Nanotechnology in Agriculture and Environmental Science

Editors

Sunil K. Deshmukh
Advisor
Greenvention Biotech Pvt. Ltd.
Pune, Maharashtra, India

Mandira Kochar
Fellow and Area Convenor
TERI-Deakin NanoBiotechnology Centre
The Energy and Resources Institute
New Delhi, India

Pawan Kaur
Associate Fellow
TERI-Deakin Nanobiotechnology Centre
The Energy and Resources Institute
New Delhi, India

Pushplata Prasad Singh
Director (Acting)
TERI-Deakin Nanobiotechnology Centre
The Energy and Resources Institute
New Delhi, India

CRC Press
Taylor & Francis Group
Boca Raton London New York

CRC Press is an imprint of the
Taylor & Francis Group, an **informa** business
A SCIENCE PUBLISHERS BOOK

First edition published 2023
by CRC Press
6000 Broken Sound Parkway NW, Suite 300, Boca Raton, FL 33487-2742

and by CRC Press
4 Park Square, Milton Park, Abingdon, Oxon, OX14 4RN

CRC Press is an imprint of Taylor & Francis Group, LLC

Library of Congress Cataloging-in-Publication Data (applied for)

ISBN: 978-1-032-34811-7 (hbk)
ISBN: 978-1-032-34812-4 (pbk)
ISBN: 978-1-003-32394-5 (ebk)

DOI: 10.1201/9781003323945

Typeset in Times New Roman
by Radiant Productions

Preface

Nanotechnology has gained tremendous attention in the last two decades due to its wide applications in several areas ranging from medicine, energy, and textiles to agriculture. These nanoparticles with smaller size with large surface area (1–100 nm) have several potential functions that could be exploited for sustainable agriculture which is the need of the hour. The development of nano chemicals has appeared as promising agents for plant growth, fertilizers, and pesticides. In recent years, the use of nanomaterials has been considered an alternative solution to control plant pests including insects, fungi, and weeds. There are several reports of nanomaterials used as antimicrobial agents for food packaging and agriculture application. Many of these nanoparticles (Ag, Fe, Cu, Si, Al, Zn, ZnO, TiO_2, CeO_2, Al_2O_3, and carbon nanotubes) have been reported to have some adverse effects on plant growth apart from their antimicrobial properties. Among these nanomaterials, Ag nanoparticles have shown great promise. In food industries, nanoparticles are currently exploited not only for increasing food quality but also for preserving its nutritive value.

This book presents applications of nanotechnology to address current problems and challenges in agriculture as well as environmental sciences. This book provides an overview of innovations in nano pesticides, nanofertilizers, bionanosensors, and nano-based delivery systems for improving different aspects of plant productivity. This includes pre-harvest and post-harvest strategies as well detection of contaminants that could be useful for enhancing soil health. Recycling of agricultural waste to beneficial products using nanotechnologies; bionanosensors; the fate of nanomaterials and the ecological consequences of their delivery into the environment; safety and nanotoxicity issues are other important issues that are dealt with in this book. Specific applications, the management of agriculture wastes and wastewater, are summarized from the viewpoint of integrated waste management in India. Book chapters have been written by experts with special reference to the innovations and latest developments in the mentioned areas of nanobiotechnology that have applications and commercial importance; especially for crop fields and post-harvest management.

Despite the research and development to promote the use of nanotechnology for agriculture and environmental issues, knowledge gaps and uncertainties about how to fill the gaps are more prevalent than scientific certainties about the public health and environmental effects of nanomaterials. So, finally, we briefly discuss the toxicity of nanomaterials to facilitate the use of agricultural nanotechnology products. This book will be highly useful for active researchers and scientists in the agricultural sector, academia as well as industry, including nanotechnologists, plant pathologists, agronomists, agro-chemists, environmental technologists, and all scientists looking for sustainability in agriculture.

We are grateful to all the contributors for their timely submission/revision of stimulating chapters in the field of Agri nanobiotechnology. We are indebted to the reviewers for value addition to the chapter by meticulous examination. The CRC Press has extended all kinds of encouragement and cooperation during the tenure of this book despite the heavy backlog, owing to the pandemic, to present this book as scheduled.

Pune, India	Sunil K. Deshmukh
New Delhi, India	Mandira Kochar
New Delhi, India	Pawan Kaur
New Delhi, India	Pushplata Prasad Singh

Contents

Section B: Environmental Nanotechnology

Recycling of Agricultural Waste

Ecosafety and Phytotoxicity

Chapter 1

Agricultural Nanobiotechnology: Current Possibilities and Constraints

Pooja and *Renu Munjal**

1. Introduction

Agriculture is backbone of most of the developing countries in the world providing food directly and indirectly to mankind, but present-day agriculture is facing challenges like: (1) **Global increase in human population.** The present global population of 7.7 billion and is projected to grow to nearly 9.8 billion by 2050. Feeding humanity will require at least a 50% increase in the production of food and other agricultural products. (2) **Hunger and extreme poverty** have decreased globally since the 1990s. Yet nearly 800 million people are chronically hungry, and 2 billion, which constitutes 27% of humanity, suffer micronutrient deficiencies. (3) **Average annual increase in crop yield** has declined since the 1960s; for example, until the mid-1980s, the rate was 3% annually for wheat but is now only 1.5%. (4) **Climate change** has reduced the potential yield of major food crops, for example, for each degree rise in average global temperature there will be reduction of 6% in wheat, 3.2% rice, 7.4% maize and 3% soya bean (FAOSTAT, 2016). Recent agricultural practices associated with the Green Revolution have greatly increased the global food supply but is not effective in the present-day scenario. For example, the alternative agriculture system like **conservation agriculture** is neither new nor practical because it works in an open system, and thereby it is thermodynamically not very tenable in such a system.

Department of Botany and Plant Physiology, CCS Haryana Agricultural University, Hisar, 125 004, Haryana, India.

* Corresponding author: munjalrenu66@gmail.com

Organic farming can neither accomplish high productivity, nor ensure a better environment and better food products. Similarly, **rainfed/dry land farming** falls short of matching the productivity that irrigated farming can provide (Mukhopadhyay, 2014). In recent years, nanobiotechnology is gaining momentum in agriculture and it is evidenced from the fact that the number of publications, the number of patents filed, and the number of patents granted in agriculture science has increased in the last two decades (Fig. 1.1). China is the leading country in filing patented products and the number of patents granted (Kah et al., 2019). The present chapter discusses various applications of nanobiotechnology in agriculture and constraints associated with it.

2. Nanobiotechnology and Nanomaterial

The term 'nanotechnology' was coined by Norio Taniguichi, a professor at Tokyo University of Science, in 1974 (Khan et al., 2014) and Richard Feynman is known as father of nanotechnology. According to the British Standard Institution (BSI) (2005), nanobiotechnology is a field of science that deals with the design, characterization, production, and application of structure, device, and system by controlling shape and size at nanoscale. International Organization for Standardization (ISO), the world's largest developer of standards, has defined nanomaterial as a material with any external dimension in the nanoscale or having internal structure or surface structure in the nanoscale, where the length range of approximately 1–100 nm is considered as nanoscale. These nanomaterials have extraordinary properties like high surface energy, high surface-to-volume ratio, high catalytic efficiency, and strong adsorption ability. Nanomaterials have different properties from their bulk counterpart because (1) Nanomaterials have a relatively larger surface area when compared to the same mass of material produced in a larger form. This can make materials more chemically reactive and affect their strength or electrical properties. (2) Quantum effects begin to dominate the behavior of matter at the nanoscale affecting the optical, electrical, and magnetic.

3. Classification, Synthesis and Characterization of Nanomaterial

Nanomaterial exists both naturally in the environment or can be fabricated artificially and are called as called anthropogenic nanoparticles, manufactured nanoparticles or engineered nanoparticles. Nanoparticles (NPs) are classified in different categories based on shape, dimension, phase composition, and nature of the material. Fundamentally, there are two approaches for the synthesis of nanomaterials (1) **top–down approach** and (2) **bottom–up approach** (Fig. 1.2).

- **Top–Down Approach**

  The top–down approach includes slicing of bulk material into nanoscale material by using micro fabrication techniques in which externally controlled tools are used to cut, mill, and shape materials into the desired size and shape.

- **Bottom–Up Approach**

  Bottom–up approach includes assembly of a defined structure by joining atom by atom, molecule by molecule, and cluster by cluster or self-organization.

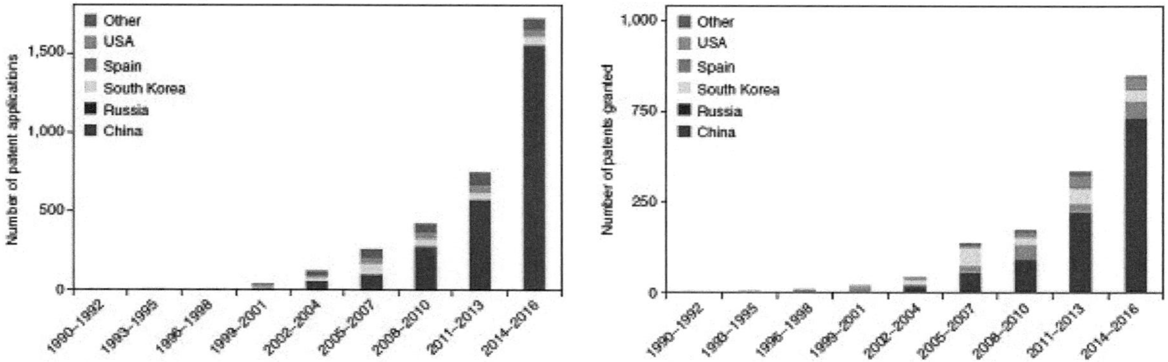

Fig. 1.1. Number of patents returned from a Google Patents search. (A) Patents in the application phase considered based on their date of filing from 1990–2016. (B) Patents granted based on their publication date from 1990–2016 (Kah et al., 2020).

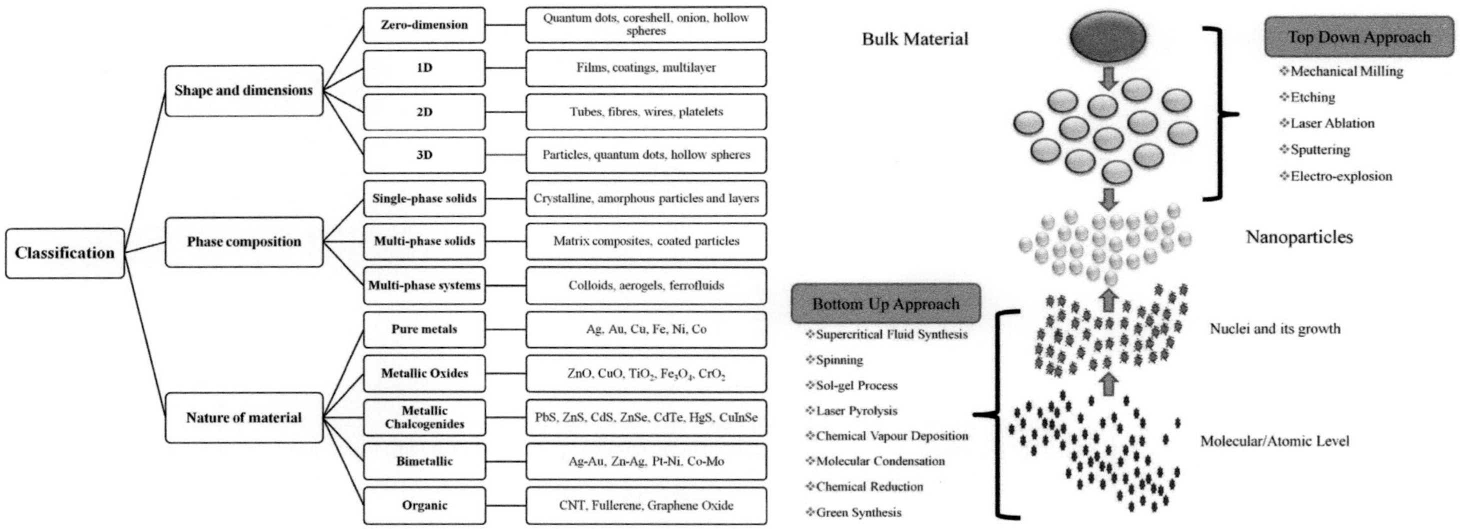

Fig. 1.2. Classification of nanomaterials and approaches for synthesis of nanomaterials (Khanna et al., 2019).

Different techniques are used for detection of nanomaterial concentration, composition, size, distribution, shape, and surface chemistry (Table 1.1). **Transmission electron microscopy** is used to trace the dynamics of an individual nanomaterial in a living cell or plant tissues. **Dynamic light scattering** (DLS) and **ultraviolet-visible spectroscopy** (UV-Vis) are not able to characterize NPs in a biological media but they provide an easy way to characterize particle-size distribution. Synchrotron-based **X-ray fluorescence microscopy** (XFM) provides a unique capability for examining the *in-situ* localization (e.g., 2D and 3D distribution) of elements in biological specimens. Other techniques are also used for detection and characterization of nanomaterial (Wang et al., 2016).

4. Application of Nanobiotechnology in Agriculture

Nanobiotechnology is running very fast in many fields of life sciences like medicines, but in the agriculture sector it is still lagging. Nanobiotechnology has various applications in the field of agriculture like **crop protection** by using various nano-agrochemicals (nanofertilizers, nanopesticides, nanoherbicides), **crop improvement** (nanoparticles as carriers of various biomolecules), **precise farming** (biosensor), **biofortification**, and **phytoremediation.**

4.1 Nanomaterial as a Magic Bullet Carrying Agrochemicals

Green Revolution have greatly increased the global food supply by use of various agrochemicals, but excessive use of these chemicals had an inadvertent, detrimental impact on the environment and on ecosystem services and are not effective in increasing the yield further highlighting the need for more sustainable agricultural methods (Mukhopadhyay, 2014).

4.1.1 Nanofertilizers

Traditional fertilizers have **low nutrient use efficiency**, for example, in China which is the world's largest consumer of nitrogen fertilizer in the world, 50% of the nitrogen applied is lost by the process of volatilization and another 5–10% by leaching. Loss of fertilizers from fields can have severe environmental consequences such as **eutrophication** which results in disturbance of aquatic flora and fauna and **unavailability of nutrients,** for example, nutrients like iron, copper or zinc, become quickly unavailable for plant uptake and therefore must be repeatedly applied; thus large doses are needed (Kah et al., 2019). Nanofertilizers have several advantages over traditional fertilizers like **more solubility** (due to smaller size and higher surface area there is high solubility in different solvents such as water), **more penetration power** (due to their small size they have more penetration power and thus can be easily taken by plants) and **slow release rate** and **target specific** (nanofertilizers with slow rate of release and target can be designed which prevent overuse of fertilizers). Nanofertilizers can be applied in three ways:

- **Direct application:** Various nanomaterials like fullerenes, carbon nanotubes, $nTiO_2$, and $nSiO_2$, ZnO, etc., can be applied directly at different growth stages of

Table 1.1. Techniques to characterize and detect nanomaterial.

Technique	Concentration	Composition/ Speciation	Size or Distribution	Shape	Surface Chemistry	Distribution
Scanning/transmission electron microscopy (SEM/TEM)	×	×	✓	✓	✓	✓
Atomic force microscopy (AFM)	×	×	✓	✓	✓	✓
Dynamic light scattering (DLS)	×	×	✓	✓	×	×
Ultraviolet-visible spectroscopy (UV-Vis)	✓	✓	×	×	×	×
Inductively coupled plasma mass spectrometry (ICP-MS)	✓	✓	×	×	×	×
Single-particle ICP-MS (SP-ICP-MS)	✓	×	✓	×	×	×
Flow field flow fractionation (FFF)-ICP-MS	✓	✓	×	U	×	×
Laser ablation ICP-MS (LA-ICP-MS)	✓	✓	×	×	×	✓
X-ray fluorescence microscopy (m-XRF)	×	×	×	×	×	✓
X-ray absorption spectroscopy (XAS)	×	✓	×	×	×	×
Transmission X-ray microscopy (TXM)	×	✓	×	×	×	✓
Nano secondary ion mass spectrometry (nano-SIMS)	×	×	×	×	×	✓
Hyperspectral microscopy	×	×	×	×	×	✓

crops at different concentrations and may or may not require traditional fertilizer practices (Millan et al., 2008).

- **Nanoencapsulations:** Fertilizers are encapsulated inside the nanostructure which are designed to allow the controlled release of nutrients in response to a specific signal, i.e., environmental factors or man-induced pulses (Aouada et al., 2015).
- **Delivered in a complex** form by nanocapsules incorporated in a matrix of organic polymers of biological or chemical origin which act as a carrier.

4.1.2 Nanopesticides

Pesticides generally used have problems due to (i) inefficiency of the pesticide. It has been estimated that 10–75% of applied pesticides do not reach the desired target; (ii) toxicity to non-target organisms; (iii) development of resistance and persistence and potential accumulation in the environment. Nanopesticides are more effective than traditional pesticides because they have more dispersion and bioavailability, improve spread and adhesion over the surface, have reduced cytotoxic and phytotoxic effects and no target effects (Huang et al., 2018). These nanopesticides can be prepared by (i) adsorption of pesticide on the surface of the nanoparticle, (ii) pesticide attachment to nanoparticle using linkers, (iii) encapsulation of pesticide in nonmaterial—made shell, and (iv) pesticide entrapped inside a nanopolymeric matrix (Fig. 1.3). These release the pesticides in response to a specific signal from the environment (Kumar et al., 2019).

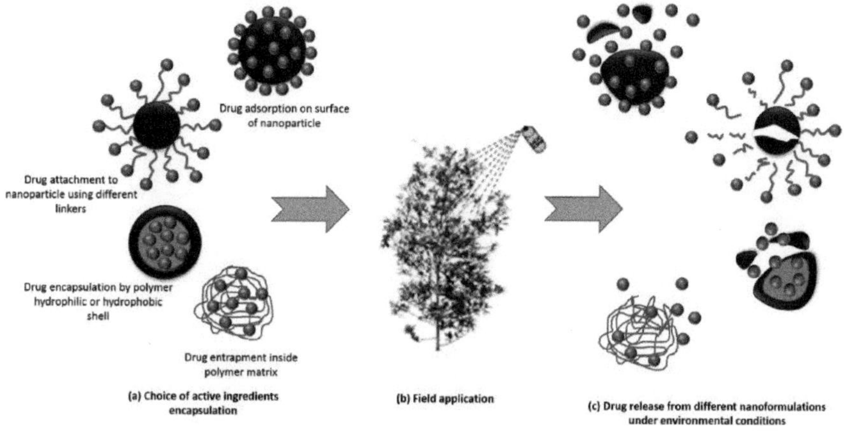

Fig. 1.3. Schematic representation of different types of nanopesticides: (a) encapsulation, entrapment, adsorption, and binding of active ingredients on the nanomaterial surface through ligands, (b) field application of pesticide nanoformulations, and (c) drug release from the nanoformulations under various environmental conditions (Kumar et al., 2019).

4.1.3 Nanoherbicides

The herbicide market in agriculture is a multi-billion-dollar industry. Conventional non-nanoherbicides have problems like dissipation through vapourisation, soil degradation, non-target effects and increased weed resistance at the topmost (Abigail and Chidambaram, 2017). Nanoherbicides development could be a new strategy to address all the problems associated with traditional herbicides as (i) a small amount of herbicide required, (ii) they protect premature degradation, (iii) they have enhanced absorption, and (iv) there is no off-target effect as target specific.

5. Nanobiotechnology as Delivery Vehicles for Biomolecules

Nanobiotechnology can be used for the efficient delivery of various biomolecules like DNA, RNA, RNP, and proteins (Fig. 1.4). All the present methods present for genetic transformations have certain drawbacks such as (i) electroporation, biolistics, agro-bacterium-mediated delivery or cationic delivery that typically target immature plant tissue, calli, meristems, or embryos, (ii) they require regeneration which is a time consuming and challenging task; (iii) being a biolistic method it often damages the target tissue and yields low level of gene expression; and (iv) agrobacterium mediated transformation is host-specific and have low levels of transformation and random integration which disturbs the expression of other genes (Cunningham et al., 2018). Nanoparticles mediated delivery is efficient and required as a nano particle (i) serves as a nonviral, biocompatible, and noncytotoxic vector. Various features like size, shape, functional group have tuned for efficient intracellular biomolecule delivery; (ii) traverses the cell wall, tuning charge, and surface properties to carry diverse cargo, and greater breadth in utility across plant species; (iii) can target chloroplast and mitochondria whereas the traditional method cannot transform the same more efficiently (Wang et al., 2016).

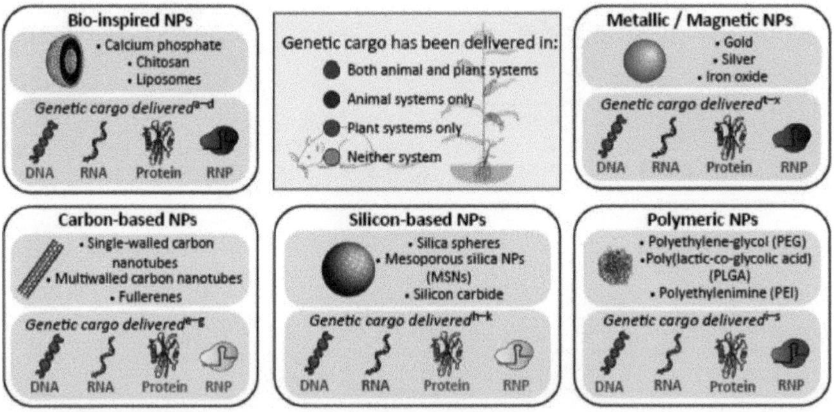

Fig. 1.4. Schematic representation of nanomaterial mediated delivery of different biomolecules (Cunningham et al., 2018).

6. Nanobiotechnology for Phytoremediation

Soil contamination by various pollutants is drawing attention of scientists worldwide. Phytoremediation is considered as a highly acceptable method to deal with these pollutants. The development of nanobiotechnology has provided efficient methods for enhancement of the phytoremediation process (Song et al., 2019). Nanobiotechnology can improve the phytoremediation process via: (i) **nanomaterial assisted removal**

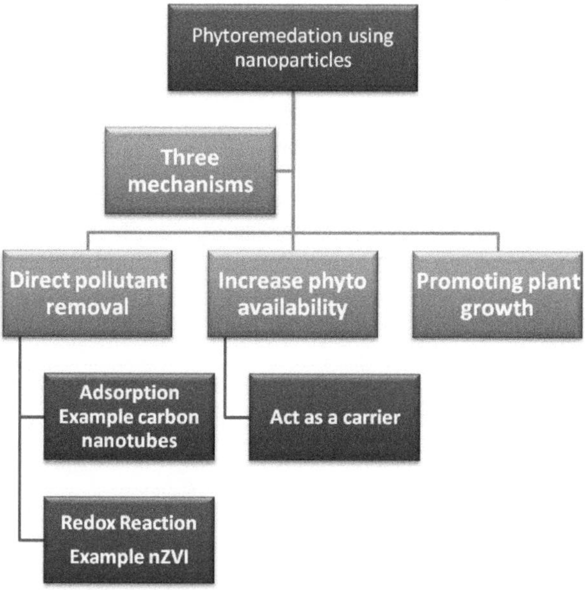

Fig. 1.5. Schematic representation mechanisms of phytoremediation using nanobiotechnology.

of pollutants, (ii) **enhanced phytoavailability**, and (iii) **promoting plant growth** (Fig. 1.5).

6.1 Nanomaterial Assisted Removal of Pollutants by Phytoremediation

Many nanomaterials can remove pollutants directly from the soil which reduces the burden of removing pollutants by plants. Nanomaterials remove pollutants via two methods: (i) via **adsorption**, for example, carbon nanotubes stabilize the organic pollutants because they have excellent adsorption capacity. They form electrostatic attraction, hydrophobic interaction, and p-p bonding with the organic pollutants that stabilize the pollutant. (ii) **Redox reactions** nZVI can be used as an electron donor for reductive degradation or stabilization of pollutants for reductive dechlorination of chlorinated organic pollutants (e.g., polychlorinated biphenyls and organochlorine pesticides) and for reductive transformation of toxic metals with high valence. nZVI can also function through adsorbing inorganic ions and coprecipitating with them.

6.2 Enhanced Phytoavailability

Phytoavailability is the main factor which influences the process of phytoremediation. Nanometerials influence phytoavailability via two methods: (i) they serve as a carrier of pollutants when they enter the cell, thereby increasing the bioavailability (Su et al., 2013) and (ii) adsorption of pollutants onto nanomaterials outside the organism may reduce the free pollutants, thereby decreasing the bioavailability (Glomstad et al., 2016). For example, by enhancing the trichloroethylene uptake by the plant. The authors explained the experimental phenomenon by co-transporting trichloroethylene with fullerene nanoparticles. The adsorbed trichloroethylene on fullerene entered the plant along with the uptake of nanoparticles.

6.3 Promoting Plant Growth

Plant biomass and growth rate are two important considerations in choosing plant species for phytoremediation, but some plants have less plant biomass and growth rate resulting in limited tolerance to pollutants in the soil. Some nonmaterials like graphene, quantum dots, carbon nanotubes, Ag nanoparticles, and ZnO nanoparticles. These nanomaterials serve as nanofertilizers or nanopesticdes thus increasing plant growth.

7. Nanobiotechnology Applications

7.1 Bioremediation

Nanoscale materials can be used for the biofortification of crops. It will solve the problem of hidden hunger and malnutrition as agriculture moves on to more marginal lands. The primary mechanism of nanomaterials' action is related to the increased activity, availability, or dissolution of materials as a function of nanoscale size. The researcher has shown that these nanomaterials can be used for biofortification. Zinc and iron deficiency is a global problem, and work such as this should be explored vigorously. For example, zinc oxide nanoparticles treated maize plant yield 15% more zinc in plant when compared with zinc sulphate (Subbaiah et al., 2016). Golubkina et al. (2017) reported that treatment of spinach with selenium under field conditions resulted in biofortification. Sundaria et al. (2018) showed that seed pre-treatment with iron oxide nanoparticles increased germination and enhanced iron in the seed.

7.2 Precise Farming through Nanobiosensors

Currently there are two classes of biosensors that are used in agriculture. First, are the **imagining, spectroscopy sensor florescence-based sensors** which present important information about the health status of a plant, but (i) are not suitable for early detection of plant stress or deficiency, (ii) lack the potential to identify specific plant stress or stressor, and (iii) are expensive for individual plant analysis. Second, there are **high throughput phenotyping devices** which can accurately generate a large amount of data in lesser time, but they also have limitations like (i) being an airborne tool it is not suitable for individual plant and crop growing in high density, and (ii) it is influenced by sunlight or shadow (Giraldo et al., 2019). Nanobiosensors are analytical devices having at least one dimension no greater than 100 nm and can overcome the limitation of a traditional biosensor as they can detect biomolecules in very minute amounts and extremely precisely. They have increased sensitivity and allow an immediate response to environmental changes once they are incorporated in the equipments. The development of these nanosensors also enhance precise farming methods. These nanobiosensors can (i) detect contaminants like viruses, bacteria, toxins bio-hazardous substances, (ii) monitor soil conditions and crop growth over vast areas. Using nanobiosensors resulted in effective detection of pathogen invasion, infection, nutrient requirement, and contamination or other biomolecules like ABA, jasmonic acid, proline, etc., which are related to biotic and abiotic stress and reduced the risk of economic losses due to these reasons (Fraceto et al., 2016).

7.3 Nanobiotechnology for Abiotic Stress Tolerance

Plants are sessile and are constantly exposed to various abiotic stressors like water, salt, and temperature, both high and low. These abiotic stresses have reduced the yield potential of many field crops. For example, for each degree rise in temperature, it is expected that there would be reduction of 6% in wheat, 3.2% rice, 7.4% maize, and 3% soya bean. It is estimated that drought and salinity may cause a 50% loss in crop production. Thus, there is an urgent need to mitigate these abiotic stresses. Many effects of nanomaterials on plants under abiotic stress have been reported. Nanomaterials mitigate abiotic stress via osmotic adjustment by synthesis of osmolytes, keeping the leaf erect and rigid to perceive more light, and prevent chlorophyll degradation, reduce the transpiration rate, stimulate the antioxidant defense system, reduce the sodium uptake and complex formation (Fig. 1.6). Zinc oxide nanoparticles increase the germination percentage and rate, decrease in seed residual fresh and dry weight in *glycine max* under drought conditions (Sedghi et al., 2013). Jaberzadeh et al., (2013) also reported that TiO_2 increased growth, yield, gluten, and starch content in wheat under drought stress. Under saline condition SiO_2, nanoparticles enhanced seed

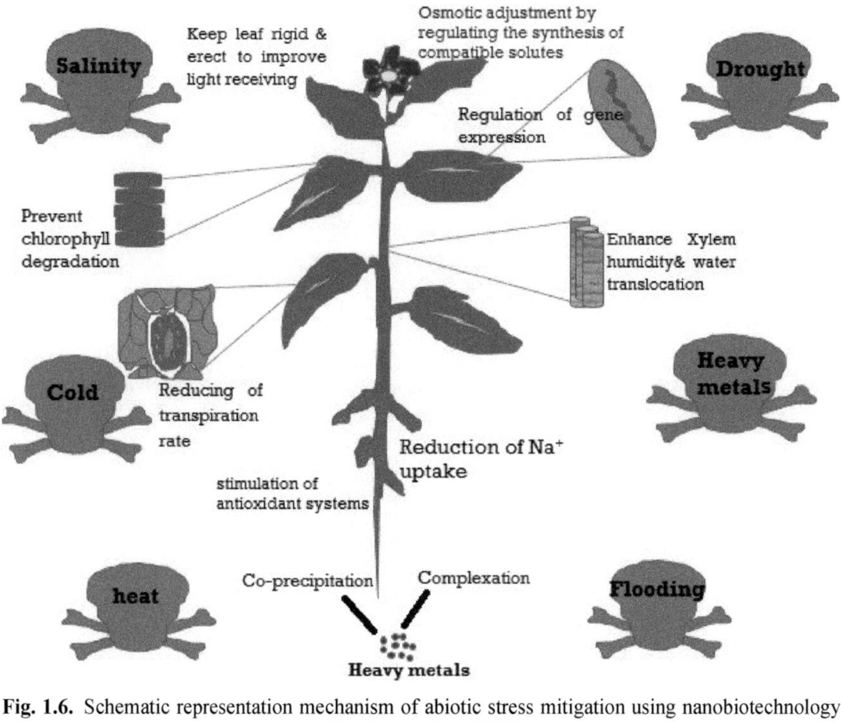

Fig. 1.6. Schematic representation mechanism of abiotic stress mitigation using nanobiotechnology (Elsakhawy et al., 2018).

germination, increased growth, activities of antioxidant enzymes relative water content and total yield of *Vicia faba* L. (Qados and Moftah, 2015). Enhanced activities of antioxidant enzymes decreased the H_2O_2 content and electrolyte leakage, enhanced the accumulation of TiO_2 by sensitive genotypes than the tolerant genotypes was reported by Mohammadi et al. (2013, 2014) in *Cicer arietinum* L. under cold stress using TiO_2 nanoparticles. Foliar application Se nanoparticles at the concentration of 10 mg L^{-1} during the booting stage of sorghum stimulate antioxidant defense system enhancing the antioxidant enzymes activity, decreased the concentration of signature oxidants, and facilitated higher levels of unsaturated phospholipids and led to a significantly increased seed yield under heat stress (Djanaguiraman et al., 2018).

8. Constraints for Nanobiotechnology

Nanobiotechnology finds application in various fields of agriculture but still there are few constraints to this new technology which does not allow it to deliver its full potential. One of the major issues associated with nanobiotechnology is nanotoxicity. Different researchers have different views about nanotoxicity. The views differ because (i) some scientists used easily scoreable parameters like germination rate, root and shoot length, etc., while others used highly complex parameters like cytotoxicity and genotoxicity; (ii) different plant species respond differently to the same nanoparticle; (iii) unrealistically high concentrations of nanoparticles; and (iv) some ignored the requirement for appropriate control. Nano quantitative structure-activity relations (nano-QSARs) predict the cytotoxicity of various metal oxides using *in vitro* cell-based assays and the adsorption of various molecules onto nanoparticles, an enthalpy descriptor or a biological surface adsorption index, and some studies used computational chemistry to simulate the translocation of nanoparticle (Wang et al., 2016).

9. Conclusion and Future Perspective

In last few decades nanotechnology has made huge progress in various fields of life sciences like medicine for diagnosis and therapy, but on the other hand applications of nanobiotechnology in the agriculture field is at an infant stage. In last decade, it has gained momentum in the field of agriculture as suggested by an increased number of publications in this field. Nanobiotechnology has different applications such as different nanoscale agrochemical (nanofertilizers, nanopesticides, and nanoherbicides), delivery of different biomolecules (DNA, RNA, RNP, and protein) for crop improvement, phytoremediation, biofortification, precise farming using nanobiosensors, and tolerance to abiotic stresses. Nanobiotechnology in agriculture has still room for improvement and it could be achieved by (i) development of specific hybrid carriers for delivering active agents which include nutrients, pesticides, and fertilizers in order to maximize their efficiency following the principles of green chemistry and environmental sustainability; (ii) design of processes which are easily upscalable at the industrial level; (iii) comparison of effects of nanoformulations/nanosystems with existing commercial products to demonstrate the real practical advantages; and (iv) acquisition of knowledge and developments of methods for risk and life-cycle assessment of nanomaterials, nanopesticides, nanofertilizers, as well as assessment of the impacts (e.g., phytotoxic effects) on non-target organisms. An important role in providing more sustainable agriculture, especially in the production of crops with nanotechnology will only be possible with the availability of nano-products with higher efficiency and lower costs.

References

Abigail, E.A. and Chidambaram, R. (2017). Nanotechnology in herbicide resistance. Nanostructured Materials-fabrication to Applications. London: IntechOpen, 207-12.

Aouada, F.A. and De Moura, M.R. (2015). Nanotechnology applied in agriculture: Controlled release of agrochemicals. pp. 103–118. *In*: *Nanotechnologies in Food and Agriculture*. Cham.: Springer.

BSI, B. (2005). A Standard for Standards-Principles of Standardization. BSI Group, London, UK.

Cunningham, F.J., Goh, N.S., Demirer, G.S., Matos, J.L. and Landry, M.P. (2018). Nanoparticle-mediated delivery towards advancing plant genetic engineering. *Trends Biotechnol.*, 36(9): 882–897.

Djanaguiraman, M., Belliraj, N., Bossmann, S.H. and Prasad, P.V. (2018). High-temperature stress alleviation by selenium nanoparticle treatment in grain sorghum. *ACS Omega*, 3(3): 2479–2491.

Elsakhawy, T., Omara, A.E.D., Alshaal, T. and El-Ramady, H. (2018). Nanomaterials and plant abiotic stress in agroecosystems. *Env. Biodivers. Soil Secur.*, 2(2018): 73–94.

FAOSTAT. (2016). FAO Statistical Database. Food and Agriculture Organization of United Nation. http://faostat.fao.org.

Fraceto, L.F., Grillo, R., de Medeiros, G.A., Scognamiglio, V., Rea, G. and Bartolucci, C. (2016). Nanotechnology in agriculture: which innovation potential does it have? *Front. Environ. Sci.*, 4: 20.

Giraldo, J.P., Wu, H., Newkirk, G.M. and Kruss, S. (2019). Nanobiotechnology approaches for engineering smart plant sensors. *Nat. Nanotechnol.*, 14(6): 541–553.

Glomstad, B., Altin, D., Sørensen, L., Liu, J., Jenssen, B.M. and Booth, A.M. (2016). Carbon nanotube properties influence adsorption of phenanthrene and subsequent bioavailability and toxicity to Pseudokirchneriella subcapitata. *Environ. Sci. Technol.*, 50(5): 2660–2668.

Golubkina, N.A., Folmanis, G.E., Tananaev, I.G., Krivenkov, L.V., Kosheleva, O.V. and Soldatenko, A.V. (2017). Comparative evaluation of spinach biofortification with selenium nanoparticles and ionic forms of the element. *Nanotechnologies Russ.*, 12(9): 569–576.

Huang, B., Chen, F., Shen, Y., Qian, K., Wang, Y., Sun, C. and Cui, H. (2018). Advances in targeted pesticides with environmentally responsive controlled release by nanotechnology. *J. Nanomater.*, 8(2): 102.

Jaberzadeh, A., Moaveni, P., Moghadam, H.R.T. and Zahedi, H. (2013). Influence of bulk and nanoparticles titanium foliar application on some agronomic traits, seed gluten and starch contents of wheat subjected to water deficit stress. *Not. Bot. Horti. Agrobot. Cluj.*, 41(1): 201–207.

Kah, M., Tufenkji, N. and White, J.C. (2019). Nano-enabled strategies to enhance crop nutrition and protection. *Nat. Nanotechnol.*, 14(6): 532–540.

Khan, M.R. and Rizvi, T.F. (2014). Nanotechnology: Scope and application in plant disease management. *Plant Pathol. J.*, 13(3): 214–231.

Khanna, P., Kaur, A. and Goyal, D. (2019). Algae-based metallic nanoparticles: Synthesis, characterization, and applications. *J. Microbiol. Methods*, 163(2019): 1–24.

Kumar, S., Nehra, M., Dilbaghi, N., Marrazza, G., Hassan, A.A. and Kim, K.H. (2019). Nano-based smart pesticide formulations: Emerging opportunities for agriculture. *J. Control. Release*, 294: 131–153.

Millán, G., Agosto, F. and Vázquez, M. (2008). Use of clinoptilolite as a carrier for nitrogen fertilizers in soils of the Pampean regions of Argentina. *Int. J. Agric.*, 35(3): 293–302.

Mohammadi, R., Maali-Amiri, R. and Abbasi, A. (2013). Effect of TiO_2 nanoparticles on chickpea response to cold stress. *Biol. Trace Elem. Res.*, 152(3): 403–410.

Mohammadi, R., Maali-Amiri, R. and Mantri, N.L. (2014). Effect of TiO_2 nanoparticles on oxidative damage and antioxidant defense systems in chickpea seedlings during cold stress. *Russ. J. Plant Physiol.*, 61(6): 768–775.

Mukhopadhyay, S.S. (2014). Nanotechnology in agriculture: Prospects and constraints. *Nanotechnol. Sci. Appl.*, 7: 63–71.

Qados, A.M.A. and Moftah, A.E. (2015). Influence of silicon and nano-silicon on germination, growth, and yield of faba bean (*Vicia faba* L.) under salt stress conditions. *J. Exp. Agric. Int.*, 509–524.

Sedghi, M., Hadi, M. and Toluie, S.G. (2013). Effect of nano zinc oxide on the germination parameters of soybean seeds under drought stress. *Ann. West Univ. Timiş., Ser. Biol.*, 16(2): 73.

Song, B., Xu, P., Chen, M., Tang, W., Zeng, G., Gong, J. and Ye, S. (2019). Using nanomaterials to facilitate the phytoremediation of contaminated soil. *Crit. Rev. Environ. Sci. Technol.*, 49(9): 791–824.

Su, Y., Yan, X., Pu, Y., Xiao, F., Wang, D. and Yang, M. (2013). Risks of single-walled carbon nanotubes acting as contaminants-carriers: Potential release of phenanthrene in Japanese medaka (*Oryzias latipes*). *Environ. Sci. Technol.*, 47(9): 4704-4710.

Subbaiah, L.V., Prasad, T.N.V.K.V., Krishna, T.G., Sudhakar, P., Reddy, B.R. and Pradeep, T. (2016). Novel effects of nanoparticulate delivery of zinc on growth, productivity, and zinc biofortification in maize (*Zea mays* L.). *J. Agric. Food Chem.*, 64(19): 3778–3788.

Sundaria, N., Singh, M., Upreti, P., Chauhan, R.P., Jaiswal, J.P. and Kumar, A. (2019). Seed priming with Iron oxide nanoparticles triggers Iron acquisition and biofortification in wheat (*Triticum aestivum* L.) grains. *J. Plant Growth Regul.*, 38(1): 122–131.

Wang, P., Lombi, E., Zhao, F.J. and Kopittke, P.M. (2016). Nanotechnology: A new opportunity in plant sciences. *Trends Plant Sci.*, 21(8): 699–712.

Section A
Agro-Nanotechnology

Nanofertilizers

Chapter 2

Application of Metallic Nanoparticles as Agri Inputs:

Modulation in Nanoparticle Design and Application Dosage Needed#

Shweta Gehlout,[1,2] *Ayushi Priyam,*[1,2] *Drishti,*[1,2]
Luis Afonso,[2] *Aaron G Schultz*[2] and
Pushplata Prasad Singh[1,2,*]

1. Introduction

Designing of ultra-small particles having exceptional properties has emerged as a promising strategy for increased plant growth and productivity in recent years (Boutchuen et al., 2019; Rizwan et al., 2019; Younes et al., 2020; Van Nguyen et al., 2021). Given this, researchers working in the arena of nanotechnology are developing nanomaterial (NMs) for agricultural applications primarily to be used as fertilizers and pesticides. The definition of nanotechnology from the United States National Nanotechnology Initiative can be quoted as "the understanding and control of matter at dimensions of roughly 1–100 nm, where unique phenomena enable novel applications; encompassing nanoscale science, engineering, and technology" (Paunovic et al., 2020). In a market research *Global Nanotechnology Market Outlook 2024* (Research and Markets, 2018), considering the research development and market trends, it was concluded that nanotechnology has a huge potential to grow as an industry especially in the areas of biomedical science, healthcare, environmental remediation, food and

[1] TERI - Deakin Nanobiotechnology Centre, The Energy and Resources Institute (TERI), Lodhi Road, New Delhi, 110003, India.
[2] School of Life and Environmental Sciences, Deakin University, Geelong, Victoria, 3217, Australia.
* Corresponding author: pushplata.singh@teri.res.in
# This project has received funding from the Department of Biotechnology, India under the Grantnumber BT/NNT/28/SP30280/2019.

agriculture, electronics, automobile, defense, and energy. The report also suggested a considerable increase in funding for nanotechnology research and development from both government and private sectors globally.

"Engineered nanomaterials (ENMs) are manufactured materials with any external dimension or internal structure or surface structure in the nanoscale (*ca.* 1–100 nm) that are designed for a specific purpose or function" (Ahamed et al., 2021). ENMs designed as nanofertilizers or nanopesticides achieve higher efficacy as compared to their conventional counterparts, which along with the smart delivery mechanisms for active ingredients can increase crop yields (Jampílek and Kráľová, 2015). However, distinct properties of nanoparticles (NPs) in NMs may increase their tendency to interact with the biological systems, which may have adverse effects on the environment and human health (Zia-ur-Rehman et al., 2018). It has also been suggested that NMs applied as agri-inputs may have a deleterious effect on soil health and plants, and may affect subsequent strata in the food chain (Theivasigamani, 2011). Thus, the development of "safe-by-design" NMs is needed for agriculture purposes.

Development of the global policy guidelines to formulate a nano-regulatory framework for the agricultural sector in the past decade by regulatory bodies, particularly, the European Union (EU) has been a brilliant initiative in this direction (Subramanian and Rajkishore, 2018). Varied outcomes have been reported in the nanotoxicity studies conducted by different research groups using the same kind of NMs (Omar et al., 2019; Shah et al., 2019), which has emphasized a need for the consideration of multiple factors during toxicity and efficacy analyses of nanoproducts. In addition, the presence of a large number of studies reporting favorable outcomes of agricultural application of NPs and an equal number of studies reporting adverse effects (Kumar et al., 2018; Yuan et al., 2018) is a major challenge that remains to be addressed for the successful development of safe nanoproducts. Inadequate understanding of mechanisms underlying phytotoxicity of NMs demands intensive investigations to understand interactions of the ENMs with plants including mechanisms involved in uptake, translocation mechanisms, biotransformation and phytotoxicity (Zhang et al., 2015).

2. The Requirement of Micronutrients for Plant Growth

Plants need various elements that are essential for their optimum functioning. Depending on the amount required for plant development, these elements are categorized as macro- and micro-nutrients. Macronutrients include primary elements: nitrogen (N), phosphorus (P), potassium (K); and secondary elements: calcium (Ca), magnesium (Mg) and sulfur (S). Iron (Fe), zinc (Zn), cobalt (Co), nickel (Ni), copper (Cu), manganese (Mn), boron (B), molybdenum (Mo), and chlorine (Cl) are the eight essential micronutrients required for plant growth. These essential elements primarily assist in the functioning of enzymes and co-enzymes that are involved in various physiological processes including photosynthesis, absorption, and utilization of other nutrients (Al-Obaidi, 2020).

Presently, most of the agricultural zones across the globe are facing a broad spectrum of challenges including declining soil organic matter, micronutrient deficiencies, low nutrient use efficiency, stagnant crop yields, and shrinking arable land (Al-Hadede et al., 2020; Zhang et al., 2020). Micronutrients' deficiencies in soil have been reported from several agro-climatic zones across the globe, which varies depending on soil type, climatic conditions, field properties and, plant types (Alloway, 2008). Erroneous management of soil resources, including cultivation at an accelerated pace without proper reloading of nutrients, finite crop rotations, and minimal addition of organic matter in the past four to five decades has led to micronutrient deficiency in soils resulting in reduced yield and deteriorated crop quality (Shukla et al., 2018). Furthermore, soil naturally contains a range of minerals and organic matter along with a wide range of micronutrients (in ion or chelated form) based on its composition. Yet the micronutrient availability in soils is dependent on different factors including, pH, redox potential, interaction with coexisting ions, organic matter dynamics, and soil microbiology (Masunaga and Fong, 2018). Under variable environmental conditions, the micronutrients present in soil may either undergo rapid reactions with phosphates and carbonates to form chemical precipitates, or interact with mineral complexes, clay, and organic matter making them unavailable to plants.

The pH of the soil is an important factor that determines the availability of mineral nutrients to plants, which in turn determines crop productivity. Depending on the soil pH, soil can be categorized into different soil types: Acidic, alkaline, and neutral. The acidic soil has a pH value of $\leq 4 - \leq 5.5$, whereas, the alkaline soils have a pH ranging between $7.3 - \geq 8.5$. The pH ~ 7 corresponds to the neutral soil (Batjes, 1995). The excessive Al^{3+} and Mn^{2+} ions present in acidic soils limit the crop production by causing low availability of K, P, Mg, Ca, or Mo to plants. On the other hand, alkaline and calcareous soil present low phytoavailability of the micronutrients Fe, Zn, Mn, and Cu (Schjoerring et al., 2019).

Nutrient use efficiency by crop plants is an important factor from both the economic and environmental point of view. On an average, the nutrient use efficiency by crop plants under all agroecological conditions has been found to be < 50%, indicating a large section of the applied nutrients being lost in the soil-plant system, resulting in an increased cost of crop production and environmental pollution (Baligar and Fageria, 2015).

Previous findings have reported the agricultural soils across the world , including India to be potentially deficient in one or more micronutrients, especially Fe, Zn, B, Mn, and Cu (Graham, 2008; Shukla et al., 2021).

Fe is the essential micronutrient present in highest concentration in the soil as well as in plants (Pingoliya et al., 2014). Deficiency of Fe leads to chlorosis in young plant leaves which alters the normal physiological function and nutritional quality in plants. Widespread deficiency of Fe can be seen among multiple crop types. Peanut (*Arachis hypogaea* L.), soybean (*Glycine max* L.), sorghum (*Sorghum bicolor* L. Moench), and upland rice (*Oryza sativa* L.) are the plant species to commonly suffer from Fe deficiency (Fageria and Moreira, 2011). Generally, the soil has high Fe content, which, however, mostly remains fixed to soil particles and, therefore is unavailable to crops. In highly aerated and alkaline soils, Fe is mainly found in its insoluble form, Fe^{3+}, making these soils deficient in the available form, Fe^{2+} (Verma et al., 2021). It is estimated that Fe deficiency is widespread occurring in about 30–50% of the cultivated soils in the world, and results in large decreases in crop production and quality. In India, one-third of the cultivated area has iron-deficient calcareous soils; most of which is usually distributed in low rainfall areas where groundnut is a major crop and suffers from Fe deficiency or interveinal chlorosis, resulting in a significant decrease in pod yield (16–32%). Fe is one of the five soil micronutrients identified as important for crop productivity in Sub-Saharan Africa (SSA) arable lands, which account for more than half of the world's potential land for cultivation (Pandey et al., 2021).

Zn is another most important micronutrient needed by plants. Deficiency of Zn can lead to various detrimental effects such as plant susceptibility to stress, a slower production rate of biomolecules like carbohydrates, cytochromes, nucleotides, auxin, and chlorophyll, and membrane disintegration in plant cells (Arunachalam et al., 2013). A number of factors influence plant-available Zn in soils, including low total Zn contents, high pH (> 7), low redox potential, prolonged flooding, rhizosphere microbial communities, high organic matter, high $CaCO_3$ and bicarbonate contents, high Fe/Mn oxide contents, high phosphorus availability, high sodium, and high exchangeable magnesium/calcium ratio (Zeng et al., 2021). It has been estimated that about 50% of the cereal-growing soils across the globe have low Zn availability (Yaseen and Hussain, 2021). The analysis of 256,000 soils and 25,000 plant samples in India revealed that 48.5% of the soils and 44% of the plant samples were deficient of bioavailable Zn (Singh, 2008). It has been suggested that by the year 2025 the Zn deficiency will range between 49–63% in different agricultural soils in India (Arunachalam et al., 2013).

B, the micronutrient involved in plant growth and sexual reproduction plays a significant role in protein and nucleic acid metabolism and helps in the maintenance of plasma membrane structural integrity. B is also involved in the regulation of stomatal pore opening and closure and production of starch for cellulose and lignin synthesis (Jehangir et al., 2017). Deficiency of B can result in altered water uptake (Wimmer and Eichert, 2013), altered expression of genes involved in Ca^{2+} signaling, and cytosolic Ca^{2+} level in plants (Quiles-Pando et al., 2013), and induction of change in cell wall components and assembly (Liu et al., 2014). On a global scale, after Zn and Fe deficiency, B is the next micronutrient deficient in crops (Ahmad et al., 2012). B deficient soils are distributed globally from tropical to temperate zones. Most prominently, sandy soils with humid climatic conditions exhibit B deficiency due to high leaching losses (Qin et al., 2021; Thapa et al., 2021). Analysis of soil and plant samples has indicated that 33% of Indian soils are deficient in B (Rai et al., 2021). A global analysis of soils has revealed that B is also deficient in Nepal, Philippines, and Thailand (Archana and Verma, 2017). On a global scale, about 31% of arable lands are estimated to be deficient in B (Hussan et al., 2021).

Another one of the most important plant micronutrients is Mn. It plays a key role in plant redox reactions during photosynthesis (Mousavi et al., 2011; Schmidt et al., 2016). The micronutrient Mn is also a part of the superoxide dismutase (SOD) enzyme and thus helps in protecting the plants against free radicals. Deficiency of Mn leads to the destabilization and disintegration of the PSII complex involved in the photosynthesis. Analysis of different soil samples collected during an analysis performed for determining the spatial spreading patterns of plant-available sulphur and micronutrients in cultivated soils of coastal districts of India showed about 23% of the soil samples to be Mn deficient (Shukla et al., 2021). An extensive deficiency of Mn (6%) has been observed in the 'rice-wheat' cropping regions of northern India, particularly in Punjab (18%) and Haryana (12%) (Shukla et al., 2012). Global data suggests that more than 10% of worldwide soils are deficient in Mn (Zulfiqar et al., 2021).

Mo plays a role in catalyzing key steps of N, C, and S metabolism in plants. Furthermore, in leguminous plants, Mo serves as a prosthetic group to nitrogenase enzyme to facilitate the symbiotic N fixation (Manuel et al., 2018). Plants that suffer from Mo deficiency have shown nitrate accumulation in seedlings and deteriorated seed quality (Kovács et al., 2015; Gopal et al., 2016). The availability of Mo in the soil is influenced by several factors, including the texture of the soil, presence of Fe, and aluminum hydroxides in the clay fraction, the organic carbon supply, the redox potential, the ionic interaction with phosphorus and sulphur, and the acidity. The pH is the key determinant of Mo availability in the soil solution. In acidic soils, the density of positive electrical charges on the surface of mineral colloids increases, which favors specific adsorption and the formation of non-labile Mo (Rosado et al., 2021). Mo is one of the most deficient micronutrients in acidic Indian soils, with its deficiency reported in 46% of the agricultural soils (Kumar, 2016). Globally, approximately 15% of the soils have been reported to be Mo deficient (Sillanpää, 1982).

Cu is an essential micronutrient for plants as it functions as a redox-active cofactor in a wide variety of plant proteins and is essential for multiple fundamental biological processes (Puig, 2014). Availability of Cu in soil is influenced by several factors such as the concentration and types of clay minerals, total Cu concentration, soil organic matter content, pH and ionic strength (Benedet et al., 2020). Analysis of a large number of soil samples during a study showed Indian soils have minimal Cu deficiency (< 1%) (Shukla et al., 2014). Analysis of soil and plant samples in a separate study showed 3% of Indian soils to be deficient in Cu (Singh, 2008). Deficiency of Cu is not much prevalent on a global scale.

3. Scope of MNPs in Agriculture

According to the Food and Agricultural Organization of the United States (FAO), "12.5 percent of the world's population (868 million people) is undernourished in terms of energy intake and 2 billion people suffer from one or more micronutrient deficiencies" (FAO/WHO). Furthermore, an approximation claims that half of the world's micronutrient deficient population lives in India (Ritchie et al., 2018). Advancement in the field of agriculture is expected to fulfill the increasing demand for food by enhancing crop production. However, degradation of arable land along with nutrient-depletion and nutrient unavailability to plants present serious challenges in the way of producing enough food agricultural products to meet the demands of the world's population through classic agricultural practices (FAO/WHO). To overcome the micronutrient deficiency in agriculture soil, excessive application of bulk fertilizers has taken place since the onset of the green revolution era. As per FAO, "chemical fertilizers are the single most important contributor to the increase in world agricultural productivity". Fertilizers containing NPK are considered as the drivers of modern agriculture (Nagendran, 2011); however, the excessive use of fertilizers in recent years has presented a serious concern regarding the environmental problems caused by these chemical entities. The act of 'fertilization' can result in the accumulation of fertilizer in soil and plant systems, and can lead to water, soil, and air pollution (Savci, 2012). Excess dosage and indiscriminate application of N and P-based fertilizers in agriculture have led to eutrophication of water bodies around the globe (Nagendran, 2011). Several processes including "leaching, drifting, hydrolysis, photolysis, and microbial degradation" lead to the loss of most of the agrochemicals applied to crops, even before reaching its target (Sabir et al., 2014). Enrichment of soil with fertilizers and pesticides can also cause alterations in soil function and properties including rhizodeposition, nutrient content, soil pH, moisture and organic carbon, as well as soil enzymatic activities, leading to a shift in the population dynamics of soil microflora with an altered soil profile (Prashar and Shah, 2016).

Nanotechnology is expected to revolutionize the agriculture sector. It has been predicted that nanotechnology may prove crucial in the advancement of sustainable agricultural practices. Due to the inherent property of nanoparticles, high surface area to volume ratio, substantial increment in quantity and quality of yield, alongside decreasing environmental contamination can be attained with the application of nanoagri inputs (Sekhon, 2014; Das et al., 2015; Liu and Lal, 2015). The small size of NPs preset in nanonutrient-based fertilizers has been reported to increase their bio-availability in plants resulting in higher nutrient uptake (Qureshi et al., 2018). With the increasing scientific pieces of evidence to support the application of MNPs for checking micronutrient deficiencies, a surge in NM-based agriproducts has been witnessed in recent years. Figure 2.1 has been reconstructed using data from a previously published report (Statnano, 2021), which presents estimates for NM-based agricultural products available in the Indian and global markets.

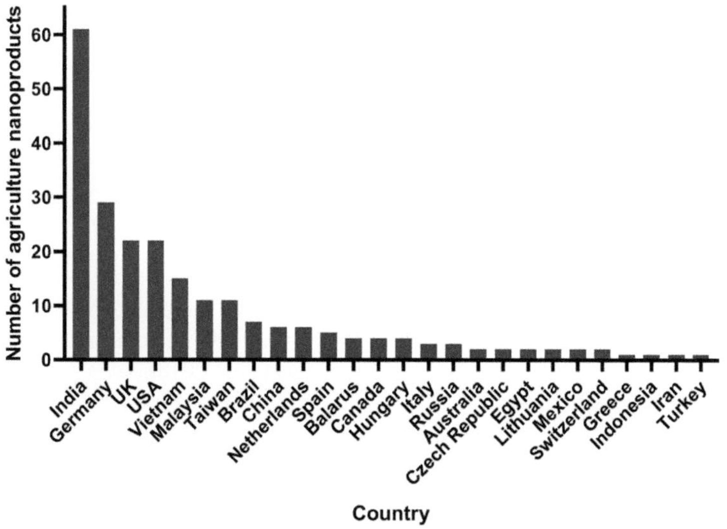

Fig. 2.1. Nanomaterials used in products by different countries for agriculture.

3.1 Effects of MNPs on Plants: *Beneficial or Adverse*

There is a vast scope for MNP application in agricultural fields, aiming mainly at the delivery of agrochemicals to the crops, detection of plant disorders, and detection and treatment of nutrient deficiency (Mehrazar et al., 2015). Currently, the most common application of MNPs in agriculture is their usage as nanofertilizers, which can be designed to release nutrients at slower rates for a longer duration (> 30 days) leading to improved nutrient use efficiency and decreased residue effects (Subramanian et al., 2015). Besides nanofertilizers, MNPs can also be used as effective nanocarriers to entrap, encapsulate, or absorb active molecules to create novel agricultural formulations (Worrall et al., 2018).

Research investigating the development of MNPs as fertilizers to deliver essential micronutrients to agricultural soils has gained increased attention in recent years (Ruttkay-Nedecky et al., 2017). The synthesis of nanofertilizers can occur via the conversion of bulk fertilizers into nanoscale particles or emulsions, surface coatings of NPs on fertilizer, nano-encapsulation of active-ingredient, and nanocarriers for targeted delivery are some of the ways to effectively deliver nutrients to crops by using nanofertilizers (Solanki et al., 2015).

3.1.1 Beneficial Effects of Different MNPs Applied to Plants

a. Iron nanoparticles

Hematite (Fe_2O_3) is the most abundant form of Fe found in soils. It is often present in a complex or chelated compound, which is in an extremely insoluble form. This leads to a decreased uptake by plants and Fe deficiency, finally causing poor plant growth and development. Fe deficiency in plants can also affect the animals that consume them, including humans, present in the food web and cause Fe deficiency (Zimmermann and Hilty, 2011; Von Moos et al., 2017). Alternative technologies improving the uptake efficiency of Fe fertilizers are therefore much desired at present (Rui et al., 2016). The beneficial effects of iron oxide nanoparticles (FeO NPs) have shown that these FeO NPs can work as an ideal substitute for traditional iron fertilizers. However, the result of the NP application largely depends on multiple factors, including the soil type and its chemistry, test plant species and its growth stage, physicochemical properties of the NPs being applied along with the NP concentration. The uptake and transportation of FeO NPs have been reported in multiple plant species (Li et al., 2013; Li et al., 2016), indicating the necessity of NP interaction studies with a large variety of crops, and in different arable soils. The subsequent effects of NP application on the food chain should also be determined (Raliya et al., 2018).

Globally, multiple research teams have investigated the effects of FeO NPs on plant growth and health, including wheat (Feizi et al., 2013), soybean (Roghayyeh et al., 2010), rice (Mankad, 2017), capsicum (Yuan et al., 2018), watermelon (Wang et al., 2016), maize (Pariona et al., 2017), and citrus plant species (Hu et al., 2017; Li et al., 2017). In the study by Wang et al. (2016), they investigated the physiological response of watermelon plants after the seeds were grown in soil supplied with a range of Fe_2O_3 NPs: 0, 20, 50, 100 milligrams per liter (mg. L^{-1}). The observations made during five successive weeks after germination of the seeds showed that 20 mg. L^{-1} Fe_2O_3 NPs does not have any adverse effect on the lipid peroxidation and the physiology of watermelon plants (Wang et al., 2016). However, at 50 mg. L^{-1} concentration of Fe_2O_3 NPs, an increase in chlorophyll, protein, and soluble sugar content was observed in treated plants. Overall, it was concluded that Fe_2O_3 NPs can facilitate improved growth of watermelon plants during Fe deficiency at proper concentrations (Wang et al., 2016). A separate field experiment investigating the effect of FeO NPs on agrarian traits of soybean in terms of yield and quality showed 0.75 grams per decimeter cube (g/dm^3) FeO NPs to increase pod and leaf dry weights in plants. A noticeable 48% increase in grain yield was observed in plants treated with 0.5 grams per liter (g. L^{-1}) FeO NPs (Roghayyeh et al., 2010). The uptake and circulation of Fe_2O_3 NPs in *Zea mays* L. has also been investigated. Treatment of the *Zea mays* L. with 20 mg. L^{-1} concentration of Fe_2O_3 NPs was found to promote root elongation by 11.5%. For the similar dosage of NPs, the germination and vigor index were found to have increased by 27.2% and 39.6%, respectively (Li et al., 2016). Similarly, exposure of *Citrus maxima* plants to a concentration of 50 mg. L^{-1} FeO NPs resulted in improvised plant health and elevated plant growth (Hu et al., 2017).

b. Zinc nanoparticles

Nano Zn is the most commonly commercialized fertilizer among all nanofertilizers (Duhan et al., 2017). Numerous Zn fertilizers are currently applied to agricultural fields to eliminate Zn deficiency in the soil, which mainly includes zinc sulfate ($ZnSO_4$) and Zinc oxide (ZnO). Besides these, other products such as synthetic chelates ($Na_2ZnEDTA$, NaZnHEDTA, NaZnNTA, and $Zn_3(C_6HSO_7)_2.2H_2O$), natural organic complexes (lignosulfonates, phenols, and polyflavonoids) and inorganic complexes (ammoniated $ZnSO_4$ solution and ammoniated $ZnCl_2$ solution) can also be used (Mortvedt and Gilkes, 1993). However, due to the restricted availability of zinc to plants, their applications as a fertilizer are limited. ZnO NPs are emerging as a possible new fertilizer product due to their ability to increase the levels of available Zn in soil due to their high surface area to volume ratio.

Nano ZnO is being tested by researchers to increase plant growth and ultimately, the yield (Shilpa and Lawre, 2014). The beneficial effect of ZnO NPs on plant growth and seed development has also been studied at the lab as well as field scale on peanut crops where treatment with a thousand ppm concentration of 25 nm sized ZnO resulted in increased seed germination and seedling vigor. The seeds treated with NPs also showed early establishment in soil followed by early flowering and successive elevation in stem and root growth. When compared to the chelated bulk $ZnSO_4$, for NP treated plants, pod yield per plant was found to have increased by 34%. A similar 29.5% and 26.3% higher pod yield was observed for the NP-treated plants in the two successive field experiments performed in two consecutive years by the same research group as compared to chelated $ZnSO_4$ (Prasad et al., 2012).

In a different study performed on chickpea plants, foliar spray with ZnO NPs resulted in a notable 21.6% increment in plant shoot growth in comparison to the control bulk $ZnSO_4$ (Burman et al., 2013). A similar study done on onion plant showed better growth and early flowering in the onion plants treated with 20 and 30 µg. mL^{-1} (microgram per milliliter) concentrations of ZnO NPs. The NP-treated plants also demonstrated higher values for seeded fruit and seed weight per umbel and 1000 seed weight over the untreated plants (Lawre and Raskar, 2014). The beneficial effects of ZnO NPs have also been studied in mung bean (*Vigna radiata* L.) seeds with a significant increase in seed germination percentage, shoot and root dry weight, shoot and root length on treatment with 50 and 100 milligrams (mg) concentrations of ZnO NPs (20 nm) (Jayarambabu et al., 2015).

c. Manganese nanoparticles

Mn is an essential micronutrient required by the plant in trace amounts and is involved in photosynthesis, respiration, and N-metabolism. It also provides protection against root and shoot pathogens (Pradhan et al., 2013; Dimkpa et al., 2018). Several research groups have shown interest in developing Mn NPs to be used as fertilizers (Pourjafar et al., 2016; Sabaghnia and Janmohammadi, 2016; El-Metwally et al., 2018). In a study conducted by Pradhan et al. (2013), Mn NPs at a concentration of 0.05 mg. L^{-1} were observed to increase photosynthetic rate in mung bean plants (*Vigna radiata*), and increase the overall growth of the plants by 52%. In this study, the root and shoot length were also found to have increased by 52% and 38%, respectively (Pradhan et al., 2013). A different study conducted by the same research group to determine the effect of Mn NPs on mung bean seeds showed increased nitrate reductase and nitrite reduction activity in root and leaf tissue of the NP-treated plants (Pradhan et al., 2014). A recent study done to evaluate the effect of nano-Mn on wheat yield and nutrient acquisition, showed nano-Mn to exhibit significant increase in plant growth with higher Mn content in plant shoot and grain, lower soil nitrate-N, and higher P content in soil and shoot, in comparison to the bulk (Mn-oxide), and ionic Mn ($MnCl_2.4H_2O$) (Dimkpa et al., 2018). Besides these, Mn NPs have also been reported to increase the seed germination (> 50%) in *Lactuca sativa* seeds when exposed to a dose of < 50 mg. L^{-1} (Liu et al., 2016).

d. Copper nanoparticles

Cu is another essential element required for plant growth and plays a major role in its many physiological processes including photosynthesis and respiration along with cell wall metabolism, ethylene sensing, molybdenum cofactor biogenesis, and protection against oxidative stress (Yruela, 2005).

Currently, there appears to be an absence of Cu NP-based fertilizers in the agricultural market as less research has been conducted on the potential beneficial effects of Cu NPs on plant growth and development. In one study, Cu NPs were added to soil and a concentration-dependent increase in the ratio of the shoot to root length was observed in lettuce plants (*Lactuca sativa*) (Shah and Belozerova, 2009). Cu NPs have also been reported to have a dose-dependent beneficial effect on germination, productivity, plant survival rate, fresh and dry weight in wheat (Pestovsky and Martinez-Antonio, 2017), and increased growth and yield nutritional characteristics of tomato plants (Juárez Maldonado et al., 2016). Further research is required to evaluate Cu NPs for suitability as fertilizer.

e. Other nanoparticles

Titanium dioxide (TiO_2) is another class of NPs which is being investigated for use in the agricultural field. In a study by Gao et al. (2006), exposure of spinach plants to nano-anatase TiO_2 was observed to increase the fresh weight and dry weight of plants by 60.36% and 71.15%, respectively. The NP-treated plants also possessed higher chlorophyll content and a 31.87% increase in net photosynthetic activity (Gao et al., 2006). Similarly, improved plant growth, increased chlorophyll content, and higher fruit yield was reported in tomato after foliar application of TiO_2 NPs (Raliya et al., 2015). TiO_2 NPs have also been reported to counteract H_2O_2 stress in common wheat and improve agronomic traits (Jaberzadeh et al., 2013).

A study done for identifying if titanium dioxide nanoparticles (TiO_2 NPs) can stimulate phytotoxicity in Canola (*Brassica napus*) at different concentration of 10–2000 mg. L^{-1} showed large radicle and plumule growth of seedling at higher concentration (1200 and 1500 mg. L^{-1}) of TiO_2 NPs (20 nm), as compared to the lower concentrations and control. Seeds treated with 2000 mg. L^{-1} TiO_2 NPs also showed significant germination percentage (75%), germination rate and seedling vigor (Mahmoodzadeh et al., 2013).

Silica (Si) NPs have also been demonstrated to improve plant growth and development (Bao-shan et al., 2004; Suriyaprabha et al., 2013; Roohizadeh et al., 2015). The addition of biologically synthesized silicon dioxide (SiO_2) NPs to soil

Table 2.1. Beneficial effects of NMs used as fertilizers.

Type of NP	Plant Species	Physiological/Biochemical Effects	Ref.
Fe	Watermelon (*Citrullus lanatus*)	Increased grain yield	(Wang et al., 2016)
	Soybean (*Glycine max* (L.) Merr.)	Increase in chlorophyll, protein, and soluble sugar content	(Roghayyeh et al., 2010)
	Corn (*Zea mays* L.)	Increase in root elongation, germination, and vigor index Increased germination rate, plant growth, and fresh weight. Increased leaf area, number of leaves, total chlorophyll content	(Li et al., 2016; de França Bettencourt et al., 2020; Hasan et al., 2020)
	Pomelo (*Citrus maxima*)	Beneficial effect on plant growth and self-protection	(Hu et al., 2017)
	Mandarin Orange (*Citrus reticulate*)	Increased agronomic productivity	(Li et al., 2017)
Zn	Peanut (*Arachis hypogaea*)	Promoted seed germination and seedling vigor	(Prasad et al., 2012)
	Chickpea plants (*Cicer arietinum* L. var. HC-1)	Increment in plant shoot growth	(Burman et al., 2013)
	Onion (*Allium cepa* L.)	Better growth and early flowering in the onion plants	(Lawre and Raskar, 2014)
	Mung bean (*Vigna radiata*)	Significant increase in seed germination percentage, shoot and root dry weight and lengths	(Jayarambabu et al., 2015)
Mn	Lettuce (*Lactuca sativa* L.)	Increased photosynthetic rate with increased root and shoot growth	(Liu et al., 2016)
	Mung bean (*Vigna radiata*)	Enhanced root elongation with increased seed germination and increased length of seedling root	(Pradhan et al., 2013)
Cu	Lettuce seeds	Increased ratio of shoot to root length in NP-treated plants	(Shah and Belozerova, 2009)
	Tomato	Positive effects on growth, yield nutritional characteristics	(Juárez Maldonado et al., 2016)
	Sweet potato (*Ipomoea batatas*)	Increased plant length	(Bonilla-Bird et al., 2020)
Ti	Spinach (*Spinacia oleracea*)	Increased fresh and dry weight of spinach with increased chlorophyll content	(Gao et al., 2006)
	Tomato (*Solanum lycopersicum*)	Enhanced plant growth with increased chlorophyll content and higher fruit yield	(Raliya et al., 2015)
	Wheat (*Triticum aestivum* L. c.v *'Pishtaz'*)	Counteracting of the H_2O stress	(Jaberzadeh et al., 2013)

Table 2.1 contd. ...

...Table 2.1 contd.

Type of NP	Plant Species	Physiological/Biochemical Effects	Ref.
Si	Maize (*Zea mays* L.)	Increase in germination percentage as indicated by increased number of germinated roots and heightened root length	(Rangaraj et al., 2013)
	Tall grasses (*Agropyron elongatum* L.)	Elevation of the tall wheatgrass seed germination	(Azimi et al., 2014)
	Tomato (*Lycopersicum esculentum*)	Improved seed germination and seed vigor index, as well as percent seed germination and mean germination time with increased fresh and dry weight of seedling	(Siddiqui and Al-Whaibi, 2014)

caused an increase in germination percentage in maize plants (Rangaraj et al., 2013). SiO_2 NPs have also been reported to improve seed germination by more than 80% in tall wheatgrass (*Agropyron elongatum* L.) (Azimi et al., 2014), and increase seed germination, mean germination time, and fresh and dry weight of seedlings in tomato plants (*Lycopersicum esculentum*) (Siddiqui and Al-Whaibi, 2014). The beneficial effects of different MNPs on different plant species are summarized in Table 2.1.

3.1.2 Adverse Effects of MNPs Applied to Plants

Assessment of nanotoxicity to plants, humans, and the environment has become one of the most extensively investigated topics in the nanoagriculture sector in recent years (Lv et al., 2018). Adverse effect of MNPs on different plant species has been recorded, generally at a higher application dosage (Ma et al., 2013; Li et al., 2015; Karimi and Mohsenzadeh, 2016; Wang et al., 2016; Rafique et al., 2018; Zhu et al., 2019). To begin with, there has been a growth in the number of studies demonstrating the benefits of Fe_2O_3 NPs for agriculture. However, the results have been variable between studies due to factors such as soil type and chemistry, crop species and age, physicochemical properties of FeO NPs of the finished product, and NP concentrations used. In a study, the effect of FeO NPs with average size 366.8 nm was tested in two plant species: cattail (*Cattail typha latifolia*) and hybrid poplars (*Populous deltoids* × *Populous nigra*). The end result of the study showed the FeO NPs to enhance the plant growth at lower concentrations of oxide but showed a significant reduction in transpiration and growth of the plants at higher concentrations (> 200 mg. L^{-1}) (Ma et al., 2013). A separate study was designed to test the effects of polyacrylic acid-coated negatively charged iron oxide nanoparticles [IONPs (IONP-NC)] and polyethyleneimine (PEI) coated positively charged IONPs (IONP-PC) on the physiology and reproductive capacity of *Arabidopsis thaliana* at two different concentrations: 3 and 25 mg. L^{-1}. Plants treated with 3 mg. L^{-1} of IONP-PC or IONP-NC did not impose any adverse effect on seedling and root length. However, treatment with 25 mg. L^{-1} of the NPs resulted in reduced seedling and root length for both the NPs (Bombin et al., 2015). An investigation carried out for assessing the effect of Fe_2O_3 NPs (20–30 nm) and bulk Fe_2O_3 at concentration: 100–1000 parts per million (ppm) on wheat (*Triticum aestivum* L.) seed germination and early growth stage showed a reduction in seed germination rate at concentration > 100 ppm (Feizi et al., 2013).

The potential toxic effects of ZnO NPs (size – 20 ± 5 nm) have been studied on ryegrass (*Lolium perenne*), wherein the end results showed the ZnO NPs to be toxic to the ryegrass seedlings in a dose-dependent manner with retardation of seedling growth with shorter roots and shoots at concentration > 50 mg.L^{-1}, in comparison to the control (Lin and Xing, 2008). On the other hand, a study determining the phytotoxicity ZnO NPs (both ~ 50 nm in diameter) on the roots of onion (*Allium cepa*) bulbs showed a significant reduction in root length on exposure to the ZnO NPs even at a lower concentration of < 20 µg. mL^{-1} (Ghodake et al., 2011).

The Food and Drug Administration (FDA) categorizes ZnO as a "GRAS" (generally recognized as safe) material. ZnO in its nano form is most commonly used in sunscreens (Smijs and Pavel, 2011). The other application areas include use in the food industry (Espitia et al., 2016), energy storage (Zhang et al., 2016), gas sensor (Chen et al., 2011) and biosensor development (Devi et al., 2011), in optical devices (Abdolmaleki et al., 2011) and solar cells (Qian et al., 2011), and in the field of medicine for drug delivery (Rasmussen et al., 2010). ZnO NPs have also been established as antimicrobial and antifungal agents (Raghupathi et al., 2011; Arciniegas-Grijalba et al., 2017; Amna et al., 2015; Jasim, 2015). These findings suggest for optimization of the dosage of ZnO NPs for agricultural applications. Large-scale toxicity and life-cycle analyses may also provide a clue for concluding the safe limits of ZnO.

Non-toxic effects of Mn NPs are the primary factors responsible for the non-availability of data regarding the phytotoxic effects of these NPs. Koppitke et al. (2010) reviewed the toxicity of Mn NPs on plants grown in solution and concluded the Mn NPs to be least toxic trace metallic elements with median phytotoxic level for the particles to be 2.5 ppm (Kopittke et al., 2010). Another study done by Pradhan et al. (2013) showed the Mn NPs to be biosafe toward the *Vigna radiata* plants and beneficial soil organisms at a concentration ranging from 0.05 mg. L^{-1} to 1 mg. L^{-1} (Pradhan et al., 2013).

Many reports regarding the determination of the potentially toxic effects of Cu NPs on plants exist in the public domain (Du et al., 2018; Da Costa and Sharma, 2016; Rajput et al., 2018). One such study performed to investigate the phytotoxicity effects of Cu NPs, with size ranging between 20–80 nm on the *Arabidopsis thaliana* roots, showed an inhibitory effect of the Cu NPs on the plant lateral root at 20 and 50 mg. L^{-1} of NP concentrations in a dose-dependent manner (Xu, 2018). A different study determining the potentially toxic effects of Cu NPs (10–30 nm) in cucumber (*Cucumis sativus*) presented an elaborate account of biological processes and changes that lead to phytotoxicity. The adverse phenotypical changes in the NP-treated plant was shown by decreased biomass followed by the reduced chlorophyll a and b content with a notable increment in hydrogen peroxide (H_2O_2) and malondialdehyde (MDA) levels (Mosa et al., 2018). The potential phytotoxic effect of CuO NPs has also been studied in coriander (*Coriandrum sativum*) (AlQuraidi et al., 2019), onion (*Allium cepa*) (Deng et al., 2016) and cauliflower (*Brassica oleracea var. botrytis*) and tomato (*Solanum lycopersicum*), etc. (Singh et al., 2017). The result for different plant studies showed small-sized CuO NPs to be more toxic to different plant species generally at higher test concentrations.

The study conducted to determine the phytotoxic effects of SiO_2 NPs (30 nm) in *Bt*-transgenic cotton showed that the application of the SiO_2 NPs significantly

Table 2.2. Toxic effects of NMs used as fertilizers.

Type of NP	Plant Species	Physiological/Biochemical Effects	Ref.
Fe	Lettuce (*Letuca sativa*)	Increased the antioxidant enzyme activities related to the changes in the mineral composition, reduced root size, lowered chlorophylls accumulation, particle aggregation on the root, affected water entrance.	(Trujillo-Reyes et al., 2014)
	Sunflower (*Helianthus annuus* L.)	Lack of uptake and translocation. Reduction of the root hydraulic conductivity	(Martínez-Fernández et al., 2016)
	Mouse ear cress (*Arabidopsis thaliana*)	Less chlorophylls, lower plant biomass	(Marusenko et al., 2013)
	Garden cress (*Lepidium sativum*) and garden pea (*Pisum sativum* L.) plants	Root accumulation due to its strong adsorption	(Bystrzejewska-Piotrowska et al., 2012)
	Onion (*Allium cepa* L.)	Increased oxidative stress and chromosomal aberration in root meristems	(Gantayat et al., 2020)
Zn	Peanut (*Arachis hypogaea*)	Promoted seed germination and seedling vigor	(Prasad et al., 2012)
	Alfalfa (*Medicago sativa* L.)	Root biomass reduction by 80%	(Bandyopadhyay et al., 2015)
	Mouse ear cress (*Arabidopsis thaliana*)	Inhibitory effects on seed germination. Inhibitory effects on root elongation and in the total number of leaves.	(Lee et al., 2010)
	Corn (*Zea mays*)	Increased Zn in roots and shoots when compared to control. Zn internalization through apoplastic pathway.	(Zhao et al., 2012)
	Phragmites australis	Reduced growth, chlorophyll content, photosynthetic efficiency, and transpiration.	(Caldelas et al., 2020)
Cu	Rice (*Oryza sativa* cv. Swarna)	Decreased root and shoot growth, weight, and germination yield	(Shaw and Hossain, 2013)
	Bean (*Phaseolus vulgaris*)	Growth inhibition, metal nutrition imbalance in shoots	(Dimkpa et al., 2015)
	Wheat (*Triticum aestivum*)	Growth inhibition (roots and shoots) and oxidative stress, Cu bioaccumulation	(Dimkpa et al., 2012)
	Radish (*Raphanus sativus*), perennial ryegrass (*Lolium perenne*)	DNA damage, growth inhibition	(Atha et al., 2012)
	Chinese elm (*Ulmus elongate*)	Reduction of net photosynthetic rate, increased carbohydrates and lipids, Cu accumulation in leaves	(Jianguo et al., 2015)

Table 2.2 contd. ...

...Table 2.2 contd.

Type of NP	Plant Species	Physiological/Biochemical Effects	Ref.
Ti	Broad bean (*Vicia faba*)	Reduction of glutathione reductase and APX activity	(Foltête et al., 2011)
	Duckweed (*Lemna minor*)	Increased SOD, CAT, and POD activity	(Song et al., 2012)
	Wheat	Biomass reduction, NP adherence to plant root surface	(Du et al., 2011)
	Bt-transgenic cotton	Significant reduction in plant height, shoot and root biomasses, reduction in Cu and Mg content in shoots and Na content in roots	(Le et al., 2014)
Si	Wheat (*Triticum aestivum* L.) and Barley (*Hordeum vulgare* L.)	Decreased seedling, shoot and root weight	(Behboudi, et al., 2017)
	Lentil (*Lens culinaris*)	Decrease in the root length, reduction of fresh weight, mitotic abnormalities	(Khan and Ansari, 2018)

decreased the plant height, shoot and root biomasses along with decreased Cu, Mg, and Na content in shoots and roots, respectively (Le et al., 2014). Another research group which aimed to investigate the effects of chitosan and SiO_2 NPs (both > 80 nm) at different doses (0, 30, 60, and 90) on wheat and barley plants showed decreased seedling, shoot and root weight, especially in wheat seeds priming, as well as direct exposure of barley seeds with the highest tested 90 ppm concentration of NPs (Behboudi et al., 2017). Significant decrease in the root and shoot length was also observed in *Lens culinaris* (Lentil) seedlings treated with > 50 $\mu g. mL^{-1}$ concentrations of 5–50 nm-sized SiNPs (Khan and Ansari, 2018). Both these studies emphasized the dosage-dependent toxic effects of these NPs. A summary of investigations carried out to determine the phytotoxic effects of MNPs in different plant species is presented in Table 2.2.

3.2 *Effects of Different MNPs on Soil Microbial Environment*

Plant microbiome is the dynamic community of microorganisms associated with the plant and has three regions: Phyllosphere, rhizosphere, and endosphere, providing a niche for the microbial community (Sharma et al., 2017). Plant microbiome are also known as the plant second genome, which plays an important role in determining plant health (Berendsen et al., 2012) and plant nutrition (Gattinger et al., 2008).

There have been studies performed in the past, demonstrating the effect of MNPs on soil microflora and soil health. The beneficial effects of FeO NPs on soil bacterial community using molecular-based methods showed that the NPs stimulate bacterial growth along with increasing the C and N cycling in soil (He et al., 2011). Another study for determining the effect of TiO_2 NPs on plants showed that the NPs enhance the root colonization of plant growth-promoting rhizobacteria (Timmusk et al., 2018). Enhanced micropropagation in pistachio plant on treatment with a formulation of nano-encapsulated plant growth-promoting rhizobacteria and their metabolites having alginate-Si NPs and carbon nanotube (Pour et al., 2019). Contrarily, the reports showing toxic effects of these NPs on the soil microflora and soil health can also be found (Ge et al., 2011; Chai et al., 2015; Shen et al., 2015). There are several factors responsible for the net effect of NPs on plant-microbe symbioses (Tian et al., 2019). One such study carried out to determine the effects of a set of five different NPs on the community of soil bacteria showed no decline in soil microbial richness. In addition, the study suggested that the environmental parameters, along with the type of microorganisms considered in the analysis, influence the behavior of NPs in the soil matrix (Concha-Guerrero et al., 2014; Shah et al., 2014). Properties of soil have also been observed to play a key role in the bioavailability of NPs (Simonin and Richaume, 2015).

Therefore, for soil health management, investigation of the potential effects of MNPs on rhizospheric microbial species becomes integral during the optimization of MNPs for crop improvement.

4. Mechanism Underlying Phytotoxicity Caused by MNPs

a. Routes of MNP exposure to plants

Direct (application into the soil/foliar spray) or Indirect (air/water stream, etc.)

A lot of information is now available for the NP uptake and localization in plants; yet, a wide research gap regarding the understanding of internalization of MNPs in the plant cells remains to be addressed in a comprehensive manner (Pérez-de-Luque, 2017; Singh et al., 2018; Shahid et al., 2019; Wagener et al., 2019).

The uptake of NPs in the plants can be governed by the two different routes—direct and indirect (Fig. 2.2). These routes facilitate uptake of MNPs in plants. In the case of direct exposure, the NPs are either sprayed on the plant shoots (Avellan et al., 2019) or are applied to the soil for their uptake and translocation (Chen, 2018). During the foliar spray, NPs can enter sub-sections of leaves and can accumulate at various locations such as epidermis, mesophyll, stroma, xylem, and phloem (Singh et al., 2018; Lv et al., 2018). Biodegradable NPs such as polymeric NPs do not pose the clearance issues from the plant systems (Pereira et al., 2019); however MNPs, if accumulated in plant tissue, can cause blocking of aquaporin (Zuverza-Mena et al., 2016), generation of reactive oxygen species (ROS) (Jiang et al., 2017), germination inhibition (Lin and Xing, 2007), reduced transpiration (Hawthorne et al., 2012), reduced root/shoot growth (Al-Huqail et al., 2018), and reduction in overall plant biomass (Adeleye et al., 2014).

In case of indirect exposure (Lv et al., 2018), the plants get exposed to nanoparticles by either using nano-contaminated water (Ma et al., 2018), nano-contaminated soil (Bao et al., 2019), or particulate matter in the air. Due to the complexity of transformation in soil or water (Pradas Del Real et al., 2016), the NPs taken up by plants may not remain in their pristine forms (Dev et al., 2018; Ma et al., 2018). A comprehensive analysis of phytotoxicity in such cases is under research and significant results are yet to be obtained (Gottschalk and Nowack, 2011; Pan and Xing, 2012).

Uptake of NPs in plants is dependent on the plant species, physicochemical characteristics of the NPs such as size (Sabo-Attwood et al., 2012) and shape of the

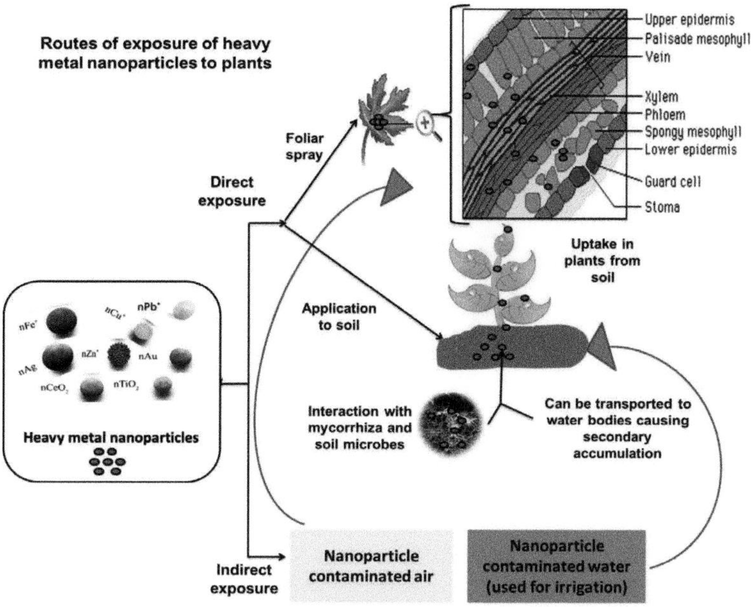

Fig. 2.2. Routes of plant exposure to NPs.

NPs (Raliya et al., 2016), and the mode of application (soil, foliar, hydroponics, or synthetic medium) (Chichiriccò and Poma, 2015).

b. Mechanism of Phytotoxicity

Despite a lot of research conducted in the arena (Yang and Watts, 2005; Lee et al., 2010; Lee et al., 2011; Geisler-Lee et al., 2012; Yang et al., 2017), mechanism of phytotoxicity caused by most NPs has not been completely understood (Yang et al., 2017). Based on the available literature data concerning the putative mechanisms underlying phytotoxicity caused by NPs, ROS generation, the effects on chloroplast, photosystems, nutrient uptake, genotoxicity, or programmed cell death (PCD) are the most prominent (Le et al., 2014; Oukarroum et al., 2015; Faisal et al., 2016; Le Van et al., 2016; Yang, Pan et al., 2017; AlQuraidi et al., 2019) (Fig. 2.3).

1. Oxidative stress

Reactive oxygen species (ROS) are generated during aerobic metabolism in plants and contain both free radicals such as superoxide anion, hydroxyl radical, and non-radical molecules like hydrogen peroxide and singlet oxygen (Sharma et al., 2012). Some of the cellular sites, including mitochondria, chloroplasts, peroxisomes, and the extracellular side of the plasma membrane are known to generate reactive molecules in plants. The normal level of ROS triggers signal transduction events, such as mitogen-activated protein (MAP) kinase cascades, which can then trigger specific cellular responses (Bailey-Serres and Mittler, 2006). However, under environmental stress, uncontrolled production of ROS can overtake cellular defense mechanisms and lead to oxidative stress. Oxidative stress can pose a threat to cells by causing lipid peroxidation, protein oxidation, enzyme inhibition, and damage to nucleic

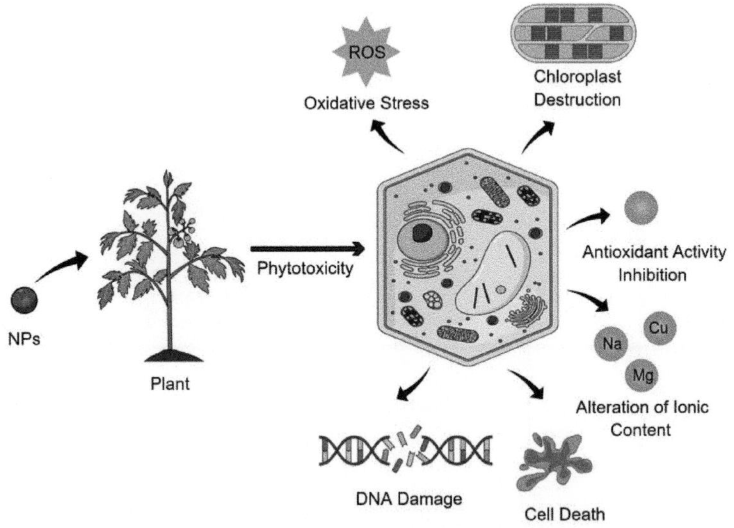

Fig. 2.3. Putative mechanisms of phytotoxicity caused by NPs. 1. ROS generation. 2. The effect on chloroplast and photosystems. 3. The effect on nutrient uptake. 4. Genotoxicity, or 5. Programmed cell death (PCD).

acid, along with activation of the programmed cell death (PCD) pathway resulting in cell death (Sharma et al., 2012). In response to oxidative damage, plants exhibit antioxidative defense mechanisms and up-regulate the production of both enzymatic and non-enzymatic components to scavenge the ROS. The enzymes include SOD, CAT (Chloramphenicol acetyltransferase), and peroxidases (POXes) including guaiacol peroxidase (GPOX), ascorbate peroxidase (APOX), glutathione reductase (GR), and dehydroascorbate reductase (DHAR). The non-enzymatic components of antioxidants include ascorbates and thiols (glutathione). In plants exposed to NPs, ROS and the peroxyl radicals are reported to be quenched by different forms of POX and CAT, whereas superoxide radicals are scavenged by SOD (Kumar et al., 2017).

Many recent studies have shown the impact of NPs on antioxidant system of plants (Qian et al., 2013; Ghosh et al., 2016; Li et al., 2016). A variety of NPs has been reported to induce ROS generation as one of the principal mechanisms of NP toxicity in plants (Nair and Chung, 2014; Vishwakarma et al., 2017; Du et al., 2018).

In a study determining the effect of CuO and TiO_2 NPs on *Brassica juncea* plants, an excess production of H_2O_2 by both the NPs was measured and it caused alterations in the activities of antioxidant enzymes: ascorbate peroxidase (APX) functions in regulating cellular levels of H_2O_2 and catalase (CAT) responsible for quenching excess H_2O_2. In this study, CAT activity was observed to decrease with increasing concentrations of CuO NPs, resulting in an increase in H_2O_2 levels in treated plants (Rao and Shekhawat, 2016). Similar to this, in a previous study, treatment of mustard seedlings with silver NPs (Ag NPs) resulted in decreased activities of antioxidants, APX and CAT (Vishwakarma et al., 2017). The authors suggested that the dis-regulation of antioxidant defense mechanisms may have been due to an interaction between the Ag NPs and the cytosolic and lipid bilayer proteins (Vishwakarma et al.,

2017). Increased cellular MDA and H_2O_2 levels in root and shoot samples, indicative of oxidative stress, have also been observed in cucumber (*Cucumis sativus*) plants, after exposure to Cu NPs (Du et al., 2018). Similarly, in a study comparing the phytotoxicity of three different metal-based NPs, Al_2O_3, ZnO, and Ag on soybean (*Glycine max*) seedlings, the ZnO NPs were observed to cause severe oxidative stress in the plants along with Ag NPs. At a concentration of 500 mg. L^{-1}, the ZnO NPs down-regulated genes involved in the oxidation-reduction cascade including, GDSL motif lipase 5, SKU5 similar 4, galactose oxidase, and quinone reductase (Hossain et al., 2016).

2. Effect on chloroplast and photosystems

Exposure to NPs has been correlated with damage to the photosynthetic system. A study performed to determine the impact of cerium dioxide (CeO_2) NPs in two separate varieties of cotton plants showed xylem sap-mediated transportation of absorbed NPs inside plants followed by the swelling and rupturing of the chloroplast (Nhan et al., 2015). Another study by the same research group observed the primary localization of iron oxide (Fe_2O_3) NPs in the endodermis and vascular cylinder of cotton plant roots, followed by transportation to shoots of both *Bt*-transgenic and non-transgenic cotton (Van Nhan et al., 2016).

Similar to these results, chloroplast thylakoid structure due to exposure to chromium oxide (Cr_2O_3) NPs has also been observed (Li et al., 2018).

NiO NPs, when studied for its effect on aquatic plant *Lemna gibba* L., showed an adverse effect on the photochemical activity of plant photosystem II (PSII) as indicated by the reduction in the quantum yield of PSII electron transport (Oukarroum et al., 2015). The observed increase in ROS formation was due to two major reasons: (i) particle effect of NiO which is enhanced in nano-form as compared to the bulk form, and (ii) dose dependency with high ROS at high concentration (1000 mg.L^{-1}) (Oukarroum et al., 2015).

3. Impact on nutrient uptake

The effect of NPs on nutrient uptake by the plants has also been studied. In a study that investigated the impacts of Ag NPs on the physiology and nutritional quality of radish (*Raphanus sativus*) sprouts, the authors reported a reduced nutritional element absorption in the radish seedlings after Ag NP exposure that correlated with reduced levels of key macro-elements (Ca and Mg) and micro-elements (B, Cu, Mn, and Zn) (Zuverza-Mena et al., 2016). A similar effect on plant nutrient uptake was observed in the transgenic cotton plant harboring the isopentenyl transferase (Ipt) gene. Treatment of these plants with elevated CuO NP concentrations resulted in reduced nutrient uptake in the shoots and roots with Fe and Zn content in the shoots significantly decreasing after treatment with 200 mg. L^{-1} CuO NP (Le Van et al., 2016).

Another study performed to understand the impact of CeO_2 NPs on growth, yield, and nutritional composition in wheat (*Triticum aestivum* L.) suggested an alteration in accumulations of different micronutrients in plant roots, leaves, and grains, leading to altered crop quality (Rico et al., 2014).

Similar dose-dependent effects of NPs on the plant nutritional quality have been observed in other plants such as radish (*Raphanus sativus*) (Zuverza-Mena et al., 2016), and Bt-transgenic cotton, etc. (Le et al., 2014).

4. Genotoxic effects of NPs

NPs, due to their ultra-small sizes, can penetrate the plant cell to reach the nuclei and cause damage to the plant at its genetic level. NPs taken up by the plants can follow either of the two pathways: apoplastic or symplastic to reach different parts of the plant (Pérez-de-Luque, 2017). Within the plant cells, accumulation of NPs may lead to the induction of genotoxic effects by their direct interaction with the genetic material or by ROS generation (Pacheco and Buzea, 2018). Genotoxic effects, due DNA damage, have also been reported for different plant species on treatment with different NPs. In a study that assessed the potential toxic effects of Cu NPs on coriander plants (*Coriandrum sativum*), grown hydroponically, results from random amplified polymorphic DNA (RAPD) analysis showed the absence of one band for the samples treated with 200 mg. L^{-1}, 400 mg. L^{-1}, and 800 mg. L^{-1} Cu NPs. An additional band at 700 bp was also observed in the plants exposed to 400 mg. L^{-1} and 800 mg. L^{-1} Cu NP highlighting the impact NPs can have on the genome of the coriander plants (AlQuraidi et al., 2019). Si NPs have also been observed to cause genotoxicity in lentils, with a dose-dependent increase in the number of chromosomal aberrations and decreased mitotic index in lentil seeds treated with Si NPs. The authors suggested that the NPs likely caused genomic instability in lentil seeds with impaired mitosis and altered DNA (Khan and Ansari, 2018). Ghosh et al. (2016) reported a significant increase in % tail DNA, leading to G2/M cell cycle arrest and apoptosis-like cell death in onion (*Allium cepa*) root nuclei after exposure to ZnO NPs. The cell death was characterized by ultrastructural alterations such as gross morphological changes, shrinkage and, vacuolation (Ghosh et al., 2016). Genotoxic effects have also been observed in root tip cells of wheat (*Triticum aestivum* L.), after exposure to both green and chemically synthesized Ag NPs (Abdelsalam et al., 2018). Irregular genome-related abnormalities including increased nuclear content and elongation, nuclear erosion and elongation were also observed in the root tip cells after exposure to the highest concentration of Ag NPs. Furthermore, the increased frequency of mitotic abnormalities and decreased mitotic index further indicated the genotoxic effects of the Ag NPs (Abdelsalam et al., 2018).

5. Programmed cell death (PCD)

NPs are believed to cause cytotoxicity of plant cells through pathways of apoptosis (programmed cell death) (Faisal et al., 2013; Ahmed et al., 2018). PCD is the genetically regulated process in multicellular eukaryotes which accounts for the elimination of undesirable and injured cells. In plants, PCD has a critical role in the control of developmental processes as well as environmental stress responses (Yanık et al., 2017). The three major forms of programmed cell death include apoptosis, autophagic cell death, and regulated necrosis (or necroptosis) (Andón and Fadeel, 2013). Among the number of studies performed to correlate MNP exposure and PCD, one includes the study done for determining the effect of single-walled carbon nanotubes (SWCNTs) on rice plants (*Oryza sativa* subsp. *japonica cv. Nipponbare*), which showed the adverse effects of the NPs on protoplasts and leaves through induction of chromatin condensation and ROS accumulation as compared to control treatments in a dose-dependent manner (Shen et al., 2010).

A different study investigating the role of cobalt oxide nanoparticles (Co_3O_4 NPs) inducing phytotoxicity, cellular DNA damage, and apoptosis in eggplant (*Solanum melongena* L. *cv. violettalunga* 2) showed NP caused apoptosis in eggplant via mitochondrial swelling and stimulation of NO signaling pathway (Faisal et al., 2016).

An earlier study done by the same research group identifying the mechanism of phytotoxicity caused by nickel oxide (NiO) nanoparticles in tomato plants showed NP-mediated induction of apoptosis in the plant. The data collected using the flow cytometry and fluorescence imaging suggested oxidative stress and mitochondrial dysfunction to be responsible for stimulating the process of apoptosis. A significant increase in caspase-3 activity by the treatment of NiO NPs proved its potential role in triggering the apoptotic pathway in the plant system (Faisal et al., 2013).

In a separate study, treatment of wheat plants with different concentrations of Al_2O_3 NPs, ranging from 5 to 50 mg. mL^{-1}, decreased the mitotic indices in wheat root cells in a dose-dependent manner and caused chromosomal abnormalities after 96 h of exposure. For all concentrations tested Al_2O_3 NP-induced caspase 3, 8, and 9-like activities were responsible for governing the PCD. Nuclear deformations and TUNEL positive reactions indicated the programmed cell death after Al_2O_3 NP induction (Yanık et al., 2017).

Despite all the progress made in recent years for determining the exact mechanism of NP interaction with plants and global regulations for the application of MNPs in agriculture, many gaps are still there in our understanding of the exact mechanism of phytotoxicity induced by NPs. As indicated by results from several studies discussed in the above sections, the phytotoxicity induced by NPs is predominantly governed by the shape, size, surface properties and the dosage to the plants. Besides this, another interesting area to explore will be to understand the effect of acute and prolonged dosages of NPs to plant systems. In addition to the current technologies being applied, advanced molecular approaches such as proteomics and genomics can assist in developing an in-depth understanding of the mechanism of phytotoxicity induced by NPs in plants (Yang et al., 2017).

5. Concerns Related to the Assessment of Potential Toxic Effects of MNPs

As evident from the published data, both beneficial and adverse effects have been observed on plant-growth when treated with MNPs. In fact, a larger number of studies have shown the MNPs to have a positive or no significant effect on plant growth (Ruttkay-Nedecky et al., 2017). The phytotoxicity has been suggested to be influenced by the environmental factors in addition to the physical and chemical nature of the MNPs and plant species (Ruttkay-Nedecky et al., 2017). Inference on toxicity has been drawn from a small number of studies conducted on a limited number of plant species. Most of the studies were focused on using a few variants of MNPs and studying their effects before fruiting, mostly at the early stages of the plant stage (Rem et al., 2012). Also, the reported phytotoxicity studies of NPs have not maintained consistency with respect to the plant system used for assessment, culturing conditions, incubation time, dosage, and physicochemical properties of ENMs. Therefore, it has become difficult to compare toxicity outcomes obtained from different studies. The relevance of the toxicity data is often defeated under such situations (Ray et al., 2009). The toxicity of nanoparticles may also be related to the size of seeds, type of plant species, plant growth stages and growth medium (Rajput et al., 2018). Furthermore, long-term exposure of plants to NPs, specifically to the crops and vegetables, has not been given much attention which are essential for generating experimental evidences to evaluate the extent that the NPs and their ionic form accumulate in the edible parts of plants, or if the NPs may get transferred along the food chain causing a threat to food safety (Lv et al., 2018). At present, insignificant evaluation of NP-contaminated food products or crop plants is performed before the commercialization of nanotechnology in multiple parts of the globe, increasing the risk of these products to be potentially toxic to the human and plant community as well as the environment (Maghsoodi et al., 2019). The studies investigating the interaction of the plant with nanoparticles at the transcriptomic and proteomic levels are scarce (Singh et al., 2017). Change in plant metabolites upon exposure to different NPs might play an important role in altering the NP speciation and thus, toxicity to plants. A mechanistic understanding of these metabolite-driven processes is however largely lacking (Ma et al., 2018). Additionally, *in vitro* studies involve assessment of phytotoxicity in modified media, synthetic or hydroponic state under a controlled environment, which may not absolutely correlate with outcomes observed under the field conditions (Dasgupta et al., 2017). Therefore, the research community across the globe must follow standard testing guidelines along with Good Laboratory Practices (GLP) norms and standard operating procedures (SOPs) for nanotoxicity evaluations. The knowledge thus obtained can be utilized to develop safe-by-design MNPs and also mitigate potential adverse effects associated with the MNPs when applied as nanonutrients and nanopesticides (Ma et al., 2018).

6. Regulatory and Commercial Aspects of NM for Agricultural Application

In order to support the development of nanotechnology in the agriculture sector, the necessity of the of nano-based products to be accepted in the regulatory arrangements have been named as the topmost preference by the international and national organizations including FDA, EU, OECD and, the Australian Pesticides and Veterinary Medicines Authority (APVMA) (Mitter and Husseyet, 2019). Amongst these organizations, EU and Switzerland have come up with specific provisions on nano-regulations. The mandatory labeling for the content of NMs has been insisted on the European Commission regulation numbered 1169/201 (Subramanian and Rajkishore, 2018). To develop an exclusive regulatory policy for NMs, an initiative was taken by the European Commission in the year 2004. Similarly, in Canada, the guidelines for the use of NMs have been specified by the Health Canada and Canadian Food Inspection Agency (CFIA), whereas, another organization, the Public Health Agency of Canada (PHAC) is responsible for food regulation. Both of these agencies are responsible for the regulation of nanoproducts in the country (Subramanian and Rajkishore, 2018). Russian corporation of nanotechnologies, National nano-safety strategic plan of South Korea, The Food and Drug Organization (FDO) of Iran, Taiwan Nanotechnology Industrialization Promotion Association (TANIDA) are the other country-based regulatory agencies for the legalization of nano products. There are three different regulatory agencies—Food Standards Australia New Zealand, the National Industrial Chemicals Notification and Assessment Scheme and APVMA—which work for the regulation of nanotechnology in the agriculture and food sectors in Australia. For nano-enabled products, there have been no specific guidelines available in New Zealand, Japan, Brazil, and African countries (Subramanian and Rajkishore, 2018). Very recently, the Department of Biotechnology, India and its allied research groups have formulated the guidelines for nano-enabled agriculture that also includes the key regulatory aspects of NPs used in agriculture (Adholeya et al., 2017).

Besides the regulatory aspects it's also worth mentioning that several of the nanoproducts are reaching the commercial markets to be used as potential fertilizers and pesticides. It can be seen from Tables 2.3 to 2.7, that NPs of B, Fe, Zn, Mo, and their composites have been produced for commercial purposes. This indicates the commercial viability of these promising NPs. It will be curious to further know about the global market acceptability and use of these products by farmers across the world.

Table 2.3. Boron-based nanofertilizer products available in global market for agriculture applications.

Commercial Name	Manufacturer	Type of Product	Applications
Nano bor 20%	Alert Biotech (India)	Boron nano particles/ nano powder	Allow crop development in regions with extreme rainfall, with acid and sandy soil and with soil with low organic matter
Nano Boron	Kanak Biotech (India)	Plant growth regulator	Helps plant's development by maintaining the balance between sugar and starch

Source: Tables 2.3–2.7: Various companies' website (Campos, 2021; He et al., 2019; Mittal et al., 2020)

Table 2.4. Iron-based nanofertilizer products available in global market for agriculture applications.

Commercial Name	Manufacturer	Type of Product	Applications
Nano iron chelate fertilizer	AFME TRADING GROUP (UK)	Iron nano particles/ nano powder	Chelated fertilizer that provide to plants essential micronutrients (Fe), resulting in plant growth acceleration/regulation
Nano Iron	Kanak Biotech (India)	Iron nano particles/ nano powder	Increases the availability of iron to plants and helps in respiration and photosynthesis process

Source: Tables 2.3–2.7: Various companies' website (Campos, 2021; He et al., 2019; Mittal et al., 2020)

7. Conclusion and Future Perspectives

Despite the potential advantages, and the growing trends in publications and patents, applications of nanotechnology in the agricultural field are still restricted to lab-scale research only. According to a search in the World Intellectual Property Organization database using the terms 'nano' and 'food', nearly 4,000 patent applications have been lodged in the past decade for these sectors. The potential benefits of nanotechnology to the agriculture and food sectors are evident from the growing number of publications and patents in the arena and yet, very few agricultural

Table 2.5. Zinc-based nanofertilizer products available in global market for agriculture applications.

Commercial Name	Manufacturer	Type of Product	Applications
Nano Zinc chelate fertilizer	AFME TRADING GROUP (UK)	Zinc nano particles/ nano powder	Chelated fertilizer that provides to plants essential micronutrients (Zn), resulting in plant growth acceleration/ regulation
Tag Nano Zinc	Tropical Agrosystem India (P) Ltd. (India)	'4G' Nano nutrient technologies (India)	Improves growth and yield, increases photosynthetic activity, and reduces chlorosis
Nano Zinc (Chelated) and Nano Zinc (Soil Application 21%)	Alert Biotech (India)	Zinc nano particles/ nano powder	Provides zinc to plants, which is a very important for plant's growth, protein synthesis, and energy production
Nano Chelated Zinc	Khazra (Iran)	Zinc nano particles/ nano powder	Slow release of chelated fertilizers for optimal use of nutrients by plants. Are resistant and stable in both alkaline and acidic soils.
Nano Zinc	Aqua-Yield® Operations, LLC (USA)	Zinc nano fertilizer	Zn absorption to roots, improved plant metabolism, seed quality, and cold tolerance
Nano Zinc	Silvertech Kimya Sanayi ve Ticaret Ltd. (Turkey)	Zinc nano fertilizer	Plant growth acceleration, photosynthesis, and crop yield enhancement, soil salinization resistant

Source: Tables 2.3–2.7: Various companies' website (Campos, 2021; He et al., 2019; Mittal et al., 2020)

Table 2.6. Molybdenum-based nanofertilizer products available in global market for agriculture applications.

Commercial Name	Manufacturer	Type of Product	Applications
Nano Molybdenum	Kanak Biotech (India)	Molybdenum nano particles/ nano powder	Provides better absorption of molybdenum by plants,which is an important component of enzymes

Source: Tables 2.3–2.7: Various companies' website (Campos, 2021; He et al., 2019; Mittal et al., 2020)

applications have been successful at commercial scale (Mitter and Hussey, 2019). This is particularly due to the high production costs of these nanoproducts that are required in large amounts in the agricultural sector, imprecise technical advantages and legislative concerns, as well as public opinion (Parisi et al., 2014). Additionally, it can be interpreted from the discussions in sections 3, 4, and 5 that the results reported from various studies conducted in the laboratory and greenhouse scale are conducted using variable experimental protocols and therefore show inconsistent results.

Table 2.7. Essential nutrient based composite nanofertilizer products available in the global market for agriculture applications.

Commercial Name	Manufacturer	Type of Product	Applications
Nova Land Nano Mn, Cu, Fe, Zn, Mo, N	Land Green & Technology Co (Taiwan)	Unspecified nano particles loading microelements (Mn, Cu, Fe, Zn, Mo, N)	Plant growth acceleration and regulation through increased chlorophyll formation and increased plant resistance against adverse conditions
Nanovec TSS 80	Laboratorios Bio-Me´dicin (Brazil)	Mn, Mo, Zn (encapsulated)	Unspecified nanoparticles loading microelements (Mn, Cu, Fe, Zn, Mo, N)
Lithocal	Litho Plant (Brazil)	Ca, Mn nano powder	Increases absorption of water and nutrients from the soil,and increases plant resistance against drought
Saula Drip, Saula Solocross, Saula Motawazen		Micro elements (Zn, Mn, Cu, B, Fe)	Fertilizer
Nubiotek Hyper Fe + Mg	Bioteksa	Fe +Mg nano particle/nano powder	Plant growth regulation, nutritional plant growth acceleration

Source: Tables 2.3–2.7: Various companies' website (Campos, 2021; He et al., 2019; Mittal et al., 2020)

Furthermore, these experiments are mostly conducted in modified media, a synthetic or hydroponic state which gives a controlled environment, which may not be able to predict precise responses that are seen under complex field conditions. Because of the lack of validated tests, the toxicology of such NPs is poorly understood. Globally accepted standard guidelines from REACH (Registration, Evaluation, Authorisation, and Restriction of Chemicals), OECD, CSIRO, EPA, AVMPA US-FDA and NIH (includes Nanoparticle Characterization Laboratory) may be followed for proper characterization of nanoparticles and the interferences induced by the nanoparticles in the available test system (Dasgupta et al., 2017).

To regulate nanotechnology in agri/feed/food sectors, different approaches are being followed in the countries which are members of the OECD and the non-member (non-OECD) countries. The REACH is the main EU regulation at present, which mainly addresses the use of NPs in plant protection products, food additives/supplements, and food contact materials (Mishra et al., 2017). Thus, similar to the specific guidelines made for testing of chemicals on terrestrial plants (OECD, 2003); specific guidelines explaining the selection and number of plant species to be tested, application of test substance, test conditions, and description of the method, should be made by OECD for toxicity studies of NPs in plants. Also, the norms of GLP and SOPs are needed to be followed for all toxic evaluations. The phytotoxicity analyses should always be conducted only by qualified and well-trained staff, following the written protocols, and using standardized equipment.

For the application of NPs to the plants, it is of great importance that the effect of physicochemical properties on phytotoxicity is established. Relevant long-term studies done using standard guidelines and GLPs are required for a crucial understanding of the effects of MNPs on plants that can focus on the following issues:

1) Characteristics of MNPs in different media (soil, hydroponics, and synthetic media) need to be conducted to understand the behavior of altered or transformed MNP on the agricultural application as compared to the pristine MNPs.
2) Application dosage needs to be optimized and defined for the application of nanoagri inputs on similar plant species.
3) Recent research has focused on the development of biogenic MNPs that are suggested to be safer to the environment and human health as compared to chemically synthesized counterparts. These MNPs have been less studied to explore their effects on plants.
4) Effect on edible parts of plants and biomagnification of these MNPs need to be explored intensively to know about their impact on successive trophic levels in the food chain.

In addition, the development of a comprehensive database is necessary for *in silico* modeling and hazard prediction of similar NMs, which will complement the evaluation procedure and help in better exploitation of nano-based agro-products (Prasad et al., 2017).

Acknowledgment

Authors acknowledge Professor Amit Kumar Dinda, All India Institute of Medical Science (AIIMS) for providing crucial discussion points regarding effects of NPs on crops.

Abbreviations/Acronyms

Ag NPs	Silver nanoparticles
Al_2O_3	Aluminum oxide
APOX	Ascorbate peroxidase
APVMA	The Australian Pesticides and Veterinary Medicines Authority
B	Boron
Ca	Calcium
CAT	Chloramphenicol acetyltransferase
CeO_2	Cerium dioxide
CFIA	the Health Canada and Canadian Food Inspection Agency
Cl	Chlorine
Co	Cobalt
Cu	Copper
DHAR	Dehydroascorbate reductase
ENMs	Engineered nanomaterials
FAO	Food and Agricultural Organization of the United States
FDA	The Food and Drug Administration

FDO	The food and drug organization
Fe	Iron
Fe_2O_3	Hematite
FeO	Iron Oxide
GLP	Good Laboratory Practices
$g.L^{-1}$	Grams per liter
g/dm^3	Grams per decimeter cube
GPOX	Guaiacol peroxidase
GR	Glutathione reductase
H_2O_2	Hydrogen peroxide
IONPs	Iron Oxide Nanoparticles
K	Potassium
MDA	Malondialdehyde
Mg	Milligram
$\mu g.mL^{-1}$	Microgram per milliliter
$mg.L^{-1}$	Milligrams per liter
Mn	Manganese
Mo	Molybdenum
N	Nitrogen
NFs	Nanofertilizers
Ni	Nickel
nm	Nanometer
NMs	Nanomaterials
NPs	Nanoparticles
OECD	Organization for Economic Co-operation and Development
P	Phosphorus
PCD	Programmed cell death
PHAC	the Public Health Agency of Canada
POXes	Peroxidases
ppm	Parts per million
REACH	Registration, Evaluation, Authorisation and Restriction of Chemicals
ROS	Reactive oxygen species
S	Sulfur
Si	Silica
SiO_2 NPs	Silicon dioxide nanoparticles
SOPs	Standard operating procedures
SOD	Superoxide dismutase
TANIDA	Taiwan Nanotechnology Industrialization Promotion Association
TiO_2	Titanium dioxide
USEPA	United States Environmental Protection Agency
US FDA	United States Food and Drug Administration
Zn	Zinc
ZnO	Zinc oxide
$ZnSO_4$	Zinc sulfate

References

Abdelsalam, N.R., Abdel-Megeed, A., Ali, H.M., Salem, M.Z.M., Al-Hayali, M.F.A. and Elshikh, M.S. (2018). Genotoxicity effects of silver nanoparticles on wheat (*Triticum aestivum* L.) root tip cells. *Ecotoxicol. Environ. Saf.*, 155: 76–85.

Abdolmaleki, A., Mallakpour, S. and Borandeh, S. (2011). Preparation, characterization and surface morphology of novel optically active poly (ester-amide)/functionalized ZnO bionanocomposites via ultrasonication assisted process. *Appl. Surf. Sci.*, 257(15): 6725–6733.

Adeleye, A.S., Conway, J.R., Perez, T., Rutten, P. and Keller, A.A. (2014). Influence of extracellular polymeric substances on the long-term fate, dissolution, and speciation of copper-based nanoparticles. *Environ. Sci. Technol.*, 48(21): 12561–12568.

Adholeya, A., Dubey, M.K., Kochar, M. and Singh, B.R. (2017). *Zero Draft Policy on Regulation of Nanoproducts in Agriculture*. The Energy and Resources Institute (TERI), New Delhi. www.teriin.org.

Ahamed, A., Liang, L., Lee, M.Y., Bobacka, J. and Lisak, G. (2021). Too small to matter? Physicochemical transformation and toxicity of engineered $nTiO_2$, $nSiO_2$, nZnO, carbon nanotubes, and nAg. *J. Hazard. Mater.*, 404: 124107.

Ahmad, W., Zia, M.S., Malhi, S., Niaz, A. and Ullah, S. (2012). Boron deficiency in soils and crops: A review. pp. 77–114. *In*: Goyal, A. (ed.). *Crop Plant* Rijeka, Croatia: InTech.

Ahmed, B., Saghir Khan, M., Saquib, Q., Al-Shaeri, M. and Musarrat, J. (2018). Interplay between engineered nanomaterials (ENMs) and edible plants: A current perspective. pp. 63–102. *In*: Faisal, M., Saquib Q., Alatar, A.A. and Al-Khedhairy, A.A. (eds.). *Phytotoxicity of Nanoparticles*. Cham: Springer.

Al-Hadede, L.T., Khaleel, S.A. and Hasan, S.K. (2020). Some applications of nanotechnology in agriculture. *Biochem. Cell. Arch.*, 20(1): 1447–1454.

Al-Huqail, A.A., Hatata, M.M., Al-Huqail, A.A. and Ibrahim, M.M. (2018). Preparation, characterization of silver phyto nanoparticles and their impact on growth potential of *Lupinus termis* L. seedlings. *Saudi J. Biol. Sci.*, 25(2): 313–319.

Al-Obaidi, J.R. (2020). Micro- and macronutrient signalling in plant cells: A proteomic standpoint under stress conditions. pp. 241–255. *In*: Aftab, T. and Hakeem, K.R. (eds.). *Plant Micronutrients: Deficiency and Toxicity Management*. Cham., Switzerland: Springer International Publishing.

Alloway, B.J. (2008). Micronutrients and crop production: An introduction. pp. 1–39. *In*: Alloway, B.J. (ed.). *Micronutrient Deficiencies in Global Crop Production*. Dordrecht: Springer.

Alquraidi, A.O., Mosa, K.A. and Ramamoorthy, K. (2019). Phytotoxic and genotoxic effects of copper nanoparticles in coriander (*Coriandrum sativum*—Apiaceae). *Plants*, 8(1): 19.

Amna, S., Shahrom, M., Azman, S., Kaus, N.H.M., Ling Chuo, A., Siti Khadijah Mohd, B., Habsah, H. and Dasmawati, M. (2015). Review on zinc oxide nanoparticles: Antibacterial activity and toxicity mechanism. *Nano-Micro Lett.*, 7: 219–242.

Andón, F.T. and Fadeel, B. (2013). Programmed cell death: Molecular mechanisms and implications for safety assessment of nanomaterials. *Acc. Chem. Res.*, 46(3): 733–742.

Archana, P.N. and Verma, P. (2017). Boron deficiency and toxicity and their tolerance in plants: A review. *J. Global Biosci.*, 6: 4958–4965.

Arciniegas-Grijalba, P.A., Patiño-Portela, M.C., Mosquera-Sánchez, L.P., Guerrero-Vargas, J.A. and Rodríguez-Páez, J.E. (2017). ZnO nanoparticles (ZnO-NPs) and their antifungal activity against coffee fungus *Erythricium salmonicolor*. *Appl. Nanosci.*, 7(5): 225–241.

Arunachalam, P., Pandian, K., Gnanasekaran, P. and Govindaraj, M. (2013). Zinc deficiency in Indian soils with special focus to enrich zinc in peanut. *Afr. J. Agri. Res.*, 8: 6681–6688.

Atha, D.H., Wang, H., Petersen, E.J., Cleveland, D., Holbrook, R.D., Jaruga, P., Dizdaroglu, M., Xing, B. and Nelson, B.C. (2012). Copper oxide nanoparticle mediated DNA damage in terrestrial plant models. *Environ. Sci. Technol.*, 46(3): 1819–1827.

Avellan, A., Yun, J., Zhang, Y., Spielman-Sun, E., Unrine, J.M., Thieme, J., Li, J., Lombi, E., Bland, G. and Lowry, G.V. (2019). Nanoparticle size and coating chemistry control foliar uptake pathways, translocation and leaf-to-rhizosphere transport in wheat. *ACS nano*, 13: 5291–5305.

Azimi, R., Farzam, M., Feizi, H. and Azimi, A. (2014). Interaction of SiO_2 nanoparticles with seed prechilling on germination and early seedling growth of tall wheatgrass (*Agropyron Elongatum* L.). *Pol. J. Chem. Technol.*, 16: 25–29.

Bailey-Serres, J. and Mittler, R. (2006). The roles of reactive oxygen species in plant cells. *Plant Physiol.*, 141(2): 311–311.

Baligar, V.C. and Fageria, N.K. (2015). Nutrient use efficiency in plants: An overview. pp. 193–206. *In*: Rakshit, A., Singh, H.B. and Sen, A. (eds.). *Nutrient Use Efficiency: From Basics to Advances*. New Delhi: Springer.

Bandyopadhyay, S., Plascencia, G., Mukherjee, A., Rico, C., José-Yacamán, M., Peralta-Videa, J. and Gardea-Torresdey, J. (2015). Comparative phytotoxicity of ZnO NPs, bulk ZnO, and ionic zinc onto the alfalfa plants symbiotically associated with *Sinorhizobium meliloti* in soil. *Sci. Total Environ.*, 515-516: 60–69.

Bao-Shan, L., Chun-Hui, L., Li-Jun, F., Shu-Chun, Q. and Min, Y. (2004). Effect of TMS (nanostructured silicon dioxide) on growth of Changbai larch seedlings. *J. For. Res.*, 15(2): 138–140.

Bao, Y., Pan, C., Liu, W., Li, Y., Ma, C. and Xing, B. (2019). Iron plaque reduces cerium uptake and translocation in rice seedlings (*Oryza sativa* L.) exposed to CeO_2 nanoparticles with different sizes. *Sci. Total Environ.*, 661: 767–777.

Batjes, N. (1995). *A Global Data Set of Soil pH Properties*. Technical Paper 27, International Soil Reference and Information Centre, Wageningen, p. 27.

Behboudi, F., Sarvestani, T., Kassaee, M., Sanavi, S. and Sorooshzadeh, A. (2017). Phytotoxicity of chitosan and SiO_2 nanoparticles to seed germination of wheat (*Triticum aestivum* L.) and barley (*Hordeum vulgare* L.) plants. *Not. Sci. Biol.*, 9(2): 242–249.

Benedet, L., Dick, D.P., Brunetto, G., Dos Santos Júnior, E., Ferreira, G.W., Lourenzi, C.R. and Comin, J.J. (2020). Copper and Zn distribution in humic substances of soil after 10 years of pig manure application in south of Santa Catarina, Brazil. *Environ. Geochem. Health*, 42(10): 3281–3301.

Berendsen, R.L., Pieterse, C.M. and Bakker, P.A. (2012). The rhizosphere microbiome and plant health. *Trends Plant Sci.*, 17(8): 478–486.

Bombin, S., Lefebvre, M., Sherwood, J., Xu, Y., Bao, Y. and Ramonell, K.M. (2015). Developmental and reproductive effects of iron oxide nanoparticles in *Arabidopsis thaliana*. *Int. J. Mol. Sci.*, 16(10): 24174–24193.

Bonilla-Bird, N., Ye, Y., Akter, T., Valdes-Bracamontes, C., Darrouzet-Nardi, A., Saupe, G., Flores-Marges, J., Ma, L., Hernandez-Viezcas, J. and Peralta-Videa, J. (2020). Effect of copper oxide nanoparticles on two varieties of sweetpotato plants. *Plant Physiol. Biochem.*, 154: 277–286.

Boutchuen, A., Zimmerman, D., Aich, N., Masud, A.M., Arabshahi, A. and Palchoudhury, S. (2019). Increased plant growth with hematite nanoparticle fertilizer drop and determining nanoparticle uptake in plants using multimodal approach. *J. Nanomater*, 2019: 6890572.

Burman, U., Saini, M. and Kumar, P. (2013). Effect of zinc oxide nanoparticles on growth and antioxidant system of chickpea seedlings. *Toxicol. Environ. Chem.*, 95(4): 605–612.

Bystrzejewska-Piotrowska, G., Asztemborska, M., Stęborowski, R., Ryniewicz, J., Polkowska-Motrenko, H. and Danko, B. (2012). Application of neutron activaton for investigation of Fe_3O_4 nanoparticles accumulation by plants. *Nukleonika*, 57: 427–430.

Caldelas, C., Poitrasson, F., Viers, J. and Araus, J. (2020). Stable Zn isotopes reveal the uptake and toxicity of zinc oxide engineered nanomaterials in *Phragmites australis*. *Environ. Sci. Nano*, 7(7): 1927–1941.

Campos, E.V.R. (2021). Commercial nanoproducts available in world market and its economic viability. pp. 561–593. *In*: Jogaiah, S., Singh, H.B., Fraceto, L.F. and Lima R.D. (eds.). *Advances in Nano-Fertilizers and Nano-Pesticides in Agriculture*. Sawston, UK: Woodhead Publishing Series in Food Science, Technology & Nutrition; Woodhead Publishing.

Chai, H., Yao, J., Sun, J., Zhang, C., Liu, W., Zhu, M. and Ceccanti, B. (2015). The effect of metal oxide nanoparticles on functional bacteria and metabolic profiles in agricultural soil. *Bull. Environ. Contam. Toxicol.*, 94: 490–495.

Chen, H. (2018). Metal-based nanoparticles in agricultural system: Behavior, transport, and interaction with plants. *Chem. Speciat. Bioavailab.*, 30(1): 123–134.

Chen, M., Wang, Z., Han, D., Gu, F. and Guo, G. (2011). Porous ZnO polygonal nanoflakes: Synthesis, use in high-sensitivity NO_2 gas sensor, and proposed mechanism of gas sensing. *J. Phys. Chem. C.*, 115(26): 12763–12773.

Chichiriccò, G. and Poma, A. (2015). Penetration and toxicity of nanomaterials in higher plants. *Nanomaterials*, 5(2): 851–873.

Concha-Guerrero, S., Brito, E., Castillo, H., Tarango-Rivero, S., Caretta, C., Luna, A., Duran, R. and Borunda, E. (2014). Effect of CuO nanoparticles over isolated bacterial strains from agricultural soil. *J. Nanomater*, 2014: 1–13.

Da Costa, M.V.J. and Sharma, P.K. (2016). Effect of copper oxide nanoparticles on growth, morphology, photosynthesis, and antioxidant response in *Oryza sativa*. *Photosynthetica*, 54(1): 110–119.

Das, S., Sen, B. and Debnath, N. (2015). Recent trends in nanomaterials applications in environmental monitoring and remediation. *Environ. Sci. Pollut. Res. Int.*, 22(23): 18333–18344.

Dasgupta, N., Ranjan, S. and Chidambaram, R. (2017). Applications of nanotechnology in agriculture and water quality management. *Environ. Chem. Lett.*, 15(4): 591–605.

De França Bettencourt, G.M., Degenhardt, J., Torres, L.A.Z., De Andrade Tanobe, V.O. and Soccol, C.R. (2020). Green biosynthesis of single and bimetallic nanoparticles of iron and manganese using bacterial auxin complex to act as plant bio-fertilizer. *Biocatal. Agric. Biotechnol.*, 30: 101822.

Deng, F., Wang, S. and Xin, H. (2016). Toxicity of CuO nanoparticles to structure and metabolic activity of Allium cepa root tips. *Bull. Environ. Contam. Toxicol.*, 97: 702–708.

Dev, A., Srivastava, A.K. and Karmakar, S. (2018). Nanomaterial toxicity for plants. *Environ. Chem. Lett.*, 16(1): 85–100.

Devi, R., Thakur, M. and Pundir, C.S. (2011). Construction and application of an amperometric xanthine biosensor based on zinc oxide nanoparticles–polypyrrole composite film. *Biosens. Bioelectron.*, 26(8): 3420–3426.

Dimkpa, C., Singh, U., Adisa, I., Bindraban, P.S., H Elmer, W., Gardea-Torresdey, J. and C White, J. (2018). Effects of manganese nanoparticle exposure on nutrient acquisition in wheat (*Triticum aestivum* L.). *Agronomy*, 8(9): 158.

Dimkpa, C.O., Mclean, J.E., Latta, D.E., Manangón, E., Britt, D.W., Johnson, W.P., Boyanov, M.I. and Anderson, A.J. (2012). CuO and ZnO nanoparticles: Phytotoxicity, metal speciation, and induction of oxidative stress in sand-grown wheat. *J. Nanopart. Res.*, 14(9): 1125.

Dimkpa, C.O., Hansen, T., Stewart, J., Mclean, J.E., Britt, D.W. and Anderson, A.J. (2015). ZnO nanoparticles and root colonization by a beneficial pseudomonad influence essential metal responses in bean (*Phaseolus vulgaris*). *Nanotoxicology*, 9(3): 271–278.

Du, W., Sun, Y., Ji, R., Zhu, J., Wu, J. and Guo, H. (2011). TiO_2 and ZnO nanoparticles negatively affect wheat growth and soil enzyme activities in agricultural soil. *J. Environ. Monit.*, 13: 822–828.

Du, W., Tan, W., Yin, Y., Ji, R., Peralta-Videa, J.R., Guo, H. and Gardea-Torresdey, J.L. (2018). Differential effects of copper nanoparticles/microparticles in agronomic and physiological parameters of oregano (*Origanum vulgare*). *Sci. Total Environ.*, 618: 306–312.

Duhan, J.S., Kumar, R., Kumar, N., Kaur, P., Nehra, K. and Duhan, S. (2017). Nanotechnology: The new perspective in precision agriculture. *Biotechnol. Rep.*, 15: 11–23.

El-Metwally, I., Abo-Basha, D.M. and El-Aziz, M.A. (2018). Response of peanut plants to different foliar applications of nano-iron, manganese, and zinc under sandy soil conditions. *Middle East J. Appl. Sci.*, 8(2): 474–482.

Espitia, P., Otoni, C. and De Fátima Ferreira Soares, N. (2016). Zinc oxide nanoparticles for food packaging applications. pp. 425–431. *In*: Barros-Velázquez, J. (ed.). *Antimicrobial Food Packaging* (1st Edn.), San Diego (2016): Elsevier.

Fageria, N.K. and Moreira, A. (2011). The role of mineral nutrition on root growth of crop plants. pp. 251–331. *In*: Sparks, D.L. (ed.). *Advances in Agronomy*. Burlington: Elsevier Inc., Academic Press.

Faisal, M., Saquib, Q., Alatar, A.A., Al-Khedhairy, A.A., Hegazy, A.K. and Musarrat, J. (2016). Cobalt oxide nanoparticles aggravate DNA damage and cell death in eggplant via mitochondrial swelling and NO signaling pathway. *Biol. Res.*, 49: 20.

Faisal, M., Saquib, Q., Alatar, A.A., Al-Khedhairy, A.A., Hegazy, A.K. and Musarrat, J. (2013). Phytotoxic hazards of NiO-nanoparticles in tomato: A study on mechanism of cell death. *J. Hazard. Mater.*, 250–251: 318–332.

FAO/WHO. (2013). *The State of Food and Agriculture, 2013*. www.fao.org/publications/sofa/2013/en/.

Feizi, H., Rezvani Moghaddam, P., Shahtahmassebi, N. and Fotovat, A. (2013). Assessment of concentrations of nano and bulk iron oxide particles on early growth of wheat (*Triticum aestivum* L.). *Annu. Rev. Res. Biol.*, 3: 752–761.

Foltête, A.S., Masfaraud, J.F., Bigorgne, E., Nahmani, J., Chaurand, P., Botta, C., Labille, J., Rose, J., Férard, J.F. and Cotelle, S. (2011). Environmental impact of sunscreen nanomaterials: Ecotoxicity and genotoxicity of altered TiO_2 nanocomposites on *Vicia Faba*. *Environ. Pollut.*, 159: 2515–2522.

Gantayat, S., Nayak, S.P., Badamali, S.K., Pradhan, C. and Das, A.B. (2020). Analysis on cytotoxicity and oxidative damage of iron nano-composite on *Allium cepa* L. root meristems. *Cytologia*, 85(4): 325–332.

Gao, F., Hong, F., Liu, C., Zheng, L., Su, M., Wu, X., Yang, F., Wu, C. and Yang, P. (2006). Mechanism of nano-anatase TiO_2 on promoting photosynthetic carbon reaction of spinach: Inducing complex of rubisco-rubisco activase. *Biol. Trace Elem. Res.*, 111(1–3): 239–253.

Gattinger, A., Palojärvi, A. and Schloter, M. (2008). Soil microbial communities and related functions. pp. 279–292. *In*: Schröder, P., Pfadenhauer, J. and Munch, J.C. (eds.). *Perspectives for Agroecosystem Management*. San Diego, CA, USA: Elsevier.

Ge, Y., Schimel, J.P. and Holden, P.A. (2011). Evidence for negative effects of TiO_2 and ZnO nanoparticles on soil bacterial communities. *Environ. Sci. Technol.*, 45(4): 1659–1664.

Geisler-Lee, J., Wang, Q., Yao, Y., Zhang, W., Geisler, M., Li, K., Huang, Y., Chen, Y., Kolmakov, A. and Ma, X. (2012). Phytotoxicity, accumulation, and transport of silver nanoparticles by *Arabidopsis thaliana*. *Nanotoxicology*, 7(3): 323–337.

Ghodake, G., Deuk Seo, Y. and Sung Lee, D. (2011). Hazardous phytotoxic nature of cobalt and zinc oxide nanoparticles assessed using *Allium cepa*. *J. Hazard. Mater.*, 186: 952–955.

Ghosh, M., Jana, A., Sinha, S., Jothiramajayam, M., Nag, A., Chakraborty, A., Mukherjee, A. and Mukherjee, A. (2016). Effects of ZnO nanoparticles in plants: Cytotoxicity, genotoxicity, deregulation of antioxidant defenses, and cell-cycle arrest. *Mutat. Res. Genet. Toxicol. Environ. Mutagen.*, 807: 25–32.

Gopal, R., Sharma, Y.K. and Shukla, A.K. (2016). Effect of molybdenum stress on growth, yield, and seed quality in black gram. *J. Plant Nutr.*, 39(4): 463–469.

Gottschalk, F. and Nowack, B. (2011). The release of engineered nanomaterials to the environment. *J. Environ. Monit. Assess.*, 13(5): 1145–1155.

Graham, R.D. (2008). Micronutrient deficiencies in crops and their global significance. pp. 41–61. *In*: Alloway, B.J. (ed.). *Micronutrient Deficiencies in Global Crop Production*. Dordrecht: Springer.

Hasan, M., Rafique, S., Zafar, A., Loomba, S., Khan, R., Hassan, S.G., Khan, M.W., Zahra, S., Zia, M. and Mustafa, G. (2020). Physiological and anti-oxidative response of biologically and chemically synthesized iron oxide: *Zea mays* a case study. *Heliyon*, 6(8): e04595.

Hawthorne, J., Musante, C., Sinha, S.K. and White, J.C. (2012). Accumulation and phytotoxicity of engineered nanoparticles to *Cucurbita pepo*. *Int. J. Phytoremediation*, 14(4): 429–442.

He, S., Feng, Y., Ren, H., Zhang, Y., Ning, G. and Lin, X. (2011). The impact of iron oxide magnetic nanoparticles on the soil bacterial community. *J. Soils Sediments*, 11: 1408–1417.

He, X., Deng, H. and Hwang, H.M. (2019). The current application of nanotechnology in food and agriculture. *J. Food Drug Anal.*, 27(1): 1–21.

Hossain, Z., Mustafa, G., Sakata, K. and Komatsu, S. (2016). Insights into the proteomic response of soybean towards Al_2O_3, ZnO, and Ag nanoparticles stress. *J. Hazard. Mater.*, 304: 291–305.

Hu, J., Guo, H., Li, J., Gan, Q., Wang, Y. and Xing, B. (2017). Comparative impacts of iron oxide nanoparticles and ferric ions on the growth of *Citrus maxima*. *Environ. Pollut.*, 221: 199–208.

Hussan, M., Hafeez, M., Saleem, M., Khan, S., Hussain, S. and Ahmad, N. (2021). Impact of soil applied humic acid, zinc and boron supplementation on the growth, yield, and zinc translocation in winter wheat. *Asian J. Agric. Biol.*, 30: 202102080.

Jaberzadeh, A., Moaveni, P., Moghadam, T. and Zahedi, H. (2013). Influence of bulk and nanoparticles titanium foliar application on some agronomic traits, seed gluten, and starch contents of wheat subjected to water deficit stress. *Not. Bot. Hortic. Agrobot.*, 41: 201–207.

Jampílek, J. and Kráľová, K. (2015). Application of nanotechnology in agriculture and food industry, its prospects and risks. *Ecol. Chem. Eng. S.*, 22(3): 321–361.

Jasim, N. (2015). Antifungal activity of zinc oxide nanoparticles on Aspergillus Fumigatus Fungus & Candida Albicans yeast. *Citeseer*, 5: 23–28.

Jayarambabu, N., Kumari, Rao, K. and Prabhu, Y. (2015). Beneficial role of zinc oxide nanoparticles on green crop production. *Int. J. Multidiscip. Adv. Res. Trends*, pp. 273–282.

Jehangir, I.A., Wani, S., Bhat, M.A., Hussain, A., Raja, W. and Athokpam, H. (2017). Micronutrients for crop production: Role of Boron. *Int. J. Curr. Microbiol.*, 6(11): 5347–5353.

Jiang, H.S., Yin, L.Y., Ren, N.N., Zhao, S.T., Li, Z., Zhi, Y., Shao, H., Li, W. and Gontero, B. (2017). Silver nanoparticles induced reactive oxygen species via photosynthetic energy transport imbalance in an aquatic plant. *Nanotoxicology*, 11(2): 157–167.

Jianguo, G., Yuhuan, W., Gendi, X., Wenqiao, L., Guohao, Y., Jun, M. and Liu, P. (2015). Effects of nano-TiO_2 on photosynthetic characteristics of Ulmus elongata seedlings. *Environ. Pollut.*, 176: 63–70.

Juárez Maldonado, A., Ortega-Ortíz, H., Pérez-Labrada, F., Cadenas-Pliego, G. and Benavides-Mendoza, A. (2016). Cu Nanoparticles absorbed on chitosan hydrogels positively alter morphological, production, and quality characteristics of tomato. *J. Appl. Bot. Food Qual.*, 89: 183–189.

Karimi, J. and Mohsenzadeh, S. (2016). Effects of silicon oxide nanoparticles on growth and physiology of wheat seedlings. *Russ. J. Plant Physiol.*, 63(1): 119–123.

Khan, Z. and Ansari, M.Y.K. (2018). Impact of engineered Si nanoparticles on seed germination, vigour index, and genotoxicity assessment via DNA damage of root tip cells in *Lens culinaris*. *J. Plant. Biochem. Physiol.*, 6: 5243–5246.

Kopittke, P.M., Blamey, F.P.C., Asher, C.J. and Menzies, N.W. (2010). Trace metal phytotoxicity in solution culture: A review. *J. Exp. Bot.*, 61(4): 945–954.

Kovács, B., Puskás-Preszner, A., Huzsvai, L., Lévai, L. and Bódi, É. (2015). Effect of molybdenum treatment on molybdenum concentration and nitrate reduction in maize seedlings. *Plant Physiol. Biochem.*, 96: 38–44.

Kumar, M. (2016). Micronutrients (B, Zn, Mo) for improving crop production on acidic soils of northeast India. *Natl. Acad. Sci. Lett.*, 39: 85–89.

Kumar, N., Tripathi, P. and Nara, S. (2018). Gold nanomaterials to plants: Impact of bioavailability, particle size, and surface coating. pp. 195–220. *In*: Tripathi, D.K., Ahmad, P., Sharma, S., Chauhan, D.K. and Dubey, N.K. (eds.). *Nanomaterials in Plants, Algae, and Microorganisms*. Academic Press, Elsevier.

Kumar, V., Khare, T., Sharma, M. and Wani, S. (2017). Impact of nanoparticles on oxidative stress and responsive antioxidative defense system of plants. pp. 393–406. *In*: Tripathi, D.K., Ahmad, P., Sharma, S., Chauhan, D.K. and Dubey, N.K. (eds.). *Nanomaterials in Plants, Algae, and Microorganisms*. Academic Press, Elsevier.

Lawre, S. and Raskar, S. (2014). Influence of zinc oxide nanoparticles on growth, flowering, and seed productivity in onion. *Int. J. Curr. Microbiol. Appl. Sci.*, 3: 874–881.

Le, N., Rui, Y., Gui, X., Li, X., Liu, S. and Han, Y. (2014). Uptake, transport, distribution, and bio-effects of SiO_2 nanoparticles in Bt-transgenic cotton. *J. Nanobiotechnol.*, 12: 50.

Le Van, N., Rui, Y., Cao, W., Shang, J., Liu, S., Nguyen Quang, T. and Liu, L. (2016). Toxicity and bio-effects of CuO nanoparticles on transgenic Ipt-cotton. *J. Plant Interact.*, 11(1): 108–116.

Le, V.N., Rui, Y., Gui, X., Li, X., Liu, S. and Han, Y. (2014). Uptake, transport, distribution and Bio-effects of SiO_2 nanoparticles in Bt-transgenic cotton. *J. Nanobiotechnology*, 12: 50.

Lee, C.W., Mahendra, S., Zodrow, K., Li, D., Tsai, Y.C., Braam, J. and Alvarez, P.J. (2010). Developmental phytotoxicity of metal oxide nanoparticles to *Arabidopsis thaliana*. *Environ. Toxicol. Chem.*, 29(3): 669–675.

Lee, W.M., Kwak, J.I. and An, Y.J. (2011). Effect of silver nanoparticles in crop plants *Phaseolus radiatus* and *Sorghum bicolor*: Media effect on phytotoxicity. *Chemosphere,* 86: 491–499.

Li, J., Chang, P., Huang, J., Wang, Y., Yuan, H. and Ren, H. (2013). Physiological effects of magnetic iron oxide nanoparticles towards watermelon. *J. Nanosci. Nanotechnol.*, 13: 5561–5567.

Li, J., Hu, J., Ma, C., Wang, Y., Wu, C., Huang, J. and Xing, B. (2016). Uptake, translocation and physiological effects of magnetic iron oxide (gamma-Fe_2O_3) nanoparticles in corn (*Zea mays* L.). *Chemosphere*, 159: 326–334.

Li, J., Hu, J., Xiao, L., Gan, Q. and Wang, Y. (2017). Physiological effects and fluorescence labeling of magnetic iron oxide nanoparticles on citrus (*Citrus reticulata*) seedlings. *Water, Air, & Soil Pollution*, 228(1): 52.

Li, J., Song, Y., Wu, K., Tao, Q., Liang, Y. and Li, T. (2018). Effects of Cr_2O_3 nanoparticles on the chlorophyll fluorescence and chloroplast ultrastructure of soybean (*Glycine max*). *Environ. Sci. Pollut. Res.*, 25(20): 19446–19457.

Li, X., Yang, Y., Gao, B. and Zhang, M. (2015). Stimulation of peanut seedling development and growth by zero-valent iron nanoparticles at low concentrations. *PLoS One*, 10(4): e0122884.

Lin, D. and Xing, B. (2008). Root uptake and phytotoxicity of ZnO nanoparticles. *Environ. Sci. Technol.*, 42(15): 5580–5585.

Lin, D.H. and Xing, B.S. (2007). Phytotoxicity of nanoparticles: Inhibition of seed germination and root elongation. *Environ. Pollut.*, 150: 243.

Liu, G., Dong, X., Liu, L., Wu, L., Peng, S.A. and Jiang, C. (2014). Boron deficiency is correlated with changes in cell wall structure that lead to growth defects in the leaves of navel orange plants. *Sci. Hortic.*, 176: 54–62.

Liu, R. and Lal, R. (2015). Potentials of engineered nanoparticles as fertilizers for increasing agronomic productions. *Sci. Total Environ.*, 514: 131–139.

Liu, R., Huiyingzhang and Lal, R. (2016). Effects of stabilized nanoparticles of copper, zinc, manganese, and iron oxides in low concentrations on lettuce (*Lactuca sativa*) seed germination: nano-toxicants or nano-nutrients? *Water Air Soil Pollut.*, 227: 42.

Lv, J., Christie, P. and Zhang, S. (2018). Uptake, translocation, and transformation of metal-based nanoparticles in plants: Recent advances and methodological challenges. *Environ. Sci. Nano*, 6(1): 41–59.

Ma, C., White, J.C., Zhao, J., Zhao, Q. and Xing, B. (2018). Uptake of engineered nanoparticles by food crops: Characterization, mechanisms, and implications. *Annu. Rev. Food Sci. Technol.*, 9(1): 129–153.

Ma, X., Gurung, A. and Deng, Y. (2013). Phytotoxicity and uptake of nanoscale zero-valent iron (nZVI) by two plant species. *Sci. Total Environ.*, 443: 844–849.

Ma, Y., Yao, Y., Yang, J., He, X., Ding, Y., Zhang, P., Zhang, J., Wang, G., Xie, C. and Luo, W. (2018). Trophic transfer and transformation of CeO_2 nanoparticles along a terrestrial food chain: Influence of exposure routes. *Environ. Sci. Technol.*, 52(14): 7921–7927.

Maghsoodi, M.R., Lajayer, B.A., Hatami, M. and Mirjalili, M.H. (2019). Challenges and opportunities of nanotechnology in plant-soil mediated systems: Beneficial role, phytotoxicity, and phytoextraction. pp. 379–404. *In*: Ghorbanpour, M. and Wani, S.H. (eds.). *Advances in Phytonanotechnology*. Academic Press, Elsevier.

Mahmoodzadeh, H., Nabavi, M. and Kashefi, H. (2013). Effect of nanoscale titanium dioxide particles on the germination and growth of canola (*Brassica napus*). *J. Ornamental Hortic. Plants*, 3: 25–32.

Mankad, M. (2017). Assessment of physiological and biochemical changes in rice seedlings exposed to bulk and nano iron particles. *Int. J. Pure Appl. Biosci.*, 5: 150–159.

Manuel, T.J., Alejandro, C.A., Angel, L., Aurora, G. and Emilio, F. (2018). Roles of molybdenum in plants and improvement of its acquisition and use efficiency. pp. 137–159. *In*: Hossain, M.A., Kamiya, T., Burritt, D.J., Phan Tran, L.S. and Fujiwara, T. (eds.). *Plant Micronutrient Use Efficiency*. Academic Press, Elsevier.

Martínez-Fernández, D., Barroso, D. and Komárek, M. (2016). Root water transport of *Helianthus annuus* L. under iron oxide nanoparticle exposure. *Environ. Sci. Pollut. Res.*, 23(2): 1732–1741.

Marusenko, Y., Shipp, J., Hamilton, G.A., Morgan, J.L., Keebaugh, M., Hill, H., Dutta, A., Zhuo, X., Upadhyay, N. and Hutchings, J. (2013). Bioavailability of nanoparticulate hematite to *Arabidopsis thaliana*. *Environ. Pollut.*, 174: 150–156.

Masunaga, T. and Fong, J. (2018). Strategies for increasing micronutrient availability in soil for plant uptake. pp. 195–208. *In*: Hossain, M.A., Kamiya, T., Burritt, D.J. Phan Tran, L.S. and Fujiwara, T. (eds.). *Plant Micronutrient Use Efficiency—Molecular and Genomic Perspectives in Crop Plants*. Academic Press, Elsevier.

Mehrazar, E., Rahaie, M. and Rahaie, S. (2015). Application of nanoparticles for pesticides, herbicides, fertilisers, and animals feed management. *Int. J. Nanopart.*, 8(1): 1–19.

Mishra, S., Keswani, C., Abhilash, P.C., Fraceto, L.F. and Singh, H.B. (2017). Integrated approach of agri-nanotechnology: Challenges and future trends. *Front. Plant Sci.*, 8: 471.

Mittal, D., Kaur, G., Singh, P., Yadav, K. and Ali, S.A. (2020). Nanoparticle-based sustainable agriculture and food science: Recent advances and future outlook. *Front. Nanotechnol.*, 2: 10.

Mitter, N. and Hussey, K. (2019). Moving policy and regulation forward for nanotechnology applications in agriculture. *Nat. Nanotechnol.*, 14(6): 508–510.

Mortvedt, J.J. and Gilkes, R.J. (1993). Zinc fertilizers. pp. 33–44. *In*: Robson, A.D. (ed.). *Developments in Plant and Soil Sciences: Zinc in Soils and Plants*. Dordrecht: Kluwer Academic Publishers, The Netherlands.

Mosa, K.A., El-Naggar, M., Ramamoorthy, K., Alawadhi, H., Elnaggar, A., Wartanian, S., Ibrahim, E. and Hani, H. (2018). Copper nanoparticles induced genotoxicty, oxidative stress, and changes in superoxide dismutase (SOD) gene expression in cucumber (*Cucumis sativus*) plants. *Front. Plant Sci.*, 9: 872–872.

Mousavi, S.R., Shahsavari, M. and Rezaei, M. (2011). A general overview on Manganese (Mn) importance for crops production. *Aust. J. Basic Appl. Sci.*, 5(9): 1799–1803.

Nagendran, R. (2011). Agricultural waste and pollution. pp. 341–355. *In*: Letcher, T.M. and Vallero, D.A. (eds.). *Waste*. Academic Press, Elsevier.

Nair, P.M. and Chung, I.M. (2014). Impact of copper oxide nanoparticles exposure on *Arabidopsis thaliana* growth, root system development, root lignificaion, and molecular level changes. *Environ. Sci. Pollut. Res. Int.*, 21(22): 12709–12722.

Nhan, L.V., Ma, C., Rui, Y., Liu, S., Li, X., Xing, B. and Liu, L. (2015). Phytotoxic mechanism of nanoparticles: Destruction of chloroplasts and vascular bundles and alteration of nutrient absorption. *Sci. Rep.*, 5: 11618.

OECD. (2003). Guideline for the Testing of Chemicals Proposal for Updating Guideline 208; Terrestrial Plant Test: Seedling Emergence and Seedling Growth Test, pp. 1–16.

Omar, R.A., Afreen, S., Talreja, N., Chauhan, D. and Ashfaq, M. (2019). Impact of nanomaterials in plant systems. pp. 117–140. *In*: Prasad, R. (ed.). *Plant Nanobionics*. Cham: Springer.

Oukarroum, A., Barhoumi, L., Samadani, M. and Dewez, D. (2015). Toxic effects of nickel oxide bulk and nanoparticles on the aquatic plant *Lemna gibba* L. *Biomed. Res. Int.*, 2015: 1–7.

Pacheco, I. and Buzea, C. (2018). Nanoparticle uptake by plants: Beneficial or detrimental? pp. 1–61. *In*: Faisal, M., Saquib, Q, Alatar, A. and Al-Khedhairy, A. (eds.). *Phytotoxicity of Nanoparticles*. Cham: Springer.

Pan, B. and Xing, B. (2012). Applications and implications of manufactured nanoparticles in soils: A review. *Eur. J. Soil Sci.*, 63(4): 437–456.

Pandey, M.K., Gangurde, S.S., Sharma, V., Pattanashetti, S.K., Naidu, G.K., Faye, I., Hamidou, F., Desmae, H., Kane, N.A., Yuan, M., Vadez, V., Nigam, S.N. and Varshney, R.K. (2021). Improved genetic map identified major QTLs for drought tolerance- and iron deficiency tolerance-related traits in groundnut. *Genes,* 12: 37.

Pariona, N., Martinez, A.I., Hdz-García, H.M., Cruz, L.A. and Hernandez-Valdes, A. (2017). Effects of hematite and ferrihydrite nanoparticles on germination and growth of maize seedlings. *Saudi J. Biol. Sci.*, 24(7): 1547–1554.

Parisi, C., Vigani, M. and Rodríguez-Cerezo, E. (2014). Agricultural nanotechnologies: What are the current possibilities? *Nano Today*, 10: 124–127.

Paunovic, J., Vucevic, D., Radosavljevic, T., Mandić-Rajčević, S. and Pantic, I. (2020). Iron-based nanoparticles and their potential toxicity: Focus on oxidative stress and apoptosis. *Chem. Biol. Interact.*, 316: 108935.

Pereira, A.D.E.S., Oliveira, H.C. and Fraceto, L.F. (2019). Polymeric nanoparticles as an alternative for application of gibberellic acid in sustainable agriculture: A field study. *Sci. Rep.*, 9(1): 7135.

Pérez-De-Luque, A. (2017). Interaction of nanomaterials with plants: What do we need for real applications in agriculture? *Front. Environ. Sci.*, 5: 12.

Pingoliya, K., Dotaniya, M. and Lata, M. (2014). Effect of iron on yield, quality and nutrient uptake of chickpea (*Cicer arietinum* L.). *Afr. J. Agric. Res.*, 9(37): 2841–2845.

Pour, M.M., Saberi-Riseh, R., Mohammadinejad, R. and Hosseini, A. (2019). Nano-encapsulation of plant growth-promoting rhizobacteria and their metabolites using alginate-silica nanoparticles and carbon nanotube improves UCB1 pistachio micropropagation. *J. Microbiol. Biotechnol.*, 29(7): 1096–1103.

Pourjafar, L., Zahedi, H. and Sharghi, Y. (2016). Effect of foliar application of nano iron and manganese chelated on yield and yield component of canola (*Brassica napus* L.) under water deficit stress at different plant growth stages. *Agric. Sci. Digest*, 36(3): 172–178.

Pradas Del Real, A.E., Castillo-Michel, H., Kaegi, R., Sinnet, B., Magnin, V., Findling, N., Villanova, J., Carrière, M., Santaella, C., Fernández-MartíNez, A., Levard, C. and Sarret, G. (2016). Fate of Ag-NPs in sewage sludge after application on agricultural soils. *Environ. Sci. Technol.*, 50(4): 1759–1768.

Pradhan, S., Patra, P., Das, S., Chandra, S., Mitra, S., Dey, K.K., Akbar, S., Palit, P. and Goswami, A. (2013). Photochemical modulation of biosafe manganese nanoparticles on *Vigna radiata*: A detailed molecular, biochemical, and biophysical study. *Environ. Sci. Technol.*, 47(22): 13122–13131.

Pradhan, S., Patra, P., Mitra, S., Dey, K.K., Jain, S., Sarkar, S., Roy, S., Palit, P. and Goswami, A. (2014). Manganese nanoparticles: Impact on non-nodulated plant as a potent enhancer in nitrogen metabolism and toxicity study both *in vivo* and *in vitro*. *J. Agric. Food Chem.*, 62(35): 8777–8785.

Prasad, R., Bhattacharyya, A. and Nguyen, Q.D. (2017). Nanotechnology in sustainable agriculture: Recent developments, challenges, and perspectives. *Front. Microbiol.*, 8: 1014.

Prasad, T.N.V.K.V., Sudhakar, P., Sreenivasulu, Y., Latha, P., Munaswamy, V., Reddy, K.R., Sreeprasad, T.S., Sajanlal, P.R. and Pradeep, T. (2012). Effect of nanoscale zinc oxide particles on the germination, growth and yield of peanut. *J. Plant Nutr.*, 35(6): 905–927.

Prashar, P. and Shah, S. (2016). Impact of fertilizers and pesticides on soil microflora in agriculture. pp. 331–361. *In*: Lichtfouse, E. (ed.). *Sustainable Agriculture Reviews* (Volume 19). Cham: Springer.

Puig, S. (2014). Function and regulation of the plant COPT family of high-affinity copper transport proteins. *Adv. Bot.*, 2014: 9.

Qian, H., Peng, X., Han, X., Ren, J., Sun, L. and Fu, Z. (2013). Comparison of the toxicity of silver nanoparticles and silver ions on the growth of terrestrial plant model *Arabidopsis thaliana*. *J. Environ. Sci.* (China), 25(9): 1947–1955.

Qian, L., Yang, J., Zhou, R., Tang, A., Zheng, Y., Tseng, T.K., Bera, D., Xue, J. and Holloway, P.H. (2011). Hybrid polymer-CdSe solar cells with a ZnO nanoparticle buffer layer for improved efficiency and lifetime. *J. Mater. Chem.*, 21(11): 3814–3817.

Qin, S., Xu, Y., Liu, H., Li, C., Yang, Y. and Zhao, P. (2021). Effect of different boron levels on yield and nutrient content of wheat based on grey relational degree analysis. *Acta Physiol. Plant.*, 43(9): 1–8.

Quiles-Pando, C., Rexach, J., Navarro-Gochicoa, M.T., Camacho-Cristóbal, J.J., Herrera-Rodríguez, M.B. and González-Fontes, A. (2013). Boron deficiency increases the levels of cytosolic Ca^{2+} and expression of Ca^{2+}-related genes in *Arabidopsis thaliana* roots. *Plant Physiol. Biochem.*, 65: 55–60.

Qureshi, A., Singh, D. and Dwivedi, S. (2018). Nano-fertilizers: A novel way for enhancing nutrient use efficiency and crop productivity. *Int. J. Curr. Microbiol. App. Sci.*, 7: 3325–3335.

Rafique, R., Zahra, Z., Virk, N., Shahid, M., Pinelli, E., Park, T.J., Kallerhoff, J. and Arshad, M. (2018). Dose-dependent physiological responses of *Triticum aestivum* L. to soil applied TiO_2 nanoparticles: Alterations in chlorophyll content, H_2O_2 production, and genotoxicity. *Agric. Ecosyst. Environ.*, 255: 95–101.

Raghupathi, K., T Koodali, R. and Manna, A. (2011). Size-dependent bacterial growth inhibition and mechanism of antibacterial activity of zinc oxide nanoparticles. *Langmuir*, 27: 4020–4028.

Rai, R., Kumar, R., Pathak, D. and Patel, K.K. (2021). Effect of zinc and Boron on potato yield and quality, nutrient uptake, and soil health. *J. Pharm. Innov.*, 10(11): 1020–1022.

Rajput, V., Minkina, T., Fedorenko, A., Sushkova, S., Mandzhieva, S., Lysenko, V., Duplii, N., Fedorenko, G., Dvadnenko, K. and Ghazaryan, K. (2018). Toxicity of copper oxide nanoparticles on spring barley (*Hordeum sativum distichum*). *Sci. Total Environ.*, 645: 1103–1113.

Rajput, V.D., Minkina, T., Fedorenko, A., Tsitsuashvili, V., Mandzhieva, S., Sushkova, S. and Azarov, A. (2018). Metal oxide nanoparticles: Applications and effects on soil ecosystems. pp. 81–106. *In*: Lund, J.E. (ed.). *Soil Contamination: Sources, Assessment and Remediation.* Hauppauge: Nova Science Publishers.

Raliya, R., Nair, R., Chavalmane, S., Wang, W.N. and Biswas, P. (2015). Mechanistic evaluation of translocation and physiological impact of titanium dioxide and zinc oxide nanoparticles on the tomato (*Solanum lycopersicum* L.) plant. *Metallomics*, 7(12): 1584–1594.

Raliya, R., Franke, C., Chavalmane, S., Nair, R., Reed, N. and Biswas, P. (2016). Quantitative understanding of nanoparticle uptake in watermelon plants. *Front. Plant Sci.*, 7: 1288.

Raliya, R., Saharan, V., Dimkpa, C. and Biswas, P. (2018). Nanofertilizer for precision and sustainable agriculture: Current state and future perspectives. *J. Agric. Food Chem.*, 66(26): 6487–6503.

Rangaraj, S., Gopalu, R.Y., Periasamy, P., Venkatachalam, R. and Kannan, N. (2013). Application of silica nanoparticles for increased silica availability in maize. *In*: Chauhan, A.K., Murli, C. and Gadkari, S.C. (eds.). *Proceedings of the 57th DAE Solid State Physics Symposium*, Mumbai, Dec. 2012. AIP Conf. Proc., 1512: 424–425.

Rao, S. and Shekhawat, G.S. (2016). Phytotoxicity and oxidative stress perspective of two selected nanoparticles in *Brassica juncea. 3 Biotech.,* 6(2): 244–244.

Rasmussen, J.W., Martinez, E., Louka, P. and Wingett, D.G. (2010). Zinc oxide nanoparticles for selective destruction of tumor cells and potential for drug delivery applications. *Expert Opin. Drug. Deliv.*, 7(9): 1063–1077.

Ray, P.C., Yu, H. and Fu, P.P. (2009). Toxicity and environmental risks of nanomaterials: Challenges and future needs. *J. Environ. Sci. Health. C. Environ. Carcinog. Ecotoxicol. Rev.*, 7(1): 1–35.

Remédios, C., Rosário, F. and Bastos, V. (2012). Environmental nanoparticles interactions with plants: Morphological, physiological, and genotoxic aspects. *J. Bot.*, 2012: 751686.

Research and Markets, 2018. *Global Nanotechnology Market Outlook 2024. Industry Research Report*. Available from https://www.researchandmarkets.com/research/9d2zws/global?w=4.

Rico, C.M., Lee, S.C., Rubenecia, R., Mukherjee, A., Hong, J., Peralta-Videa, J.R. and Gardea-Torresdey, J.L. (2014). Cerium oxide nanoparticles impact yield and modify nutritional parameters in wheat (*Triticum aestivum* L.). *J. Agric. Food Chem.*, 62(40): 9669–9675.

Ritchie, H., Reay, D.S. and Higgins, P. (2018). Quantifying, projecting, and addressing India's hidden hunger. *Front. Sustain. Food Syst.*, 2: 11.

Rizwan, M., Ali, S., Ali, B., Adrees, M., Arshad, M., Hussain, A., Ur Rehman, M.Z. and Waris, A.A. (2019). Zinc and iron oxide nanoparticles improved the plant growth and reduced the oxidative stress and cadmium concentration in wheat. *Chemosphere*, 214: 269–277.

Roghayyeh, S., Mehdi, T. and Rauf, S. (2010). Effects of nano-iron oxide particles on agronomic traits of soybean. *Notulae Sci. Biol.*, 2: 112–113.

Roohizadeh, G., Majd, A. and Arbabian, S. (2015). The effect of sodium silicate and silica nanoparticles on seed germination and growth in the *Vicia faba* L. *Trop. Plant Res.*, 2(2): 85–89.

Rosado, T.L., Freitas, M.S.M., Carvalho, A.J.C.D., Gontijo, I., Pires, A.A., Vieira, H.D. and Barcellos, R. (2021). Soil chemical properties and nutrition of conilon coffee fertilized with molybdenum and nitrogen. *Rev. Bras. Cienc. Solo.*, 45: e0210034.

Rui, M., Ma, C., Hao, Y., Guo, J., Rui, Y., Tang, X., Zhao, Q., Fan, X., Zhang, Z., Hou, T. and Zhu, S. (2016). Iron oxide nanoparticles as a potential iron fertilizer for peanut (*Arachis hypogaea*). *Front. Plant. Sci.*, 7: 815.

Ruttkay-Nedecky, B., Krystofova, O., Nejdl, L. and Adam, V. (2017). Nanoparticles based on essential metals and their phytotoxicity. *J. Nanobiotechnology*, 15(1): 33.

S Pestovsky, Y. and Martinez-Antonio, A. (2017). The use of nanoparticles and nanoformulations in agriculture. *J. Nanosci. Nanotechnol.*, 17(12): 8699–8730.

Sabaghnia, N. and Janmohammadi, M. (2016). Analysis of the impact of nano-zinc, nano-iron, and nano-manganese fertilizers on chickpea under rain-fed conditions. *Ann. Univ. Mariae Curie-Sklodowska, Sect. C Biol.*, 70: 43–45.

Sabir, S., Arshad, M. and Chaudhari, S.K. (2014). Zinc oxide nanoparticles for revolutionizing agriculture: Synthesis and applications. *Sci. World J.*, 2014: 925494.

Sabo-Attwood, T., Unrine, J.M., Stone, J.W., Murphy, C.J., Ghoshroy, S., Blom, D., Bertsch, P.M. and Newman, L.A. (2012). Uptake, distribution, and toxicity of gold nanoparticles in tobacco (*Nicotiana xanthi*) seedlings. *Nanotoxicology*, 6(4): 353–360.

Savci, S. (2012). Investigation of effect of chemical fertilizers on environment. *APCBEE Proc.*, 1: 287–292.

Schjoerring, J.K., Cakmak, I. and White, P.J. (2019). Plant nutrition and soil fertility: synergies for acquiring global green growth and sustainable development. *Plant Soil*, 434(1): 1–6.

Schmidt, S.B., Jensen, P.E. and Husted, S. (2016). Manganese deficiency in plants: The impact on photosystem II. *Trends Plant Sci.*, 21(7): 622–632.

Sekhon, B.S. (2014). Nanotechnology in agri-food production: An overview. *Nanotechnol. Sci Appl.*, 7: 31–53.

Shah, T., Xu, J., Zou, X., Cheng, Y., Zhang, X., Hussain, Q. and Gill, R.A. (2019). Impact of nanomaterials on plant economic yield and next generation. pp. 203–214. *In*: Ghorbanpour, M. and Wani, S.H. (eds.). *Advances in Phytonanotechnology*. Academic Press, Elsevier.

Shah, V. and Belozerova, I. (2009). Influence of metal nanoparticles on the soil microbial community and germination of lettuce seeds. *Water Air Soil Pollut.*, 197(1): 143–148.

Shah, V., Collins, D., Walker, V.K. and Shah, S. (2014). The impact of engineered cobalt, iron, nickel and silver nanoparticles on soil bacterial diversity under field conditions. *Environ. Res. Lett.*, 9(2): 024001.

Shahid, M., Dumat, C., Khalid, S., Rabbani, F., Farooq, A.B.U., Amjad, M., Abbas, G. and Niazi, N.K. (2019). Foliar uptake of arsenic nanoparticles by spinach: An assessment of physiological and human health risk implications. *Environ. Sci. Pollut. Res.*, 26(20): 20121–20131.

Sharma, P., Jha, A.B., Dubey, R.S. and Pessarakli, M. (2012). Reactive oxygen species, oxidative damage, and antioxidative defense mechanism in plants under stressful conditions. *J. Bot.*, 2012: 26.

Sharma, R., Singh, M. and Chauhan, A. (2017). Rhizosphere microbiome and its role in plant growth promotion. pp. 29–56. *In*: Kalia, V., Shouche, Y., Purohit, H. and Rahi, P. (eds.). *Mining of Microbial Wealth and MetaGenomics*. Singapore: Springer.

Shaw, A.K. and Hossain, Z. (2013). Impact of nano-CuO stress on rice (*Oryza sativa* L.) seedlings. *Chemosphere*, 93(6): 906–915.

Shen, C.X., Zhang, Q.F., Li, J., Bi, F.C. and Yao, N. (2010). Induction of programmed cell death in Arabidopsis and rice by single-wall carbon nanotubes. *Am. J. Bot.*, 97(10): 1602–1609.

Shen, Z., Chen, Z., Hou, Z., Li, T. and Lu, X. (2015). Ecotoxicological effect of zinc oxide nanoparticles on soil microorganisms. *Front. Environ. Sci. Eng.*, 9: 912–918.

Shilpa, R. and Lawre, S. (2014). Effect of zinc oxide nanoparticles on cytology and seed germination in onion. *Int. J. Curr. Microbiol. App. Sci.*, 3: 467–473.

Shukla, A., Behera, S., Shivay, Y., Singh, M.P. and Singh, A.K. (2012). Micronutrients and field crop production in India: A review. *Indian J. Agron.*, 57: 123–130.

Shukla, A., Tiwari, P. and Prakash, C. (2014). Micronutrients deficiencies vis-a-vis food and nutritional security of India. *Indian J. Fert.*, 10(12): 94–112.

Shukla, A., Behera, S., Pakhre, A. and Chaudhary, K.S. (2018). Micronutrients in soils, plants, animals and humans. *Indian J. Fertil.*, 14(4): 30–54.

Shukla, A., Behera, S., Tripathi, R., Prakash, C., Nayak, A., Kumar, P.S., Chitdeshwari, T., Kumar, D., Nayak, R. and Babu, P.S. (2021). Evaluation of spatial spreading of phyto-available sulphur and micronutrients in cultivated coastal soils. *PLoS One,* 16(10): e0258166.

Shukla, A.K., Behera, S.K., Prakash, C., Tripathi, A., Patra, A.K., Dwivedi, B.S., Trivedi, V., Rao, C.S., Chaudhari, S.K., Das, S. and Singh, A.K. (2021). Deficiency of phyto-available sulphur, zinc, boron, iron, copper and manganese in soils of India. *Sci. Rep.*, 11(1): 19760.

Siddiqui, M.H. and Al-Whaibi, M.H. (2014). Role of nano-SiO_2 in germination of tomato (*Lycopersicum esculentum seeds Mill.*). *Saudi J. Biol. Sci.*, 21(1): 13–17.

Sillanpää, M. (1982). *Micronutrients and the Nutrient Status of Soils: A Global Study.* FAO Soils Bulletin No. 48, FAO/Finnish International Development Agency, Rome, Italy.

Simonin, M. and Richaume, A. (2015). Impact of engineered nanoparticles on the activity, abundance, and diversity of soil microbial communities: A review. *Environ. Sci. Pollut. Res.*, 22: 13710–13723.

Singh, A., Singh, N.B., Hussain, I. and Singh, H. (2017). Effect of biologically synthesized copper oxide nanoparticles on metabolism and antioxidant activity to the crop plants *Solanum lycopersicum* and *Brassica oleracea var. botrytis*. *J. Biotechnol.*, 262: 11–27.

Singh, A., Singh, N., Afzal, S., Singh, T. and Hussain, I. (2018). Zinc oxide nanoparticles: a review of their biological synthesis, antimicrobial activity, uptake, translocation and biotransformation in plants. *J. Mater. Sci.*, 53(1): 185–201.

Singh, M.V. (2008). Micronutrient deficiencies in crops and soils in India. pp. 93–125. *In*: Alloway, B.J. (ed.). *Micronutrient Deficiencies in Global Crop Production.* Dordrecht: Springer.

Singh, S., Vishwakarma, K., Singh, S., Sharma, S., Dubey, N.K., Singh, V.K., Liu, S., Tripathi and Chauhan, D.K. (2017). Understanding the plant and nanoparticle interface at transcriptomic and proteomic level: A concentric overview. *Plant Gene*, 11: 265–272.

Smijs, T.G. and Pavel, S. (2011). Titanium dioxide and zinc oxide nanoparticles in sunscreens: Focus on their safety and effectiveness. *Nanotechnol. Sci. Appl.*, 4: 95–112.

Solanki, P., Bhargava, A., Chhipa, H., Jain, N. and Panwar, J. (2015). Nano-fertilizers and their smart delivery system. pp. 81–101. *In*: Rai, M., Ribeiro, C., Mattoso, L. and Duran, N. (eds.). *Nanotechnologies in Food and Agriculture.* Cham: Springer International Publishing.

Song, G., Gao, Y., Wu, H., Hou, W., Zhang, C. and Ma, H. (2012). Physiological effect of anatase TiO_2 nanoparticles on *Lemna minor*. *Environ. Toxicol. Chem.*, 31: 2147–2152.

Statnano. (2021). *Nanotechnology Products Database*. https://product.statnano.com/.

Subramanian, K. and Rajkishore, S. (2018). Regulatory framework for nanomaterials in agri-food systems. pp. 319–342. *In*: Rai, M. and Biswas, J. (eds.). *Nanomaterials: Ecotoxicity, Safety, and Public Perception.* Cham: Springer International Publishing.

Subramanian, K.S., Manikandan, A., Thirunavukkarasu, M. and Rahale, C.S. (2015). Nano-fertilizers for balanced crop nutrition. pp. 69–80. *In*: Rai, M., Ribeiro, C., Mattoso, L. and Duran N. (eds.). *Nanotechnologies in Food and Agriculture*. Cham: Springer International Publishing.

Subramanian, K.S. and Rajkishore, S.K. (2018). Regulatory framework for nanomaterials in agri-food systems. pp. 319–342. *In*: Rai, M. and Biswas, J. (eds.). *Nanomaterials: Ecotoxicity, Safety, and Public Perception.* Cham:.Springer International Publishing.

Suriyaprabha, R., Karunakaran, G., Yuvakkumar, R., Prabu, P., Rajendran, V. and Kannan, N. (2013). Application of silica nanoparticles for increased silica availability in maize. *Solid State Phys.*, 1512: 424–425.

Thapa, S., Bhandari, A., Ghimire, R., Xue, Q., Kidwaro, F., Ghatrehsamani, S., Maharjan, B. and Goodwin, M. (2021). Managing micronutrients for improving soil fertility, health, and soybean yield. *Sustainability,* 13(21): 11766.

Theivasigamani, P. (2011). Phytotoxicity of nanoparticles in agricultural crops. pp. 51–60. *In*: *Green Technology and Environmental Conservation (GTEC 2011). IEEE*, Chennai.

Tian, H., Kah, M. and Kariman, K. (2019). Are nanoparticles a threat to mycorrhizal and rhizobial symbioses?: A critical review. *Front. Microbiol.*, 10: 1660.

Timmusk, S., Seisenbaeva, G. and Behers, L. (2018). Titania (TiO_2) nanoparticles enhance the performance of growth-promoting rhizobacteria. *Sci. Rep.*, 8(1): 617.

Trujillo-Reyes, J., Majumdar, S., Botez, C., Peralta-Videa, J. and Gardea-Torresdey, J. (2014). Exposure studies of core–shell Fe/Fe_3O_4 and Cu/CuO NPs to lettuce (*Lactuca sativa*) plants: Are they a potential physiological and nutritional hazard? *J. Hazard. Mater.*, 267: 255–263.

Van Nguyen, D., Nguyen, H.M., Le, N.T., Nguyen, K.H., Nguyen, H.T., Le, H.M., Nguyen, A.T., Dinh, N.T.T., Hoang, S.A. and Van Ha, C. (2021). Copper nanoparticle application enhances plant growth and grain yield in maize under drought stress conditions. *J. Plant Growth Regul.*, pp. 1–12.

Van Nhan, L., Ma, C., Rui, Y., Cao, W., Deng, Y., Liu, L. and Xing, B. (2016). The effects of Fe_2O_3 nanoparticles on physiology and insecticide activity in non-transgenic and Bt-transgenic cotton. *Front. Plant Sci.*, 6: 1263.

Verma, H., Jindal, M. and Rather, S.A. (2021). Bacterial siderophores for enhanced plant growth. pp. 314–331. *In*: Malik, A.J. (ed.). *Handbook of Research on Microbial Remediation and Microbial Biotechnology for Sustainable Soil*. Hershey: IGI Global.

Vishwakarma, K., Shweta, Upadhyay, N., Singh, J., Liu, S., Singh, V.P., Prasad, S.M., Chauhan, D.K., Tripathi, D.K. and Sharma, S. (2017). Differential phytotoxic impact of plant mediated silver nanoparticles (AgNPs) and silver nitrate ($AgNO_3$) on Brassica sp. *Front. Plant Sci.*, 8: 1501.

Von Moos, L.M., Schneider, M., Hilty, F.M., Hilbe, M., Arnold, M., Ziegler, N., Mato, D.S., Winkler, H., Tarik, M., Ludwig, C., Naegeli, H., Langhans, W., Zimmermann, M.B., Sturla, S.J. and Trantakis, I.A. (2017). Iron phosphate nanoparticles for food fortification: Biological effects in rats and human cell lines. *Nanotoxicology*, 11(4): 496–506.

Wagener, S., Jungnickel, H., Dommershausen, N., Fischer, T., Laux, P. and Luch, A. (2019). Determination of nanoparticle uptake, distribution, and characterization in plant root tissue after realistic long-term exposure to sewage sludge using information from Mass Spectrometry. *Environ. Sci. Technol.*, 53(9): 5416–5426.

Wang, F., Liu, X., Shi, Z., Tong, R., Adams, C.A. and Shi, X. (2016). Arbuscular mycorrhizae alleviate negative effects of zinc oxide nanoparticle and zinc accumulation in maize plants: A soil microcosm experiment. *Chemosphere*, 147: 88–97.

Wang, Y., Hu, J., Dai, Z., Li, J. and Huang, J. (2016). *In vitro* assessment of physiological changes of watermelon (*Citrullus lanatus*) upon iron oxide nanoparticles exposure. *Plant Physiol. Biochem.*, 108: 353–360.

Wimmer, M.A. and Eichert, T. (2013). Review: Mechanisms for boron deficiency-mediated changes in plant water relations. *Plant Sci.*, 203–204: 25–32.

Worrall, E.A., Hamid, A., Mody, K.T., Mitter, N. and Pappu, H.R. (2018). Nanotechnology for plant disease management. *Agronomy*, 8(12): 285.

Xu, L. (2018). Adsorption and inhibition of CuO nanoparticles on *Arabidopsis thaliana* root. *IOP Conference Series: Earth and Environmental Science*, 113: 012230.

Yang, J., Cao, W. and Rui, Y. (2017). Interactions between nanoparticles and plants: Phytotoxicity and defense mechanisms. *J. Plant Interact.*, 12(1): 158–169.

Yang, L. and Watts, D.J. (2005). Particle surface characteristics may play an important role in phytotoxicity of alumina nanoparticles. *Toxicol. Lett.*, 158: 122.

Yang, X., Pan, H., Wang, P. and Zhao, F.J. (2017). Particle-specific toxicity and bioavailability of cerium oxide (CeO_2) nanoparticles to *Arabidopsis thaliana*. *J. Hazard Mater.*, 322(Pt A): 292–300.

Yanık, F., Aytürk, Ö. and Vardar, F. (2017). Programmed cell death evidence in wheat (*Triticum aestivum* L.) roots induced by aluminum oxide (Al_2O_3) nanoparticles. *Caryologia*, 70(2): 112–119.

Yaseen, M.K. and Hussain, S. (2021). Zinc-biofortified wheat required only a medium rate of soil zinc application to attain the targets of zinc biofortification. *Arch. Agron. Soil Sci.*, 67(4): 551–562.

Younes, N., Hassan, H.S., Elkady, M.F., Hamed, A. and Dawood, M.F. (2020). Impact of synthesized metal oxide nanomaterials on seedlings production of three *Solanaceae* crops. *Heliyon*, 6(1): e03188.

Yruela, I. (2005). Copper in plants. *Braz. J. Plant Physiol.*, 17: 145–156.

Yuan, J., Chen, Y., Li, H., Lu, J., Zhao, H., Liu, M., Nechitaylo, G.S. and Glushchenko, N.N. (2018). New insights into the cellular responses to iron nanoparticles in *Capsicum annuum*. *Sci. Rep.*, 8(1): 3228.

Yuan, L., Richardson, C.J., Ho, M., Willis, C.W., Colman, B.P. and Wiesner, M.R. (2018). Stress responses of aquatic plants to silver nanoparticles. *Environ. Sci. Technol.*, 52(5): 2558–2565.

Zeng, H., Wu, H., Yan, F., Yi, K. and Zhu, Y. (2021). Molecular regulation of zinc deficiency responses in plants. *J. Plant Physiol.*, 261: 153419.

Zhang, J., Gu, P., Xu, J., Xue, H. and Pang, H. (2016). High performance of electrochemical lithium storage batteries: ZnO-based nanomaterials for lithium-ion and lithium–sulfur batteries. *Nanoscale*, 8(44): 18578–18595.

Zhang, P., Ma, Y. and Zhang, Z. (2015). Interactions between engineered nanomaterials and plants: phytotoxicity, uptake, translocation, and biotransformation. pp. 77–99. *In*: Siddiqui, M., Al-Whaibi, M. and Mohammad, F. (eds.). *Nanotechnology and Plant Sciences*. Cham: Springer.

Zhang, P., Guo, Z., Zhang, Z., Fu, H., White, J.C. and Lynch, I. (2020). Nanomaterial transformation in the soil–plant system: Implications for food safety and application in agriculture. *Small*, 16(21): 2000705.

Zhao, L., Hernandez-Viezcas, J.A., Peralta-Videa, J.R., Bandyopadhyay, S., Peng, B., Munoz, B., Keller, A.A. and Gardea-Torresdey, J.L. (2012). ZnO nanoparticle fate in soil and zinc bioaccumulation in corn plants (*Zea mays*) influenced by alginate. *Environ. Sci. Process Impacts*, 15(1): 260–266.

Zhu, J., Zou, Z., Shen, Y., Li, J., Shi, S., Han, S. and Zhan, X. (2019). Increased ZnO nanoparticle toxicity to wheat upon co-exposure to phenanthrene. *Environ. Pollut.*, 247: 108–117.

Zia-Ur-Rehman, M., Qayyum, M.F., Akmal, F., Maqsood, M.A., Rizwan, M., Waqar, M. and Azhar, M. (2018). Recent progress of nanotoxicology in plants. pp. 143–174. *In*: Tripathi, D.K., Ahmad, P., Sharma, S., Chauhan, D.K. and Dubey, N.K. (eds.). *Nanomaterials in Plants, Algae, and Microorganisms*. Academic Press, Elsevier.

Zimmermann, M.B. and Hilty, F.M. (2011). Nanocompounds of iron and zinc: Their potential in nutrition. *Nanoscale*, 3(6): 2390–2398.

Zulfiqar, U., Hussain, S., Ishfaq, M., Ali, N., Yasin, M.U. and Ali, M.A. (2021). Foliar manganese supply enhances crop productivity, net benefits, and grain manganese accumulation in direct-seeded and puddled transplanted rice. *J. Plant Growth Regul.*, 40(4): 1539–1556.

Zuverza-Mena, N., Armendariz, R., Peralta-Videa, J.R. and Gardea-Torresdey, J.L. (2016). Effects of silver nanoparticles on radish sprouts: Root growth reduction and modifications in the nutritional value. *Front. Plant Sci.*, 7: 90.

Chapter 3

Recent Advances in Nanofertilizer Development

Ankita Bedi and *Braj Raj Singh**

1. Introduction

With the unbridled increase in population, there is a constant demand for increased food production. This in turn is creating stress on agricultural land owing to a reduction in available cultivable land and water resources. Moreover, the present exploitive agricultural practices are causing soil degradation and macro and micronutrient deficiencies, which are proving to be major constraints in the path towards sustainable food and crop production. Fertilizer is a key component in increasing food production and feeding the increasing population. According to a report by the Food and Agriculture Organization (FAO) of United States 2017, the total fertilizer demand globally in terms of macronutrients, i.e., N + P_2O_5 + K_2O was around 184.67 million tonnes in 2014 and reached about 186.6 million tonnes in 2015 indicating a subsequent increase of 1.6% per annum. This shows that fertilizer consumption of fertilizers is forecast to reach about 201.66 million tonnes by 2020.[1]

This alarming increase in the consumption of chemical fertilizers poses a serious threat to the environment. It will cause irretrievable damage to the soil microbial flora, chemical ecology, soil structure, and food chain existing in the ecosystem (Chinnamuthu and Boopathi, 2009; Conley et al., 2009; Sekhon, 2014; Chhipa, 2017; Prasad et al., 2017). The plausible solution to the above-mentioned problems is the development and use of smart fertilizers that are environment friendly, cost-effective, and less toxic.

Over the last few decades, many technologies are being worked upon to improve crop productivity. Among them, nanotechnology is a promising alternate

TERI-Deakin Nanobiotechnology Centre, The Energy and Resources Institute (TERI), Lodhi Road, New Delhi, 110003, India.
* Corresponding author: brajnano99@gmail.com
[1] Details available at http://www.fao.org/3/a-i6895e.pdf, last accessed on 19 March 2020.

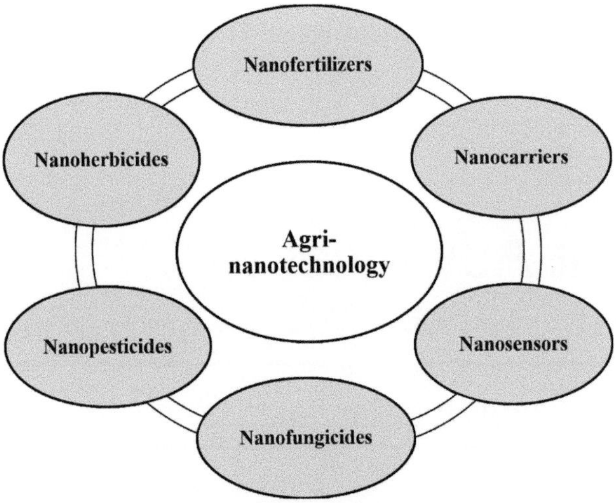

Fig. 3.1. Applications of nanotechnology in agriculture.

solution. The term 'nanotechnology' was first coined by Professor Norio Taniguichi from Tokyo University in 1974 (Gogos et al., 2012). Nanoparticles, the leading edge of nanotechnology, are quite versatile with a wide range of applications in the agricultural sector (Fig. 3.1). They have proven to bridge the gap between bulk and macro materials.

Nanoparticles, because of their high surface area to volume ratio, have higher reactivity as compared to their bulk counterpart, i.e., chemical fertilizers (Naderi and Danesh-Shahraki, 2013). This, in turn, enhances their absorption as fertilizers, thus minimizing the loss due to leaching. Therefore, nanofertilizers entail being a sustainable approach to reduce dependency on chemical fertilizers, increasing nutrient quality of crops, and protecting the environment.

2. Conventional Fertilizer v/s Nanofertilizer

2.1 Challenges with Conventional Fertilizers

For decades, conventional fertilizers have been used to boost crop productivity. To feed the ever-increasing population, exhaustive farming practices are being followed which in turn are resulting in a vicious circle of land degradation and thus deteriorating the crop yield. Fertilizers help in replacing nutrients in their chemical form to the plants which are not completely accessible by them. This loss of nutrients is due to many reasons such as chemical leaching, run-off, decomposition, hydrolysis by soil moisture even due to degradation by microbes (Sabir et al., 2014; Bindraban et al., 2020). There have been reports indicating that around 40–70% of nitrogen, 80–90% of phosphorus, and 50–90% of potassium content of applied fertilizers are in the chemically bound form or washed away in the environment, thus becoming unavailable to plants (Ombodi et al., 2000; El-Aila et al., 2015; Liu and Lal, 2015; Miao et al., 2015; Wang et al., 2015; Yang et al., 2016; Bouwman et al., 2017; Duhan et al., 2017; Giroto et al., 2017; Bindraban et al., 2020). It also results in environmental pollution because of leaching or binding in the soil. It is quite well known that loss of nutrients results in groundwater pollution and eutrophication in aquatic ecosystems, thus affecting the flora and fauna[2] (Davies et al., 2001). Thus, to overcome nutrient losses, repeated application of fertilizers is practiced. As per the statistical data by International Fertilizer Industry Association (IFIA), fertilizer consumption is increasing on an average by 1.5% per annum, and by 2021–22, 199 million tonnes will be used (Heffer and Homme, 2017). Wastage of fertilizers means huge economic losses and financial encumbrance both for developing and for developed nations. Taking these facts into consideration, the application of conventional fertilizers on a large scale is not a sustainable solution for the future.

Consequently, there is a dire need to develop eco-friendly fertilizers that can enhance crop productivity by expediting maximum nutrient uptake. Nanofertilizers are a sustainable solution that can be applied in lower dosages and on specified targeted sites; thus, preventing the loss of nutrients and in turn reducing environmental pollution (Naderi and Danesh-Shahraki, 2013; Sekhon, 2014; Siddiqui et al., 2015; Kopittke et al., 2019). Mortvedt (1992) observed that there is an increase in the suspension rate of less soluble fertilizers such as ZnO in water due to the increased surface area of particles. Reduction in particle size also increases the number of particles per unit weight of applied nutrient, thus preventing the repeated application of fertilizer. Nanofertilizers are being applied at a comparatively lower dosage, thus providing an economical solution (Prasad et al., 2012).

2.2 *Nanofertilizers: A Sustainable Solution*

Nanofertilizers are new-age fertilizers that can be used in lower dosages for slow/controlled release of nutrients in comparison to bulk fertilizers (Reynolds, 2002; Batsmanova et al., 2013; Subramanian et al., 2015; Adisa et al., 2019). They thus enhance their efficacy in terms of plant growth and yield. The nanoparticles are synthesized using various physical and chemical processes. However, these processes are complex involving high temperature, pressure conditions along with the use of toxic chemicals thus making them harmful to the environment, whereas biological routes of synthesis are environment friendly as well as cost-effective. Biological entities have a unique potential to synthesize molecules with selective properties, thus becoming a potential tool for nanoparticles synthesis (Yadav et al., 2008). Biosynthesis has been carried out by exploiting various microbes and plant extracts (Bharde et al., 2008; Ghormade et al., 2011; Iravani et al., 2011; Mazumdar et al., 2011; Waghmare et al., 2011; Srivastava et al., 2012; Jayaseelan et al., 2012; Jain et al., 2013; Byrne et al., 2014; Sarkar et al., 2014; Singh et al., 2014; Panpatte et al., 2016; Bedi et al., 2018a; 2018b; Rajesh et al., 2018).

Nanomaterials, in this case, are targeted towards reduction in usage of chemical products by providing smart delivery of key elements, minimizing the loss of nutrients in fertilization, and lastly enhancing the plant productivity via optimized water management (Fig. 3.2).

[2] Details available at https://www.bio-fit.eu/q1/lo4-nano-fertilizers-and-genetically-engineered-microbes; last accessed on 19 March 2020.

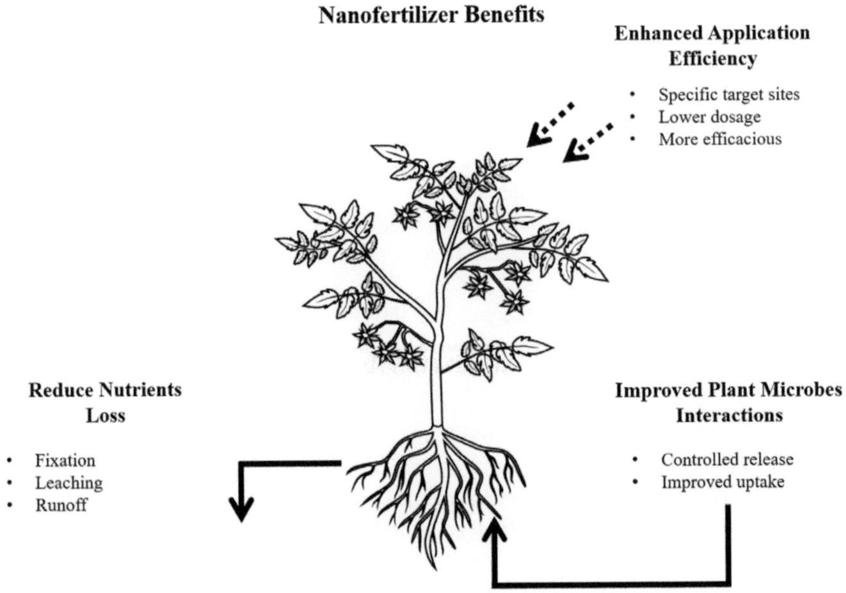

Fig. 3.2. Benefits of nanofertilizers.

To enhance plant yield more effectively, as compared to their conventional variants, nanonutrients need to be released in a controlled phase which depends on various environmental factors such as temperature, soil pH, moisture, and type of soil (Ghormade et al., 2011; Wang et al., 2012; Naderi and Danesh-Shahraki, 2013; Dapkekar et al., 2018). Nano-sized particles improve the photosynthesis activity, absorption capacity, and increase the plant biomass and its leaf surface area. Besides, they are also known to contribute towards preventing eutrophication and water pollution (Solanki et al., 2015; Shang et al., 2019).

Porous nanomaterials such as zeolites, chitosan, and clay have proven to increase crop productivity and reduce nutrient losses because of their slow and demand-based release of nutrients in plants (Millan et al., 2008; Abdel-Aziz et al., 2016; Panpatte et al., 2016; Medina-Velo et al., 2017). Lateef et al. (2016) tested nano-zeolite and zeolite-based nanocomposite with various micro and macronutrients for their activity as a slow-release fertilizer. Promising results were obtained when tested in water and soil systems for 7 and 14 days, respectively. Abdel-Aziz et al. (2016) have tested the effect of NPK-nanochitosan composite in the range of 10–100 mgL^{-1} through the foliar application on wheat. Grain yield was recorded to enhance by 51% and 56% in comparison to control and conventional NPK, respectively. However, the comparison with pure nano chitosan as control was missing.

Laware and Raskar (2014) demonstrated an increase in the germination of onion seeds at a lower concentration of 20 $\mu g\ mL^{-1}$ and 30 $\mu g\ mL^{-1}$ of nano-ZnO. Bedi et al. (2018a) indicated that Zn–Fe nanoparticles at a concentration of 20 ppm showed enhanced germination when applied to wheat. Kisan et al. (2015) reported that 500 and 1000 ppm of nano-ZnO increased the protein content and nutritional value of spinach. Groundnut seeds have also been reported to show enhanced growth when treated with 1000 ppm of ZnO nanoparticles (Prasad et al., 2012). Dimkpa et al.

(2017) have studied the efficacy of nanocomposite, i.e., nanoparticles composite (ZnO, CuO, and B_2O_3) through the soil as well as a foliar application on soybean under drought conditions. They claimed that these nanocomposite showed similar effects as micronutrient salts in enhancing the grain yield and plant uptake of N and P. Even enhancement in the growth of wheat in terms of increased chlorophyll content, shoot Zn uptake, and Zn content in grain was observed under drought conditions on the application of nano Zn as compared to its bulk counterpart, i.e., ZnO (Dimkpa et al., 2020a; 2020b). TiO_2 nanoparticles exhibit an increase in plant growth when applied to mung bean through the foliar application (Raliya et al., 2015). An increase of 46.4% and 94% in chlorophyll and total soluble leaf protein content, respectively, was also observed at a concentration of 10 mgL^{-1}. Singh et al. (2016) tested the effects of $ZnSO_4$ and nano-Zn on seeds of tomato (*Solanum lycopersicum* L.). The nanoparticles showed increased seed germination and vigor at lower concentrations in comparison to the bulk supplement. As per the observations, nanoparticles possibly penetrate the seed coat thus improving the development of the embryo. The protein and sugar content were also reported to increase in the case of nano-Zn. Dapkekar et al. (2018) reported the ferti-fortification of wheat using zinc complex with nanochitosan (Zn-CNP). The four-year field trials indicated that the results were comparable with the bulk $ZnSO_4$ at a concentration 10 times lower than the latter. The results were recorded on the plots with varying zinc concentrations. Sun et al. (2020) have also indicated the enhanced growth of winter wheat (*Triticum aestivum* L.) when treated with ZnO nanoparticles. Synchrotron-based X-ray fluorescence microscopy (μ-XRF) and laser ablation inductively coupled plasma mass spectrometry (LA-ICP-MS) indicated a 30-fold increase in the concentration of zinc present in grain endosperm in comparison to control.

Iannone et al. (2016) described the effect of Fe_3O_4 nanoparticles in wheat (*Triticum aestivum* L.). Approximately, 20% increase in root biomass was observed at the concentration of 20 mgL^{-1} in comparison to control, whereas no translocation was recorded in the case of aerial parts. An increase in the activity of oxidative stress enzymes was also recorded in the case of iron nanoparticles. The authors indicated that this enhanced antioxidant activity could have prevented oxidative cell damage. Suresh et al. (2016) have reported that peanut seeds (*Arachis hypogea* L.) when subjected to seed priming with Fe_3O_4 nanoparticles at a concentration of 500 ppm showed an increase in protein and carbohydrate content in the leaves. The Fe_3O_4 nanoparticles were found to be toxic at higher concentrations. Li et al. (2016) indicated that there was an increase of about 10% in the root growth of corn seeds when subjected to Fe_2O_3 nanoparticles. The authors reported that nanoparticles were accumulated in the vacuoles, while nothing was observed in the shoot. Sheykhbaglou et al. (2010) also showed the enhanced plant yield in soybean on the application of nano iron-oxide at 0.75 gmL^{-1}. The application of nano-iron fertilizer increased the agronomic traits and concentration of essential oils in dragonhead (*Dracocephalum moldavica*) (Yousefzadeh and Sabaghnia, 2016).

Reports are available that indicate the effect of nanofertilizers on various fruits. Davarpanah et al. (2018) have shown that the foliar application of nano-Ca decreased the fruit cracking in the case of pomegranate in both seasons as compared to calcium chloride. On the other hand, the application of combined dosage of nano-Zn and nano-B on pomegranate enhanced the fruit yield and quality by affecting pollen germination, tube elongation, and flowering (Davarpanah et al., 2016). Wang et al. (2016) observed that the growth and chlorophyll content of watermelon (*Citrullus lanatus*) enhanced after treatment with γ-Fe_2O_3 nanoparticles (soil treatment), thus treating iron-chlorosis. Lopez-Vargas et al. (2018) studied the effect of Cu nanoparticles (50 nm) on fruit yield and nutrient content. It was reported that the concentration of 250 mgL^{-1} increased the concentration of bioactive compounds such as vitamin C, lycopene, and flavonoids in the fruit. The shelf-life was also predicted to enhance due to the better firmness of the fruit. On the other hand, concentration as high as 500 mgL^{-1} was found to be toxic.

Carbon nanoparticles have shown promising results in increasing the grain yield of various crops such as *Oryza sativa* (10.3%), *Triticum aestivum* (28.8%), *Zea mays* L. (10.9%), *Glycine max* (16.7%), and vegetables (12.3–19.8%) (Liu et al., 2009). Many reports indicated the enhanced germination of crops such as wheat, barley, tomato, soybean, corn, garlic, etc., on the application of multiwalled carbon nanotubes (MWCNTs) (Lahiani et al., 2013; Srivastava and Rao, 2014; Khodakovskaya et al., 2016; Joshi et al., 2018).

Several publications have also investigated the use of Mn nanoparticles on a variety of crops such as lettuce (Liu et al., 2016), wheat (Dimkpa et al., 2018), mung bean (Pradhan et al., 2014), tomato, and eggplant (Elmer and White, 2016).

Literature highlights the efficacious properties of these nanomaterials resulting in an increase in agricultural productivity in terms of both yield and quality. Although there are still loopholes in understanding the exact mechanism responsible, a few reports indicate the enhancement of root vigor because of increased enzyme activity (Dubey et al., 2016; Shang et al., 2019; Shojaei et al., 2019). Efforts are being made to study the physicochemical parameters that influence the plant–nanoparticle interactions, adsorption, and translocation studies. Nanomaterials are claimed to stimulate plant growth by facilitating complexation with root exudates as well as at the molecular level by creating new pores on the leaf surface and root or by endocytosis/ion channels (Mastronardi et al., 2015). The increased surface area to volume ratio facilitates a greater number of nanonutrients being slowly adsorbed and assimilated in plants with time, thus ensuring a balanced growth cycle in terms of nutrition (Subramanian et al., 2015; Monreal et al., 2016; Lowry et al., 2019).

3. Market Scenario of Nanofertilizers

These smart fertilizers, though being in their relative infant stage, are gaining popularity globally because of their endless benefits. Many countries are, thus, readily investing in their development and application. Various products are already available in the market with the majority being composite/formulation based. In many cases, they may not have provided the complete information related to their being 'nano'. Some of the examples are being discussed here.

Nano-Gro by Agro Nanotechnology Corp., U.S.A., is a composite of active ingredients (micronutrients in nanomolecular scale) in the form of coded sugar pellets less than one-eighth ins in diameter, dissolved in ordinary tap water. It is a plant growth regulator and immunity enhancer.[3] Tropical Agro, India has developed nano 4G fertilizers (combination of protein-lacto-gluconate formulation with various nutrients) using 4G nanotechnology in collaboration with ICAR. They are designed

[3] Details available at http://www.agronano.com/nanogro.htm, last accessed on 21 September 2019.

Table 3.1. List of some commercially available nanofertilizers.

Product Name	**Company**
Nanofertilizer Products Available in the Indian Market	
4G" Nutrients (nano fertilizers and nano micronutrients with Proteino - Lacto - Gluconates) • '4G' complete fertilizer (Organic N-P-K) - Complex • Prathista Phosphate – DAP replacement • Prathista Potash – MoP replacement • Megacal – Secondary Nutrients • Prathista Zinc – Zinc fertilizer • Prathista Aishwarya – Complete Organic Fertilizer • New Suryamin – Complete Nutritional liquid fertilizer	Prathista Industries Limited, Hyderabad
'4G' Nutrients (Nano Fertilizers & Nano micronutrients with Proteino - Lacto - Gluconates) • Tag Nano NPK • Tag Nano Phos • Tag Nano Potash • Tag Nano Cal • Tag Nano Zinc	Tropical AgroSystem (I) Pvt. Ltd, Chennai, Tamil Nadu
'4G' Nutrients (Nano Fertilizers & Nano micronutrients with Proteino - Lacto - Gluconates) • Nano Max Potash • Nano Max NPK • Nano Max Cal • Nano max Zinc	JU Agri Sciences Pvt. Ltd., Janakpuri, Delhi
'4G' Nutrients (Nano Fertilizers & Nano micronutrients with Proteino - Lacto - Gluconates) • Nano Zn • Nano Mg • Nano N • Nano K • Nano P • Nano S	Kanak Biotech Pvt. Ltd., Patparganj Industrial Area, New Delhi
A nanonutrient with 12 essential nutrients loaded on a nano-silica • Nualgi foliar spray (NFS)	Nualgi Nanobiotech, Jayanagar, Bangalore
• Rich Herba Green Plant Nutrient (Calcium - 44.08%, Magnesium - 2.02%, Silicon - 10.98%, and trace elements: Fe, Zn, Cu, Mo, Co)	Richfield Fertiliser Private Limited, Nashik, Maharashtra
Nanoparticles liquid formulation of micronutrients • SURPLUS	Rallis India Limited, Mumbai
• Nano Urea • Nano Zinc • Nano Copper	Indian Farmers Fertiliser Cooperative Limited, New Delhi
Nanofertilizer Products Available in the Market outside India	
Micronutrients in nanomolecular scale • Nano Gro	Agro Nanotechnology Corp., U.S.A.

Table 3.1 contd. ...

...Table 3.1 contd.

Product Name	Company
NPK fertilizer with a blend of billions of microbes, sea-kelp, and mineral electrolytes • Nano-Ag Answer	Urth Agriculture, U.S.A.
A biodegradable formulation • Nano Green	Nano Green Sciences Inc., U.S.A.
Formulation of specific nanonutrients as oxygen nutrients, glucose in water • Nutri Brix	Nutri Brix, Canada
Dispersion of biological humus with purified water. • GreenEarth-NanoPlant Ready to Use Spray	GreenEarth - Nanoplant, LLC Florida
Nanofertilizer Products Ready for Launch in the Market	
• Nano Zn-Fe • Nanophosphorus	TERI-Deakin Nanobiotechnology Centre, The Energy and Resources Institute, New Delhi
Nanofertilizer Products under Field Trials-based Performance Validation Stage	
• Nano Nitrogen • Nano Zinc • Nano Copper	Indian Farmers Fertiliser Cooperative Limited, New Delhi
• Nano NPK • Nano Sulphur • Nano Magnesium • Nano Boron • Nano Copper • Nano Molybdenum	TERI-Deakin Nanobiotechnology Centre, The Energy and Resources Institute, New Delhi

for foliar application with 3–4 ml L^{-1} dosage).[4] Nano Green also referred to as Plant tonic a product by Nano Green Sciences Inc. is a foliar-based fertilizer that is made up of colloidal micelles in the size range of 1–4 nm. It is claimed to be a biodegradable formulation.[5] The Nano-Ag Answer from Urth Agriculture, U.S.A. is an NPK fertilizer with a blend of billions of microbes, sea-kelp, and mineral electrolytes. In this case, the word 'nano' focuses on the small amount being applied.[6] Nutri Brix, with offices in Canada and U.S.A., has patented an innovative process using molecular nanotechnology where they have developed a formulation of specific nanonutrients as oxygen nutrients, and glucose in water. The formulation itself is known as NutriBrix.[7] Along with these, there are also other nanoproducts commercially available claiming to enhance crop yield (Table 3.1).

[4] Details available at http://www.tropicalagro.in/products, last accessed on 21 September 2019.

[5] Details available at http://www.nanogreensciences.com/index.html, last accessed on 21 September 2019.

[6] Details available at http://www.urthagriculture.com/nano-ag-fertilizer accessed on 21/9/19, last accessed on 21 September 2019.

[7] Details available at https://nutribrix.com/, last accessed on 21 September 2019.

4. Health Risks and Regulatory Concerns

It is well documented that recent advances in nanotechnology, are likely to transform various aspects of human life. However, precautionary measures concerning practical applications of this interdisciplinary research field still need to be considered. This concern has led to the development of the term 'nanotoxicology' which focuses on the assessment of toxicity of nanoparticles, thus ensuring their innocuous design and application (Oberdorster et al., 2005).

Standardized protocols need to be designed concerning the physicochemical characterization of nanoparticles. The prescribed exposure level to the environment also has to be demarcated on utmost priority. To date, no such protocols/parameters have been specified (Dhawan et al., 2009; Solanki et al., 2015). The absence of such guidelines makes it difficult to authenticate the protocols developed by different research groups. These are the fundamental points that are required to identify the threat caused by nanoparticles exposure to the environment as well as to the biological system. Nanoparticles, other than their size and shape, vary at different levels such as their surface functionalities, chemical composition, mode of synthesis, stability, and dissolution behaviour. The risk assessment criteria are, thus, quite subjective because of such a vast heterogeneity[8] and it must be calculated and determined on a case-to-case basis (Bryksa et al., 2012; Xiaojia et al., 2019).

Even though research indicating the direct effect of nanomaterials on human health is missing but one cannot overlook the fact that such minuscule size particles can enter the human being's system via any of the oral, respiratory, or intracutaneous pathways. There are reports, though limited in number, of scientists working on finding the effect of nanomaterials at the cellular and sub-cellular levels (Zhu et al., 2013; Kah, 2015; Prasad et al., 2017).

Even the European Food Security Authority (EFSA) committee has suggested that "the risk assessment paradigm (hazard identification, hazard characterization, exposure assessment, and risk characterization) is applicable for nanoparticles" (EFSA Scientific Committee, 2011). Thus, extensive research to study the complete life cycle assessment/toxicity is the need of the hour to develop a regulatory framework.

5. Conclusion and Future Prospects

The extensive occurrence of soil nutrient deficiency worldwide has led to a major reduction in crop yield thus leading to economic losses. Though usage of chemical fertilizers has been a common practice for ages, the need for a more sustainable approach resulted in the introduction of 'agri-nanotechnology'. Nanofertilizers, because of their low dosage, specific target sites, and slow release, are proving to be economically feasible and environmentally friendly. However, there are some major concerns because of the absence of studies on the interaction of nanoparticles with biota of soil and environment, toxicity, uptake, and pathways in plant systems as well as humans/animals, and safe disposal. These life-cycle assessment (LCA) questions should be addressed to develop the regulatory framework for the safe applications of these nanofertilizers and their commercialization. Further research is, therefore,

[8] Details available at https://www.bio-fit.eu/q1/lo4-nano-fertilizers-and-genetically-engineered-microbes, last accessed on 19 March 2020.

required to elucidate their fate and action mechanism in the biological systems, i.e., their effect on soil characteristics, physiological and biochemical responses, growth, and crop yield under different agronomic conditions. Eventually, researchers, industries, and regulators need to take joint responsibility for providing the advantage of nanofertilizers to farmers to cultivate crops under diverse climatic vulnerable conditions with minimum environmental hazards and risks.

Acknowledgment

This activity was supported by the Centre of Excellence for Advanced Research in Agricultural Nanotechnology supported by the Department of Biotechnology, Govt. of India (Grant No. BT/NNT/28/SP30280/2019) and TERI-Deakin Nano Biotechnology Centre (TDNBC), India. We thank Dr. Alok Adholeya, Project Coordinator and former Director, TDNBC who provided insight and expertise that greatly assisted this chapter.

References

Adisa, I.O., Pullagurala, V.L.R., Peralta-Videa, J.R., Dimkpa, C.O., Elmer, W.H., Gardea-Torresdey, J.L. and White, J.C. (2019). Recent advances in nano-enabled fertilizers and pesticides: A critical review of mechanisms of action. *Environ. Sci. Nano*, 6(7): 2002–2030.

Aziz, H.M.A., Hasaneen, M.N. and Omer, A.M. (2016). Nano chitosan-NPK fertilizer enhances the growth and productivity of wheat plants grown in sandy soil. *Span. J. Agric. Res.*, 14(1): 17.

Batsmanova, L.M., Gonchar, L.M., Taran, N.Y. and Okanenko, A.A. (2013). *Using a Colloidal Solution of Metal Nanoparticles as Micronutrient Fertiliser for Cereals*. Doctoral Dissertation, Sumy State University.

Bedi, A., Singh, B.R., Deshmukh, S.K., Aggarwal, N., Barrow, C.J. and Adholeya, A. (2018a). Development of a novel myconanomining approach for the recovery of agriculturally important elements from jarosite waste. *J. Environ. Sci.*, 67: 356–367.

Bedi, A., Singh, B.R., Deshmukh, S.K., Adholeya, A. and Barrow, C.J. (2018b). An *Aspergillus aculateus* strain was capable of producing agriculturally useful nanoparticles via bioremediation of iron ore tailings. *J. Environ. Manage.*, 215: 100–107.

Bharde, A.A., Parikh, R.Y., Baidakova, M., Jouen, S., Hannoyer, B., Enoki, T., Prasad, B.L.V., Shouche, Y.S., Ogale S. and Sastry, M. (2008). Bacteria-mediated precursor-dependent biosynthesis of superparamagnetic iron oxide and iron sulfide nanoparticles. *Langmuir*, 24(11): 5787–5794.

Bindraban, P.S., Dimkpa, C.O., White, J.C., Franklin, F.A., Melse-Boonstra, A., Koele, N., Pandey, R., Rodenburg, J., Senthilkumar, K, Demokritou, P. and Schmidt, S. (2020). Safeguarding human and planetary health demands a fertilizer sector transformation. *Plants People Planet*, 2(4): 302–309.

Bouwman, A.F., Beusen, A.H.W., Lassaletta, L., Van Apeldoorn, D.F., Van Grinsven, H.J.M. and Zhang, J. (2017). Lessons from temporal and spatial patterns in global use of N and P fertilizer on cropland. *Sci. Rep.*, 7(1): 1–11.

Bryksa, B.C. and Yada, R.Y. (2012). Challenges in food nanoscale science and technology. *J. Food Drug Anal.*, 20.

Byrne, J.M., Coker, V.S., Cespedes, E., Wincott, P.L., Vaughan, D.J., Pattrick, R.A., van der Laan, G., Arenholz, E., Tuna, F., Bencsik, M. and Lloyd, J.R. (2014). Biosynthesis of zinc substituted magnetite nanoparticles with enhanced magnetic properties. *Adv. Funct. Mater.*, 24(17): 2518–2529.

Chhipa, H. (2017). Nanofertilizers and nanopesticides for agriculture. *Environ. Chem. Lett.*, 15(1): 15–22.

Chinnamuthu, C.R. and Boopathi, P.M. (2009). Nanotechnology and agroecosystem. *Madras Agric. J.*, 96(1/6): 17–31.

Conley, D.J., Paerl, H.W., Howarth, R.W., Boesch, D.F., Seitzinger, S.P., Havens, K.E., Lancelot, C. and Likens, G.E. (2009). Controlling eutrophication: Nitrogen and phosphorus. *Science*, 323(5917): 1014–1015.

Dapkekar, A., Deshpande, P., Oak, M.D., Paknikar, K.M. and Rajwade, J.M. (2018). Zinc use efficiency is enhanced in wheat through nanofertilization. *Sci. Rep.*, 8(1): 1–7.

Davarpanah, S., Tehranifar, A., Davarynejad, G., Abadía, J. and Khorasani, R. (2016). Effects of foliar applications of zinc and boron nano-fertilizers on pomegranate (*Punica granatum* cv. Ardestani) fruit yield and quality. *Sci. Hortic.*, 210: 57–64.

Davarpanah, S., Tehranifar, A., Abadía, J., Val, J., Davarynejad, G., Aran, M. and Khorassani, R. (2018). Foliar calcium fertilization reduces fruit cracking in pomegranate (*Punica granatum* cv. Ardestani). *Sci. Hortic.*, 230: 86–91.

Davies, S., Reed, M. and O'Brien, S. (2001). *Vermont Legislative Research Shop*. The University of Vermont.

Dhawan, A., Sharma, V. and Parmar, D. (2009). Nanomaterials: A challenge for toxicologists. *Nanotoxicology*, 3(1): 1–9.

Dimkpa, C.O., Bindraban, P.S., Fugice, J., Agyin-Birikorang, S., Singh, U. and Hellums, D. (2017). Composite micronutrient nanoparticles and salts decrease drought stress in soybean. *Agron. Sustain. Dev.*, 37(1): 5.

Dimkpa, C.O., Singh, U., Adisa, I.O., Bindraban, P.S., Elmer, W.H., Gardea-Torresdey, J.L. and White, J.C. (2018). Effects of manganese nanoparticle exposure on nutrient acquisition in wheat (*Triticum aestivum* L.). *Agronomy*, 8(9): 158.

Dimkpa, C.O., Andrews, J., Sanabria, J., Bindraban, P.S., Singh, U., Elmer, W.H., Gardea-Torresdey J.L. and White, J.C. (2020a). Interactive effects of drought, organic fertilizer, and zinc oxide nanoscale and bulk particles on wheat performance and grain nutrient accumulation. *Sci. Total Environ.*, 722: 137808.

Dimkpa, C., Andrews, J., Fugice, J., Singh, U., Bindraban, P.S., Elmer, W.H., Gardea-Torresdey, J.L. and White, J.C. (2020b). Facile coating of urea with low-dose ZnO nanoparticles promotes wheat performance and enhances Zn uptake under drought stress. *Front. Plant Sci.*, 11: 168.

Dubey, A. and Mailapalli, D.R. (2016). Nanofertilisers, nanopesticides, nanosensors of pest and nanotoxicity in agriculture. pp. 307–330. *In*: *Sustainable Agriculture Reviews*. Cham: Springer.

Duhan, J.S., Kumar, R., Kumar, N., Kaur, P., Nehra, K. and Duhan, S. (2017). Nanotechnology: The new perspective in precision agriculture. *Biotechnol. Rep.*, 15: 11–23.

EFSA Scientific Committee. (2011). Guidance on the risk assessment of the application of nanoscience and nanotechnologies in the food and feed chain. *EFSA Journal*, 9(5): 2140.

El-Aila, H.I., El-Sayed, S.A.A. and Yassen, A.A. (2015). Response of spinach plants to nanoparticles fertilizer and foliar application of iron. *Int. J. Environ.*, 4(3): 181–185.

Elmer, W.H. and White, J.C. (2016). The use of metallic oxide nanoparticles to enhance growth of tomatoes and eggplants in disease infested soil or soilless medium. *Environ. Sci. Nano*, 3(5): 1072–1079.

Ghormade, V., Deshpande, M.V. and Paknikar. K.M. (2011). Perspectives for nano-biotechnology enabled protection and nutrition of plants. *Biotechnol. Adv.*, 29(6): 792–803.

Giroto, A.S., Guimarães, G.G.F., Foschini, M. and Ribeiro, C. (2017). Role of slow-release nanocomposite fertilizers on nitrogen and phosphate availability in soil. *Sci. Rep.*, 7: 46032.

Gogos, A., Knauer, K. and Bucheli, T.D. (2012). Nanomaterials in plant protection and fertilization: current state, foreseen applications, and research priorities. *J. Agric. Food Chem.*, 60(39): 9781–9792.

He, X., Deng, H. and Hwang, H.M. (2019). The current application of nanotechnology in food and agriculture. *J. Food Drug Anal.*, 27(1): 1–21.

Heffer, P. and Prud'homme, M. (2017) (May). Fertilizer Outlook 2017–2021. pp. 22–24. *In*: 85th *International Fertilizer Industry Association Annual Conference*.

Iannone, M.F., Groppa, M.D., de Sousa, M.E., van Raap, M.B.F. and Benavides, M.P. (2016). Impact of magnetite iron oxide nanoparticles on wheat (*Triticum aestivum* L.) development: Evaluation of oxidative damage. *Environ. Exp. Bot.*, 131: 77–88.

Iravani, S. (2011). Green synthesis of metal nanoparticles using plants. *Green Chem.*, 13(10): 2638–2650.

Jain, N., Bhargava, A., Tarafdar, J.C., Singh, S.K. and Panwar, J. (2013). A biomimetic approach towards synthesis of zinc oxide nanoparticles. *Appl. Microbiol. Biotechnol.*, 97(2): 859–869.

Jayaseelan, C., Rahuman, A.A., Kirthi, A.V., Marimuthu, S., Santhoshkumar, T., Bagavan, A., Gaurav, K., Karthik, L. and Rao, K.B. (2012). Novel microbial route to synthesize ZnO nanoparticles using *Aeromonas hydrophila* and their activity against pathogenic bacteria and fungi. *Spectrochim. Acta A Mol. Biomol.*, 90: 78–84.

Joshi, A., Kaur, S., Dharamvir, K., Nayyar, H. and Verma, G. (2018). Multi-walled carbon nanotubes applied through seed-priming influence early germination, root hair, growth, and yield of bread wheat (*Triticum aestivum* L.). *J. Sci. Food Agric.*, 98(8): 3148–3160.

Kah, M. (2015). Nanopesticides and nanofertilizers: Emerging contaminants or opportunities for risk mitigation? *Front. Chem.*, 3: 64.

Khodakovskaya, M.V. and Biris, A.S., University of Arkansas. (2016). Method of using carbon nanotubes to affect seed germination and plant growth. U.S. Patent 9,364,004.

Kisan, B., Shruthi, H., Sharanagouda, H., Revanappa, S.B. and Pramod, N.K. (2015). Effect of nano-zinc oxide on the leaf physical and nutritional quality of spinach. *Agrotech.*, 5(1): 135.

Kopittke, P.M., Lombi, E., Wang, P., Schjoerring, J.K. and Husted, S. (2019). Nanomaterials as fertilizers for improving plant mineral nutrition and environmental outcomes. *Environ. Sci. Nano*, 6(12): 3513–3524.

Lahiani, M.H., Dervishi, E., Chen, J., Nima, Z., Gaume, A., Biris, A.S. and Khodakovskaya, M.V. (2013). Impact of carbon nanotube exposure to seeds of valuable crops. *ACS Appl. Mater. Interfaces*, 5(16): 7965–7973.

Lateef, A., Nazir, R., Jamil, N., Alam, S., Shah, R., Khan, M.N. and Saleem, M. (2016). Synthesis and characterization of zeolite-based nano-composite: An environment friendly slow-release fertilizer. *Microporous Mesoporous Mater.*, 232: 174–183.

Laware, S.L. and Raskar, S. (2014). Influence of zinc oxide nanoparticles on growth, flowering, and seed productivity in onion. *Int. J. Curr. Microbiol. Appl. Sci.*, 3(7): 874–881.

Li, J., Hu, J., Ma, C., Wang, Y., Wu, C., Huang, J. and Xing, B. (2016). Uptake, translocation, and physiological effects of magnetic iron oxide (γ-Fe_2O_3) nanoparticles in corn (*Zea mays* L.). *Chemosphere*, 159: 326–334.

Liu, J., Zhang, Y.D. and Zhang, Z.M. (2009). The application research of nano-biotechnology to promote increasing of vegetable production. *Hubei Agricultural Sciences*, 1: 20–25.

Liu, R. and Lal, R. (2015). Potentials of engineered nanoparticles as fertilizers for increasing agronomic productions. *Sci. Total Environ.*, 514: 131–139.

Liu, R., Zhang, H. and Lal, R. (2016). Effects of stabilized nanoparticles of copper, zinc, manganese, and iron oxides in low concentrations on lettuce (*Lactuca sativa*) seed germination: Nanotoxicants or nanonutrients? *Water Air Soil Pollut.*, 227(1): 42.

López-Vargas, E.R., Ortega-Ortíz, H., Cadenas-Pliego, G., de Alba Romenus, K., Cabrera de la Fuente, M., Benavides-Mendoza, A. and Juárez-Maldonado, A. (2018). Foliar application of copper nanoparticles increases the fruit quality and the content of bioactive compounds in tomatoes. *Appl. Sci.*, 8(7): 1020.

Lowry, G.V., Avellan, A. and Gilbertson, L.M. (2019). Opportunities and challenges for nanotechnology in the agri-tech revolution. *Nat. Nanotechnol.*, 14(6): 517–522.

Mastronardi, E., Tsae, P., Zhang, X., Monreal, C. and DeRosa, M.C. (2015). Strategic role of nanotechnology in fertilizers: Potential and limitations. pp. 25–67. *In*: *Nanotechnologies in Food and Agriculture*. Cham: Springer.

Mazumdar, H. and Haloi, N. (2011). A study on biosynthesis of iron nanoparticles by *Pleurotus* sp. *J. Microbiol. Biotechnol. Res.*, 1(3): 39–49.

Medina-Velo, I.A., Dominguez, O.E., Ochoa, L., Barrios, A.C., Hernández-Viezcas, J.A., White, J.C., Peralta-Videa, J.R. and Gardea-Torresdey, J.L. (2017). Nutritional quality of bean seeds harvested from plants grown in different soils amended with coated and uncoated zinc oxide nanomaterials. *Environ. Sci. Nano*, 4(12): 2336–2347.

Miao, Y.F., Wang, Z.H. and Li, S.X. (2015). Relation of nitrate N accumulation in dryland soil with wheat response to N fertilizer. *Field Crops Res.*, 170: 119–130.

Millán, G., Agosto, F. and Vázquez, M. (2008). Use of clinoptilolite as a carrier for nitrogen fertilizers in soils of the Pampean regions of Argentina. *Int. J. Agric. Nat. Resour.*, 35(3): 293–302.

Monreal, C.M., DeRosa, M., Mallubhotla, S.C., Bindraban, P.S. and Dimkpa, C. (2016). Nanotechnologies for increasing the crop use efficiency of fertilizer-micronutrients. *Biol. Fertil. Soils*, 52(3): 423–437.

Mortvedt, J.J. (1992). Crop response to level of water-soluble zinc in granular zinc fertilizers. *Fertil. Res.*, 33(3): 249–255.

Naderi, M.R. and Danesh-Shahraki, A. (2013). Nanofertilizers and their roles in sustainable agriculture. *Int. J. Agric. Crop Sci.*, 5(19): 2229–2232.

Oberdörster, G., Oberdörster, E. and Oberdörster, J. (2005). Nanotoxicology: an emerging discipline evolving from studies of ultrafine particles. *Environ. Health Perspect.*, 113: 823–839.

Ombódi, A. and Saigusa, M. (2000). Broadcast application versus band application of polyolefin-coated fertilizer on green peppers grown on andisol. *J. Plant Nutr.*, 23(10): 1485–1493.

Panpatte, D.G., Jhala, Y.K., Shelat, H.N. and Vyas, R.V. (2016). Nanoparticles: The next generation technology for sustainable agriculture. pp. 289–300. *In*: *Microbial Inoculants in Sustainable Agricultural Productivity*. New Delhi: Springer.

Pradhan, S., Patra, P., Mitra, S., Dey, K.K., Jain, S., Sarkar, S., Roy, S., Palit, P. and Goswami, A. (2014). Manganese nanoparticles: Impact on non-nodulated plant as a potent enhancer in nitrogen metabolism and toxicity study both *in vivo* and *in vitro*. *J. Agric. Food Chem.*, 62(35): 8777–8785.

Prasad, R., Bhattacharyya, A. and Nguyen, Q.D. (2017). Nanotechnology in sustainable agriculture: Recent developments, challenges, and perspectives. *Front. Microbiol.*, 8: 1014.

Prasad, T.N.V.K.V., Sudhakar, P., Sreenivasulu, Y., Latha, P., Munaswamy, V., Reddy, K.R., Sreeprasad, T.S., Sajanlal, P.R. and Pradeep, T. (2012). Effect of nanoscale zinc oxide particles on the germination, growth, and yield of peanut. *J. Plant Nutr.*, 35(6): 905–927.

Rajesh, K.M., Ajitha, B., Reddy, Y.A.K., Suneetha, Y., Reddy, P.S. and Ahn, C.W. (2018). A facile biosynthesis of copper nanoparticles using Cuminum cyminum seed extract: antimicrobial studies. *Adv. Nat. Sci.: Nanosci. Nanotechnol.*, 9(3): 035005.

Raliya, R., Biswas, P. and Tarafdar, J.C. (2015). TiO_2 nanoparticle biosynthesis and its physiological effect on mung bean (*Vigna radiata* L.). *Biotech. Rep.*, 5: 22–26.

Reynolds, G.H. (2002). *Forward to the Future: Nanotechnology and Regulatory Policy*. Pacific Research Institute.

Sabir, A., Yazar, K., Sabir, F., Kara, Z., Yazici, M.A. and Goksu, N. (2014). Vine growth, yield, berry quality attributes, and leaf nutrient content of grapevines as influenced by seaweed extract (*Ascophyllum nodosum*) and nanosize fertilizer pulverizations. *Sci. Hortic.*, 175: 1–8.

Sarkar, J., Ghosh, M., Mukherjee, A., Chattopadhyay, D. and Acharya, K. (2014). Biosynthesis and safety evaluation of ZnO nanoparticles. *Bioprocess Biosyst. Eng.*, 37(2): 165–171.

Sekhon, B.S. (2014). Nanotechnology in agri-food production: an overview. *Nanotechnol. Sci. Appl.*, 7: 31.

Shang, Y., Hasan, M., Ahammed, G.J., Li, M., Yin, H. and Zhou, J. (2019). Applications of nanotechnology in plant growth and crop protection: A review. *Molecules*, 24(14): 2558.

Sheykhbaglou, R., Sedghi, M., Shishevan, M.T. and Sharifi, R.S. (2010). Effects of nano-iron oxide particles on agronomic traits of soybean. *Not. Sci. Biol.*, 2(2): 112–113.

Shojaei, T.R., Salleh, M.-A.M., Tabatabaei, M., Mobli, H., Aghbashlo, M., Rashid, S.A. and Tan, T. (2019). Applications of nanotechnology and carbon nanoparticles in agriculture. pp. 247–277. *In*: *Synthesis, Technology, and Applications of Carbon Nanomaterials*. Elsevier.

Siddiqui, M.H., Al-Whaibi, M.H., Firoz, M. and Al-Khaishany, M.Y. (2015). Role of nanoparticles in plants. pp. 19–35. *In*: *Nanotechnology and Plant Sciences*. Cham: Springer.

Singh, A., Singh, N.B., Hussain, I., Singh, H., Yadav, V. and Singh, S.C. (2016). Green synthesis of nano zinc oxide and evaluation of its impact on germination and metabolic activity of *Solanum lycopersicum*. *J. Biotechnol.*, 233: 84–94.

Singh, B.N., Rawat, A.K.S., Khan, W., Naqvi, A.H. and Singh, B.R. (2014). Biosynthesis of stable antioxidant ZnO nanoparticles by *Pseudomonas aeruginosa* rhamnolipids. *PLoS One*, 9(9): 106937.

Solanki, P., Bhargava, A., Chhipa, H., Jain, N. and Panwar, J. (2015). Nano-fertilizers and their smart delivery system. pp. 81–101. *In*: *Nanotechnologies in Food and Agriculture*. Cham: Springer.

Srivastava, A. and Rao, D.P. (2014). Enhancement of seed germination and plant growth of wheat, maize, peanut, and garlic using multiwalled carbon nanotubes. *Eur. Chem. Bull.*, 3 (5): 502–504.

Srivastava, S.K. and Constanti, M. (2012). Room temperature biogenic synthesis of multiple nanoparticles (Ag, Pd, Fe, Rh, Ni, Ru, Pt, Co, and Li) by *Pseudomonas aeruginosa* SM1. *J. Nanopart. Res.*, 14(4): 831.

Subramanian, K.S., Manikandan, A., Thirunavukkarasu, M. and Rahale, C.S. (2015). Nano-fertilizers for balanced crop nutrition. pp. 69–80. *In*: *Nanotechnologies in Food and Agriculture*. Cham: Springer.

Sun, H., Du, W., Peng, Q., Lv, Z., Mao, H. and Kopittke, P.M. (2020). Development of ZnO nanoparticles as an efficient Zn fertilizer: Using synchrotron-based techniques and laser ablation to examine elemental distribution in wheat grain. *J. Agric. Food Chem.*, 68(18): 5068–5075.

Suresh, S., Karthikeyan, S. and Jayamoorthy, K. (2016). Effect of bulk and nano-Fe_2O_3 particles on peanut plant leaves studied by Fourier transform infrared spectral studies. *J. Adv. Res.*, 7(5): 739–747.

Waghmare, S.S., Deshmukh, A.M., Kulkarni, S.W. and Oswaldo, L.A. (2011). Biosynthesis and characterization of manganese and zinc nanoparticles. *Univers. J. Environ. Res. Technol.*, 1(1): 64–69.

Wang, J., Mao, H., Zhao, H., Huang, D. and Wang, Z. (2012). Different increases in maize and wheat grain zinc concentrations caused by soil and foliar applications of zinc in Loess Plateau, China. *Field Crops Res.*, 135: 89–96.

Wang, Y., Hu, J., Dai, Z., Li, J. and Huang, J. (2016). *In vitro* assessment of physiological changes of watermelon (*Citrullus lanatus*) upon iron oxide nanoparticles exposure. *Plant Physiol. Biochem.*, 108: 353–360.

Wang, Z.H., Miao, Y.F. and Li, S.X. (2015). Effect of ammonium and nitrate nitrogen fertilizers on wheat yield in relation to accumulated nitrate at different depths of soil in drylands of China. *Field Crops Res.*, 183: 211–224.

Yadav, V., Sharma, N., Prakash, R., KK, R., Bharadwaj, L.M. and Prakash, N.T. (2008). Generation of selenium containing nano-structures by soil bacterium *Pseudomonas aeruginosa*. *Biotechnology*, 7(2): 299–304.

Yang, H., Xu, M., Koide, R.T., Liu, Q., Dai, Y., Liu, L. and Bian, X. (2016). Effects of ditch-buried straw return on water percolation, nitrogen leaching, and crop yields in a rice–wheat rotation system. *J. Sci. Food Agric.*, 96(4): 1141–1149.

Yousefzadeh, S. and Sabaghnia, N. (2016). Nano-iron fertilizer effects on some plant traits of dragonhead (*Dracocephalum moldavica* L.) under different sowing densities. *Acta Agric. Slov.*, 107(2): 429–437.

Zhu, M., Nie, G., Meng, H., Xia, T., Nel, A. and Zhao, Y. (2013). Physicochemical properties determine nanomaterial cellular uptake, transport, and fate. *Acc. Chem. Res.*, 46(3): 622–631.

Chapter 4

Nanofertilizers: Importance in Nutrient Management

Mona Nagargade,[1,*] *Vishal Tyagi,*[2] *Dileep Kumar,*[1] *SK Shukla*[3] and *AD Pathak*[4]

1. Introduction

The world population is projected to reach 9.7 billion in 2050 and 10.87 billion by 2100 (UN DESA, 2019). With the alarming rise in population, particularly in less-developed regions, the demand for food will increase manifold. Even if the total demand for food and feed were to grow rapidly, meeting the expected food and feed demand will require a substantial increase in the global food production at 70% by 2050 (FAO, 2009). About 80% of the required increase will need to come from the higher yields and 10% from the increase in cropping intensity (Alexandratos and Bruinsma, 2012). Though this demand for increased production of food must be achieved, our stock of natural resources is on the verge of depletion and climate change risks are getting severe with each passing day. Therefore, feeding the future population without environmental degradation and depletion of natural resources is a major challenge that researchers and policymakers face across the world. This objective can only be met by adopting sustainable agricultural practices. Growth factors, such as light, oxygen, CO_2, water, and nutrients determine the performance of the crop. If one factor is limiting, crop performance is negatively affected. Among all, optimal crop nutrition is a fundamental requirement to ensure food and nutritional security

[1] Scientist, ICAR – Indian Institute of Sugarcane Research, Lucknow, 226002, Uttar Pradesh, India.
[2] Scientist, ICAR – Indian Institute of Seed Science, Mau, 275103, Uttar Pradesh, India.
[3] Principal Scientist and Head, Division of Crop Production, ICAR – Indian Institute of Sugarcane Research, Lucknow, 226002, Uttar Pradesh, India.
[4] Director, ICAR – Indian Institute of Sugarcane Research, Lucknow, 226002, Uttar Pradesh, India.
* Corresponding author: monanagargade@gmail.com

in the present scenario. The food grain production is directly influenced by fertilizer application. Fertilizer applications, based on the crop nutrient demand, enhance crop yield which further ensure food, nutritional, and livelihood security. Crop nutrient demand and soil nutrient reserve are maintained by the application of conventional fertilizers. According to estimates, the use of fertilizers contributes to almost 50% of crop production in the agriculture sector (Qureshi et al., 2018). Higher production may be achieved due to nanofertilizer application in agriculture. Therefore, to achieve higher yields, farmers rely on heavy amounts of chemical fertilizers all over the world. By 2050, the projected consumption of nitrogen (N), phosphorous (P), and potassium (K) fertilizers will increase by 172%, 175%, and 150%, respectively as compared to the current levels (Khan et al., 2018), and the total global fertilizer consumption is expected to be reach a staggering 263 million MT by 2050 (Fig. 4.1). As observed in Fig. 4.2, India is also showing an increasing pattern of fertilizer use.

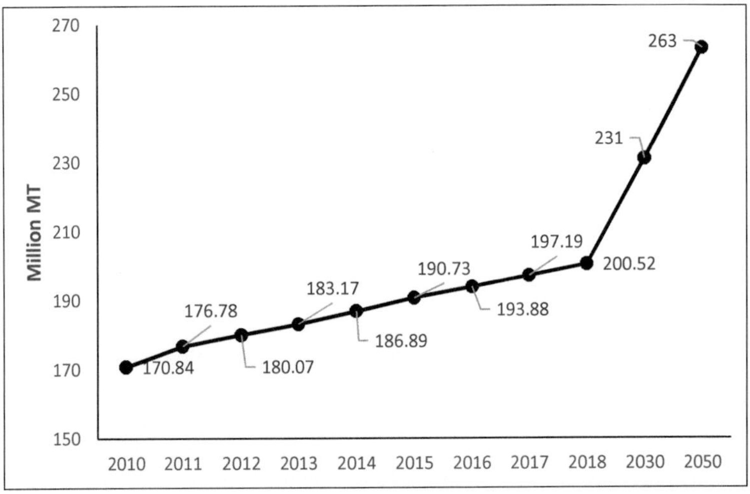

Fig. 4.1. Actual and projected global fertilizer consumption (Source: FAO, 2018).

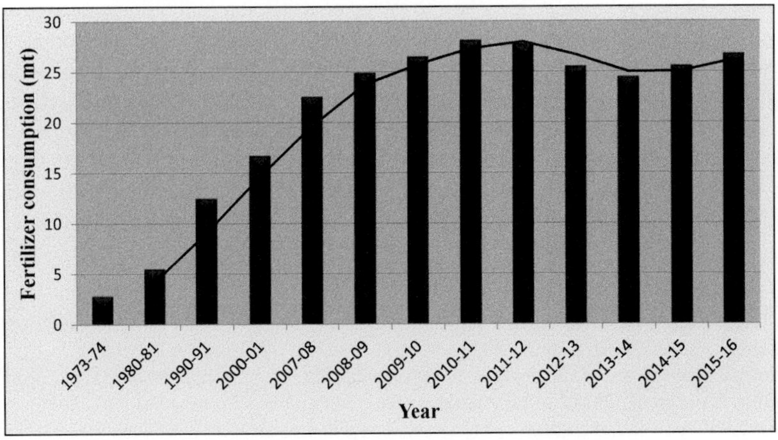

Fig. 4.2. Fertilizer consumption in India (Source: Indiastat.com).

After the Green Revolution, the introduction of high-yielding nutrient-responsive varieties of rice and wheat accelerated the rate of fertilizer use in India. No doubt, agro-chemicals have contributed significantly to increasing food grain production (El-Ghamry et al., 2018) in the country but their indiscriminate use, especially fertilizers and pesticides, is causing soil and water pollution, climate change, and human health problems among others (Leon-Silva et al., 2018). For supplementing nutrients to the crop, farmers mostly use conventional fertilizers, but these have poor nutrient use efficiency (Table 4.1), which means that major part of applied fertilizers are lost in the environment. Reduced fertilizer use efficiency is the result of leaching, decomposition, hydrolysis, volatilization, or denitrification losses (El-Azeim et al., 2020). Thus, more fertilizers are required to produce the same quantity of food grain (Fig. 4.3). Further, an improper and imbalanced use of conventional fertilizers causes several problems, such as significant nutrient losses during delivery, environmental pollution, shorter duration of nutrient release, more solubility, and climate change risks due to greenhouse gas emissions. For example, an excessive use of nitrogen leads to eutrophication and air pollution is caused by nitrogen oxides (NO, N_2O, NO_2) emissions and soil degradation (Savci, 2012). Similarly, phosphorus fertilizers cause eutrophication problems in surface freshwater bodies and coastal ecosystems (Liu and Lal, 2015). Fertilizers reach the water bodies through run-off, further allowing the growth of aquatic plants and algal blooms that cause depletion of dissolved oxygen, thereby negatively affecting aquatic life. The soluble forms of these nutrients reach the groundwater through deep percolation after dissolution with irrigation and/or rainfall where these contaminate the drinking water supply, making it unsuitable for both human and livestock consumption (Khan et al., 2018). Additionally, due to a rise in costs to acquire higher doses of these fertilizers, farmers find themselves at the brunt of a reducing profit margin (Zulfiqar et al., 2019).

As per the previous discussion, it could be inferred that ecosystem-based crop management practices can play an instrumental role in management of growth, environment, and enhancement of crop productivity. Low nutrient use efficiency (NUE) is typically the result of a higher release rate of fertilizers than the absorption rate by plants, and/or transformation of fertilizers/nutrients to forms that are not

Table 4.1. Fertilizers and their nutrient-use efficiency.

S. No.	Category of Fertilizer	Nutrient-use Efficiency (%)	Causes of Low Efficiency	References
1.	Nitrogenous	30–50	Immobilization, volatilization, denitrification, leaching	Linquist et al., 2012; Dimkpa et al., 2020
2.	Phosphatic	15–20	Fixation in soils Al – P, Fe – P, Ca – P	Suh and Lee, 2011
3.	Potassic	70–80	Fixation in clay – lattices	El-Azeim et al., 2020
4.	Sulphur-based	8–10	Immobilization, leaching	Havlin et al., 2012
5.	Micronutrients-based (Zn, Fe, Cu, Mn, B)	< 5	Fixation in soils	Monreal et al., 2015

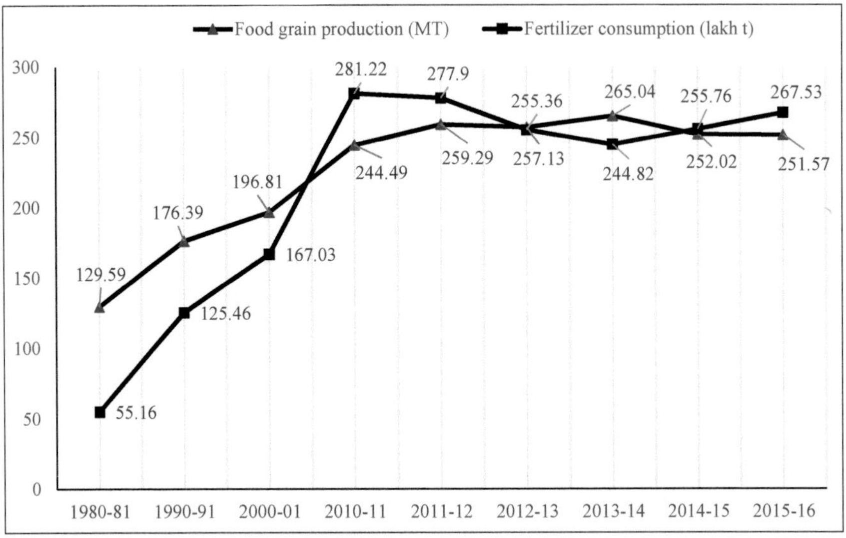

Fig. 4.3. Food grain and fertilizer consumption in India (Ministry of Agriculture and Farmers Welfare, 2017).

bio-available to the crops (Guo et al., 2018). In this situation, there is an urgent need to develop smart materials that can systematically release nutrients to specific targeted sites in plants (Solanki et al., 2015; Calabi-Floody et al., 2018). The introduction of some nutrient-efficient technologies could not only enhance nutrient use efficiency but also improve crop production (Lateef et al., 2019). Higher nutrient-use efficiency is mainly associated with more plant nutrient uptake from soil system and efficient conversion of nutrients into vegetative biomass and ultimately yields. The main aim of fertilizer use should be to "feed the soil and let the soil feed the plant" (Jat et al., 2011). The key principle of crop fertilization is to avoid nutrient losses and establish synchronization between nutrient release and crop demand. There are two strategies to enhance nutrient use efficiency, which are listed as follows:

1. **Product-based:** It is mainly based on enhancing the efficiency of fertilizers, i.e., nanofertilizers, slow-release fertilizers, nitrification inhibitors, urease inhibitors, biofertilizers, etc.
2. **Process-based:** This includes enhanced efficiency of fertilization through site-specific nutrient management, integrated nutrient management, split application, band application, etc.

2. Nutrient Management

Nutrient management implies that managing all nutrient sources (organic, chemical, and biofertilizers) in a way to maximize crop yield, restore/maintain/increase soil fertility and minimize loss of nutrients to the environment. Maintain or enhance soil productivity through a balanced use of local and external sources of plant nutrients is the prime objective of nutrient management. Concept of nutrient management mainly relies on below equation:

Nutrient management = Maintenance of soil fertility + Increase Crop productivity

Nutrient management of crop should be based on following points (Singh and Singh, 2001):

1. Nutrient removed must be returned to the soil.
2. Organic carbon levels should be maintained and enhanced.
3. Soil physical conditions should be maintained and upgraded.
4. The buildup of abiotic stress should be minimal.
5. The soil quality with respect to soil acidity/toxic elements build up must be sustained.
6. Degradation of land due to soil erosion must be controlled.
7. The efficiency of plant nutrients must be improved, thus limiting losses to the environment.

The intervention of nanotechnology in agriculture facilitates ample opportunity for nutrient management. It facilitates the smart delivery of inputs (slow and target delivery) in the agroecosystem. In 2002, the United States Department of Agriculture (USDA) drafted the world's first roadmap for application of nanotechnology in agriculture (Manjunatha, 2016). In the Indian context, former President late Dr. APJ Abdul Kalam emphasized the application of nanotechnology in agriculture and implemented it, through the development of at least 10 nanotechnology-based products in different sectors such as water, energy, agriculture, healthcare, space, and defence through public-private partnership.

3. Nanofertilizers

Nanofertilizers are those organic or inorganic materials that fall into the size range of 1–100 nm and supply as well as make available the essential elements to crop plants. They may improve the performance of conventional fertilizers through modification in the form of nanomaterials (Liu and Lal, 2017). They may work as a reservoir of nutrients and release them slowly and steadily, based on crop demand. Application of nanofertilizer increases effective duration of nutrient release and increases nutrient-use efficiency and crop productivity. Nano-based materials are required in lesser amounts due to their improved effectiveness, reducing the dosage and frequency of fertilizer use. Different nanomaterials associated with plant nutrition such as macronutrient nanofertilizers, micronutrient nanofertilizers, green nanofertilizers, biofertilizers, nanonutrient-engineered nanomaterials, chitosan-based nanomaterials, nutrient-loaded nanofertilizers, and plant growth-enhancing nanomaterials must be explored.

Nanocarriers act as a delivery platform that can deliver nutrients to the plants and these materials are user- and environmental-friendly. Naturally derived nanocarriers provide a platform for better utilization of biomass or organic waste. Nanoclays, hydroxyapatite nanoparticles, mesoporous silica, carbon-based nanomaterials, polymeric nanomaterials, and others pine oleoresin [POR], and nanoscale zinc oxide (ZnO) or rock phosphate [RP] as carriers for urea, nano-sized manganese carbonate

($MnCO_3$), hollow core-shell loaded with zinc sulphate ($ZnSO_4$) can be used as nanocarriers for efficient delivery of nutrients (Guo et al., 2018). Nanofertilizers are highly reactive and catalytic due to their high surface/volume ratio. The faster dissolution kinetics of nanomaterials may pass the nutrients through the cell membrane and reduce the problem of leaching (Arruda et al., 2015; Mastronardi et al., 2015). The use of nanobiosensors with nanomaterial coating facilitates the release of nutrients in response to signals received from soil microbes in the plant's rhizospheric environment under the nutrient deficient condition. Nanofertilizers have a plethora of advantages towards high production and effective utilization of natural resources in contrary to uses and application of conventional fertilizers. These nanomaterials help improve the overall productivity through enhanced seed germination, better seedling growth, and other physiological and biochemical activities, such as photosynthesis, nitrogen metabolism pathway, protein, and carbohydrate metabolism, along with biotic and abiotic stress tolerance. Simultaneously, it has other advantages, namely, It is required in lesser quantity which ultimately reduces the cost of transportation and ease of application.

3.1 Properties of Nanofertilizers

1. Small size and large surface area
2. More surface area-to-volume ratio
3. More reactivity with other compounds
4. High solubility in different solvents
5. More penetration
6. Slow release.

3.2 Classification of Nanofertilizers

Nanofertilizers are categorized into different groups based on their functions, which are listed as follows:

3.2.1 Macronutrient-based Nanofertilizers

Plants require macronutrients in large amounts for their growth and development and macronutrient nanofertilizers supply all the necessary macronutrients (N, P, K, Ca, Mg, and S). According to estimates, the global demand for macronutrient fertilizers will reach 263 MT by 2050 (Alexandratos and Bruinsma, 2012). Engineered nanomaterials show potentially higher nutrient-use efficiency than conventional fertilizers. The crop nutrient demand for nitrogen is mainly fulfilled by the application of urea (NH_2-CO-NH_2), diammonium phosphate ($(NH_4)_2HPO_4$), and ammonium sulphate ($(NH_4)_2SO_4$) (rice and tea) nitrogenous fertilizers. These fertilizers are prone to volatilization, denitrification, leaching, and run-off losses. Therefore, the slow release of nitrogen is accomplished using nanomaterials, such as nano-assisted-coated urea and urea-modified hydroxyapetite (HA) (Subbaiya et al., 2012). The nanocomposites show a bi-phasic nutrient release pattern with the rapid release of nitrogen at an early stage and later, with a successive steady release over 60 days (Kottegoda et al., 2017). The nanomaterial releases 78% more nitrogen than conventional fertilizers on 60 days after application in a sandy soil (pH 7). This temporal release pattern could effectively increase nitrogen-use efficiency and ultimately crop yield. Urea hydroxyapatite nanohybrids facilitate the slow and steady release of nitrogen and phosphorus. Therefore, the application of phosphorus with calcium formulation in a controlled manner contributes to more plant growth and improved crop yield.

Examples

1. Foliar application of NPK-nanochitosan composite (10–100 mg/L) on wheat crop significantly reduced the plant life cycle by 40 days and increased the yield by 51% and 56% as compared to control and conventional NPK fertilizers, respectively (Abdel-Aziz et al., 2016).
2. N_2O emission is reduced by the application of fortified urea with 2% nano ZnO and 35% nano rock phosphate ($Ca_{10}(PO_4)_6F_2$) particle using (pine oleo resin) as a binding agent (Kundu et al., 2016)

Advantages of Macronutrient Nanofertilizers

1. Increases crop productivity through direct, enhanced availability, and target delivery of nutrients both spatially and temporally.
2. Efficient uptake and utilization of nutrients facilitates plants to stand under biotic and abiotic stress condition.,
3. Reduces greenhouse gas (N_2O) emission which ultimately reduces the problem of climate change.
4. Reduces frequency and amount of chemicals used in agriculture.
5. Ultimately reduces cost of production and increases income of farmers.

3.2.2 Micronutrient Nanofertilizers

Micronutrient fertilizers supply essential nutrients required by plants in relatively small amounts, usually less than 10 mg/kg of soil. Micronutrients on nanoscale can increase the availability of those elements such as Fe, Zn, Cu, B, Mn that promote plant metabolism, enhance growth, development, and nutritional quality. Root or foliar applications of Zn nanomaterial or a composite of Zn, Cu, and B nanomaterial can increase grain yield as well as N and K accumulation. The fortification of N and K macronutrient fertilizers with micronutrients at nanoscale can increase the overall nutrient-use efficiency (Dimpka and Brindaban, 2017).

3.2.3 Green Nanofertilizers

Products developed by using bacteria, fungus, yeast, actinomycetes, and plant extract could be considered because the of biosynthesis of green nanomaterials (El-Ghamry et al., 2018). The development of new nanoproducts to use microorganisms must have properties of being eco-friendly, acting as capping and reducing agents. The microbes can be utilized to produce biofertilizers using indirect means of microbial activity through redox reaction and metabolic energy.

3.2.4 Nanobiofertilizers

Biofertilizers are substances containing live or latent cells of microorganisms (rhizobium, phosphorus solubilizing bacteria, azotobactor) of efficient strain in sufficient number used with the objective of increasing their number and to accelerate those activities of the microorganism which augment the availability of nutrients that can be easily available to crops. The use of biofertilizers along with nanomolecules to enhance production and productivity of crops may be considered nanobiofertilizers. There are several factors that makes a slight to quite a lot more difference between nanofertilizers and nanobiofertilizers, such as the application route, i.e., soil or plant, application methods, application dose, and its further impact on the environment (Simarmata et al., 2016). The application of formulated nanomolecules helps to improve the efficiency of biofertilizer. Numerous nano-coated materials are used with biofertilizers such as polymeric nanoparticles, and hydrophobic silica which increase the effectiveness and shelf life of biofertilizers (Kaushik et al., 2017). Delivery mechanism of biofertilizers can be enhanced with the use of nanomaterials either by foliar or rhizospheric application. An exhaustive field research is required to study the interaction between nanofertilizers and nanobiofertilizers.

3.2.5 Non-nutrient-engineered Nanomaterials (ENMs) with Fertilizer Potential

There are a few ENMs that affect plant growth and development indirectly. These materials, such as carbon nanotubes (CNT) CeO_2, SiO_2, and TiO_2, are not essential for the plant to complete its life cycle though they increase plant growth that finally converts to yields.

3.2.6 Chitosan-based Nanomaterials as Nanofertilizers

Chitosan is a decomposable biopolymer, which is natural and in an abundant source. Chitosan has multipronged effects on plant growth, development, antimicrobial effects, source of micro-nutrients, and pesticides. Chitosan nanomolecules are highly reactive and pose high positive charges on their surfaces. The other chemical property of these materials is high water solubility. In contrary to bulk chitosan, the engineered nanomaterials attract both organic and inorganic materials and macro, micronutrients. Chitosan is also a good source of nitrogen nutrient and contains about 9–10% N. There are several physical factors affecting plant response to chitosan nanomaterials, such as particle size, ionic charges, and polydispersity index (Rai et al., 2012).

3.2.7 Nutrient-loaded Nanofertilizers

Nutrient-loaded nanofertilizers are those nanomaterials which are capable of loading essential nutrients inside the pores of a plant, e.g., zeolite, mesoporous silica. Nutrient-augmented zeolites are the best example of such kinds of materials. The three-dimensional positioning structure of Al and Si as AlO_4 and SiO_4 provide sufficient porosity within the nanoscale (0.3–10 nm). Zeolite possesses the characteristics of nanostructure due to its size, which is less than nanoscale. Zeolites with high specific surface area and high cation exchange capacity (CEC) are capable of supplying more nutrients to plants (Liu and Lal, 2015).

3.2.8 Plant-growth-enhancing Nanomaterials

There are some other nanomaterials (Table 4.2) which influence plant growth indirectly. These materials do not contain essential nutrients but affect enzymes' activity and ultimately the plant growth.

Table 4.2. Plant-growth-enhancing nanomaterials.

S. No.	Nanomaterial	Effects on Crop Growth	References
1.	TiO_2	Increased plant photosynthetic activity, enhanced plant enzymatic reaction, fixation of atmospheric nitrogen	(Khodakovskaya et al., 2013; Villagarcia et al., 2012).
2.	CNT at 50 mg/liter	Increased tomato yield and water use efficiency	

3.2.9 Nanotechnological Approaches to Enhance Nutrient-use Efficiency (NUE)

- Encapsulated by nanomaterials or coated with a thin protective film (Encapsulation): It is packaged fertilizers in small protected coated shells, i.e., nanotubes or nanoporous materials. Such materials enhance the following activities of fertilizer granules:
  - Decrease solubility
  - Higher use efficiency
  - Environmentally safe
- Slow release: In this process, soluble fertilizers coated with certain materials counter the exposure of fertilizers to water or release the resultant nutrient to solution by diffusion. The release of nutrients from coated capsules is governed by nanomaterials and sub-nanomaterials.
- Nanobiosensors: While being exposed to biotic and abiotic stresses, plants secrete definite exudates from the roots in rhizospheric environments. These agrochemicals activate the microbes that are responsible for biotic mineralization, and they enhance the mineralization of nanoparticles either from soil organic matter or soil organic colloids. Such exudates (i.e., organic acid, phenolic acid, fatty acid, enzymes, etc.) may be categorized as an environmental indicator and it can be specified as a nanobiosensor, and release of the nutrient takes place which fulfils the needs of the plants. However, at several instances, mismatch of the root exudates signals the soil microbes and plant rhizosphere yielding without any nutrient release or a meagre amount of release (Suppan, 2017) affects the efficiency of nano-biosensors.

3.3 Factors Affecting Nanofertilizers

There are three factors which affect the uptake, assimilation, and utilization of nanomaterials used as fertilizers

1. Plant factor: Type of crop, varieties, age of plant, type of plant part exposed to the fertilizers and leaf properties.
2. Edaphic factor: Organic matter content, pH, electrical conductivity, soil texture, soil structure, soil moisture, and microbial population.
3. Environmental factor: Temperature, rainfall, humidity, wind velocity, and solar radiation.

3.4 Limitations of Nanofertilizers

1. Inadequate information related to environmental safety and human health may limit their usage.
2. Some nanomaterials have characteristics of phytotoxicity with respect to the dose application on the plant.
3. Nanomaterials are highly reactive and variable during application due to their microscopic size and high surface area. These two physical characteristics pose threat for farmers (respiratory disease, skin problems) who will ultimately handle and apply these materials to the field. This could raise serious concerns at both the industry and farm sectors [Ultimate stakeholder of fertilizers are farmers and in the Indian context, more than 70% farmers are small and marginal. They utilize the product at the field manually (either broadcasting, band placement, or foliar application) for crop growth, therefore it should be safe for users].
4. Nanomaterials pose health risks. Therefore, with reference to fertilizer cycle, the assessment of risks and hazards are critical to identify priorities for future toxicological research.

4. The Way Forward

1. The application of nanotechnology is in the initial phase to address the global problem of low fertilizer-use efficiency. The use of nanotechnology is inevitable, which demands intensive field studies to match the needs of farmers.
2. Identification of type of nanomaterials, the amount to be applied, and the stage of application must be highly precise to get better results.
3. Based on soil testing and deficiency of a particular nutrient, a target-based delivery mechanism must be strengthened for high input use efficiency.
4. Extensive field trials, along with laboratory testing, are required to generate enough information and an efficient nano product for exemplary output.
5. Nanoproducts developed based on the studies discussed earlier must have wider acceptance while having undergone all regulatory compliances before being promoted on a commercial scale for agricultural use.
6. Nanofertilizers developed at the commercial scale must adhere to all environmental norms and have eco-friendly characteristics.

References

Abdel-Aziz, H.M.M., Hasaneen, M.N.A. and Omer, A.M. (2016). Nano chitosan-NPK fertilizer enhances the growth and productivity of wheat plants grown in sandy soil. *Spanish Journal of Agricultural Research*, 14(1): e0902.

Alexandratos, N. and Bruinsma, J. (2012). *World Agriculture towards 2030/2050: The 2012 Revision.* ESA Working Paper No. 12-03. FAO, Rome.

Arruda, S.C.C., Silva, A.L.D., Galazzi, R.M., Azevedo, R.A. and Arruda, M.A.Z. (2015). Nanoparticles applied to plant science: A review. *Talanta*, 131: 693–705.

Calabi-Floody, M., Medina, J., Rumpel, C., Condron, L.M., Hernandez, M., Dumont M. and Mora, M.L. (2018). Smart Fertilizers as a strategy for sustainable agriculture. *Advances in Agronomy*, 147: 119–157.

Dimkpa, C.O. and Bindraban, P.S. (2017). Nanofertilizers: New products for the industry? *Journal of Agriculture Food Chemistry*, 66: 6462–6473. https://doi.org/10.1021/acs.jafc.7b02150.

Dimkpa, C.O., Fugice, J., Singh, U. and Lewis T.D. (2020). Development of fertilizers for enhanced nitrogen use efficiency: Trends and perspectives. *Science of the Total Environment*, 731: 139113.

El-Azeim, M.M.A., Sherif, M.A., Hussien, M.S., Tantawy, I.A.A. and Bashandy, S.O. (2020). Impacts of nano- and non-nanofertilizers on potato quality and productivity. *Acta Ecologica Sinica* (Article in press). doi.org/10.1016/j.chnaes.2019.12.007.

El-Ghamry, A.M., Mosa, A.A., Alshaal, T.A. and El-Ramady. H.R. (2018). Nanofertilizers vs. Biofertilizers: New insights. *Environment, Biodiversity and Soil Security*, 2: 51–72.

FAO. (2018). Modified from FAO, *World Fertilizer Trends and Outlook.*

Guo, H., White, J.C., Wang, Z. and Xing, B. (2018). Nano-enabled fertilizers to control the release and use efficiency of nutrients. *Current Opinion in Environmental Science & Health*, 6: 77–83.

Havlin, J.L., Beaton, J.M., Tisdale, S.L. and Nelson, W.L. (2012). *Soil Fertility and Fertilizers an Introduction to Nutrient Management* (7th Edn.). N.J: PHI Learning Private Limited Publishers, pp. 1–515.

Jat, M.L, Saharawat, Y.S. and Gupta, R. (2011). Conservation agriculture in cereal systems of South Asia: Nutrient management perspectives. *Karnataka Journal of Agricultural Sciences*, 24(1): 100–105.

Kaushik, S. and Djiwanti, S.R. (2017). Nanotechnology for enhancing crop productivity. pp. 249–262. *In*: Prasad, R., Kumar, M., Kumar, V. (eds.). *Nanotechnology: An Agricultural Paradigm.* Singapore: Springer.

Khan, M.N., Mobin, M., Abbas, Z.K. and Alamri, S.A. (2018). Fertilizers and their contaminants in soils, surface and groundwater. *In*: Dominick A. DellaSala and Michael I. Goldstein (eds.). *The Encyclopedia of the Anthropocene*, 5: 225–240. Oxford: Elsevier.

Khodakovskaya, M.V., Kim, B.S., Kim, J.N., Alimohammadi, M., Dervishi, E., Mustafa, T. and Cernigla, C.E. (2013). Carbon nanotubes as plant growth regulators: Effects on tomato growth, reproductive system, and soil microbial community. *Small*, 9: 115–123.

Kottegoda, N., Sandaruwan, C. and Priyadarshana, G. (2017). Urea–hydroxyapatite nanohybrids for slow release of nitrogen. *ACS Nano*, 11: 1214–1221. https://doi.org/10.1021/acsnano.6b07781.

Kundu, S., Adhikari, T., Mohanty, S.R., Rajendiran, S., Vassanda Coumar, M., Sahaj, K. and Patra A.K. (2016). Reduction in nitrous oxide emission from nano zinc oxide and nano rock phosphate coated urea. *Agrochimica*, 60(2): 59–70.

Lateef, A., Nazir, R., Jamil, N., Alam, S., Shah, R., Khan, M.N., Saleem, M. and Shafiq-ur-Rehman. (2019). Synthesis and characterization of environmental-friendly Corncob biochar-based nano-composite: A potential slow-release nano-fertilizer for sustainable agriculture. *Environmental Nanotechnology, Monitoring & Management*, 11: 1–12.

Leon-Silva, S., Arrieta-Cortes, R., Fernandez-Luqueno, F. and Lopez-Valdez, F. (2018). Design and production of nanofertilizers. *Agricultural Nanobiotechnology*, 17–31.

Linquist, B.A., Adviento-Borbe, M.A., Pittelkow, C.M., Van Kessel, C. and Van Groeningen, K.J. (2012). Fertilizer management practices and greenhouse gas emissions from rice systems: A quantitative review and analysis. *Field Crop Research*, 135: 10–21.

Liu, R. and Lal, R. (2015). Potentials of engineered nanoparticles as fertilizers for increasing agronomic productions. *Science of the Total Environment*, 514: 131–139.

Liu, R. and Lal, R. (2017). Nanofertilizers. pp. 1511–1515. *In*: *Encyclopedia of Soil Science* (3rd Edn.). DOI: 10.1081/E-ESS3-120053199.

Manjunatha, S.B., Biradar, D.P. and Aladakatti, Y.R. (2016). Nanotechnology and its applications in agriculture: A review. *Journal of Farm Science*, 29(1): 1–13.

Mastronardi, E., Tsae, P., Zhang, X., Monreal, C. and DeRosa, M.C. (2015). Strategic role of nanotechnology in fertilizers: Potential and limitations. pp. 25–67. *In*: *Nanotechnologies in Food and Agriculture*. N.Y.: Springer International Publishing.

Monreal, C.M., DeRosa, M., Mallubhotla, S.C., Bindraban, P.S. and Dimkpa, C. (2015). Nanotechnologies for increasing the crop use efficiency of fertilizer-micronutrients. *Biology and Fertility of Soils*, 52(3): 423–437.

Qureshi, A., Singh, D.K. and Dwivedi, S. (2018). Nano-fertilizers: A novel way for enhancing nutrient use efficiency and crop productivity. *International Journal of Current Microbiology and Applied Sciences*, 7(2): 3325–3335.

Rai, V., Acharya, S. and Dey, N. (2012). Implications of nanobiosensors in agriculture. *J. Biomater. Nanobiotechnol.*, 3: 315–324.

Savci, S. (2012). Investigation of effect of chemical fertilizers on environment. *APCBEE Procedia*, 1: 287–292.

Simarmata, T., Hersanti, T., Turmuktini, N., Betty Fitriatin, R. and Mieke Setiawati, Purwanto. (2016). Application of bioameliorant and biofertilizers to increase the soil health and rice productivity. *Hayati J. Biosci.*, 23: 181–184.

Singh, B. and Singh, Y.S. (2001). Concepts in nutrient management. pp. 92–109. *In*: *Recent Advances in Agronomy*. Indian Society of Agronomy, New Delhi, India.

Solanki, P., Bhargava, A., Chhipa, H., Jain, N. and Panwar, J. (2015). Nano-fertilizers and their smart delivery system. pp. 81–101. *In*: Rai, M., Ribeiro, C., Mattos, L. and Duran, N. (eds.). *Nanotechnologies in Food and Agriculture*. Springer.

Subbaiya, R., Priyanka, M. and Selvam, M.M. (2012). Formulation of green nano-fertilizer to enhance the plant growth through slow and sustained release of nitrogen. *J. Pharm. Res.*, 5: 5178–5183.

Suh, S. and Lee, S. (2011). Phosphorus use-efficiency of agriculture and food system in the US. *Chemosphere*, 84(6): 806–813.

Suppan, S. (2017). Applying Nanotechnology to Fertilizer: Rationales, research, risks, and regulatory challenges. (The Institute for Agriculture and Trade Policy works locally and globally at the intersection of policy and practice to ensure fair and sustainable food, farm, and trade systems.) The Institute for Agriculture and Trade Policy, pp. 1–21.

UN DESA (United Nations, Department of Economic and Social Affairs, Population Division). (2019). *World Population Prospects 2019*.

Villagarcia, H., Dervishi, E., de Silva, K., Biris, A.S. and Khodakovskaya, M.V. (2012). Surface chemistry of carbon nanotubes impacts the growth and expression of water channel protein in tomato plants. *Small*, 8: 2328–2334.

Zulfiqara, F., Navarro, M., Ashraf, M., Akram, N.A. and Munne-Bosch, S. (2019). Nanofertilizer use for sustainable agriculture: Advantages and limitations. *Plant Science*, 289: 1–11.

Plant Disease Management

Chapter 5

Polymeric Nano-fungicides for the Management of Fungal Diseases in Crops

Ruma Rani,[1] and *Pawan Kaur*[2,*]

1. Introduction

Agriculture is the most significant and stable division since it produces and delivers raw materials for food and feed productions. Worldwide, people are trying to increase the crop production for feeding appetite of the increasing population, which is one of the foremost challenges. For the last five to six decades, pesticides, chemical fertilizers, genetic modified plants, and disease-resistant crops have been used in the farming sector. Consequently, in the present scenario, pesticides (insecticides, fungicides and herbicides) have drawn attention as they are extensively used in numerous agricultural applications, which result in contamination of soil and groundwater. Constant use of pesticides, finally leads it to the food or food chain, affecting human health as they are carcinogenic and cytotoxic in nature. No doubt, augmentation of chemical fertilizers and pesticides have been very popular in agriculture in terms of enhanced food production for the rapidly growing human population, while it lowers down the quality of food that we eat and the fertility of the soil (Gianessi, 2013; Linquist et al., 2013; Hossain et al., 2015). Nevertheless, use of pesticides is one of best option for the protection of crop/plants from pests, fungi, and weeds as their use reduces the labor and improves life quality.

Unprecedented change in the global market of pesticides is due to their use in pest management as they have played a significant role to increase the agricultural yield by controlling pests and plant diseases. Usage of pesticides (herbicides, insecticides, fungicides, and fumigants) in the world was nearly 6 billion pounds in both 2011 and

[1] ICAR-National Research Centre on Equines, Hisar-125001, India.
[2] TERI-Deakin Nanobiotechnology Centre, The Energy and Resources Institute (TERI), Lodhi Road, New Delhi, 110003, India.
* Corresponding author: pawankaurnano@gmail.com

2012. U.S. pesticide usage was over 1.1 billion pounds in both 2011 and 2012 (Atwood and Paisley-Jones, 2017). The global crop protection chemicals market was estimated at USD 58.38 billion in 2019 and projected to expand at a Compound Annual Growth Rate (CAGR) of 3.3% from 2020 to 2027 (Global Industry Report, 2027). An estimated loss caused due to insects, plant diseases, and weeds was found to be 14%, 13%, and 13% respectively (Pimentel et al., 2009; Abd-Elsalam and Alghuthaymi, 2015). The plant pathogens, such as various fungus like Ascomycetes (genera: *Alternaria, Fusarium, Verticillium*) and Basidiomycetes (genera: *Rhizoctonia, Sclerotium*) caused contamination in the plant and fruit tissues. These group of plant pathogens causes losses in crop every year and responsible for damage during fruit transport and storage. A fungicide is a categorized under pesticides that have been used to controls fungal disease by preventing, impeding, or killing of the fungus responsible for spread of the disease. Fungicides work by contact (barrier on the plant that is not absorbed) or by penetration (absorbed by the plant at various levels of systemic movement).

Pesticide nanoformulations provide a wide variety of benefits that include (i) improved effectiveness and robustness, (ii) biodegradability in soil and environment, (iii) dispersibility and wettability, (iv) less cytotoxicity due to lesser dose of active ingredients (Kah and Hofmann, 2014; Chauhan et al., 2017; Adisa et al., 2019). Use of nanoencapsulation in the field of agriculture enhanced the pest control efficacy by encapsulation of active ingredients within the polymer matrix and averting the early degradation of active ingredients under adverse environment conditions (Kumar et al., 2019). Various reports have been reviewed the application nanotechnology in the field of agriculture include seed germination, plant growth and progression, fertilizer delivery, pesticides delivery, pathogen detection, etc. (Nair et al., 2010; Ghormade et al., 2011; Gogos et al., 2012; Goswami et al., 2012; Sekhon, 2014; Nuruzzaman et al., 2016; Worrall et al., 2018). Nanofungicides have been used to deliver any fungicide formulation in nanometer size range and owing to their vast physiochemical and functionalization properties, nanofungicides can be used for combating the plant disease either by delivery of active ingredients in controlled manner to increase the bioavailability of the active ingredient at the site of action (Choudhury et al., 2011; Alghuthaymi et al., 2015; Bhattacharyya et al., 2016; Haq and Ijaz, 2019).

The chapter has summarized the issues with conventional pesticides and the advantages of nanoformulations over the regular pesticides. Different polymeric nanocarriers of fungicide formulations for fungal infection in crops are also included. Lastly, we have drawn a conclusion and future perspective for the use of nanofungicides.

2. Conventional vs. Nanopesticides

Conventional pesticides are all active ingredients other than biological pesticides and antimicrobial pesticides. These have been a worldwide popular agrochemical system that brings about increase in crop yields, food production, and food quality (Gianessi, 2013; Linquist et al., 2013). Along with improved crop yield, these also have sustenance quality and safety of food against weed, insect, fungus, etc. Despite the many benefits of using pesticides in crop production, inappropriate use results in several unwanted effects that cause risk for humans and the environment. Pesticides applied to crops can cause water run-off and get leached to the groundwater. Also, they can be harmful to non-target organisms like frogs, bees, etc. and can pollute the

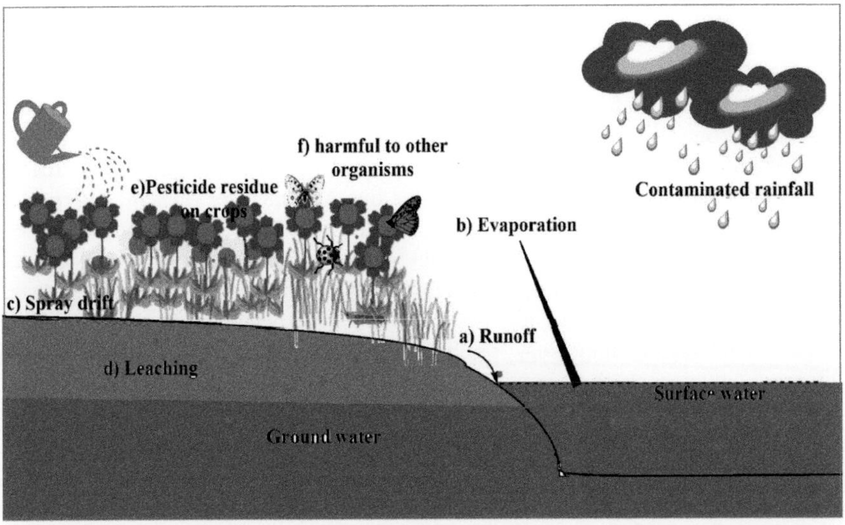

Fig. 5.1. Depiction of probable ways of environment pollution due to application of pesticides in the fields. (a) Pesticide run-off with rainfall, (b) evaporation of pesticide can leads contaminated rainfall, (c) pesticide spray drift into soil, (d) leaching of pesticide can cause contamination of ground water, (e) pesticide residue on crops leads contamination of food, (f) harmful to other organisms.

soil and air (Zhang et al., 2015). Only a part (less than 0.1%) of pesticide formulation is utilized by biological target uptake and the rest is loss in the form of run-off, spray/dust drift, rolling down, and rainwater leaching (He et al., 2016; Nuruzzaman et al., 2016; Massinon et al., 2017; Song et al., 2017) as shown in Fig. 5.1.

The use of conventional methods has been reported to influence the environment as well as on the economy of farmers, as about 90% of applied pesticide is lost during application. Thus, it is an alarming issue for the public as well as for the researchers that the overuse of pesticides harmful for the environment. Shahid et al. (2018) assessed the morpho-anatomical, biological, and chemical changes of three fungicides, namely carbendazim, hexaconazole, and kitazin, on *Pisum sativum* (peas) and reported a dose-dependent effect of fungicide application, specially carbendazim that showed a gradual decrease in development, and crop yield. Researchers have also reported the adverse effects such as cellular/ultrastructural damage, reduced viability, decreased plant nutrient, induced nodule variation, and oxidative damage, etc., caused by incessant use of pesticides (Anuradha et al., 2016; Shahid et al., 2018a, 2018b, 2019a, 2019b). The inappropriate delivery of pesticides resulted in numerous detrimental effects like water contamination, soil contamination, air pollution, increased pest and pathogen resistivity, obliteration of various species (e.g., earthworms, frog, bees, a non-target species) (Goulson et al., 2015; Goswami et al., 2017). Humans are also exposed to pesticides through many routes such as direct route (oral, inhalation, dermal) and indirect route (food delivery and consumption) (Kennedy et al., 2015; Smith et al., 2017) that leads to several health problems like burning eyes, dizziness wheezing, cancer, obesity, endocrine disorders, neurological disorders, etc. (Hernández et al., 2013; Rossignol et al., 2014; Richendrfer and Creton, 2015; Lammoglia et al., 2017; Rostami et al., 2019).

In nanoformulations of fungicides, nanoparticles are united with active ingredients and with other inorganic and organic compounds that must be in a nanometric size. Owing to its extremely small nanosize, the nanoformulation can be permeable through cellular membranes and help in the transfer of the active ingredient. The nanofungicides are more stable and effective at lower amounts of fungicidal compound as compared to conventional fungicides (Shyla et al., 2018). Therefore, these nonfungicides are smart nanocarriers of conventional fungicides that produce a significant effect at its minimum dose and is also economic for the farmers. Due to the high reactivity of materials at the nanoscale level support, the use of nanofungicides at lesser quantities has significant impacts on crop protection. Moreover, administration of nanofungicides at lower doses helps in decreasing the toxicity level while increasing the efficacy. The nanoformulations of fungicides enhance the solubility of active ingredients and support the slow release of compound at the target sites (Kumar et al., 2012; Roy et al., 2014; Liu et al., 2016; Volova et al., 2019). Thus, the bioavailability of agrochemicals for a longer period makes them eco-friendly as compared to conventional formulations (Kumar et al., 2013; Mattos et al., 2017; Shang et al., 2019).

The mechanism of killing a fungal cell was the same as that of the convention fungicides. But the rapidity and greater efficacy exhibited by the nanoformulations may be due to their small size and more permeability through the cell membrane of a fungal cell. The likely antifungal mechanism of nanoparticles on a fungal cell may be that when nanoparticles are taken by the fungal cell, the nanoparticles cause interruption in the DNA replication process. Moreover, the nanoparticles interact with enzymes present in the mitochondria that is involved in the process of cellular respiration which further promotes the reactive oxygen species (ROS) and severe oxidative damage to lipids, proteins, carbohydrates, and DNA that leads to necrosis or apoptotic cell death (Reidy et al., 2013; Haq and Ijaz, 2019), as depicted in Fig. 5.2.

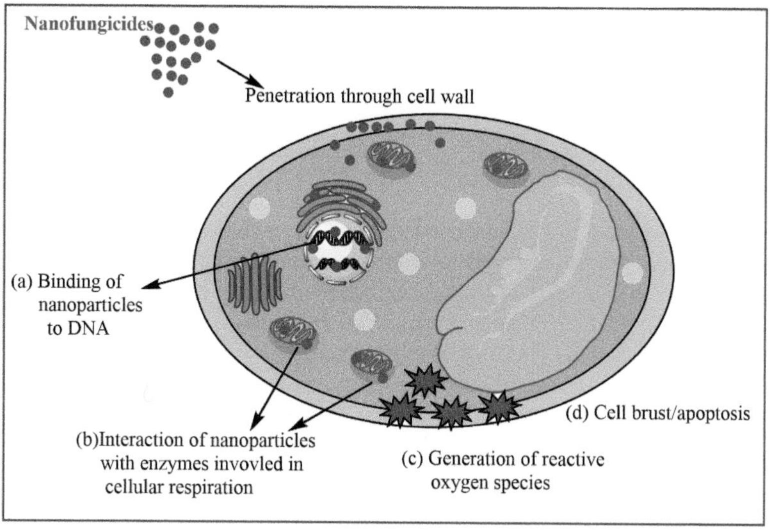

Fig. 5.2. Illustration of probable fungicidal mechanism nanoparticles. (a) Intake of nanoparticles, binding with DNA and stop DNA replication; (b) Interaction of nanoparticles with enzymes involved in cellular respiration; (c) Generation of ROS, and (d) apoptosis of fungal cell.

3. Nanocarriers for Management of Plant Disease

Nanocarriers are very beneficial for the encapsulation and protection of active ingredients from contrary environmental conditions (high temperature, pH, radiation), and increase their chemical stability. Smart nano-based carriers are prepared to deliver enough amount of active ingredients based on a stimulus and they act as prompts for targeted and controlled release of the active ingredient. Controlled release nanoformulations mitigate the harmful effects of agrochemicals by decrease of phytotoxicity, leaching, volatilization, drifting, and soil degradation, as well as increase and promote care on application (Camara et al., 2019). Different types of nanoformulations for agrochemicals have been proposed and developed like nanoparticles, nanocapsules, nanoemulsions, nanosuspensions, and nanocomposites (Alghuthaymi et al., 2015; Díaz-Blancas et al., 2016; Duhan et al., 2017; Majeed et al., 2017; Worrall et al., 2018; Kumar et al., 2019). Basis on the physico-chemical nature of nanoformulations, they are classified as follows: (1) polymer nanoformulations, (2) metallic nanoformulations, (3) lipid-based nanoformulations (4) carbon-based nanoformulations, (5) semiconductor nanoformulations, and (6) ceramic nanoformulations.

The main advantages of the nanoencapsulation of active ingredients in the form of nanocarriers are:

- Larger surface to volume ratio, biodegradable in nature and target-specific
- Improvement in the solubility and permeability of the active ingredient
- Protection and control release of active ingredients (stimuli-responsive nanoformulations like slow, quick, specific, moisture, heat, pH, magnetic, ultrasound release) to achieve target-specific release to maximize the utilization efficacy of pesticides
- Enhancement of dispersion of pesticides, lowering the run-off in the surrounding area
- Enhancement of biological efficacy due to prolonged/sustained release of active ingredients
- Reduction in the application rate and the various cytotoxic effects caused by active ingredients.

Among various nanoformulation strategies, polymeric nanomaterials and engineered nanoparticles are most used for the entrapment of active ingredient/s, as they seem to be biocompatible and biodegradable in nature (Ramasamy et al., 2017; Kumar et al., 2019). Polymers are categorized as either (1) natural and synthetic or (2) biodegradable and nonbiodegradable, based on controlled drug-delivery approach. Natural and biodegradable polymers (include cellulose, gelatin, pullulan, chitosan, alginate, and gliadin) and synthetic biodegradable polymers include poly-(lactide-co-glycolide) (PLGA), polylactic acid (PLA), polyanhydrides (PAHs), poly-ε-caprolactone (PCL), polyhydroxyalkanoates (PHAs), and polyphosphazene have been reported as nanocarriers for encapsulation of active ingredients (Chopra et al., 2014; Rani et al., 2015; Kumar et al., 2016; Wang et al., 2018; Hasheminejad et al., 2019). Figure 5.3 depicts the different types of nanoparticles as protectants and carriers for the benefits in the field of pest protection and pest management.

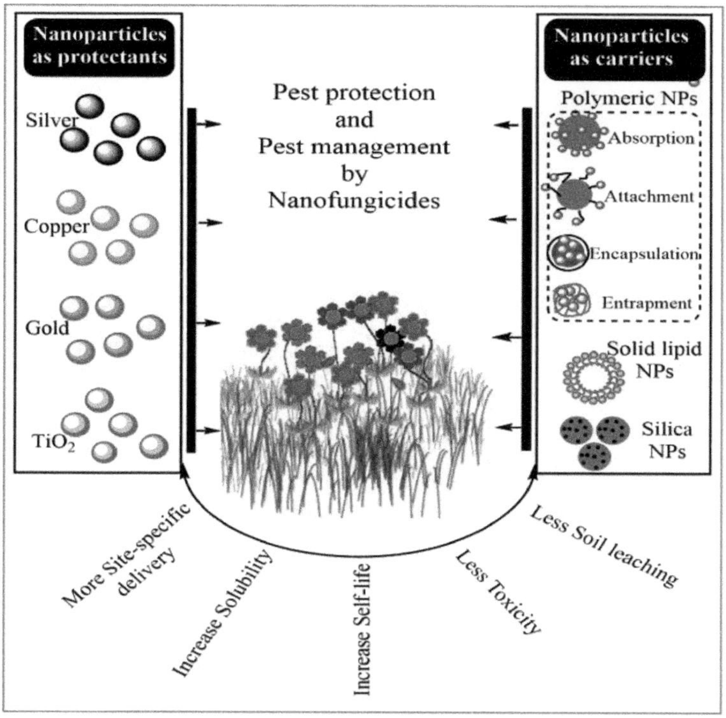

Fig. 5.3. Depicts different nanoformulations used for encapsulation of fungicides for pest protection and management.

4. Chitosan as Polymeric Nanoformulation

Polymeric nanoformulations can offer resourceful applications for the delivery of agrochemicals by inhibiting the harmful effects on the human health and environment because of their direct use. Many studies have been done for the optimization of the exact concentration of polymers and other excipients required in the preparation of nanoformulations. Various organic solvents and surfactants used in the nanoformulations are the main source of harmful effects (Katagi, 2008; Chen and Mullin, 2013; Chen and Mullin, 2015). Therefore, researchers have tried for the preparation of different nanoformulations with or without surfactant. Chitosan is one of the most significant materials used as a carrier for agrochemical as it is biodegradable, biocompatible, non-toxic, permeable, and has a film-forming property. Moreover, it has been reported to have many properties to carry out antifungal, antibacterial, antiviral, antiparasitic, and insecticidal activities (Rabea et al., 2005; Said et al., 2011; Ayaz et al., 2016; Kim, 2018). Improvement in the solubility of pesticide formulation can also be done by using chitosan. Chitosan has two reactive groups, that allow modification, graft reactions, and ionic interactions, enabling improvement of chitosan properties (Argüelles-Monal et al., 2018). Moreover, it adheres to the plant parts and prolong the contact time which might facilitate the uptake of active ingredients from the nanocarriers (Zhang et al., 2013; Malerba and Cerana, 2016). Chitosan, alone or in combination with others and its derivatives have already been used for the controlled release of fungicides (Mujtaba et al., 2020).

Metal nanoparticles such as silver, zinc, silica, copper, gold, iron, etc., have also become prevalent and putative alternatives to agrochemicals (Haq and Ijaz, 2019). They have been reported to have a potential to eradicate annoying and fatal microbes from different infected places like plants, soils, and even from water bodies (Sharma et al., 2012). Various reports have been published on preparation and assessment of antifungal activity of chitosan nanoparticles and chitosan-metal nanocomposites against various phytopathogenic fungi. Chowdappa et al. (2014) reported the synthesis of chitosan-silver nanocomposite against the fungus, *Colletotrichum gloeosporioides* and observed inhibitory effects. Saharan et al. (2013) prepared different type of nanoparticles of chitosan in combination with saponin and copper using the ionic gelation method and evaluated against three phytopathogenic fungi, namely *Rhizoctonia solani*, *Alternaria alternata*, and *Macrophomina phaseolina*, and observed more effective inhibitory effects shown by chitosan-copper nanoparticles, among the three nanoformulations. Likewise, the effect of chitosan nanoparticles has been evaluated on protection of rice plant from the blast fungus (*Pyricularia grisea*) (Manikandan and Sathiyabama, 2016). Sathiyabama and Parthasarathy (2016) reported the biological preparation of chitosan nanoparticles by adding isolated protein from the *Penicillium oxalicum* culture to chitosan solutions and *in vitro* evaluation of their potential toward growth inhibition of phytopathogenic fungi, *Pyricularia grisea*, *Fusarium oxysporum*, and *Alternaria solani*. Rubina et al. (2017) reported the preparation of copper nanoparticles using the metal-vapor synthesis method and further used the preparation of chitosan-copper nanocomposites. Moreover, the research group demonstrated a significant inhibitory effect of chitosan-copper nanocomposites against two sclerotia-forming pathogenic fungi *Rhizoctonia solani* and *Sclerotium rolfsii*. Thereafter, bimetallic blend-chitosan nanocomposite using copper and zinc nanoparticles instead of single metal nanocomposite were prepared. The bimetallic blend, chitosan nanocomposite, has been evaluated *in vitro* and *in vivo* against *Rhizoctonia solani* and has been reported to have found the synergetic effect against the fungi (Abd-Elsalam et al. 2018). Evaluated the synergistic effect of chitosan-silver nanoparticles against a fungus, causative agent for rice blast disease *Pyricularia oryzae*. Likewise, Kaur et al. (2018) reported preparation of chitosan nanoparticles, chitosan-silver nanoparticles, and a chitosan nanocomposite with copper oxide and zinc oxide and their evaluation against fusarium wilt of chickpea (*Cicer arietinum* L.). The highest reduction in wild disease was observed in the chitosan-copper oxide nanocomposite followed by chitosan-zinc oxide nanocomposites, chitosan-silver, and chitosan. Hashim et al. (2019) investigated different type of nanoparticles, chitosan, copper, and silica against *Botrytis cinerea*, the causative agent of gray mold on table grapes and reported that chitosan and silica nanoparticles have more potential as an antifungal agent. Thereafter, the same research group demonstrated the antifungal effect of different nanoparticles by themselves (chitosan nanoparticles, silica nanoparticles) and in combined form (chitosan-silica nanocomposites) against *Botrytis cinereal*. From both *in vitro* and *in vivo* evaluations, the researchers concluded the synergistic effect of chitosan-silica nanocomposites as compared to the other two nanoformulations (Youssef et al., 2019).

Various reports on encapsulation of fungicides in chitosan-based matrix and other nanomaterial-based are given in Table 5.1. Moreover, the release of the active ingredient from the matrix can be controlled based on different stimuli like light, pH,

Table 5.1. Different fungicides encapsulated in chitosan-based polymer.

S. No.	Fungicide	Material and Technique Used for Encapsulation	Achievements	References
1.	Tebuconazole or chlorothalonil	Polyvinylpyridine (PVP), PVP copolymer and polymer blend for nanoparticles	Delivered to southern yellow pine wood for protection against fungal attack of *Gloeophyllum trabeum*	Liu et al., 2002a
2.	Chlorothalonil, tebuconazole, and KATHON 930-fungicides; chlorpyrifos-insecticide	PVP, PVP copolymer and polymer blend for nanoparticles	Delivered to birch and southern yellow pine for protection against fungal attack by *Gloeophyllum trabeum* and *Trametes versicolor*	Liu et al., 2002b
3.	Tebuconazole and chlorothalonil	PVP, PVP copolymer and polymer blend for nanoparticles	Delivered to birch and southern yellow pine for protection against fungal attack by *Gloeophyllum trabeum* and *Trametes versicolor*	Liu et al., 2002c
4.	Pyoluteorin	Nanophase material of silicon dioxide loaded with drug synthesized by a highly ordered monolith method	Prolonged the antifungal effects	Chen et al., 2010
5.	Validamycin	Nano-sized calcium carbonate prepared by reversed-phase microemulsion method	Better germicidal efficiency against *R. solani* in comparison to pure fungicide	Qian et al., 2011
6.	Tebuconazole	Incorporated in porous hollow silica nanospheres (PHSN) using mini-emulsion method	Controlled release nanoformulation with release due to diffusion process and showed more antifungal activity as compared to pure	Qian et al., 2013
7.	Metalaxyl fungicide	Mesoporous silica nanoparticles prepared by sol-gel process and drug introduced by rotary evaporation method.	Sustained release of antifungal drug into soil	Wanyika, 2013
8.	Volatile essential oil components from plants- carvacrol, allyl isothiocyanate, eugenol, thymol, thymoquinone and cinnamaldehyde, diallyl disulfide	Encapsulated into mesoporous silica material	Encapsulated compounds tested against *Aspergillus niger* and found controlled release with more antifungal activity as compared to pure	Janatova et al., 2015

Table 5.1 contd. ...

...Table 5.1 contd.

S. No.	Fungicide	Material and Technique Used for Encapsulation	Achievements	References
9.	Carbendazim and Tebuconazole	Solid lipid nanoparticles and polymeric nanocapsules	More antifungal activity and less toxicity	Campos et al., 2015
10.	Pyraclostrobin	Water soluble chitosan derivative-capped mesoporous silica nanoparticles	Equal antifungal activity against *Phomopsis asparagi* (*Sacc.*) even at one-half of nanoparticles dose	Cao et al., 2016
11.	Essential oil from *Zataria multiflora*	Solid lipid nanoparticles (SLNs) prepared by ultra-sound technique and high shear homogenization	SLNs more effective against fungal pathogens as compared to pure oil.	Nasseri et al., 2016
12.	Eugenol and thymol	Immobilized on mesoporous silica microparticles	Enhanced antifungal activity in strawberry jam	Ribes et al., 2017
13.	Pyrimethanil	Mesoporous silica nanoparticles	Applied on cucumber leaves and found low risk of fungicide accumulation on the edible part of the plant	Zhao et al., 2017
14.	Oregano (*Origanum vulgare*) essential oil	Nanoemulsion prepared by phase inversion temperature method	Tested against *Cladosporium* sp., *Fusarium* sp., and *Penicillium* sp. Genera and found more *in vitro* antifungal activity	Bedoya-Serna et al., 2018
15.	Tebuconazole, propineb, fludioxonil	Silver nanoparticles synthesized by ginkgo fruit extract	Synergistic antifungal activity of AgNPs and three different fungicides	Huang, 2018
16.	Pyraclostrobin	Biodegradable chitosan-lactide copolymer prepared for encapsulation by nano-precipitation method	Nanoparticles showed sustained release and pH-responsive controlled release profile with better antifungal activity against *Colletotrichum gossypii Southw*	Xu et al., 2014
17.	Flusilazole	An amphiphilic chitosan-polylactide graft copolymer for encapsulation using nano-precipitation method	Controlled release and more permeability of flusilazole	Mei et al., 2014
18.	Hexaconazone	Hexaconazone-loaded chitosan-TPP nanocapsules has been prepared by ionotropic gelation method	The nanoformulation reported controlled delivery of hexaconazole against *Rhizoctonia solani.* and their less cytotoxic effect as compared to pure hexaconazole.	Chauhan et al., 2017

19.	Kaempferol	Loaded into lecithin/chitosan nanoparticles	Enhanced antifungal activity of encapsulated kaempferol against the phytopathogenic fungus *Fusarium oxysporium* as compared to pure one	Ilk et al., 2017
20.	*Cymbopogon artini* Essential Oil	Chitosan nanoparticles	Increased antifungal activity against *Fusarium graminearum*	Kalagatur et al., 2018
21.	HarpinPss, an elicitor from *Pseudomonas syringae* pv. *syringae*	HarpinPss-loaded chitosan nanoparticles	Improved foliar permeability and bioavailability of HarpinPss against *Rhizoctonia solani* infection.	Nadendla et al., 2018
22.	Azoxystrobin	Emulsion-based synchronous pesticide loading using Mesoporous silica/carboxymethyl chitosan	More loading capacity, pH-sensitive release, and enhanced antifungal activity against the late tomato pest, *Phytophthora infestans*.	Xu et al., 2018
23.	Clove essential oil	Chitosan nanoparticles by emulsion-ionic gelation	Controlled release nanoformulation with more antifungal efficacy	Hasheminejad et al., 2019
24.	Hexaconazole and dazomet	Chitosan nanoparticles encapsulated both drugs	Improved fungicidal potential against *Ganoderma boninense*	Maluin et al., 2019a, and 2019b
25.	Tebuconazole	Tebuconazole-loaded metal-organic frameworks using meso-tetra (4-carboxyphenyl) porphine, metal ions, and layer-by-layer deposition of pectin and chitosan	pH-dependent controlled release and showed dual microbicidal effect on plant pathogens	Tang et al., 2019

temperature, redox, enzymes, magnetic field, etc. Stimuli-based controlled active ingredients delivery systems can be prepared by means of natural polymers, namely, chitosan, alginate, cyclodextrin, starch, carboxymethylcellulose, etc., and silica due to their extensive availability and biodegradability. The use of stimuli-based polymers/carriers in nanoformulations can empower the release of different active ingredients at a specific time in such a manner to enhance the bioefficacy and reduce the adverse effects, dosage requirement, and time-to-time application (Zhao et al., 2018; Neri-Badang et al., 2019). Xu et al. (2014) used biodegradable chitosan-lactide copolymer matrix for the encapsulation of pyraclostrobin using the nano-precipitation method and reported pH-responsive controlled release and sustained release profile with better antifungal activity against tested fungus. An amphiphilic chitosan-polylactide graft copolymer has been used for encapsulation of the fungicide compound, flusilazole, using the nanoprecipitation method and reported enhanced antifungal activity through a controlled release approach as compared to the classical lusilazole emulsifiable concentrate (Mei et al., 2014). Sandhya et al. (2016) reported controlled release of carbendazim from chitosan-pectin loaded polymeric nanoparticles at variable pH ranges. The research group also evaluated and reported more bioefficacy of carbendazim nanoformulation against *Aspergillus parasiticus* and *Fusarium oxysporum* as related to the pure drug. Moreover, they also confirmed from observations of phytotoxicity studies that carbendazim nanoformulation was safe towards seed germination and root growth of different seeds.

Ketoconazole loaded chitosan-gellan gum nanoparticles exhibited significantly antifungal activity against *Aspergillus niger* in comparison to pure drug (Kumar et al., 2016). Furthermore, encapsulation of flavonoid compounds has also been reported for the enhancement of their potential antioxidant and antifungal activities, which can be done by nanotechnology approach to improve the limited poor dissolution and bioavailability. Kaempferol-loaded lecithin/chitosan nanoparticles showed improved antifungal potential against *Fusarium oxysporium*, a phytopathogenic fungus and reported for improved bioavailability of kaempferol as liken to kaempferol (Ilk et al., 2017). Hexaconazone-loaded chitosan-TPP nanocapsules has been prepared by ionotropic gelation method and reported for controlled delivery of hexaconazole against *Rhizoctonia solani.* Moreover, the nanoformulation reported for their less cytotoxic effect as compared to pure hexaconazole (Chauhan et al., 2017). Recently, Maluin et al. (2019) reported the targeted and sustained nano delivery of hexaconazole-loaded chitosan-TPP nanoparticles up to 86 h against a pathogenic fungus, *Ganoderma boninense*, caused Ganoderma Disease in Oil Palm. Other reports regarding the encapsulation of fungicides in chitosan-based polymer are summarized in Table 5.1.

5. Conclusion

In the field of agriculture, nanotechnology in association with biotechnology has prolonged the competence of nanomaterials in protection of crop plants. Nanomaterials considered to increase the efficacy and efficiency of fungicides, pesticides, and herbicides to combat against bacteria, viruses, fungi, and any other crop pathogens. For plant protection and pest management, controlled and targeted delivery of agrochemicals have great potential in the agriculture industry. Nanoencapsulation carriers like polymers, lipids, metal nanoparticles, clay, metal organic frameworks, and green nanoformulations are projected to endorse targeted delivery of agrochemicals while reducing degradation because of environmental effects. Additionally, nanoparticulate-based agrochemicals have many advantages over the conventional ones such as being environment-friendly, having fewer toxic effects, requiring reduced doses of agrichemicals, are controlled release and highly efficient. Therefore, agrochemical nanoformulations are effective and are an advanced approach to overcome the current problems of environment and unmanageable practices of land reuse. Nano-based agrichemical products have various advantages as well as constraints in their use due to the long-time presence of these nanomaterials in the environment. Consequently, it is the need of hour to resolve the issue related to the interaction of nanoformulations with environmental entities such as soil, water bodies, etc., and ensure their eco-friendly and socio-friendly use.

References

Abd-Elsalam, K.A. and Alghuthaymi, M.A. (2015). Nanobiofungicides: Is it the nextgeneration of fungicides? *J. Nanotech Material Sci.*, 2(2): 38–40.

Abd-Elsalam, K.A., Vasilkov, A.Y., Said-Galiev, E.E., Rubina, M.S., Khokhlov, A.R., Naumkin, A.V., Shtykova, E.V. and Alghuthaymi, M.A. (2018). Bimetallic blends and chitosan nanocomposites: Novel antifungal agents against cotton seedling damping-off. *Euro J. Plant Pathol.*, 151(1): 57–72.

Adisa, I.O., Pullagurala, V.L.R., Peralta-Videa, J.R., Dimkpa, C.O., Elmer, W.H., Gardea-Torresdey, J. and White, J. (2019). Recent advances in nano-enabled fertilizers and pesticides: A critical review of mechanisms of action. *Environ Sci: Nano*, 6: 2002–2030.

Alghuthaymi, M.A., Almoammar, H., Rai, M., Said-Galiev, E. and Abd-Elsalam, K.A. (2015). Myconanoparticles: Synthesis and their role in phytopathogens management. *Biotech Biotechnologic Equip.*, 29(2): 221–236.

Anuradha, B., Rekhapadmini, A. and Rangaswamy, V. (2016). Influence of tebuconazole and copper hydroxide on phosphatase and urease activities in red sandy loam and black clay soils. *3 Biotech.*, 6(1): 78.

Arguelles-Monal, W., Lizardi-Mendoza, J., Fernandez-Quiroz, D., Recillas-Mota, M. and Montiel-Herrera, M. (2018). Chitosan derivatives: Introducing new functionalities with a controlled molecular architecture for innovative materials. *Polym.*, 10(3): 342.

Atwood, D. and Paisley-Jones, C. (2017). *Pesticides Industry Sales and Usage: 2008–2012 Market Estimates.* United States Environmental Protection Agency: Washington, DC, USA.

Ayaz, M., Junaid, M., Ullah, F., Sadiq, A., Ovais, M., Ahmad, W. and Zeb, A. (2016). Chemical profiling, antimicrobial, and insecticidal evaluations of *Polygonum hydropiper* L. *BMC Complement Altern. Med.*, 16(1): 502.

Bedoya-Serna, C.M., Dacanal, G.C., Fernandes, A.M. and Pinho, S.C. (2018). Antifungal activity of nanoemulsions encapsulating oregano (*Origanum vulgare*) essential oil: *In vitro* study and application in *Minas P*adrao cheese. *Brazilian J. Microbiol.*, 49(4): 929–935.

Camara, M.C., Campos, E.V.R., Monteiro, R.A., Santo-Pereira, A.D.E., de Freitas Proenca, P.L. and Fraceto, L.F. (2019). Development of stimuli-responsive nano-based pesticides: Emerging opportunities for agriculture. *J. Nanobiotechnol.*, 17(1): 100.

Campos, E.V., De Oliveira, J.L., Da Silva, C.M., Pascoli, M., Pasquoto, T., Lima, R., Abhilash, P.C. and Fraceto, L.F. (2015). Polymeric and solid lipid nanoparticles for sustained release of carbendazim and tebuconazole in agricultural applications. *Sci. Rep.*, 5: 13809.

Cao, L., Zhang, H., Cao, C., Zhang, J., Li, F. and Huang, Q. (2016). Quaternized chitosan-capped mesoporous silica nanoparticles as nanocarriers for controlled pesticide release. *Nanomater*, 6: 126.

Chauhan, N., Dilbaghi, N., Gopal, M., Kumar, R., Kim, K.H. and Kumar, S. (2017). Development of chitosan nanocapsules for the controlled release of hexaconazole. *Int. J. Biol. Macromol.*, 97: 616–624.

Chen, J., Wang, W., Xu, Y. and Zhang, X. (2010). Slow-release formulation of a new biological pesticide, pyoluteorin, with mesoporous silica. *J. Agric. Food Chem.*, 59(1): 307–311.

Chen, J. and Mullin, C.A. (2013). Quantitative determination of trisiloxane surfactants in beehive environments based on liquid chromatography coupled to mass spectrometry. *Environ. Sci. Technol.*, 47(16): 9317–9323.

Chen, J. and Mullin, C.A. (2015). Characterization of trisiloxane surfactants from agrochemical adjuvants and pollinator-related matrices using liquid chromatography coupled to mass spectrometry. *J. Agric. Food Chem.*, 63(21): 5120–5125.

Chopra, M., Kaur, P., Bernela, M. and Thakur, R. (2014). Surfactant assisted nisin loaded chitosan-carageenan nanocapsule synthesis for controlling food pathogens. *Food Control*, 37: 158–164.

Choudhury, S.R., Ghosh, M., Mandal, A., Chakravorty, D., Pal, M., Pradhan, S. and Goswami, A. (2011). Surface-modified sulfur nanoparticles: An effective antifungal agent against *Aspergillus niger* and *Fusarium oxysporum*. *Appl. Microbiol. Biotechnol.*, 90(2): 733–743.

Chowdappa, P., Gowda, S., Chethana, S. and Madhura, S. (2014). Antifungal activity of chitosan-silver nanoparticle composite against Colletotrichum gloeosporioides associated with mango anthracnose. *African J. Microbio. Res.*, 8: 1803–1812.

Diaz-Blancas, V., Medina, D., Padilla-Ortega, E., Bortolini-Zavala, R., Olvera-Romero, M. and Luna-Barcenas, G. (2016). Nanoemulsion formulations of fungicide tebuconazole for agricultural applications. *Molecul.*, 21(10): 1271.

Duhan, J.S., Kumar, R., Kumar, N., Kaur, P., Nehra, K. and Duhan, S. (2017). Nanotechnology: The new perspective in precision agriculture. *Biotechnol. Rep.*, 15: 11–23.

Ghormade, V., Deshpande, M.V. and Paknikar, K.M. (2011). Perspectives for nanobiotechnology enabled protection and nutrition of plants. *Biotechnol. Adv.*, 29: 792–803.

Gianessi, L.P. (2013). The increasing importance of herbicides in worldwide crop production. *Pest Managem. Sci.*, 69(10): 1099–1105.

Global Industry Report. *Crop Protection Chemicals Market Size, Crop Protection Chemicals Market Size, Share & Trends Analysis Report by Product (Herbicides, Fungicides, Insecticides, Biopesticides), by Application, by Region and Segment Forecasts, 2020–2027*. Report ID: 978-1-68038-567-0, March, 2020.

Gogos, A., Knauer, K. and Bucheli, T.D. (2012). Nanomaterials in plant protection and fertilization: Current state, foreseen applications, and research priorities. *J. Agric. Food Chem.*, 60: 9781–9792.

Goswami, A. and Bandyopadh, A. (2012). Contribution of nanobiotechnology in Indian agriculture: Future prospects. *J. Indian Inst. Sci.*, 92: 219–232.

Goswami, L., Kim, K.H., Deep, A., Das, P., Bhattacharya, S.S., Kumar, S. and Adelodun, A.A. (2017). Engineered nano particles: Nature, behaviour, and effect on the environment. *J. Environ. Manage.*, 196: 297–315.

Goulson, D., Nicholls, E., Botias, C. and Rotheray, E.L. (2015). Bee declines driven by combined stress from parasites, pesticides, and lack of flowers. *Sci.*, 347: 1255957.

Haq, I.U. and Ijaz, S. (2019). Use of metallic nanoparticles and nanoformulations as nanofungicides for sustainable disease management in plants. pp. 289–316. *In*: *Nanobiotechnology in Bioformulations*. Cham: Springer.

Hasheminejad, N., Khodaiyan, F. and Safari, M. (2019). Improving the antifungal activity of clove essential oil encapsulated by chitosan nanoparticles. *Food Chem.*, 275: 113–122.

Hashim, A.F., Youssef, K. and Abd-Elsalam, K.A. (2019). Ecofriendly nanomaterials for controlling gray mold of table grapes and maintaining postharvest quality. *Eur. J. Plant Pathol.*, 154(2): 377–388.

He, Y., Zhao, B. and Yu, Y. (2016). Effect, comparison and analysis of pesticide electrostatic spraying and traditional spraying. *Bulg. Chem. Commun.*, 48: 340–344.

Hernandez, A.F., Parron, T., Tsatsakis, A.M., Requena, M., Alarcon, R. and Lopez-Guarnido, O. (2013). Toxic effects of pesticide mixtures at a molecular level: Their relevance to human health. *Toxicol.*, 307: 136–45.

Hossain, M.M. (2015). Recent perspective of herbicide: Review of demand and adoption in world agriculture. *J. Bangladesh Agri. Univ.*, 13: 13–24.

Huang, W., Wang. C., Duan, H., Bi, Y., Wu, D., Du, J. and Yu, H. (2018). Synergistic antifungal effect of biosynthesized silver nanoparticles combined with fungicides. *Int. J. Agric. Biol.*, 20: 1225–1229.

Ilk, S., Saglam, N. and Ozgen, M. (2017). Kaempferol loaded lecithin/chitosan nanoparticles: Preparation, characterization, and their potential applications as a sustainable antifungal agent. *Artif. Cells Nanomed. Biotechnol.*, 45(5): 907–916.

Janatova, A., Bernardos, A., Smid, J., Frankova, A., Lhotka, M., Kourimska, L., Pulkrabek, J. and Kloucek, P. (2015). Long-term antifungal activity of volatile essential oil components released from mesoporous silica materials. *Ind. Crop Prod.*, 67: 216–220.

Kah, M., Beulke, S., Tiede, K. and Hofmann, T. (2013). Nanopesticides: State of knowledge, environmental fate, and exposure modeling. *Crit. Rev. Env. Sci. Tech.*, 43: 1823–1867.

Kah, M. and Hofmann, T. (2014). Nanopesticide research: Current trends and future priorities. *Environ. Int.*, 63: 224–235.

Kalagatur, N.K., Ghosh, S.N., Sundararaj, N. and Mudili, V. (2018). Antifungal activity of chitosan nanoparticles encapsulated with *Cymbopogon martinii* essential oil on plant pathogenic fungi *Fusarium graminearum*. *Front. Pharmacol.*, 9: 610.

Katagi, T. (2008). Surfactant effects on environmental behavior of pesticides. pp. 71–177. *In*: *Reviews of Environmental Contamination and Toxicology*. New York, NY: Springer.

Kaur, P., Duhan, J.S. and Thakur, S. (2018). Comparative pot studies of chitosan and chitosan-metal nanocomposites as nano-agrochemicals against fusarium wilt of chickpea (Cicer arietinum L.). *Biocatalys. Agri. Biotech.*, 14: 466–471.

Kennedy, M.C., Glass, C.R., Bokkers, B., Hart, A.D., Hamey, P.Y., Kruisselbrink, J.W., de Boer, W.J., van der Voet, H., Garthwaite, D.G. and van Klaveren, J.D. (2015). A European model and case studies for aggregate exposure assessment of pesticides. *Food Chem. Toxicol.*, 79: 32–44.

Kim, S. (2018). Competitive biological activities of chitosan and its derivatives: Antimicrobial, antioxidant, anticancer, and anti-inflammatory activities. *Int. J. Polym. Sci.*, 2018: 1–13.

Kumar, S., Dilbaghi, N., Saharan, R. and Bhanjana, G. (2012). Nanotechnology as emerging tool for enhancing solubility of poorly water-soluble drugs. *Bio. Nano Sci.*, 2(4): 227–250.

Kumar, S., Dilbaghi, N., Rani, R., Bhanjana, G. and Umar, A. (2013). Novel approaches for enhancement of drug bioavailability. *Reviews in Advanced Sciences and Engineering*, 2(2): 133–154.

Kumar, S., Kaur, P., Bernela, M., Rani, R. and Thakur, R. (2016). Ketoconazole encapsulated in chitosan-gellan gum nanocomplexes exhibits prolonged antifungal activity. *Int. J. Biol. Macromol.*, 93: 988–994.

Kumar, S., Nehra, M., Dilbaghi, N., Marrazza, G., Hassan, A.A. and Kim, K.H. (2019). Nano-based smart pesticide formulations: Emerging opportunities for agriculture. *J. Control Release*, 294: 131–153.

Lammoglia, S.K., Kennedy, M.C., Barriuso, E., Alletto, L., Justes, E., Munier-Jolain, N. and Mamy, L. (2017). Assessing human health risks from pesticide use in conventional and innovative cropping systems with the BROWSE model. *Environ. Int.*, 105: 66–78.

Linquist, B.A., Liu, L., van Kessel, C. and van Groenigen, K.J. (2013). Enhanced efficiency nitrogen fertilizers for rice systems: Meta-analysis of yield and nitrogen uptake. *Field Crops Res.*, 154: 246–254.

Liu, B., Wang, Y., Yang, F., Wang, X., Shen, H., Cui, H. and Wu, D. (2016). Construction of a controlled-release delivery system for pesticides using biodegradable PLA-based microcapsules. *Colloids Surf B: Biointerfac.*, 144: 38–45.

Liu, Y., Laks, P. and Heiden, P. (2002a). Controlled release of biocides in solid wood. I. Efficacy against brown rot wood decay fungus (*Gloeophyllum trabeum*). *J. Appl. Polym. Sci.*, 86: 596–607.

Liu, Y., Laks, P. and Heiden, P. (2002b). Controlled release of biocides in solid wood. II. Efficacy against *Trametes versicolor* and *Gloeophyllum trabeum* wood decay fungi. *J. Appl. Polym. Sci.*, 86: 608–614.

Liu, Y., Laks, P. and Heiden, P. (2002c). Controlled release of biocides in solid wood. III. Preparation and characterization of surfactant-free nanoparticles. *J. Appl. Polym. Sci.*, 86: 615–621.

Majeed, A., Najar, R.A., Choudhary, S., Rehman, W.U., Singh, A., Thakur, S. and Bhardwaj, P. (2017). Practical and plausible implications of chitin and chitosan-based nanocomposites in agriculture. Chitosan: Derivatives, Composites, and Applications, 411.

Malerba, M. and Cerana, R. (2016). Chitosan effects on plant systems. *Int. J. Mol. Sci.*, 17: 996.

Maluin, F.N., Hussein, M.Z., Yusof, N.A., Fakurazi, S., Idris, A.S., Hilmi, Z., Hailini, N. and Jeffery Daim, L.D. (2019a). Preparation of chitosan–hexaconazole nanoparticles as fungicide nano delivery system for combating ganoderma disease in oil palm. *Molecules*, 24(13): 2498.

Maluin, F.N., Hussein, M.Z., Yusof, N.A., Fakurazi, S., Seman, I.A., Hilmi, N.H.Z. and Daim, L.D.J. (2019b). Enhanced fungicidal efficacy on *Ganoderma boninense* by simultaneous co-delivery of hexaconazole and dazomet from their chitosan nanoparticles. *RSC Adv.*, 9(46): 27083–27095.

Manikandan, A. and Sathiyabama, M. (2016). Preparation of chitosan nanoparticles and its effect on detached rice leaves infected with *Pyricularia grisea. Int. J. Biol. Macromol.*, 84(5): 8–61.

Massinon, M., Cock, N.D., Forster, W.A., Nairn, J.J., Mccue, S.W., Zabkiewicz, J.A. and Lebeau, F. (2017). Spray droplet impaction outcomes for different plant species and spray formulations. *Crop Prot.*, 99: 65–75.

Mattos, B.D., Tardy, B.L., Magalhaes, W.L. and Rojas, O.J. (2017). Controlled release for crop and wood protection: Recent progress toward sustainable and safe nanostructured biocidal systems. *J. Control Release*, 262: 139–150.

Mei, X.D., Liang, Y.H., Zhang, T., Ning, J. and Wang, Z.Y. (2014). An amphiphilic chitosan-polylactide graft copolymer and its nanoparticles as fungicide carriers. *Proc. Adv. Mater. Res.*, 1051: 21–28.

Mujtaba, M., Khawar, K.M., Camara, M.C., Carvalho, L.B., Fraceto, L.F., Morsi, R.E., Elsabee, M.Z., Kaya, M., Labidi, J., Ullah, H. and Wang, D. (2020). Chitosan-based delivery systems for plants: A brief overview of recent advances and future directions. *Int. J. Biol. Macromol.*, 154: 683–697.

Nadendla, S.R., Rani, T.S., Vaikuntapu, P.R., Maddu, R.R. and Podile, A.R. (2018). HarpinPss encapsulation in chitosan nanoparticles for improved bioavailability and disease resistance in tomato. *Carbohydr. Polym.*, 199: 11–19.

Nair, R., Varghese, S.H., Nair, B.G., Maekawa, T., Yoshida, Y. and Kumar, D.S. (2010). Nanoparticulate material delivery to plants. *Plant Sci.*, 179: 154–163.

Nasseri, M., Golmohammadzadeh, S., Arouiee, H., Jaafari, M.R. and Neamati, H. (2016). Antifungal activity of *Zataria multiflora* essential oil-loaded solid lipid nanoparticles *in-vitro* condition. *Iran J. Basic Med. Sci.*, 19: 1231–1237.

Neri-Badang, M.C. and Chakraborty, S. (2019). Carbohydrate polymers as controlled release devices for pesticides. *J. Carbohydr. Chem.*, 38: 67–85.

Nuruzzaman, M., Rahman, M.M., Liu, Y. and Naidu, R. (2016). Nanoencapsulation, nano-guard for pesticides: A new window for safe application. *J. Agric. Food Chem.*, 64(7): 1447–1483.

Pimentel, D. (2009). Pesticide and pest control. pp. 83–87. *In*: Peshin, P. and Dhawan, A.K. (eds.). *Integrated Pest Management: Innovation-development Process*. Dordrecht, Netherlands: Springer.

Qian, K., Shi, T., Tang, T., Zhang, S., Liu, X. and Cao, Y. (2011). Preparation and characterization of nano-sized calcium carbonate as controlled release pesticide carrier for validamycin against *Rhizoctonia solani. Microchimica. Acta*, 173(1–2): 51–57.

Qian, K., Shi, T., He, S., Luo, L. and Cao, Y. (2013). Release kinetics of tebuconazole from porous hollow silica nanospheres prepared by miniemulsion method. *Microporo. Mesoporo. Material*, 169: 1–6.

Rabea, E.I., Badawy, M.E., Rogge, T.M., Stevens, C.V., Hofte, M., Steurbaut, W. and Smagghe, G. (2005). Insecticidal and fungicidal activity of new synthesized chitosan derivatives. *Pest Managem. Sci.*, 61(10): 951–960.

Ramasamy, T., Ruttala, H.B., Gupta, B., Poudel, B.K., Choi, H.G., Yong, C.S. and Kim, J.O. (2017). Smart chemistry-based nanosized drug delivery systems for systemic applications: A comprehensive review. *J. Control Release*, 258: 226–253.

Rani, R., Dilbaghi, N., Dhingra, D. and Kumar, S. (2015). Optimization and evaluation of bioactive drug-loaded polymeric nanoparticles for drug delivery. *Int. J. Biol. Macromol.*, 78: 173–179.

Reidy, B., Andrea, H., Luch, A., Dawson, K. and Lynch, I. (2013). Mechanisms of silver nanoparticle release, transformation, and toxicity: A critical review of current knowledge and recommendations for future studies and applications. *Materials*, 6: 2295–2350.

Ribes, S., Ruiz-Rico, M., Perez-Esteve, E., Fuentes, A., Talens, P., Martinez-Manez, R. and Barat, J.M. (2017). Eugenol and thymol immobilised on mesoporous silica-based material as an innovative antifungal system: Application in strawberry jam. *Food Control*, 81: 181–188.

Richendrfer, H. and Creton, R. (2015). Chlorpyrifos and malathion have opposite effects on behaviors and brain size that are not correlated to changes in AChE activity. *Neurotoxicol.*, 49: 50–58.

Rossignol, D.A., Genuis, S.J. and Frye, R.E. (2014). Environmental toxicants and autism spectrum disorders: A systematic review. *Transl. Psychiatry*, 4: e360.

Roy, A., Singh, S.K., Bajpai, J. and Bajpai, A.K. (2014). Controlled pesticide release from biodegradable polymers. *Cent. Eur. J. Chem.*, 12(4): 453–469.

Rubina, M.S., Vasilkov, A.Y., Naumkin, A.V., Shtykova, E.V., Abramchuk, S.S., Alghuthaymi, M.A. and Abd-Elsalam, K.A. (2017). Synthesis and characterization of chitosan–copper nanocomposites and their fungicidal activity against two sclerotia-forming plant pathogenic fungi. *J. Nanostructure Chem.*, 7(3): 249–258.

Saharan, V., Mehrotra, A., Khatik, R., Rawal, P., Sharma, S.S. and Pal, A. (2013). Synthesis of chitosan-based nanoparticles and their *in vitro* evaluation against phytopathogenic fungi. *Int. J. Biol. Macromol.*, 62: 677–683.

Saharan, V., Sharma, G., Yadav, M., Choudhary, M.K., Sharma, S.S., Pal, A., Raliya, R. and Biswas, P. (2015). Synthesis and *in vitro* antifungal efficacy of Cu–chitosan nanoparticles against pathogenic fungi of tomato. *Int. J. Biol. Macromol.*, 75: 346–353.

Said, S.M., EL-Sayed, S.M., Farid, H.E.A. and Abozid, M.M. (2011). Insecticidal effect of chitosan prepared by different chemical processing sequences against *Galleria melleonella*. Az. *J. Pharm. Sci.*, 43: 123–132.

Sandhya, K., Kumar, S. and Dilbaghi, N. (2017). Preparation, characterization, and bio-efficacy evaluation of controlled release carbendazim-loaded polymeric nanoparticles. *Environ. Sci. Pollut. Res.*, 24(1): 926–937.

Sathiyabama, M. and Parthasarathy, R. (2016). Biological preparation of chitosan nanoparticles and its *in vitro* antifungal efficacy against some phytopathogenic fungi. *Carbohydr. Polym.*, 151: 321–325.

Sekhon, B.S. (2014). Nanotechnology in agri-food production: An overview. *Nanotechnol. Sci. Appl.*, 7: 31–53.

Shahid, M., Ahmed, B. and Khan, M.S. (2018a). Evaluation of microbiological management strategy of herbicide toxicity to green gram plants. *Biocatalyst Agri. Biotechnol.*, 14: 96–108.

Shahid, M., Ahmed, B., Zaidi, A. and Khan, M.S. (2018b). Toxicity of fungicides to *Pisum sativum*: A study of oxidative damage, growth suppression, cellular death, and morpho-anatomical changes. *RSC Adv.*, 8(67): 38483–38498.

Shahid, M. and Khan, M.S. (2018). Cellular destruction, phytohormones and growth modulating enzymes production by *Bacillus subtilis* strain BC8 impacted by fungicides. *Pesticide Biochemistry and Physiology*, 149: 8–19.

Shahid, M., Khan, M.S. and Kumar, M. (2019a). Kitazin-pea interaction: understanding the fungicide induced nodule alteration, cytotoxicity, oxidative damage, and toxicity alleviation by *Rhizobium leguminosarum*. *RSC Adv.*, 9(30): 16929–16947.

Shahid, M., Zaidi, A., Ehtram, A. and Khan, M.S. (2019b). *In vitro* investigation to explore the toxicity of different groups of pesticides for an agronomically important rhizosphere isolate *Azotobacter vinelandii*. *Pesticide Biochemistry and Physiology*, 157: 33–44.

Shang, Y., Hasan, M., Ahammed, G.J., Li, M., Yin, H. and Zhou, J. (2019). Applications of nanotechnology in plant growth and crop protection: A review. *Molecul.*, 24(14): 2558.

Sharma, K., Sharma, R., Shit, S. and Gupata, S. (2012). Nanotechnological application on diagnosis of a plant disease. *In*: *Proceedings of the 2012 International Conference on Advances in Biological and Medical Sciences*. Singapore.

Shyla, K.K., Natarajan, N. and Nakkeeran, S. (2018). Antifungal activity of zinc oxide, silver and titanium dioxide nanoparticles against Macrophomina phaseolina. *J. Mycol. Plant Pathol.*, 44: 268–273.

Smith, M.N., Workman, T., McDonald, K.M., Vredevoogd, M.A., Vigoren, E.M., Griffith, W.C., Thompson, B., Coronado, G.D., Barr, D. and Faustman, E.M. (2017). Seasonal and occupational trends of five organophosphate pesticides in house dust. *J. Exposure Sci. Environ. Epidemiol.*, 27: 372–378.

Song, M., Ju, J., Luo, S., Han, Y., Dong, Z., Wang, Y., Gu, Z., Zhang, L., Hao, R. and Jiang, L. (2017). Controlling liquid splash on superhydrophobic surfaces by a vesicle surfactant. *Sci Adv.*, 3: e1602188.

Tang, J., Ding, G., Niu, J., Zhang, W., Tang, G., Liang, Y., Fan, C., Dong, H., Yang, J., Li, J. and Cao, Y. (2019). Preparation and characterization of tebuconazole metal-organic framework-based microcapsules with dual-microbicidal activity. *Chem. Eng. J.*, 359: 225–232.

Volova, T., Prudnikova, S., Boyandin, A., Zhila, N., Kiselev, E., Shumilova, A., Baranovskiy, S., Demidenko, A., Shishatskaya, E. and Thomas, S. (2019). Constructing slow-release fungicide formulations based on poly (3-hydroxybutyrate) and natural materials as a degradable matrix. *J. Agr. Food Chem.*, 67(33): 9220–9231.

Wang, Y., Zhang, S. and Benoit, D.S. (2018). Degradable poly (ethylene glycol) (PEG)-based hydrogels for spatiotemporal control of siRNA/nanoparticle delivery. *J. Control Release*, 287: 58–66.

Wanyika, H. (2013). Sustained release of fungicide metalaxyl by mesoporous silica nanospheres. pp. 321–329. *In*: *Nanotechnology for Sustainable Development*. Cham: Springer.

Worrall, E., Hamid, A., Mody, K., Mitter, N. and Pappu, H. (2018). Nanotechnology for plant disease management. *Agronom.*, 8(12): 285.

Xu, C., Cao, L., Zhao, P., Zhou, Z., Cao, C., Li, F. and Huang, Q. (2018). Emulsion-based synchronous pesticide encapsulation and surface modification of mesoporous silica nanoparticles with carboxymethyl chitosan for controlled azoxystrobin release. *Chem. Eng. J.*, 348: 244–254.

Xu, L., Cao, L.D., Li, F.M., Wang, X.J. and Huang, Q.L. (2014). Utilization of chitosan-lactide copolymer nanoparticles as controlled release pesticide carrier for pyraclostrobin against *Colletotrichum gossypii*. *Southw. J. Disper. Sci. Technol.*, 35(4): 544–550.

Youssef, K., de Oliveira, A.G., Tischer, C.A., Hussain, I. and Roberto, S.R. (2019). Synergistic effect of a novel chitosan/silica nanocomposites-based formulation against gray mold of table grapes and its possible mode of action. *Int. J. Biol. Macromol.*, 141: 247–258.

Zhang, J., Li, M., Fan, T., Xu, Q., Wu, Y., Chen, C. and Huang, Q. (2013). Construction of novel amphiphilic chitosan copolymer nanoparticles for chlorpyrifos delivery. *J. Polym. Res.*, 20(3): 107.

Zhang, M., Zeiss, M.R. and Shu, G.S. (2015). Agricultural pesticide use, and food safety: California's model. *J. Integr. Agric.*, 14(11): 2340–2357.

Zhao, P., Cao, L., Ma, D., Zhou, Z., Huang, Q. and Pan, C. (2017). Synthesis of pyrimethanil-loaded mesoporous silica nanoparticles and its distribution and dissipation in cucumber plants. *Molecules*, 22: E817.

Zhao, X., Cui, H., Wang, Y., Sun, C., Cui, B. and Zeng, Z. (2018). Development strategies and prospects of nano-based smart pesticide formulation. *J. Agric. Food Chem.*, 66: 6504–6612.

Chapter 6

Nano-enabled Pesticides: Status and Perspectives

CC Sheeja, Damodaran Arun and *Lekha Divya**

1. Introduction

The chemicals used in farm fields to protect crops from pests, weeds, and plant diseases are known as agrochemicals. Pesticides and fertilizers are the commonly used agrochemicals. The use of agrochemicals started around 4500 years ago as a significant pest eradication measure (IUPAC, 2010). It facilitated increased yield and ensured improved crop varieties. The regular use of agrochemicals became one of the major causes of pollution, mostly water and soil contamination. Within the context of agricultural practices, pest management is not a new concern. The low solubility resulted in long-term persistence of pesticides and biomagnification in non-target organisms. In parallel, the demand for food and agricultural products multiplied due to an alarming rise in the human population. Such a situation intensified the search for alternative strategies that ensured minimum environmental degradation and maximum productivity (Fig. 6.1). To ensure judicious use of pesticides with minimal risks to the ecosystem, various approaches, such as integrated pest management (IPM) were implemented.

The primary ingredients in pesticides are water-insoluble organic compounds (Nuruzzaman et al., 2016). These ingredients are processed as either emulsifiers or dispersants to obtain a suitable pesticide formulation (Ghormade et al., 2011). The final form of pesticides is generally facilitated by spray application. The excess of pesticides falls on plant foliar parts that run-off and mix up with natural water resources, thereby causing water pollution. The loss of pesticides is one of the major issues during field application. The loss rate is approximately calculated at 70% and the useful share of pesticides is estimated at 0.1% (He et al., 2011; Kah et al., 2013; Nuruzzaman et al., 2016; Zhao et al., 2018). The non-target loss is one of the

Department of Zoology, University of Calicut, Thenhipalam, Malappuram, 673 635, Kerala, India.
* Corresponding author: divyacuk@gmail.com

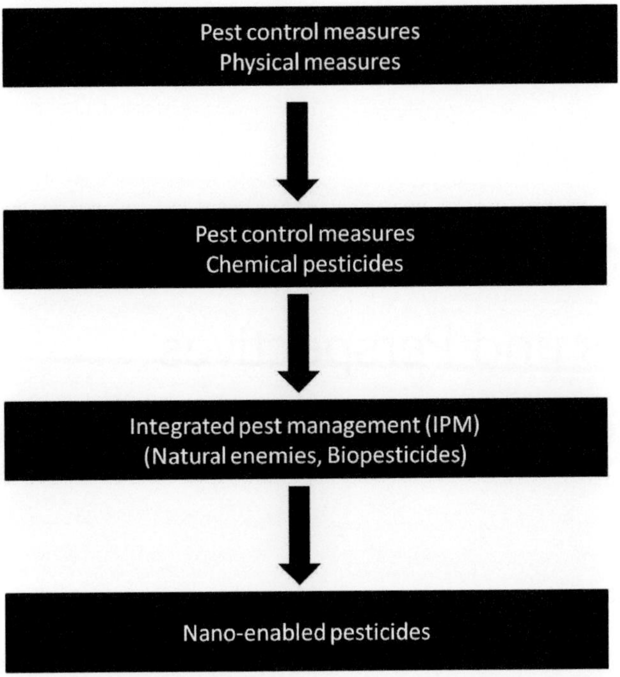

Fig. 6.1. The advancement of pest control strategies.

crises with conventional pesticide formulations (He et al., 2016). The emergence of resistant strains of target organisms is another concern. These disadvantages limit the use of conventional pesticides. However, the controlled use of chemical pesticides is essential for better agricultural productivity.

The development and advancement of technology have revived the field of agriculture. The emerging nanotechnology has opened new avenues in the targeted delivery of pesticides. The unique chemical and physical properties of nanomaterials extensively benefit both the industrial and agricultural sectors. In agriculture, nanomaterials gained much prominence due to their role in the formulation of agrochemicals. The nanoagrochemicals so far synthesized comprise nanofertilizers and nanopesticides.

Nanopesticides are those formulations that include nanosized entities with unique properties stemming from their size (Kah et al., 2013). Nano-enabled pesticides may reduce environmental pollution by lowering the number and quantity of pesticides used, thereby minimizing pesticide losses. However, nanopesticides may harmfully affect the aquatic and soil microbiota due to their higher environmental retention, rapid transport, and increased toxicity. This chapter deals with the status currently prevailing and the perspectives on nanopesticides, premised on their formulation, mechanism of action, and environmental impact.

2. Nano-enabled Pesticides: Formulation and Mechanism of Action

A pesticide formulation is the process of formulating an end-user product using active and inert ingredients. The formulation of nano-enabled pesticides or nanopesticides achieves targeted delivery of active ingredients and controlled release of the core materials (Kawabata et al., 2011; Huang et al., 2018). Nanocarriers such as nanoemulsions, nanocapsules and polymerized metal nanoparticles (NPs) are efficient modes for nano-enabled pesticide delivery (Fig. 6.2). These formulations can effectively achieve pest control by altering the pH and the inorganic salt concentration (Zhao et al., 2017).

Among the above-mentioned nanocarriers, nanoemulsions are cost-effective and the safest way of delivery (Knowles, 2008). A nanoemulsion is a nano-sized colloid dispersion of one or more liquids that are immiscible in nature. Nanoemulsions possess uniform and extremely small droplet sizes, ranging from 20 nm to 200 nm (Forgiarini et al., 2000; Wang et al., 2007). Formulations of nanoemulsion require typically 3–10% less organic solvents and surfactants (McClements, 2012). Besides that, the miniature droplets ensure uniform distribution of the pesticide on the surface of plant

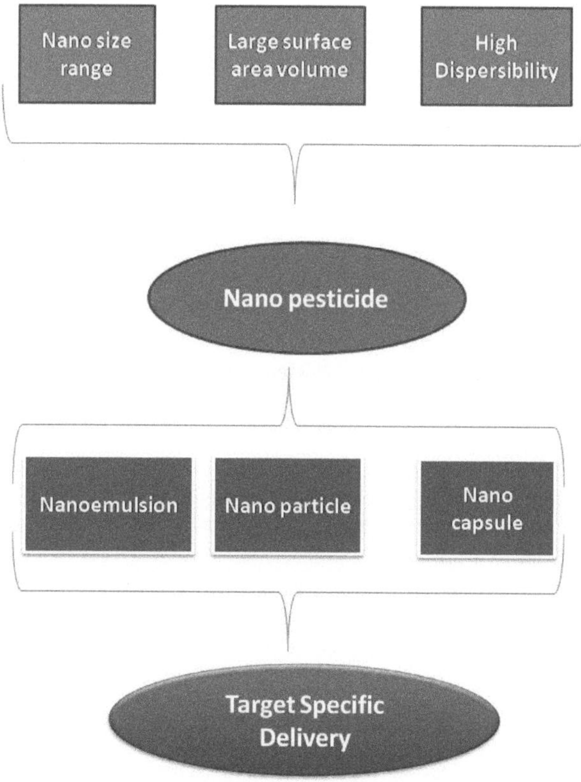

Fig. 6.2. Nanopesticides ensure the active ingredients' target-specific action through efficient delivery modes such as nanoemulsions, nanoparticles, and nanocapsules.

leaves. Nanoemulsions augment the bioavailability, thereby improving the efficacy of pesticides. Most of the pesticides are highly hydrophobic with low water solubility. Emulsifiable concentrates of such pesticides can be nano-formulated for efficient action. Reduced toxic effects along with efficient pest control are the major advantages of nanoemulsions (Suresh Kumar et al., 2013). Nanoemulsion of tebucanazol, a triazole fungicide, is found to be less toxic when compared to its conventional form (Díaz-Blancas et al., 2016). In a similar study, the insecticide beta cypermethrin was found to possess excellent spreading ability in two different nanoemulsion systems (Du et al., 2016). More importantly, to achieve sustainable environmental protection, the oil phase of the emulsion should be non-toxic, and the surfactants should be eco-friendly in nature (Du et al., 2016). Eco-friendly nanoemulsions based on plant oils provide a new perspective for insect pest control. Many studies proved the potential insecticidal activity of plant-based nanoemulsions against cotton pests. Peppermint oil nanoemulsion is one of the best examples for such formulations that showed insecticidal activity against cotton aphid (Heydari et al., 2020).

Another method of nanoformulation is nanoencapsulation. Nanoencapsulation is the method of packing nanoparticles of solid, liquid, or gas, also known as the core or active, within a secondary material, named as the matrix or shell (Augustin and Hemar, 2009). Nano-encapsulated pesticides can be formulated by coating the pesticide using an appropriate nanomaterial in its nano-size range. It has high selectivity when compared to other methods. Nanoencapsulation enhances the dispersion for the targeted release of the active ingredient in a controlled way (Peteu et al., 2010). Solid lipid NPs, polymer-based nanomaterials, inorganic porous nanomaterials, nanoclays and layered double hydroxides have been found as excellent encapsulation materials (Ghormade et al., 2011). Nano-sized polymers are found to be suitable for pesticide delivery. Naturally derived polymers are ecologically best because of their biodegradable nature. Nanomicelle, nanosphere, nanogel, nanocapsule are a few examples of the nano-encapsulated mode of delivery (Nuruzzaman et al., 2016).

Moreover, an array of engineered NPs is used in pesticide nanoformulations. NPs of metal and metal oxides of Ag^{+} (silver), Cu^{2+} (copper), Zn^{2+} (zinc), Mn^{2+} (manganese), Ti^{2+} titanium, Ce^{4+} (cerium), chitosan, and β-D-glycan are the most widely studied NPs (Niemeyer et al., 2001; Khot et al., 2012; Rico et al., 2014; Anusuya and Sathiyabama, 2015; Kumaraswamy et al., 2018). Biosynthesis of active nanopesticides from natural sources like plants has also been emphasized and is a trendy topic in the field of NP research (Pariona et al., 2019). Several *in vitro* studies have demonstrated the antimicrobial activity of nanomaterials (Dutta and Kaman, 2017). The antimicrobial property of nanopesticides can be utilized for crop improvement by developing target nanopesticides with specific inhibitory activity. Previous studies reported the *in vitro* inhibitory activity of NPs against the growth of different pathogens (Loo et al., 2018; Qais et al., 2019). Ag NPs cause membrane disruption by binding to cysteine-containing proteins of the plasma membrane of the pathogen (Dutta and Kaman, 2017). A wide spectrum of antimicrobial and antifungal efficacy has been improved by green synthesized ZnO NPs (He et al., 2011; Gunalan et al., 2012; Jayaseelan et al., 2012; Dimkpa et al., 2013).

The mechanism of action against pests is more-or-less similar in all the nanomaterials. Nanomaterials may induce reactive oxygen species (ROS) generation that eventually leads to several molecular changes in pathogenic organisms. Several studies have demonstrated that Nps disrupt cell membranes and cause cell death.

Cations such as Ag^+, Cu^{2+}, Zn^{2+}, and Ti^{4+} bind to functional groups, such as sulfhydryl (R-SH) in proteins and damage major pathways (El-Argawy et al., 2017). We had studied the effects of Molybdenum disulfide NPs (MoS_2 NPs) in the foraging ant pests. The study demonstrated that the MoS_2 NPs exert ROS-induced cellular effects in the immune cells of the ants (Sheeja et al., 2020). It is suggested that the MoS_2 NPs may exert similar effects in other insect pests too. Having insecticidal properties, MoS_2 NPs can be formulated into nanocarriers. Further studies are needed for validation. There are several recent paradigms of successful implementation of NPs as pesticides, specifically, fungicides and insecticides. Fungicides are crucial in fruit plantation. In the fruit plantation of mango, papaya, citrus, etc., the fungal stain *Colletotrichum gloeosporioides* causes Anthracnose disease. Benzimidazole has been being applied as a fungicide. But the severe use of benzimidazole resulted in a resistant strain of the pathogen. The potential use of nanocomposites as insecticides and fungicides has gained attention in the recent times. Insecticidal effects of various NPs were studied in honeybees (Milivojević et al., 2015), drosophila (Pompa et al., 2011) and other insect pests (Yasur and Rani, 2015). The use of Ag NPs was found to be effective against *C. gloeosporioides* (Park et al., 2006; Lamsal et al., 2011; Kim et al., 2012).

3. Compatibility of Nano-encapsulated Pesticides with the Environment

Nano-enabled pesticides containing active ingredients with minimal risk profile assure several benefits over conventional pesticides. This may enable improved behavior of the active ingredients, target specific delivery, increased efficacy, and better environmental compatibility. Nanotechnology helps to design a formulation that targets only the specific pest while it ensures safety for handlers and non-target organisms. It also provides targeted action on the plant or the pest via better interaction (Pérez-de-Luque and Rubiales, 2009).

The *in vitro* and *in vivo* study of polyethylene glycol (PEG) encapsulated acephate on murine model was reported by Pradhan et al. (2013). Cytotoxicity-related abnormality was not observed in the selected cell population. Both the previously discussed studies provided evidence of a broad spectrum of activities against agricultural pests (Pradhan et al., 2013). A nano-encapsulated imidacloprid coated with sodium alginate retained efficacy against pests although cytotoxicity was low (Kumar et al., 2019). Therefore, nano-enabled pesticides are considered safer and more effective compared to their commercial counterparts with reported toxicity.

However, the current state of knowledge is insufficient for dependably assessing the environmental risk of nanopesticides. Along with the affirmative sides, the possible harmful impacts of nano dimensions of the nanoscale formulation should also considered. In an assessment of the copper(II) hydroxide ($Cu(OH)_2$) nanopesticide, it was found that $Cu(OH)_2$ nanopesticides affected the total soil microbial communities (Zhang et al., 2019).

Similar studies suggest proper reviews of nanopesticides and comprehensive research studies are the need of the hour to confirm the toxicity of the nanoscale formulations and understand the influence of size, charge, shape, and chemistry of these NPs on toxicity. Hence, future studies are essential to check the toxicity aspects of various physicochemical parameters of these formulations.

Registered nano-enabled agrochemical products are a handful. To make the nanoagrochemical innovations bench to market, an iterative problem formulation and risk assessment (United States Environmental Protection Agency (USEPA), 1998) is required regarding the market release of the nano-enabled pesticides. A major criterion for risk assessment is the durability or stability after field applications of these nanoformulations. In a risk assessment study of the nano-formulated herbicide, pendimethalin, the modification of the fate and behavior of the active ingredients were observed. Additionally, the depth of penetration and the rate of release of the active ingredients were found to be significant when compared to their conventional form (Walker et al., 2018).

Nanoencapsulation technology has great potential, but more investigations are required to make pesticides more eco-friendly. The high surface/mass ratio of nanoformulations is double-edged on the target species, promoting rapid action while affecting entry routes via the skin or inhalation. Nano-encapsulated pesticides are known for their controlled release and improved efficacy. By analysing the various nano-sized capsules, their studied effects on the target and non-target species open paths to innovation in pesticide formulations. Moreover, these formulations offer more advantages than conventional pesticides. The practical field implementation of nanopesticides needs thorough evaluation to ensure their compatibility with the environment.

4. Conclusion

Over the years, unsustainable agrochemicals, especially pesticides have found their way into our ecosystems and contaminated soils and waterways. The emergence of novel pesticides, nanopesticides, offers an eco-friendlier approach. Nanopesticides emerged in the agricultural sector as an alternative to conventional pesticides that adversely affect human health and the environment. The pesticide formulation and application require continued systematic research to ensure environmentally sustainable targeted, controlled release. As a contradictory, innovation always results in both negative and positive effects on human and environmental health. Technological advancement could be able to balance both agricultural and environmental sustainability. The synthesis of research, innovation, and sustainability would help achieve the goal of sustainable development. Currently, there are no clear assessment reports available on the use of nanopesticides. Hence, before the implementation and release of nanopesticides into an open market, a thorough evaluation of the fate of applied nanopesticides is a must. Therefore, a guideline for their evaluation is required and the suitability of the present regulation must be revisited.

References

Anusuya, S. and Sathiyabama, M. (2015). Foliar application of β-d-Glucan nanoparticles to control rhizome rot disease of turmeric. *Int. J. Biol. Macromol.*, 72: 1205–1212.

Augustin, M.A. and Hemar, Y. (2009). Nano- and micro-structured assemblies for encapsulation of food ingredients. *Chem. Soc. Rev.*, 38: 902–912.

Díaz-Blancas, V., Medina, D.I., Padilla-Ortega, E., Bortolini-Zavala, R., Olvera-Romero, M. and Luna-Bárcenas, G. (2016). Nanoemulsion formulations of fungicide tebuconazole for agricultural applications. *Molecules*, 21(10): 1271.

Dimkpa, C.O., McLean, J.E., Britt, D.W. and Anderson, A.J. (2013). Antifungal activity of ZnO nanoparticles and their interactive effect with a biocontrol bacterium on growth antagonism of the plant pathogen *Fusarium Graminearum*. *BioMetals*, 26 (6): 913–924.

Du, Z., Wang, C., Tai, X., Wang, G. and Liu, X. (2016). Optimization and Characterization of Biocompatible oil-in-water nanoemulsion for pesticide delivery. *ACS Sustain. Chem. Eng.*, 4(3): 983–991.

Dutta, P. and Kaman, P.K. (2017). Nanocentric plant health management with special reference to silver. *Int. J. Curr. Microbiol. Appl. Sci.*, 6(6): 2821–2830.

El-Argawy, E., Rahhal, M.M.H., El-Korany, A., Elshabrawy, E.M. and Eltahan, R.M. (2017). Efficacy of some nanoparticles to control damping-off and root rot of sugar beet in El-Behiera Governorate. *Asian J. Plant Pathol.*, 11: 35–47.

Forgiarini, A., Esquena, J., González, C. and Solans, C. (2000). Studies of the relation between phase behavior and emulsification methods with nanoemulsion formation. *Progr. Colloid and Polymer Sci.*, 115: 36–39.

Ghormade, V., Deshpande, M.V. and Paknikar, K.M. (2011). Perspectives for nano-biotechnology enabled protection and nutrition of plants. *Biotechnol. Adv.*, 29(6): 792–803.

Gunalan, S., Sivaraj, R. and Rajendran, V. (2012). Green synthesized ZnO nanoparticles against bacterial and fungal pathogens. *Prog. Nat. Sci.: Mater. Int.*, 22(6): 693–700.

He, L., Liu, Y., Mustapha, A. and Lin, M. (2011). Antifungal activity of zinc oxide nanoparticles against *Botrytis Cinerea* and *Penicillium Expansum*. *Microbiol Res.*, 166(3): 207–215.

He, Y., Zhao, B. and Yu, Y. (2016). Effect, comparison and analysis of pesticide electrostatic spraying and traditional spraying. *Bulg. Chem. Commun.*, 48(D): 340–44.

Heydari, M., Amirjani, A., Bagheri, M., Sharifian, I. and Sabahi, Q. (2020). Eco-friendly pesticide based on peppermint oil nanoemulsion: Preparation, physicochemical properties, and its aphicidal activity against cotton aphid. *Environ. Sci. Pollut. Res.*, 27: 6667–6679.

Huang, B., Chen, F., Shen, Y., Qian, K., Wang, Y., Sun, C., Zhao, X., Cui, B., Gao, F., Zeng, Z. and Cui, H. (2018). Advances in targeted pesticides with environmentally responsive controlled release by nanotechnology. *Nanomaterials*. https://doi.org/10.3390/nano8020102.

IUPAC. (2010). *History of Pesticide Use*. Available at: http://agrochemicals.iupac.org/index.php?option=com_sobi2&sobi2Task=sobi2Details&catid=3&sobi2Id=31.

Jayaseelan, C., Rahuman, A.A., Kirthi, A.V., Marimuthu, S., Santhoshkumar, T., Bagavan, A., Gaurav, K., Karthik, L. and Rao, K.B. (2012). Novel microbial route to synthesize ZnO nanoparticles using *Aeromonas hydrophila* and their activity against pathogenic bacteria and fungi. *Spectrochim. Acta A Mol. Biomol. Spectrosc.*, 90: 78–84.

Kah, M., Beulke, S., Tiede, K. and Hofmann, T. (2013). Nanopesticides: State of knowledge, environmental fate, and exposure modeling. *Crit. Rev. Environ. Sci. Technol.*, 43(16): 1823–1867.

Kawabata, Y., Wada, K., Nakatani, M., Yamada, S. and Onoue, S. (2011). Formulation design for poorly water-soluble drugs based on Biopharmaceutics Classification System: Basic approaches and practical applications. *Int. J. Pharm.*, 420: 1–10.

Khot, L.R., Sankaran, S., Maja, J.M., Ehsani, R. and Schuster, E.W. (2012). Applications of nanomaterials in agricultural production and crop protection: A review. *Crop Protection*, 35: 64–70.

Kim, S.W., Jung, J.H., Lamsal, K., Kim, Y.S., Min, J.S. and Lee, Y.S. (2012). Antifungal effects of silver nanoparticles (AgNPs) against various plant pathogenic fungi. *Mycobiology*, 40(1): 53–58.

Knowles, A. (2008). Recent developments of safer formulations of agrochemicals. *Environmentalist*, 28(1): 35–44.

Kumar, S., Nehra, M., Dilbaghi, N., Marrazza, G., Hassan, A.A. and Kim, K.H. (2019). Nano-based smart pesticide formulations: Emerging opportunities for agriculture. *J. Control. Release*, 294: 131–153.

Kumaraswamy, R.V., Kumari, S., Choudhary, R.C., Pal, A., Raliya, R., Biswas, P. and Saharan, V. (2018). Engineered chitosan-based nanomaterials: Bioactivities, mechanisms, and perspectives in plant protection and growth. *Int. J. Biol. Macromol.*, 113: 494–506.

Lamsal, K., Kim, S.W., Jung, J.H., Kim, Y.S., Kim, K.S. and Lee, Y.S. (2011). Application of silver nanoparticles for the control of Colletotrichum species *in vitro* and pepper anthracnose disease in field. *Mycobiology*, 39(3): 194–199.

Loo, Y.Y., Rukayadi, Y., Nor-Khaizura, M.A.R., Kuan, C.H., Chieng, B.W., Nishibuchi, M. and Radu, S. (2018). *In vitro* antimicrobial activity of green synthesized silver nanoparticles against selected gram-negative foodborne pathogens. *Front. Microbiol.*, 9: 1555. https://www.frontiersin.org/article/10.3389/ fmicb.2018.01555.

McClements, D.J. (2012). Nanoemulsions versus microemulsions: Terminology, differences, and similarities. *Soft Matter*, 8(6): 1719–1729.

Milivojević, T., Glavan, G., Božič, J., Sepčić, K., Mesarič, T. and Drobne, D. (2015). Neurotoxic potential of ingested ZnO nanomaterials on bees. *Chemosphere*, 120: 547–554.

Niemeyer, B.A., Bergs, C., Wissenbach, U., Flockerzi, V. and Trost, C. (2001). Competitive regulation of CaT-like-mediated Ca^{2+} entry by protein kinase C and calmodulin. *Proceedings of the National Academy of Sciences*, 98(6): 3600–3605.

Nuruzzaman, M., Rahman, M.M., Liu, Y. and Naidu, R. (2016). Nanoencapsulation, nano-guard for pesticides: A new window for safe application. *J. Agric Food Chem.*, 64(7): 1447–1483.

Pariona, N., Mtz-Enriquez, A.I., Sánchez-Rangel, D., Carrión, G., Paraguay-Delgado, F. and Rosas-Saito, G. (2019). Green-synthesized copper nanoparticles as a potential antifungal against plant pathogens. *RSC Adv.*, 9(33): 18835–18843.

Park, H.J., Kim, S.H., Kim, H.J. and Choi, S.H. (2006). A new composition of nanosized silica-silver for control of various plant diseases. *Plant Pathol. J.*, 22(3): 295–302.

Peteu, S.F., Oancea, F., Sicuia, O.A., Constantinescu, F. and Dinu, S. (2010). Responsive polymers for crop protection. *Polymers*, 2(3): 229–251.

Pompa, P.P., Vecchio, G., Galeone, A., Brunetti, V., Sabella, S., Maiorano, G., Falqui, A., Bertoni, G. and Cingolani, R. (2011). *In vivo* toxicity assessment of gold nanoparticles in *Drosophila melanogaster*. *Nano Res.*, 4(4): 405–413.

Pradhan, S., Roy, I., Lodh, G., Patra, P., Choudhury, S.R., Samanta, A. and Goswami, A. (2013). Entomotoxicity and biosafety assessment of PEGylated acephate nanoparticles: A biologically safe alternative to neurotoxic pesticides. *J. Environ. Sci. Health, Part B*, 48(7): 559–569.

Pérez-de-Luque, A. and Rubiales, D. (2009). Nanotechnology for parasitic plant control. *Pest Manag. Sci.*, 65(5): 540–545.

Qais, F.A., Shafiq, A., Khan, H.M., Husain, F.M., Khan, R.A., Alenazi, B., Alsalme, A. and Ahmad, I. (2019). Antibacterial effect of silver nanoparticles synthesized using *Murraya koenigii* L. against multidrug-resistant pathogens. *Bioinorg. Chem. Appl.*, Vol. 2019. https://doi.org/10.1155/2019/4649506.

Rico, C.M., Lee, S.C., Rubenecia, R., Mukherjee, A., Hong, J., Peralta-Videa, J.R. and Gardea-Torresdey, J.L. (2014). Cerium oxide nanoparticles impact yield and modify nutritional parameters in wheat (*Triticum aestivum* L.). *J. Agric. Food Chem.*, 62(40): 9669–9675.

Sheeja, C.C., Anusri, A., Levna, C., Aneesh, P.M. and Lekha, D. (2020). MoS2 nanoparticles induce behavioral alteration and oxidative stress-ediated cellular toxicity in the social insect *Oecophylla Smaragdina* (Asian Weaver Ant). *J. Hazard Mater.*, 385: 121624. https://doi.org/10.1016/j.jhazmat.2019.121624.

Suresh Kumar, R.S., Shiny, P.J., Anjali, C.H., Jerobin, J., Goshen, K.M., Magdassi, S., Mukherjee, A. and Chandrasekaran, N. (2013). Distinctive effects of nano-sized permethrin in the environment. *Environ. Sci. Pollut. Res. Int.*, 20(4): 2593–2602.

United States Environmental Protection Agency (USEPA). (1998) *Guidelines for Ecological Risk Assessment*. Washington, D.C.:EPA/630/R 95/002F.

Walker, G.W., Kookana, R.S., Smith, N.E., Kah, M., Doolette, C.L., Reeves, P.T., Lovell, W., Anderson, D.J., Turney, T.W. and Navarro, D.A. (2018). Ecological risk assessment of nano-enabled pesticides: A perspective on problem formulation. *J. Agric. Food Chem.*, 66(26): 6480–6486.

Wang, L., Li, X., Zhang, G., Dong, J. and Eastoe, J. (2007). Oil-in-water nanoemulsions for pesticide formulations. *J. Colloid Interface Sci.*, 314(1): 230–235.

Yasur, J. and Rani, P.U. (2015). Lepidopteran insect susceptibility to silver nanoparticles and measurement of changes in their growth, development, and physiology. *Chemosphere*, 124: 92–102.

Zhang, X., Xu, Z., Wu, M., Qian, X., Lin, D., Zhang, H., Tang, J., Zeng, T., Yao, W., Filser, J. and Li, L. (2019). Potential environmental risks of nanopesticides: Application of $Cu(OH)_2$ nanopesticides to soil mitigates the degradation of neonicotinoid thiacloprid. *Environ. Int.*, 129: 42–50.

Zhao, X., Zhu, Y., Zhang, C., Lei, J., Ma, Y. and Du, F. (2017). Positive charge pesticide nanoemulsions prepared by the phase inversion composition method with ionic liquids. *RSC Adv.*, 48586–48596.

Zhao, X., Cui, H., Wang, Y., Sun, C., Cui, B. and Zeng, Z. (2018). Development strategies and prospects of nano-based smart pesticide formulation. *J. Agric. Food Chem.*, 66(26): 6504–6512.

Chapter 7

Chitosan Nanomaterials for Post Flowering Stalk Rot Control in Maize

Garima Sharma,[1] *Damyanti Prajapati,*[1] *Ajay Pal*[2] and *Vinod Saharan*[1,*]

1. Introduction

In agriculture, pathogens cause numerous plant diseases and are responsible for losses in crop yield and quality. Post-flowering stalk rot (PFSR) complex is a wide-spread group of diseases in maize that causes internal decay and discoloration of stalk tissues. It blocks the translocation of water and nutrients resulting in death and lodging of plant, and ultimately reducing the crop yield. This disease is prevalent in areas where there is scarcity of irrigation, especially after the post-flowering stage of crop growth. PFSR complex is caused by *Fusarium verticillioides* (Fusarium stalk rot), *Macrophomina phaseolina* (charcoal rot), and *Cephalosporium maydis* (late wilt). Among these three species, the former is the most accountable pathogen of PFSR which can be found in association with maize as a pathogen or as a symptomless intercellular endophyte (Pereira et al., 2011). Fusarium stalk rot causes 38% loss in plant in the rain-fed northern, central, and southern regions in India (AICRP, 2014). Crop protection has heavily relied on synthetic pesticides over the past few years, but their applications are under strong regulation with the implementation of new legislation and resistance evolution in pest populations (Chandler et al., 2011). Pesticide application, worth ~ 3 million tonne per annum, is used to control pathogens and pests in agriculture, of which 90% get run-

[1] Department of Molecular Biology and Biotechnology, Rajasthan College of Agriculture, Maharana Pratap University of Agriculture and Technology, Udaipur, Rajasthan 313001, India.

[2] Department of Biochemistry, College of Basic Sciences and Humanities, Chaudhary Charan Singh Haryana Agricultural University, Hisar, Haryana 125004, India.

* Corresponding author: vinodsaharan@gmail.com

off to air, water, and soil during application (Pimentel and Burgess, 2014). Chemical pesticide is a double-edged sword where its application helps the farmers to raise the productivity but harms the ecosystem. It also leads to the development of resistant plant pathogenic strains. Plant diseases, therefore, need to be controlled in a regulated manner to maintain the quality and safety of agricultural produces (Xing et al., 2015). It is also pertinent to develop environment friendly pesticides and techniques which can be used to reduce pesticide usage while ensuring vigorous sustainable agriculture to feed the ever-increasing population. Application of nanotechnologies in agriculture can potentially contribute to address the issue of sustainability. Another core objective of nanotechnology in agriculture is the efficient use of fertilizers and pesticides using nanoscale carriers and compounds, reducing the amount of agrochemicals without impairing productivity (Fraceto et al., 2016). Reports indicate that metal, metal oxide, and carbon- and polymer-based nanomaterials are under various stages of research and application in agriculture. Among the nanomaterials, 60% of them are metal-based and intended for use in disease management and plant growth. But uses of metal-based nanomaterials have raised issues of metal toxicity in living organisms; their fates in soil, water, and air are also uncertain. The situation has further become critical as there are contradictory reports on toxicity of metal nanomaterials in end results and majority of them suggest further research for concrete conclusion. Therefore, researchers are searching for novel approaches through nanotechnology to reorganize the existing agrochemicals and/or design novel eco-friendly agro-active components for agriculture while protecting the environment. To reduce the negative impacts to human health and environment, natural products are excellent alternatives to synthetic pesticides. The interplay of antimicrobial and eliciting properties makes chitosan a potential antimicrobial agent to control plant disease caused by pathogens. Chitosan induces host defense responses through lignification, ion flux variations, cytoplasmic acidification, membrane depolarization, chitinase glucanase activation, phytoalexin biosynthesis, and generation of reactive oxygen species, etc. (Pichyangkura and Chadchawan, 2015). The immunomodulatory role of chitosan is well demonstrated in plants, whilst its nanoparticles have mostly been examined for biomedical applications (Chandra et al., 2015; Saharan and Pal, 2016). Recent studies have explored the positive effect of chitosan-based nanomaterials on the plant's immune system and growth (Chandra et al., 2015; Anusuya and Banu, 2016).

Salicylic acid (SA) is classified as a phenolic growth regulator. In plants, it is a major signaling molecule to amend responses of plants to environmental stresses (Hovarth et al., 2007). Traditionally, it has been used in agriculture for seed treatment and foliar application to induce plant innate immunity and increase plant growth under stress conditions (Vicente and Placensia, 2011). SA has not only significantly controlled the Basal Stem Rot (BSR) progression caused by *Ganoderma boninense* in many industrially important crops, including oil palm seedlings, but has also promoted plant growth (Surendran et al., 2018). However, it is unstable in different environments during field application, and at certain concentrations, exerts phytotoxicity by reducing seed germination, respiration, and photosynthesis (Pancheva et al., 1996). Therefore, application of SA in agriculture needs to be strictly controlled to avoid its toxicity and improve efficiency. Development of slow, systemic, targeted, and protective release methods of SA through chitosan nanoformulation could enhance its efficacy and reduce its phytotoxicity. Copper (Cu) is a conversant metal used as an antimicrobial agent in addition to its use as a micronutrient, but toxic effects are also observed when it is applied in excess amount. This shortcoming could effectively be sublimed by providing slow and sustained release of metal from porous nanostructure of natural biopolymer such as chitosan (Choudhary et al., 2017). It was witnessed that SA and Cu played crucial roles in disease control as well as in enhancing crop yield through their precise application (Saharan et al., 2015; Choudhary et al., 2019; Kumaraswamy et al., 2019). This chapter has reviewed recent insights of chitosan nanotechnology to control plant disease with special reference to PFSR disease of maize.

2. Application of Nanotechnology in Crop Protection and Growth

Current challenges of sustainability, food security, and climate change are engaging researchers in exploring the field of nanotechnology as a new source of key improvements for the agricultural sector (Parisi et al., 2015). The agronomic application of nanotechnology in plants (phytonanotechnology) has the potential to alter conventional plant production systems, allowing for the controlled release of agrochemicals (fertilizers, pesticides, and herbicides) and target-specific delivery of biomolecules (nucleotides, proteins, and activators). Successful applications of agricultural nanotechnology both at small scale and research and development stage include plant protection products (Anjali et al., 2012), fertilizers (Milani et al., 2015), soil improvement (http://www.geohumus.com/us/products.html), water purification (McMurray et al., 2006), diagnostics (Vamvakaki and Chaniotakis, 2007), and plant breeding (Torney et al., 2007). Nanosensor technology has emerged as one of the emerging technologies in aiding decision-making in crop monitoring, accurate analysis of nutrients and pesticides in soil, or for maximizing the efficiency of water use for a smart agriculture (Fraceto et al., 2016). Metallic nanomaterials have various applications in agriculture. For example, silver nanoparticles (AgNPs) have been proved effective against seed-borne fungal pathogen *Gibberella fujikuroi* in rice (Jo et al., 2015), gold (Au) NPs have shown antifungal activity against *Candida albicans* (Wani and Ahmed, 2013), Zinc oxide (ZnO) NPs are effective against *Botrytis cinerea* and *Penicillium expansum* (He et al., 2011), Karnal bunt disease in wheat was detected by nanogold-based immunosensors (Singh et al., 2010), oxidized multi-walled carbon nanotube (MWCNT) enhanced germination and increased root and shoot growth in mustard. But manufactured NPs (MNPs) can have a detrimental effect on the environment such as air and soil because of their handling, result of accidents, etc. (Gottschalk et al., 2009; Cornelis et al., 2014; Aziz et al., 2015; Prasad et al., 2016) and adversely affect the soil organisms, plants, and environment. The growing interest and demand for organic or non-polluted food/crop from the more health-conscious consumers and the unsteady and pernicious nature of metal nanoparticles have raised serious concerns with respect to their use (Saharan et al., 2013; Saharan et al., 2015). So, there is a need to explore bio-based materials for plant growth and protection.

3. Exploitation of Chitosan in Agriculture

Chitosan is a biopolymer formed from deacetylation of chitin. Chitosan, chemically a co-polymer of N-acetyl-D-glucosamine and D-glucosamine, is a biocompatible, biodegradable, highly permeable, cost-effective, non-toxic, and plant growth and defense-stimulating biomaterial. It has emerged as a most promising polymer (Shukla et al., 2013). These properties make chitosan-based materials applicable in many fields including agriculture, where they are used as biostimulants (Pichyangkura and Chadchawan, 2015). Chitosan has been reported to induce turmeric rhizome development as well as curcumin bioaccumulator in rhizome roots by 38% and 56%, respectively. It induces lipoxygenase activity in excised grapevine leaves which leads to lipid peroxidation and ultimately in systemic acquired resistance (Hu et al., 2015). Chitosan can induce multifaceted disease resistance in plants and is known for its broad spectrum of antimicrobial and insecticidal activities (Rabea et al., 2003; Hadrami et al., 2010). It is easily biodegraded through enzymes present in living organisms predominantly by lysozyme in vertebrates and bacterial enzymes in colon (Kean and Thanou, 2010). Therefore, its use in various fields as a bioactive compound could be recommended. Chitosan traditionally acts as an elicitor to induce the plant immune system and, thus, provides immunity to plants against various pathogens by triggering various defense-related reactions. Chitosan induces the innate immune responses in plants by producing reactive oxygen species (ROS), inducing defense-related enzymes (superoxide dismutase, catalase, and peroxidase), several defensive genes such as pathogenesis-related genes (glucanase and chitinase), biosynthesis of jasmonic acid, lignifications, ion flux, etc. (Sathiyabama and Balasubramanian, 1998; Prapagdee et al., 2007; Bueter et al., 2013; Hadwiger, 2013; Pichyangkura and Chadchawan, 2015; Sathiyabama et al., 2016).

Several reports have been describing the use of chitosan for biotic and abiotic stress management in agriculture due to its multi-dynamic bioactivities in plants (Sharp, 2013; Katiyar et al., 2014; Handwiger, 2013; Wang et al., 2015). Some important applications of chitosan involve antifungal activity in papaya (Hewajulige et al., 2009) and grape (Reglinski et al., 2010). Chitosan has been reported to induce defense response in tomato (Benhamou and Thériault, 1992; Benhamou and Nicole, 1994) and chilli seeds (Photchanachai et al., 2006), strawberry fruit (Ghouth et al., 1992), and rose shrubs (Wojdyla, 2004). Chitosan triggered innate immunity in rice by stimulating hydrogen peroxide (H_2O_2) production (Lin et al., 2005; Li et al., 2009), induced defense response by nitric oxide (NO) pathways in tobacco (Zhao et al., 2007; Zhang et al., 2011). Despite such fascinating properties, the insolubility of bulk chitosan in aqueous media limits its wide spectrum of application (Saharan et al., 2013; Sathiyabama and Charles, 2015). In addition, applications of chitosan in acidic aqueous media also exhibit toxicity to the environment. Practically, acid soluble chitosan has also lesser biological activity as compared to its dispersed form (Saharan and Pal, 2016).

4. Application of Chitosan Nanomaterial in Crop Disease Management

Due to specific physico-chemical properties of chitosan, this biopolymer is not soluble in water and moreover cannot be dispersed in water for various applications, especially in seed treatment and foliar application (Saharan and Pal, 2016). However, chitosan is considered as a highly tenable biopolymer because of its functional groups such as

–NH_2 and –OH. Numerous chemical modifications have been introduced in chitosan to exhibit different levels of solubility in water as well as other biophysical characteristics (Kean and Thanou, 2010). However, chemical modifications of chitosan may affect its inherent properties such as biodegradability, biocompatibility, and non-toxicity. The chelating property of chitosan towards organic and inorganic compounds renders it suitable for improvement in solubility and biocidal activity (Shukla et al., 2013), and has led the researchers in preparing chitosan NPs. Chitosan-based NPs are more effective than bulk material (Du et al., 2009; Smitha et al., 2015) and used world-wide for various applications including agriculture due to their biodegradability, solubility, high permeability, non-toxicity, and cost-efficacy (Ali et al., 2011; Brunel et al., 2013; Manikandan and Sathiyabama, 2016). Chitosan-based nanomaterials have been also developed and tested for antifungal efficacy (Saharan et al., 2013; Saharan et al., 2015; Sathiyabama and Parthasarathy, 2016), plant defense boosting activity (Chandra et al., 2015) and as a carrier system for plant growth hormone (Periera et al., 2017) which proved their superiority in results as compared to their bulk counterpart (Table 7.1).

Recent studies have investigated the use of chitosan NPs loaded with Cu as an antimicrobial agent (Saharan et al., 2013; Saharan et al., 2015) (Fig. 7.1). Application

Fig. 7.1. TEM micrographs of (a) Cu–chitosan NPs in inset showing porous structures. (b) Red arrow denoted Cu embedded and blue arrow denoted chitosan porous network. SEM images of (c) Cu–chitosan nanoparticles, and (d) porous Cu–chitosan showed nano in red rectangular and micro size pores in blue rectangular. Source: Saharan et al., 2015; copyright permission from Elsevier.

Table 7.1. Chitosan nanomaterials exploited in agriculture for plant growth and protection.

Nanoparticles	**Purpose**	**References**
Cu–chitosan nanomaterials	To control the growth of *Fusarium graminerarium*	(Brunel et al., 2013)
	To evaluate NPs against phytopthogenic fungi *Alternaria alternate, Macrophomina phaseolina,* and *Rhizoctonia solani*	(Saharan et al., 2013)
	To evaluate the antifungal activity of nanoparticles on *Alternaria solani* and *Fusarium oxysporum* and their substantial growth promotory effect on tomato seedling	(Saharan et al., 2015)
	To enhance seedling growth via reserve food mobilization in maize	(Saharan et al., 2016)
	To boost defence responses against Curvularia leaf spot (CLS) disease of maize and plant growth promotory activity	(Choudhary et al., 2017)
Chitosan nanomaterials	To evaluate nanoparticles against *A. alternate, M. phaseolina,* and *R. Solani*	(Saharan et al., 2013)
Chitosan nanomaterials	To enhance antifungal activity against *Pyricularia grisea, A. solani, F. oxysporum,* and also promote growth of chickpea seedlings	(Sathiyabama and Parthasarathy, 2016)
	To check antifungal activity against *Pyricularia grisea*	(Manikandan and Sathiyabama, 2016)
	To provide increased defence responses against blast disease in finger millet	(Sathiyabama and Manikandan, 2016)
	To check antifungal activity against *Fusarium graminerarium*	(Kheiri et al., 2016)
Chitosan-saponin nanomaterials	To evaluate nanoparticles against *A. alternate, M. phaseolina,* and *R. Solani*	(Saharan et al., 2013)
Cell wall polymer-based nanomaterials	Antifungal activity improvement against *F. oxysporum f.* sp. *Lycopersici*	(Sathiyabama and Charles, 2015)
Ag–Chitosan nanomaterials	To promote the growth of chickpea and improve defence-related enzymes	(Anusuya and Banu, 2016)
Chitosan nanomaterials loaded with essential oil	To improve the application potential of summer savoury essential oil by allowing its dispersibility in water	(Feyzioglu and Tornuk, 2016)
NO donor S-nitroso-mercaptosuccinic acid encapsulating chitosan nanoparticles	To alleviate the effect of salt stress in maize plant	(Oliveira et al., 2016)
Chitosan nanoparticles	Growth inhibition of *Pyricularia grisea, Alternaria solani, Fusarium oxysporum*	(Sathiyabama and Parthasarathy, 2016)

Table 7.1 contd. ...

...Table 7.1 contd.

Nanoparticles	Purpose	References
Chitosan nanomaterials as carriers for plant growth hormone	To encapsulate GA_3 to improve its efficiency in terms of leaf development and level of carotenoid content and higher economic value of agricultural products	(Periera et al., 2017)
Zn complexed chitosan nanomaterials	To encapsulate the promising micronutrient for foliar application	(Deshpande et al., 2017)
Ag nanomaterials immobilized in chitosan nanocarrier	To check antibacterial activity against Gram +ve and –ve bacteria	(Sharma, 2017)
Thymol loaded chitosan silver nanomaterials	To check antibacterial activity against multiple food borne pathogens	(Manukumar et al., 2017)
Lanthanum-modified chitosan oligosaccharide nanoparticles	To load avermectin and provide defence response against rice blast	(Liang et al., 2018)
Salicylic acid encapsulated chitosan nanomaterials	To use salicylic acid as a biostimulant for plant defence and growth	(Kumaraswamy et al., 2019)
Salicylic acid and copper encapsulated chitosan nanomaterials	To increase the source activity in maize	(Sharma et al., 2020)

of Cu–chitosan NPs on maize seedling led to superior values of % germination, root and shoot length, root number, seedling length, fresh and dry weight, and seed vigur index as compared to water, $CuSO_4$, and bulk chitosan treatment. Same treatments also increased the activity of α-amylase and protease which corroborated with decreased content of starch and protein, respectively (Saharan et al., 2016). One recent study has demonstrated the successful application of S-nitroso mercaptosuccinic acid (MSA) encapsulated chitosan NPs in the amelioration of deleterious effects of salinity on photosystem II activity, chlorophyll content, and growth of maize plants (Oliveira et al., 2016). In another study, chitosan NPs showed antifungal activity against four plant pathogenic fungi, namely *Rhizoctonia solani*, *Colletotrichum accutatum*, *Phytophthora infestens*, and *F. oxysporum.* It also showed antioxidant and cytotoxic activity. A further study proved that chitosan NPs induced defence responses in finger millet against blast disease (Sathiyabama and Manikandan, 2016) (Table 7.1).

From the above-cited literature, it can be confirmed that chitosan nanomaterials have been used in two ways: (1) used alone for plant protection or/and growth and (2) used as a carrier/or encapsulating unit for organic and inorganic materials. The core objective of encapsulation of organic/inorganic materials into chitosan nanostructures is to increase the efficacy of the encapsulated material by slow, systemic, targeted, and protected release and to avoid its toxicity to plants and the environment.

5. Role of Salicylic Acid and Copper in Crop Protection: Its Prospects with Nanotechnology

In recent years, the focus has shifted on regulating the plant endogenous defence and growth regulatory system through various signaling molecules. Therefore, researchers have been working on various signaling molecules, especially SA owing to its function as a local and systemic plant defence eliciting agent against pathogens as well as its role during plant response to abiotic stresses such as drought, chilling, heavy metal toxicity, heat, and osmotic stress. Adding to its role in biotic and abiotic stresses, SA also plays an essential role in plant growth regulation. SA, chemically ortho-hydroxybenzoic acid, belongs to plant phenolic compound, which has diverse effects on various processes such as photosynthesis, ion uptake, different plant growth aspects, membrane permeability, defence, and antioxidant enzyme activities related to biotic and abiotic stresses (Hayat et al., 2008; Vicente and Placencia, 2011).

For disease control, strengthening the immune system, and promotion of plant growth, SA has been applied to plants via seed treatment (Horvath et al., 2002) and foliar application (Zhang et al., 2016). Exogenous-applied SA controlled disease caused by *Rhizoctonia solani* through rapid induction of antioxidant and defence enzymes including phenylalanine ammonia lyase (PAL), polyphenol oxidase (PPO), and peroxidase (POD) enzymes in cowpea plant (Chandra et al., 2007). Foliar application of SA on apple cultivar (*Malus domestica* Borkh cv. 'Gala') induced disease resistance against Glomerella leaf spot (GST) caused by *Glomerella cingulata* (Zhang et al., 2016). Application of SA in maize crop delayed initial infection of maize dwarf mosaic virus and prevented the damaging effect of this deadly virus (Cueto-Ginzo et al., 2016). Therefore, it is proved that the signalling molecule SA is directly involved in plant defence system to protect plants from diverse biotic stresses. Recent reports revealed the crucial role of SA in mitigating biotic stresses in various crop plants. Application of SA can abate the water scarcity damages and improve plant efficiency by altered activities of enzymatic antioxidants and reduced generation of ROS (Dianat et al., 2016). SA provides protection to plants from several abiotic stresses by regulating important physiological processes such as photosynthesis, nitrogen metabolism, proline metabolism, production of glycine-betaine (GB), antioxidant defence system, and plant–water relations under stress conditions (Khan et al., 2010, 2012, 2014). SA application promoted flower bud formation in tobacco (Lee and Skoog, 1965) and induced flowering in various genera of the Lemnaceae family including long day (LD), short day (SD) and photoperiod insensitive types (Khurana and Clelend, 1992). Exogenous application of SA significantly regulated the vegetative growth in a concentration-dependent manner. In soybean plants treated with 10 nM, 100 μM, and up to 10 mM SA, the shoot and root growth increased by 20% and 45% with 10 nM and 100 μM, respectively. Likewise leaf rosettes and chamomile plant root growth were stimulated by 32% and 65%, respectively, with 50 μM SA treatment but the same study with higher concentrations have shown opposite effects. In each case the evidence confirmed that concentration is an important aspect in the execution of SA as a growth regulator.

With the intervention of nanoencapsulation, there is scope to develop a delivery system of SA for its slow, systemic, targeted, and protected release to enhance the

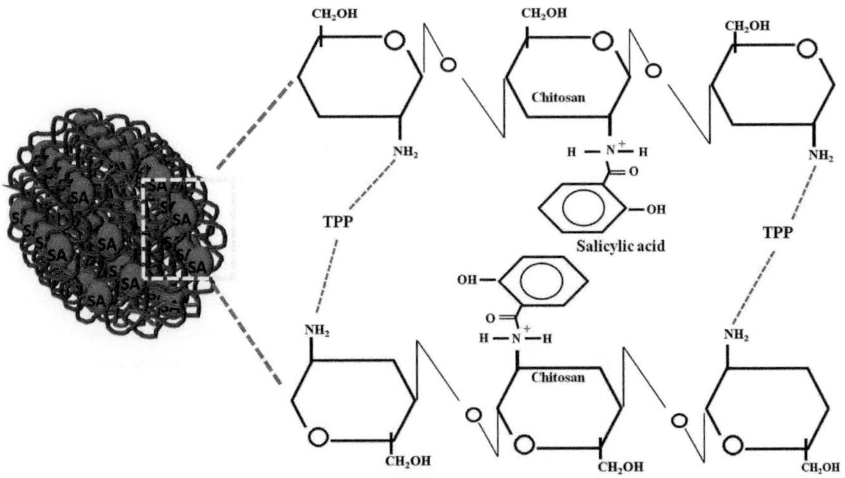

Fig. 7.2. Salicylic acid-chitosan NMs. Source: Kumaraswamy et al., 2019; copyright permission from Elsevier.

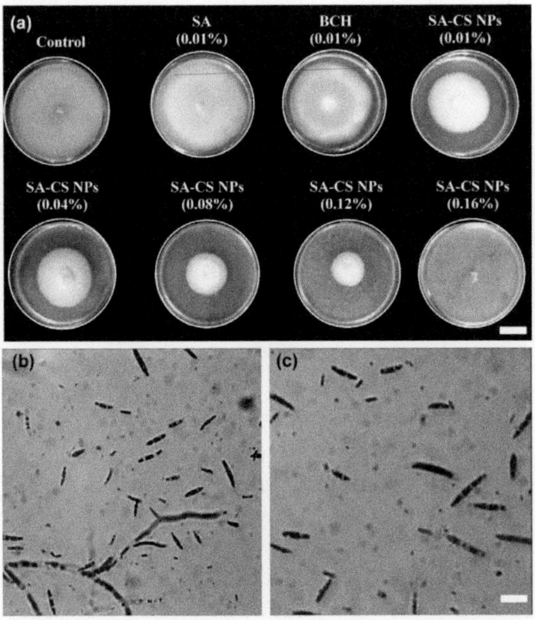

Fig. 7.3. Effect of SA-CS NPs on (a) mycelial growth (Bar = 3.0 cm) and spore germination of *F. verticillioides* in (b) control, and (c) in SA-CS NPs (0.01%) (Bar = 50 μm). Source: Kumaraswamy et al., 2019; copyright permission from Elsevier.

efficacy and avoid toxicity to plants and the environment. Recent invention signified and proved that SA could be encapsulated into chitosan at nanoscale (Fig. 7.2) and its sustained and protected release for higher activity against *Fusarium verticillioides* causing PFSR disease in maize (Fig. 7.3) (Kumaraswamy et al., 2019).

The US Environmental Protection Agency (EPA) has recognized Cu as the first metallic antimicrobial agent (Grass et al., 2011). It is a prevalent micronutrient in the plant system and is an essential component of many important enzymes of plant metabolism. However, application of Cu in crops has certain disadvantages. Uncontrolled exposure to plant cells shows toxic effects and severely expresses an antagonistic effect. A recent study by Choudhary et al. (2017) has come up with slow and sustained application of Cu through nanotechnology. The use of Cu-encapsulated nanochitosan (360.3 nm mean hydrodynamic diameter, 0.48 PDI, and + 32.6 mV zeta potential) provided effective control of tomato fungal pathogen (Saharan et al., 2015). Further, successive study on chitosan NPs encapsulating Cu confirmed the enhanced growth of seedling by mobilizing the reserve food in maize (Saharan et al., 2016). In maize plants, Cu–chitosan nanoparticles raised defence response against CLS disease via increased activity of antioxidant and defense enzymes in addition to increased plant growth (Choudhary et al., 2017). Additionally, studies on Cu–chitosan NPs claimed slow and sustained release of Cu from NPs (Fig. 7.4) which was the main reason for its effectiveness (Saharan et al., 2015; Choudhary et al., 2017).

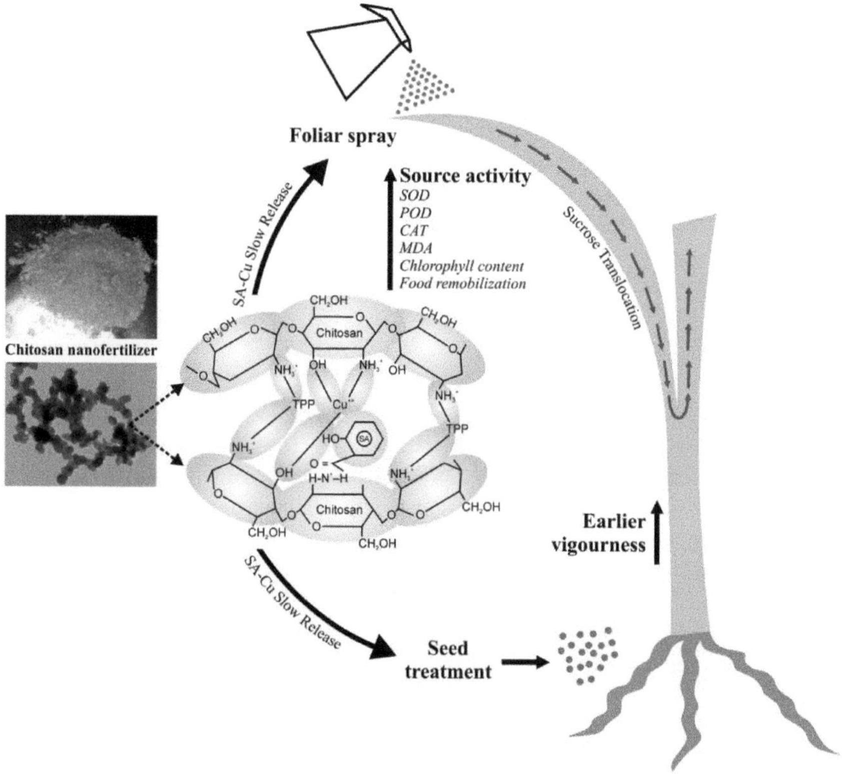

Fig. 7.4. Working of Cu/SA co-encapsulated chitosan nanomaterial to foster source activity in maize. Source: Sharma et al., 2020; copyright permission from Elsevier.

6. Post-flowering Stalk Rot of Maize

Post-flowering Stalk Rot (PFSR) complex is one of the most serious, destructive, and wide-spread group of fungal diseases in maize accounting for yield loss ranging from 10% to 42% and sometimes as high as 100% in some areas. Fusarium stalk rot has been observed in the plant age group of 55–65 days which coincides with tasselling and silking and immediately followed by grain formation stage. The stalk rot usually occurs after the flowering stage and prior to physiological maturity, which reduces yields in two ways: (1) affected plants die prematurely, thereby producing lightweight ears having poorly filled kernels and (2) plants with stalk rot easily lodge which makes harvesting difficult and ears are left in the field during harvesting (Singh et al., 2012). Stalk rot reduces maize yield directly by affecting the physiological activity of the plants (Ledencan et al., 2003). The disease becomes apparent when the crop enters the senescence phase and severity increases during the grain-filling stage. Stalk rot symptoms are observed during post-flowering and pre-harvest stages (Lal and Singh, 1984). Rotting extends from infected roots to the stalk and causes premature drying, stalk breakage and ear dropping, thus significantly reducing the yields (Colbert et al., 1987). The disease causes internal decay and discoloration of stalk tissues, directly reduces yield by blocking the translocation of water and nutrients, thus resulting in lodging and death of the plants. Symptom development depends on several stress factors including an excess or lack of moisture, heavy and continuous cloudiness, high plant density, foliar diseases, and corn borer infestation (Parry et al., 1995). As the stalk rot of maize is a complex disease involving more than one organism, it is very difficult to manage it with a single control measure. Hence, efforts are needed to explore the feasibility of a combination of various control measures for integrated management of stalk rots (Kulkarni and Anahosur, 2011).

7. Agrochemical and Biocontrol Measures for PFSR Disease Control

Besides the global population explosion of human beings, climate change is another issue which affects the future agriculture scenario. To feed the world, stringent assessment is needed to find out the possibility of alternative technology for sustainable food production (Chen and Yada, 2011). Use of agrochemicals seems the only effective solution to sustain crop productivity. In this regard, various systemic and non-systemic fungicides, namely Bavistin, Dithane M-45, Blitox 50, Hexacap-75, TMTD, Topsin-M, and apron are generally recommended to control PFSR in maize (Khokhar et al., 2014). Among the other agrochemicals, fungicides, such as tebuconazole and thiabendazole were evaluated for their ability to inhibit the growth of toxigenic *F. verticillioides* and it was found that 5% aqueous solution of tebuconazole effectively reduced ear rot disease and accumulation of fumonisins to the maximum extent as compared with other fungicides (Chandra et al., 2008). The agro-chemical residues left to a variable extent in the food materials after harvesting are beyond the control of the consumer and have deleterious effects on human health. The presence of chemical residues is a major bottleneck in international trade of food commodities (Bajwa and Sandhu, 2014). Bioagents may be a useful and important alternative for the effective management of fungal pathogens in maize (Khokhar et al., 2014). Growth suppression of *F. verticillioides* by *Trichoderma pseudokoningii* and *Trichoderma harzianum* has been reported in maize for management of PFSR diseases (Sobowale et al., 2005; Shekhar and Kumar, 2010). Application of biopesticides appear more pivotal to overcome the side effects of massive application of synthetic pesticides. To control PFSR disease in maize, recent efforts had emphasized on the use of a combination of biopesticides with synthetic fungicides. Biopesticides are bioformulations of living organisms which have different behaviour at different ecosystems. Hence, the performance of biopesticides towards disease control is inconsistent and needs intensified research (Ghormade et al., 2011).

8. Recent Advancements in Chitosan Nanotechnology to Control PFSR Disease of Maize

Slow release of SA from salicylic acid–chitosan NPs has significantly amended the physiological and biochemical responses in maize plant for commendable disease control, plant growth, and yield compared to sole SA application. *F. verticillioides* is an intercellular endophytic pathogen and symptoms of PFSR appear at the flowering stage. At this late stage, most of the approaches of disease control may not be effective. Indeed, application of salicylic acid–chitosan NPs as seed treatment and foliar application before the flowering stage has been an effective and preventive approach as it boosted the plant's innate immunity before onset of pathogen infection. The study further deemed that SA functionalized chitosan nanoparticles could further customize its physicochemical properties such as size, porosity, encapsulation capacity, and slow-release pattern to widen its application in disease control and higher yield in different crop plants. Strong evidence inferred from review literature show that the SA-functionalized chitosan NPs could act as an effective biostimulant for plants and have immense potential for commercialization (Kumaraswamy et al., 2019).

A co-encapsulation approach has so far not been attempted for the synthesis of chitosan nanomaterials; however, various compounds such as Cu, Zn, and SA have been individually encapsulated in nanoscale chitosan and applied solely to control plant disease and yield improvement. The core objective of the recent study was to deliver a robust biodegradable chitosan-based nanomaterial which can be applied to up-regulate the source activity to support sink strength for higher crop yield in maize and to strengthen the innate immunity (Fig. 7.4).

The developed nanomaterials were characteristically novel in the physico-chemical properties such as size, monodispersity, stability, and symmetric porosity of its particulates, and provides Cu as well as SA (a biostimulant strengthening the plant signal transduction) to harness their synergistic effect on plants (Fig. 7.5). Foliar application of chitosan (SA and Cu encapsulated) nanomaterials significantly invigorated the activities of antioxidant enzymes which contribute to the maintenance of cellular redox homeostasis and plant innate immunity. Among all, the activities of POD (peroxidase) and CAT (catalase) were elusively induced by chitosan nanomaterials followed by SOD (superoxide dismutase). SOD catalyses the dismutation of superoxide (O_2^{-1}) into H_2O_2 and further H_2O_2 is swapped into H_2O and O_2 by POD and CAT. The MDA (malondiadehyde) content was lower in plant leaves treated with chitosan nanomaterials which signify the stability of cell membranes. To study the effect of nanomaterials on reserve food mobilization towards maize

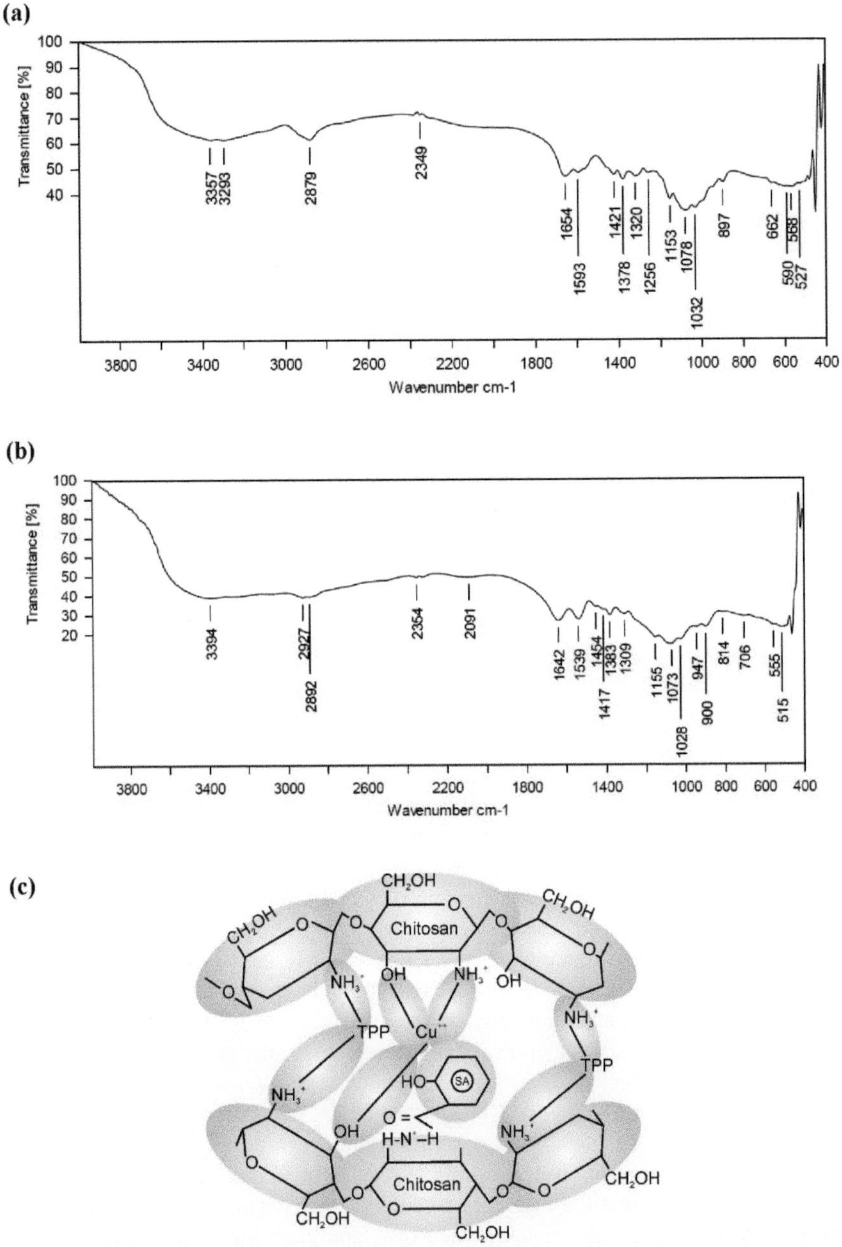

Fig. 7.5. FTIR spectra analysis of (a) bulk chitosan, (b) chitosan nanofertilizer, and (c) hypothetical representation of chitosan nanofertilizer showing interactions of sodium tripolyphosphate (TPP), –COOH group of salicylic acid (SA) and Cu ions with primary amide of chitosan in chitosan nanomaterial. Source: Sharma et al., 2020; copyright permission from Elsevier.

grains, a second internode was used to quantify the sucrose (non-reducing sugar). The study deduced that the treatment with chitosan nanomaterials (0.01%–0.16%) endowed higher mobilization of sucrose in internode which is a crucial factor for higher grain yield. Besides, higher bioactivities of Cu and SA through their slow release and synergistic effects, the presence of 9%–10% nitrogen (N) in chitosan and 24% phosphorous (P) in TPP is another factor that cannot be overlooked. The study speculated that N and P present in chitosan nanomaterials might be used by plants for growth and development which needs to be studied in future. Chitosan nanofertilizer application in the maize field claimed for significant control of PFSR disease in pot and field conditions by direct antifungal activity and by boosting maize crop immunity by balancing the ROS.

9. Conclusion

The elusive bioactivities of copper and salicylic acid encapsulating chitosan nanofertilizer attributed to slow release and synergistic effects of both these ingredients. We further conclude that chitosan, being a biodegradable biopolymer, which easily degrades by enzymes released by soil microorganisms, makes the developed nanomaterials environment friendly. We expect that the co-encapsulation approach used in the previous study can further be expanded for synthesis of nanofertilizers having more than one micronutrient to deliver a consortium of micronutrients to the crop. However, to realize the claims made in that study, field experiments need to be performed for translation/validation of technology from the laboratory to the field conditions.

References

AICRP. (2014). *Annual Report of AICRP Maize Pathology.* Udaipur.

Ali, S.W., Rajendran, S. and Joshi, M. (2011). Synthesis and characterization of chitosan and silver loaded chitosan nanoparticles for bioactive polyester. *Carbohydr. Polym.*, 83(2): 438–446.

Anjali, C.H., Sharma, Y., Mukherjee, A. and Chandrasekaran, N. (2012). Neem oil (*Azadirachta indica*) nanoemulsion: A potent larvicidal agent against *Culex quinquefasciatus. Pest Manag. Sci.*, 68(2): 158–163.

Anusuya, S. and Banu, K.N. (2016). Silver-chitosan nanoparticles induced biochemical variations of chickpea (*Cicer arietinum* L.). *Biocatal. Agric. Biotechnol.*, 8: 39–44.

Aziz, N., Faraz, M., Pandey, R., Shakir, M., Fatma, T., Varma, A. and Prasad, I. (2015). Facile algae-derived route to biogenic silver nanoparticles: synthesis, antibacterial, and photocatalytic properties. *Langmuir*, 31(42): 11605–11612.

Bajwa, U. and Sandhu, K.S. (2014). Effect of handling and processing on pesticide residues in food: A review. *J. Food Sci. Technol.*, 51(2): 201–220.

Benhamou, N. and Thériault, G. (1992). Treatment with chitosan enhances resistance of tomato plant to the crown and root rot pathogen *Fusarium oxysporum* sp. *radicis lycopersici. Physiol. Mol. Plant Pathol.*, 41: 33–52.

Benhamou, N., Lafontaine, P.J. and Nicole, M. (1994). Induction of systemic resistance to Fusarium crown and root rot in tomato plant by seed treatment with chitosan. *Phytopathology*, 84(12): 1432–1444.

Brunel, F.N., Gueddari, E.E. and Moercshbacher, B.M. (2013). Complexation of copper (II) with chitosan nanogels: Toward control of microbial growth. *Carbohydr. Polym.*, 92: 1348–1356.

Bueter, C.L., Specht, C.A. and Levitz, S.M. (2013). Innate sensing of chitin and chitosan. *PLoS Pathog.*, 9(1): 1–3.

Chandler, D., Bailey, A.S., Mark Tatchell, G., Davidson, G., Greaves, J. and Grant, W.P. (2011). The development, regulation, and use of biopesticides for integrated pest management. *Philos. Trans. R. Soc. B: Biol. Sci.*, 366(1573): 1987–1998.

Chandra, A., Anand, A. and Dubey, A. (2007). Effect of salicylic acid on morphological and biochemical attributes in cowpea. *J. Environ. Biol.*, 28(2): 193–196.

Chandra, N.S., UdayaShankar, A.C., Niranjan, R.S., Niranjana, S.R. and Prakash. H.S. (2008). Tebuconazole and Thiabendazole-novel fungicides to control toxigenic *Fusarium verticilloides* and fumonisin in maize. *J. Mycol. Plant Pathol.*, 38(3): 430–436.

Chandra, S., Chakraborty, N., Dasgupta, A., Sarkar, J., Panda, K. and Acharya, K. (2015). Chitosan nanoparticles: A positive modulator of innate immune responses in plants. *Sci. Rep.*, 5: 1–13.

Chen, H. and Yada, R. (2011). Nanotechnologies in agriculture: New tools for sustainable development. *Trends Food Sci. Technol.*, 22(11): 585–594.

Choudhary, R.C., Kumaraswamy, R.V., Kumari, S., Sharma, S.S., Pal, A., Raliya, R., Biswas, P. and Saharan, V. (2017). Cu-chitosan nanoparticle boost defense responses and plant growth in maize (*Zea mays* L.). *Sci. Rep.*, 7(1): 1–11.

Choudhary, R.C., Kumaraswamy, R.V., Kumari, S., Sharma, S.S., Pal, A., Raliya, R., Biswas, P. and Saharan, V. (2019). Zinc encapsulated chitosan nanoparticle to promote maize crop yield. *Int. J. Biol. Macromol.*, 127: 126–135.

Colbert, T.R., Kang, M.S., Myers, O. and Zuber, M.S. (1987). General and specific combining ability estimates for pith cell death in stalk internodes of maize. *Field Crops Res.*, 17(2): 155–161.

Cornelis, G., Hund-Rinke, K., Kuhlbusch, T., Van den Brink, N. and Nickel, C. (2014). Fate and bioavailability of engineered nanoparticles in soils: a review. *Crit. Rev. Environ. Sci. Technol.*, 44(24): 2720–2764.

Cueto-Ginzo, I.A., Serrano, L., Sin, E., Rodríguez, R., Morales, J.G., Lade, S.B., Medina, V. and Achon, M.A. (2016). Exogenous salicylic acid treatment delays initial infection and counteracts alterations induced by maize dwarf mosaic virus in the maize proteome. *Physiol. Mol. Plant Pathol.*, 96: 47–59.

Deshpande, P., Dapkekar, A., Oak, M.D., Paknikar, K.M. and Rajwade, J.M. (2017). Zinc complexed chitosan/TPP nanoparticles: A promising micronutrient nanocarrier suited for foliar application. *Carbohydr. polym.*, 165: 394–401.

Dianat, M., Saharkhiz, M.J. and Tavassolian, I. (2016). Salicylic acid mitigates drought stress in *Lippia citriodora* L.: Effects on biochemical traits and essential oil yield. *Biocatal. Agric. Biotechnol.*, 8: 286–293.

Du, W.L., Niu, S.S., Xu, Y.L., Xu, Z.R. and Fan, C.L. (2009). Antibacterial activity of chitosan tripolyphosphate nanoparticles loaded with various metal ions. *Carbohydr. Polym.*, 75: 385–389.

Feyzioglu, G.C. and Tornuk, F. (2016). Development of chitosan nanoparticles loaded with summer savory (*Satureja hortensis* L.) essential oil for antimicrobial and antioxidant delivery applications. *LWT - Food Science and Technology*, 70: 104–110.

Fraceto, L.F., Grillo, R., de Medeiros, G.A., Scognamiglio, V., Rea, G. and Bartolucci, C. (2016). Nanotechnology in agriculture: Which innovation potential does it have? *Front. Environ. Sci.*, 4: 20.

Ghaouth, A.E., Arul, J., Grenier, J. and Asselin, A. (1992). Antifungal activity of chitosan on two post harvest pathogens of strawberry fruits. *Phytopathology*, 82: 398–402.

Ghormade, V., Deshpande, M.V. and Paknikar, K.M. (2011). Perspectives for nano-biotechnology enabled protection and nutrition of plants. *Biotechnol. Adv.*, 29(6): 792–803.

Grass, G., Rensing, C. and Solioz. M. (2011). Metallic copper as an antimicrobial surface. *Appl. Environ. Microbiol.*, 77(5): 1541–1547.

Hadrami, A.E.I., Adam, L.R., Hadrami, I.E. and Daayf, F. (2010). Chitosan in plant protection. *Mar. Drugs*, 8: 968–987.

Hadwiger, L.A. (2013). Multiple effects of chitosan on plant systems: Solid science or hype. *Plant Sci.*, 208: 42–49.

Hayat, S., Hasan, S.A., Fariduddin, Q. and Ahmad, A. (2008). Growth of tomato (*Lycopersicon esculentum*) in response to salicylic acid under water stress. *J. Plant Interact.*, 3: 297–304.

He, L., Liu, Y., Mustapha, A. and Lin. M. (2011). Antifungal activity of zinc oxide nanoparticles against *Botrytis cinerea* and *Penicillium expansum*. *Microbiol. Res.*, 166(3): 207–215.

Hewajulige, I.G.N., Sultanbawa, Y., Wijeratnam, R.S.W. and Wijesundara, R.L.C. (2009). Mode of action of chitosan coating on anthracnose disease control in papaya. *Phytoparasitica*, 37: 437–444.

Horvath, E., Janda, T., Szalai, G. and Paldi, E. (2002). *In vitro* salicylic acid inhibition of catalase activity in maize: Differences between the isozymes and a possible role in the induction of chilling tolerance. *Plant Sci.*, 163: 112–1135.

Horvath, E., Szalai, G. and Janda, T. (2007). Induction of abiotic stress tolerance by salicylic acid signaling. *J. Plant Growth Regul.*, 26: 290–300.

http://www.geohumus.com/us/products.html.

Hu, T., Hu, Z., Zeng, H., Qv, X. and Chen, G. (2015). Tomato lipoxygenase D involved in the biosynthesis of jasmonic acid and tolerance to abiotic and biotic stress in tomato. *Plant Biotechnol. Rep.*, 9(1): 37–45.

Jo, Y.K., Cromwell, W., Jeong, H.K., Thorkelson, J., Roh, J.H. and Shin, D.B. (2015). Use of silver nanoparticles for managing *Gibberella fujikuroi* on rice seedlings. *Crop Prot.*, 74: 65–69.

Katiyar, D., Hemantaranjan, A., Singh, B. and Bhanu, A.N. (2014). A future perspective in crop protection: chitosan and its oligosaccharides. *Adv. Plants Agric. Res.*, 1(1): 23–30.

Kean, T. and Thanou, M. (2010). Biodegradation, biodistribution, and toxicity of chitosan. *Adv. Drug Deliv. Rev.*, 62(1): 3–11.

Khan, M.I.R., Syeed, S., Nazar, R. and Anjum, N.A. (eds.). (2012). An insight into the role of salicylic acid and jasmonic acid in salt stress tolerance. pp. 277–300. *In*: *Phytohormones and Abiotic Stress Tolerance in Plants.* Berlin: Springer.

Khan, M.I.R. and Khan, N.A. (2014). Ethylene reverses photosynthetic inhibition by nickel and zinc in mustard through changes in PSII activity, photosynthetic nitrogen use efficiency, and antioxidant metabolism. *Protoplasma*, 251: 1007–1019.

Khan, N., Syeed, S., Masood, A., Nazar, R. and Iqbal, N. (2010). Application of salicylic acid increases contents of nutrients and antioxidative metabolism in mung bean and alleviates adverse effects of salinity stress. *Int. J. Plant Biol.*, 1(1): e1–e1.

Kheiri, A., Jorf, S.A.M., Malihipour, A., Saremi, H. and Nikkhah, M. (2016). Application of chitosan and chitosan nanoparticles for the control of Fusarium head blight of wheat (*Fusarium graminearum*) *in vitro* and green house. *Int. J. Biol. Macromol.*, 93: 1261–1272.

Khokhar, M.K., Sharma, S.S. and Gupta, R. (2014). Influence of sowing dates on incidence and severity of post flowering stalk rot of maize caused by Fusarium verticillioides. *J. Mycol. Plant Pathol.*, 44(2): 205–208.

Khurana, J.P. and Cleland, C.F. (1992). Role of salicylic acid and benzoid acid in flowering of a photoperiod-insensitive strain, *Lemna paucicostata* LP6. *Plant Physiol.*, 100: 1541–1546.

Kulkarni, S. and Anahosur. K.H. (2011). Integrated management of dry stalk rot disease of maize. *J. Plant Dis. Sci.* 6(2): 99–106.

Kumaraswamy, R.V., Kumari, S., Choudhary, R.C., Sharma, S.S., Pal, A., Raliya, R., Biswas, P. and Saharan, V. (2019). Salicylic acid functionalized chitosan nanoparticle: A sustainable biostimulant for plant. *Int. J. Biol. Macromol.*, 123: 59–69.

Lal, S. and Singh, I.S. (1984). Breeding for resistance to downy mildews and stalk rots in maize. *Theor. Appl. Genetics*, 69 (2): 111–119.

Ledencan, T., Simic, D., Brkic, I., Jambrovic, A. and Zdunic, Z. (2003). Resistance of maize inbreds and their hybrids to Fusarium stalk rot. *Czech J. Genet. Plant Breed., UZPI (Czech Republic)* 39: 15–20.

Lee, T.T. and Skoog, F. (1965). Effects of substituted phenols on bud formation and growth of tobacco tissue cultures. *Physiol. Plantarum*, 18(2): 386–402.

Li, Y., Yin, H., Wang, Q., Zhao, X.M., Du, Y.G. and Li, F.L. (2009). Oligochitosan induced *Brassica napus* L. production of NO and H_2O_2 and their physiological function. *Carbohydr. Polym.*, 75: 612–617.

Liang, W., Yu, A., Wang, G., Zheng, F., Hu, P., Jia, J. and Xu, H. (2018). A novel water-based chitosan-La pesticide nanocarrier enhancing defense responses in rice (*Oryza sativa* L.) growth. *Carbohydr. Polym.*, 199: 437–444.

Lin, W., Hu, X., Zhang, W., Rogers, W.J. and Cai, W. (2005). Hydrogen peroxide mediates defence responses induced by chitosan of different molecular weights in rice. *J. Plant Physiol.*, 162: 937–944.

Manikandan, A. and Sathiyabama, M. (2016). Preparation of chitosan nanoparticles and its effect on detached rice leaves infected with *Pyricularia grisea. Int. J. Biol. Macromol.*, 84: 58–61.

Manukumar, H.M., Umesha, S. and NaveenKumar, H.N.N. (2017). Promising biocidal activity of thymol loaded chitosan silver nanoparticles (T-C@AgNPs) as anti-infective agents against perilous pathogens. *Int. J. Biol. Macromol.*, 102: 1257–1265.

McMurray, T.A., Dunlop, P.S.M. and Byrne, J.A. (2006). The photocatalytic degradation of atrazine on nanoparticulate TiO_2 films. *J. Photochem. Photobiol.*, 182(1): 43–51.

Milani, N., Hettiarachchi, G.M., Kirby, J.K., Beak, D.G., Stacey, S.P. and McLaughlin, M.J. (2015). Fate of zinc oxide nanoparticles coated onto macronutrient fertilizers in an alkaline calcareous soil. *PLoS One*, 10(5): 1–16.

Oliveira, H.C., Gomes, B.C.R., Pelegrino, M.T. and Seabra, A.B. (2016). Nitric oxide-releasing chitosan nanoparticles alleviate the effects of salt stress in maize plants. *Nitric Oxide*, 61: 10–19.

Pancheva, T.V., Popova, L.P. and Uzunova, A.N. (1996). Effects of salicylic acid on growth and photosynthesis in barley plants. *J. plant Physiol.*, 149: 57–63.

Parisi, C., Vigani, M. and RCerezo, E. (2015). Agricultural nanotechnologies: What are the current possibilities? *Nano Today*, 10(2): 124–127.

Parry, D.W., Jenkinson, P. and McLeod, L. (1995). Fusarium ear blight (scab) in small grain cereals: A review. *Plant Pathol.*, 44(2): 207–238.

Pereira, A.E.S., Silva, P.M., Oliveira, J.L., Oliveirac, H.C. and Fraceto, L.F. (2017). Chitosan nanoparticles as carrier systems for the plant growth hormone gibberellic acid. *Colloids Surf. B.*, 150: 141–152.

Photchanachai, S., Singkaew, J. and Thamthong, J. (2006). Effects of chitosan seed treatment on *Colletotrichum* sp. and seedling growth of chili cv. jinda. pp. 585–590. *In*: *IV International Conference on Managing Quality in Chains*: *The Integrated View on Fruits and Vegetables Quality 712*. Bangkok, Thailand: IIEE.

Pichyangkura, R. and Chadchawan, S. (2015). Biostimulant activity of chitosan in horticulture. *Sci. Hortic.*, 196: 49–65.

Pimentel, D. and Burgess, M. (2014). Pesticides applied worldwide to combat pests. pp. 1–12. *In*: *Integrated Pest Management*. Springer.

Prapagdee, B., Kotchadat, K., Kumsopa, A. and Visarathanonth, N. (2007). The role of chitosan in protection of soybean from sudden death syndrome caused by *Fusarium solani* f. sp. *glycines*. *Bioresour. Technol.*, 98: 1353–1358.

Prasad, R., Pandey, R. and Barman, I. (2016). Engineering tailored nanoparticles with microbes: Quovadis? *Wiley Interdisciplinary Reviews*: *Nanomed. Nanobiotechnol.*, 8: 316–330.

Rabea, E.I., Badawy, M.E.T., Stevens, C.V., Smagghe, C.V. and Steurbaut, W. (2003). Chitosan as antimicrobial agent: applications and mode of action. *Biomacromol.*, 4(6): 1457–1465.

Reglinski, T., Elmer, P.A.G., Taylor, J.T., Wood, P.N. and Hoyte, S.M. (2010). Inhibition of Botrytis cinerea growth and suppression of botrytis bunch rot in grapes using chitosan. *Plant Pathol.*, 59(5): 882–890.

Saharan, V., Mehrotra, A., Khatik, R., Rawal, P., Sharma, S.S. and Pal, A. (2013). Synthesis of chitosan-based nanoparticles and their *in vitro* evaluation against phytopathogenic fungi. *Int. J. Biol. Macromol.*, 62: 677–683.

Saharan, V., Sharma, G., Yadav, M., Choudhary, M.K., Sharma, S.S., Pal, A., Raliya, R. and Biswas, P. (2015). Synthesis and *in vitro* antifungal efficacy of Cu–chitosan nanoparticles against pathogenic fungi of tomato. *Int. J. Biol. Macromol.*, 75: 346–353.

Saharan, V. and Pal, A. (2016). Chitosan-based nanomaterials in plant growth and protection. Springer.

Saharan, V., Kumaraswamy, R.V., Choudhary, R.C., Kumari, S., Pal, A., Raliya, R. and Biswas, P. (2016). Cu-chitosan nanoparticle mediated sustainable approach to enhance seedling growth in maize by mobilizing reserved food. *J. Agricu. Food Chem.*, 64(31): 6148–6155.

Sathiyabama, M. and Balasubramanian, R. (1998). Chitosan induces resistance components in Arachis hypogaea against leaf rust caused by *Puccinia arachidis* Speg. *Crop Prot.*, 17(4): 307–313.

Sathiyabama, M. and Charles, R.E. (2015). Fungal cell wall polymer-based nanoparticles in protection of tomato plants from wilt disease caused by *Fusarium oxysporum* f. sp. lycopersici. *Carbohydr. Polym.*, 133: 400–407.

Sathiyabama, M. and Parthasarathy, R. (2016). Biological preparation of chitosan nanoparticles and its *in vitro* antifungal efficacy against some phytopathogenic fungi. *Carbohydr. Polym.*, 151: 321–325.

Sathiyabama, M. and Manikandan, A. (2016). Chitosan nanoparticle induced defense responses in fingermillet plants against blast disease caused by *Pyricularia grisea* (Cke.) Sacc. *Carbohydr. Polym.*, 154: 241–246.

Sharma, G., Kumar, A., Devi, K.A., Prajapati, D., Bhagat, D., Pal, A., Raliya, R., Biswas, P. and Saharan, V. (2020). Chitosan nanofertilizer to foster source activity in maize. *Int. J. Biol. Macromol.*, 145: 226–234.

Sharma, S. (2017). Enhanced antibacterial efficacy of silver nanoparticles immobilized in a chitosan nanocarrier. *Int. J. Biol. Macromol.*, 104: 1740–1745.

Sharp, R.G. (2013). A review of the applications of chitin and its derivatives in agriculture to modify plant-microbial interactions and improve crop yields. *Agronomy*, 3(4): 757–793.

Shekhar, M. and Kumar, S. (2010). Potential biocontrol agents for the management of Macrophomina phaseolina, incitant of charcoal rot in maize. *Arch. Phytopathol. Plant Prot.*, 43(4): 379–383.

Shukla, S.K., Mishra, A.K., Arotiba, O.A. and Mamba, B.B. (2013). Chitosan-based nanomaterials: A state-of-the-art review. *Int. J. Biol. Macromol.*, 59: 46–58.

Singh, N., Rajendran, A., Shekhar, M. and Mittal, G. (2012). Biochemical response and host-pathogen relation of stalk rot fungi in early stages of maize (*Zea mays* L.). *Afr. J. Biotechnol.*, 11(82): 14837–14843.

Singh, S., Singh, M., Agrawal, V.V. and Kumar, A. (2010). An attempt to develop surface plasmon resonance based immunosensor for Karnal bunt (*Tilletia indica*) diagnosis based on the experience of nano-gold based lateral flow immuno-dipstick test. *Thin Solid Films*, 519(3): 1156–1159.

Sobowale, A.A, Cardwell, K.F., Odebode, A.C., Bandyopadhyay, R. and Jonathan, S.G. (2005). Growth inhibition of *Fusarium verticillioides* (Sacc.) Nirenberg by isolates of *Trichoderma pseudokoningii* strains from maize plant parts and its rhizosphere. *J. Plant Prot. Res.*, 45(4): 249–266.

Surendran, A., Siddiqui, Y., Manickam, S. and Ali, A. (2018). Role of benzoic and salicylic acids in the immunization of oil palm seedlings challenged by Ganoderma boninense. *Ind. Crops Prod.*, 122: 358–365.

Torney, F., Trewyn, B.G., Lin, V.S.Y. and Wang, K. (2007). Mesoporous silica nanoparticles deliver DNA and chemicals into plants. *Nature Nanotechnol.*, 2(5): 295–300.

Vamvakaki, V. and Chaniotakis, N.A. (2007). Pesticide detection with a liposome-based nano-biosensor. *Biosens. Bioelectron.*, 22(12): 2848–2853.

Vicente, M.R.S. and Plasencia, J. (2011). Salicylic acid beyond defence: Its role in plant growth and development. *J. Exp. Bot.*, 62: 3321–3338.

Wang, M., Chen, Y., Zhang, R., Wang, W., Zhao, X., Du, Y. and Yin, H. (2015). Effects of chitosan oligosaccharides on the yield components and production quality of different wheat cultivars (*Triticum aestivum* L.) in Northwest China. *Field Crops, Res.*, 172: 11–20.

Wani, I.A. and Ahmad, T. (2013). Size and shape dependant antifungal activity of gold nanoparticles: A case study of Candida. *Colloids Surf.*, 101: 162–170.

Wojdyla., A.T. (2004). Chitosan in the control of some ornamental foliage diseases. *Commun. Agric. Appl. Biol. Sci.*, 69: 705–715.

Xing, K., Xiao, Z., Peng, X. and Qin, X. (2015). Chitosan antimicrobial and eliciting properties for pest control in agriculture: A review. *Agron. Sustain. Dev.*, 35(2): 569–588.

Zhang, H., Zhao, X., Yang, J., Yin, H., Wang, W., Lu, H. and Du, Y. (2011). Nitric oxide production and its functional link with OIPK in tobacco defense response elicited by chitooligosaccharide. *Plant Cell Rep.*, 30: 1153–1162.

Zhang, Y., Shi, X., Li, B., Zhang, Q., Liang, W. and Wang, C. (2016). Salicylic acid confers enhanced resistance to *Glomerella* leaf spot in apple. *Plant Physiol. Biochem.*, 106: 64–72.

Zhao, X.M., She, X.P., Yu, W., Liang, X.M. and Du, Y.G. (2007). Effects of oligochitosans on tobacco cells and role of endogenous nitric oxide burst in the resistance of tobaccoto tobacco mosaic virus. *J. Plant Pathol.*, 89: 55–65.

Chapter 8

Nanophytovirology Approach to Combat Plant Viral Diseases

Sanjana Varma,[1,2] *Neha Jaiswal,*[1,2] *Niraj Vyawahare,*[3] *Anil T Pawar,*[4] *Rashmi S Tupe,*[5] *Varsha Wankhade,*[6] *Koteswara Rao Vamkudoth*[1,2] and *Bhushan P Chaudhari*[1,2,*]

1. Introduction

It has been estimated that by 2050, global population growth will exceed 9 billion people. Globally, one of the major challenges facing humanity is food security. The various abiotic, like climate change, and biotic factors (pests) are the reasons for the occurrence of challenges in food security (Elmer and White, 2018; Farooq et al., 2021). Plant pathogens are the major contributing factor to the food security challenge because they are responsible for substantial quantitative and qualitative loss of the crops (Farooq et al., 2021). These phytopathogens are accountable for about 10–40% losses in the quality and yield of food crops and horticultural produce. The major causative agents for plant diseases are viruses, fungi, bacteria, and nematodes (Rajwade et al., 2020). Plant viruses or phytoviruses are responsible for the austere negative effect on agriculture due to their rapid mutation, genetic diversity, and shortage of their controlling ways (Farooq et al., 2021). One evaluation says that around 50% of emerging plant diseases were due to viruses (Vargas-Hernandez et al., 2020). Phytoviruses are the major threat to crops, globally, and are responsible

[1] Biochemical Sciences Division, CSIR-National Chemical Laboratory, Pashan, Pune-411008, India.
[2] Academy of Scientific and Innovative Research (AcSIR), Ghaziabad-201002, India.
[3] Dr. D. Y. Patil College of Pharmacy, Akurdi, Pune-411044, India.
[4] School of Pharmacy, Dr. Vishwanath Karad MIT World Peace University, Pune-411038, India.
[5] Symbiosis School of Biological Sciences (SSBS), Symbiosis International (Deemed University) (SIU), Lavale, Pune-412115, India.
[6] Department of Zoology, Savitribai Phule Pune University, Pune-411007, India.
* Corresponding author: bp.chaudhari@ncl.res.in

for an annual economic effect of over $30 billion (Farooq et al., 2021). It has been also reported that greater than 700 crop species are infected by around 900 species of phytoviruses. Around 20% loss of cultivation worldwide is due to viruses. Various factors like viral concentration, stage of infection, physiological factors, type of host, time of replication and strain of virus decide the infectivity of plant viral infections. Commonly visible symptoms of viral infections in plants are dry leaves, mottling, necrosis, malformation of fruits, plant stunting, also death (Vargas-Hernandez et al., 2020). Phytoviruses comprise either simple DNA (deoxyribonucleic acid) or RNA (ribonucleic acid) as their genetic material. Plant viruses transmission mainly occurs through vectors, i.e., insect pests and also through human activities, although non-vector routes of transmission also exist. The attributes like genomic diversity, great potential to adapt to the environment, transmission through various vectors, rapid evolution, and dynamic genetic structure owes to the difficulty of their control among all the plant disease causing pathogens (Farooq et al., 2021). The major economically and scientifically significant top 10 plant viruses are tobacco mosaic virus (TMV), tomato spotted wilt virus, tomato yellow leaf curl virus, cucumber mosaic virus, potato virus Y, cauliflower mosaic virus, African cassavamosaic virus, Plum pox virus, Bromemosaic virus, and Potato virus X (Khater et al., 2017; Rajwade et al., 2020; Tortella et al., 2021).

The conventional ways to combat plant viral infections are not effective and come with many drawbacks. The common conventional practices are the use of chemical pesticides and other relevant chemicals, thermotherapy, crop rotation (Tortella et al., 2021; Rajwade et al., 2020), executing good farming practices and proper disposal of infected plant parts to avoid transmission of infection (Rajwade et al., 2020). Apart from these traditional ways, the cultivation of genetically modified diseases resistant plant varieties can also help to combat plant diseases (Rajwade et al., 2020; Tortella et al., 2021). Genetically modified plants are not cultivated widely due to their unacceptance by the consumer and certain risks involved with them (Rajwade et al., 2020). Therefore, a novel way to combat plant diseases is the need of the hour. The control of viral infection is difficult as compared to infections caused by other pathogens due to the dependence of viruses on host cell machinery and these are detected after the appearance of symptoms (Tortella et al., 2021).

The use of nanoparticles (NPs) to control or cure plant viral diseases is gaining interest day by day. This has been also termed nanophytovirology. Nanophytovirology is the emerging field that involves the use of various engineered NPs and non-engineered NPs to combat plant viral diseases. Although there is very little research exploring the field to date (Farooq et al., 2021). The unique physiochemical properties like the high surface area to volume ratio, great reactivity, ease of functionalization, and small size in the range of 1–100 nm make them to interact with phytoviruses, host plants, and viral vectors in specific and unique ways which are useful for the treatment of plant viral infections (Farooq et al., 2021). Different types of nanomaterials have a role in nanophytovirology like metallic NPs, metal oxide NPs, organic NPs, etc., of varied shapes and sizes (Vargas-Hernandez et al., 2020). Nanomaterials are advantageous over conventional plant disease management ways due to their properties which lead to enhanced efficiency, reduce environmental toxicity, and lower inputs (Farooq et al., 2021). The different strategies by which NPs can be used to control plant diseases are as antimicrobial agents that can be directly applied to seeds, foliage, or roots (Worrall et al., 2018), as nanocarrier mainly to carry pesticides, other antimicrobial chemicals (Farooq et al., 2021) and genetic materials (Elmer and White, 2018) and also helps in their controlled and sustainable release, as a biostimulant which induces plant immunity, as a nanofertilizers (Farooq et al., 2021; Duhan et al., 2017) and as a biosensing probe for early detection of plant infections before the appearance of symptoms (Duhan et al., 2017; Rai and Ingle, 2012). Nanotechnology is the field that provides solutions to various human, animal, environmental, and agriculture-related problems. This chapter will further illustrate the role of nanotechnology in controlling plant viral infections.

2. General about Nanotechnology

The concept of nanotechnology was seeded by Richard Feynman, a physicist (Misra et al., 2013). 'Nano' is the Greek word meaning dwarf and nano is also considered as one billionth part of the meter (10^{-9}) (Rai and Ingle, 2012). Nanotechnology is a multidisciplinary field that involves physics, chemistry, biology, material science, medicine, agriculture, and pharmaceutical sciences and deals with materials that have one or more dimensions in the size range of 1–100 nm (Duhan et al., 2017; Elmer and White, 2018). Due to their size, nanomaterials has unique properties from their bulk material (Rai and Ingle, 2012; Elmer and White, 2018; Worrall et al., 2018). NPs consist of properties like high surface area to volume ratio, high reactivity, tunability, and flexibility of coating, etc. (Elmer and White, 2018; Farooq et al., 2021). Nanostructures can be zero dimensional (e.g., NPs), one-dimensional (e.g., nanowires), two-dimensional (e.g., nanofilms), or three-dimensional (e.g., arrays, hierarchical structures) (Misra et al., 2013). Naturally occurring NPs are oceanic salt spray, volcanic dust, many viral and viroid particles, but except viral particles other naturally present NPs have irregularity in shape and size. On the other hand, engineered NPs have uniform shape and size like spherical, rods, sheets, multiwalled tubes, balls, and bifurcating tree-like structures (Elmer and White, 2018). Different types of NPs are inorganic like metallic (silver, platinum, gold), metal oxide NPs (ferric oxide, zinc oxide, aluminium oxide, silica oxide, etc.); organic-like liposomes, chitosan, polymeric NPs and carbon-based like carbon nanotubes (single-walled and multiwalled), graphene, etc. (Hajong et al., 2019).

Nanoparticles are synthesized by three methods which are chemical, physical, and biological. These synthesis methods are categorized into two approaches that are top–down approaches and bottom–up approaches. Top–down means synthesizing nanoscale structures from the large-size structure while the bottom–up approaches involve altering individual atoms and molecules to form nanostructures (Banerjee et al., 2021; Hajong et al., 2019). The various bottom–up approaches are the green synthesis of NPs using microorganisms and plants, laser pyrolysis, atomic condensation, and chemical vapour deposition whereas top–down approaches include etching, laser ablation, sputtering, mechanical milling, and electro explosion (Banerjee et al., 2021). Green synthesis of NPs by using plants and microorganisms is gaining more attention due to its cost-effective, eco-friendly, and simple nature (Banerjee et al., 2021; Hajong et al., 2019), contrararily, physical and chemical methods are expensive and have harmful effects on the environment and human health (Hajong et al., 2019). The synthesis of desirable size and properties of nanomaterials can be achieved by tuning the reaction conditions like temperature, pH, time, concentrations of chemicals, etc.

After synthesis, nanomaterials can be characterized for various properties by employing high-end characterizations techniques. Techniques like transmission electron microscopy (TEM), scanning transmission X-Ray microscopy (STXM), atomic force microscopy (AFM), scanning electron microscopy (SEM), and X-Ray fluorescence microscopy (XFM) reveal information about the morphology and size of nanomaterials. The electrostatic stability and hydrodynamic radius of NPs can be determined by zeta potential analysis and dynamic light scattering (DLS) techniques, respectively. Nuclear magnetic resonance (NMR) helps in gaining information on structural attributes of NPs while optical properties and stability of NPs in colloidal suspension can be studied by UV-visible spectroscopy. X-ray diffraction (XRD) analysis gives an idea about the crystallinity of NPs. In view of the increasing interest of researchers in nanotechnology-based applications, other novel, cost-effective, and simple techniques should be developed for the characterization of nanomaterials (Banerjee et al., 2021).

3. Limitations of Conventional Methods to Combat Plant Viral Diseases

The commonly employed conventional way to control plant disease is the use of chemical pesticides (Ahsan and Yuanhua, 2021). However, the toxic nature of chemical pesticides is very harmful to humans, animals, and the environment (Rai and Ingle, 2012; Mokrini and Bouharroud, 2019; Ahsan and Yuanhua, 2021). Excessive use of chemical pesticides is paving way for the development of pest and plant pathogen resistance (Mokrini and Bouharroud, 2019) which will continue the demand for new products repeatedly (Ahsan and Yuanhua, 2021) and also causes biomagnification of pesticides, loss of natural habitat of useful animals in farming and pollinator decline (Duhan et al., 2017). In addition to these, chemical pesticides kill useful microbes leading to loss of soil fertility (Duhan et al., 2017; Rai and Ingle, 2012) and causing ecological imbalance (Ahsan and Yuanhua, 2021). Chemical antivirals that are used to treat plant viral infections at issue culture levels are 2-thiouracil azidothymidine, acycloguanosine (Acyclovir), 5-azacytidine (Zidovudine), 2, 4-dioxohexahydro-2, 5-triazine (DHT), etc. The drawbacks associated with these antivirals are that these are not that selectively useful for crop viral disease, have a different mode of action, and their *ex vitro* application is ineffective (Vargas-Hernandez et al., 2020). Another way to control plant diseases is their detection by using laboratory techniques such as PCR (Polymerase Chain Reaction), ELISA (Enzyme-Linked Immunosorbent Assay), microscopic techniques, and other cultural ways. However, these techniques for detecting the causing agents of various plant diseases are time consuming, require skilled persons, are complex in sample handling, and cannot be applied in the field directly at the site of infection. Further, the reagents involved in the above techniques are expensive (Hong and Lee, 2018; Hajong et al., 2019; Mokrini and Bouharroud, 2019; Rajwade et al., 2020). Further, the use of genetically modified disease-resistant varieties of plants and crops is one of the promising ways to combat plant diseases. However, there are many limitations associated with the use of genetically modified crops. Firstly, this is time-consuming and requires skilled individuals; secondly, only plant populations which have abundant genetic variation can be modified; thirdly, undesirable traits can be introduced with the desirable one; and lastly, this way is restricted to some plants only that can mate with each other (Ahsan and Yuanhua, 2021). Also this way leads to the development of toxicity; emergence on new resistant viral strains can occur and this is expensive (Akhtar et al., 2020). Another risk with the usage of genetically modified disease-resistant plants is the unacceptance of the consumer (Rajwade et al., 2020). Employing nanotechnology is the efficient alternative to all these conventional ways of combating plant viral diseases due to the unique and significant properties of NPs. Also, the use of nanomaterials has the potential to overcome the limitations of conventional techniques.

4. Nanomaterials to Combat Plant Viral Diseases

Owing to properties like high stability, low cost, ability to incorporate hydrophobic and hydrophilic substances, renewable nature, and secure handing (Vargas-Hernandez et al., 2020), different kinds of NPs can be used per se as an antiviral: (1) as a nanocarrier to carry pesticides and other antiviral agrochemicals, (2) as a biosensing material for sensitive and early detection of phytoviruses so that they can be treated before their vast spread, and (3) as a biostimulant to enhance plant immunity. These various roles of nanomaterials for controlling plant viral diseases are illustrated further. Figure 8.1 gives a general idea about the different roles of nanomaterials in the field nanophytovirology.

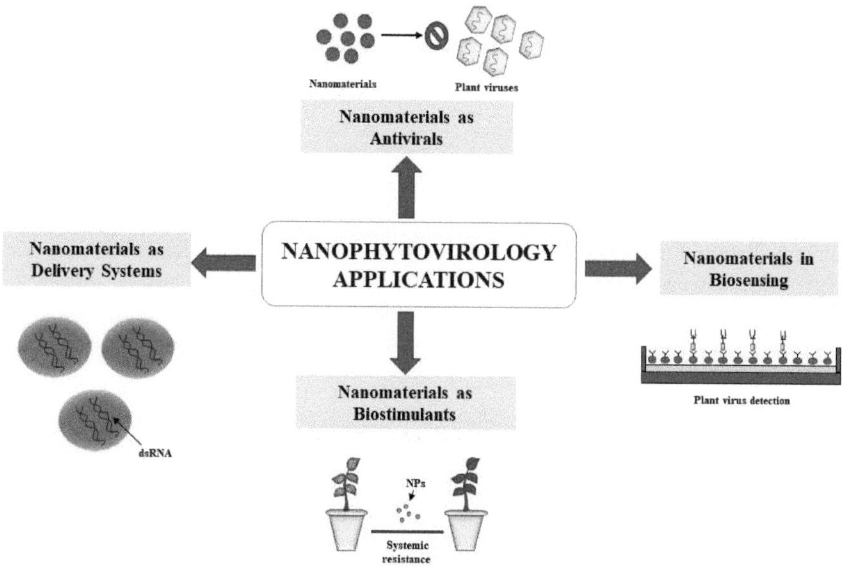

Fig. 8.1. Different applications of nanomaterials in the field nanophytovirology.

4.1 Nanomaterials as Antivirals

Nanomaterials exhibit efficient antimicrobial activity as compared to their bulk counterparts due to their high surface area to volume ratio which helps them to interact with microbes (Rajwade et al., 2020). Mainly metallic and metal oxides NPs had been studied for their antiviral potential against different types of plant viruses

(Rajwade et al., 2020; Vargas-Hernandez et al., 2020). However, carbon-based NPs were also explored for their antiviral potential (Adeel et al., 2021). Looking into the difficulty to control phytoviruses, this novel strategy of using NPs as antivirals will help in food security. NPs per se can efficiently inhibit various phytoviruses. NPs also inhibit viruses indirectly by inducing plant defence mechanisms in different ways. In a stress situation the plant responds by increasing reactive oxygen species (ROS) levels which restricts the pathogen entry and spread and also generates local and systemic defence responses like pathogenesis-related genes activation. NPs application leads to the induction of the antioxidant system and repression of oxidative stress shown in various reports (Vargas-Hernandez et al., 2020). Tobacco leaves treated with iron oxide NPs had shown production of ROS which might be the reason for the development of resistance against the virus in the tobacco plant (Cai et al., 2020). Silica oxide NPs, when applied on cucumber plat to treat papaya ringspot virus (PRSV) infection, induced the expression of *pox* and *pal* gene (Elsharkawy and Mousa, 2015). Similarly, the activity of antioxidant enzymes polyphenol oxidase (PPO) and peroxide (POD) increased when silver NPs were used in the tomato plant to treat tomato mosaic virus and potato virus Y (Noha et al., 2018). Cucumber plants with PRSV infection, when treated with silica oxide NPs, showed induction of the expression of pathogenesis related gene PR1; this gene helps in plant immunity (Elsharkawy and Mousa, 2015). Another way of induction of plant defences so that it can develop resistance against pathogens are by plant hormones; the signalling via hormones regulates the defence mechanism and plant growth. Various plant hormones have the ability to trigger appropriate defence mechanisms in plants and NPs application influences the up or down regulation of the plant hormones leading to development of their defence mechanism. This depends on the dose of NPs used as well as the time of application (Vargas-Hernandez et al., 2020). Zinc oxide NPs have been reported for upregulating plant stress hormone salicylic acid (SA) and abscisic acid (ABA) (Vargas-Hernandez et al., 2020). Tobacco plant infected with the turnip mosaic virus, when treated with iron oxide and titanium dioxide NPs, showed an impact on the levels of various phytohormones (Hao et al., 2018). Furthermore, infected tomato plants when treated with silver NPs had shown an increase in the proline content (Noha et al., 2018). Also, plant secondary metabolites help in their defence mechanisms. Table 8.1 gives a more detailed idea about NPs that have been explored for their antiviral activity against phytoviruses.

4.1.1 Silver Nanoparticles (AgNPs)

Silver NPs show a broad-spectrum antimicrobial potential due to which it is gaining more interest in plant infection management (Hajong et al., 2019; Rajwade et al., 2020; Banerjee et al., 2021). AgNPs were the first NPs to be used in agriculture to control various plant pathogens (Elmer and White, 2018; Banerjee et al., 2021). AgNPs were most widely studied for their antiviral effects against phytoviruses. They can inactivate the microbes of both prokaryotic and eukaryotic organisms (Ramezani et al., 2019). AgNPs showed antiviral properties against the Sunhemp rosette virus which had infected *Cymopsis tetragonaloba* (guar beans). When biosynthesized AgNPs of size 15 nm were sprayed on the leaves of guar beans in concentration 50 mg L^{-1}, it was observed that after 3–4 days AgNPs treated plant did not show lesions while the untreated ones had shown lesions. This infers that treatment with AgNPs led

Table 8.1. Different types of nanomaterials used to combat plant viral diseases.

Sr. No.	Nanoparticles	Virus to be Treated/Detected	Experimental Plant/Host Plant	Applications	References
1	Silicon nanowires	*Cucumber Mosaic Virus* (*CMV*), *Papaya Ring Spot Virus* (*PRSV*)	Cucumber and Papaya	Biosensor	(Ariffin et al., 2014)
2	Gold nanoparticles	*Plum pox virus*	Plum (*Prunus domestica*) & Tobacco (*Nicotiana benthamiana*) leaves	Biosensor	(Jarocka et al., 2011)
3	Gold nanoparticles (AuNPs)	*Potato virus X* (*PVX*)	Potato leaf tissue and sprout	Molecular diagnostic	(Drygin et al., 2012)
4	Gold nanoparticles (AuNPs)	*Tomato yellow leaf curl virus* (*TYLCV*)	Tomato	Biosensor	(Razmi et al., 2019)
5	Carbon nanotubes & zinc nnaocomposite (MWCNTs-ZnNPs)	*Cognate begomovirus*	Chili leaf curl betasatellite (ChLCB)	Biosensor	(Tahir et al., 2017)
6	Gold nanoparticles (GNP)	*Grapevine leafroll-associated virus 3* (*GLRaV-3*)	Grapes	Biosensor	(Byzova et al., 2018)
7	Gold nanoparticles (GNPs)	*Potato virus Y (PVY)*	Potato	Biosensor	(Razo et al., 2018)
8	Gold nanorods (AuNRs)	*Cymbidium mosaic virus (CymMV) or Odontoglossum ringspot virus* (*ORSV*)	Orchid	Biosensor	(Lin et al., 2014)
9	Silver nanoparticles (AgNPs)	*Tomato mosaic virus* (*ToMV*) *and Potato virus Y (PVY)*	Tomato	Treatment	(Noha et al., 2018)
10	Silver nanoparticles (AgNPs)	*Tomato spotted wilt virus (TSWV)*	Potato plants	Treatment	(Shafie et al., 2018)
11	Gold nanoparticle (AuNP)	*Citrus tristeza virus* (CTV)	Citrus	Biosensor	(Khater et al,. 2019b)
12	Graphene-based silver nanocomposites (GO-AgNps)	*Tomato Bushy Stunt Virus* (*TBSV*)	*Capsicum annuum* (pepper)	Treatment	(Elazzazy et al., 2017)

Table 8.1 contd. ...

...Table 8.1 contd.

Sr. No.	Nanoparticles	Virus to be Treated/Detected	Experimental Plant/Host Plant	Applications	References
13	Gold nanoparticles	*Horsegram yellow mosaic viruses (HgYMV)*	*Nicotiana benthamiana*	Biosensor	(Abubakar et al., 2018)
14	Gold nanoparticles (AuNPs)	*Barley yellow dwarf virus-PAV (BYDV-PAV)*	Barley (*Hordeum vulgare*)	Treatment/ Biostimulant	(Alkubaisi and Aref, 2017)
15	Gold nanoparticles (AuNPs)	*Banana bunchy top virus (BBTV)*	Banana	Biosensor	(Wei et al., 2014)
16	Cadmium-telluride quantum dots (QDs) & Carbon nanoparticles (CNPs)	*Citrus tristeza virus (CTV)*	Citrus	Biosensor	(Shojaei et al., 2016a)
17	Gold nanoparticles (AuNPs)	*Citrus tristeza virus (CTV)*	Citrus	Biosensor	(Khater et al., 2019a)
18	Zinc oxide nanoparticles (ZnONPs) and Silica nanoparticles (SiO_2NPs)	*Tobacco mosaic virus (TMV)*	*Nicotiana benthamiana*	Treatment/ Biostimulant	(Cai et al., 2019)
19	Titanium dioxide nanostructure (TDNS)	*Broad bean stain virus (BBSV)*	Faba bean plants	Treatment	(Elsharkawy and Derbalah, 2019)
20	Cadmium-telluride quantum dots (QDs) & Gold nanoparticles (AuNPs)	*Citrus tristeza virus (CTV)*	Citrus	Biosensor	(Shojaei et al., 2016b)
21	Silica (SiO_2) nanoparticles	*Papaya ring spot virus (PRSV)*	Cucumber	Treatment/ Biostimulant	(Elsharkawy and Mousa, 2015)
22	Nickel oxide nanostructures (NONS)	*Cucumber mosaic virus (CMV)*	Cucumber	Treatment/ Biostimulant	(Derbalah and Elsharkawy, 2019)
23	Carbon nanotubes (CNTs) & C_{60} fullerenes	*Tobacco Mosaic Virus (TMV)*	*Nicotiana benthamiana*	Treatment/ Biostimulant	(Adeel et al., 2021)
24	Gold nanoparticles (AuNPs)	*Tomato leaf curl New Delhi virus (ToLCNDV)*	Tomato	Biosensor	(Dharanivasan et al., 2019)

Table 8.1 contd. ...

...Table 8.1 contd.

Sr. No.	Nanoparticles	Virus to be Treated/Detected	Experimental Plant/Host Plant	Applications	References
25	Gold nanoparticles (AuNPs)	*Potato virus X*	Potato	Biosensor	(Panferov et al., 2018)
26	Gold nanoparticles (AuNPs)	*Tomato leaf curl New Delhi virus (ToLCNDV)*	Tomato	Biosensor	(Dharanivasan et al., 2016)
27	Carbon nanotube-based copper nanoparticle composite (MWCNTs-Cu NPs)	*Begomovirus (CLCuKoV-Bur)*	Cotton leaves	Biosensor	(Tahir et al., 2018)
28	Fe_3O_4 nanoparticles	*Tobacco mosaic virus (TMV)*	*Nicotiana benthamiana*	Treatment/ Biostimulant	(Cai et al., 2020)
29	Silver nanoparticles	*Sunhemp rosette virus (SHRV)*	Cluster bean (*Cymopsis tetragonaloba*)	Treatment	(Jain and Kothari, 2014)
30	Fe_2O_3, TiO_2, MWCNTs & C_{60}	*Turnip mosaic virus*	Tobacco (*Nicotiana benthamiana*)	Treatment/ Biostimulant	(Hao et al., 2018)
31	Silver nanoparticles	*Potato virus Y (PVY)*	Potato	Treatment/ Biostimulant	(El-Shazly et al., 2017)
32	Zinc nanoparticles (ZnNPs)	*Cucumber mosaic virus (CMV)*	Eggplant	Treatment	(El-Sawy et al., 2017b)
33	Cerium dioxide nanoparticles	*Tobacco mosaic virus (TMV)*	Tobacco	Treatment	(Eugene and Zholobak, 2016)
34	Silica nanoparticles	*Tomato Yellow Leaf Curl Virus (TYLCV)*	Tomato	Treatment/ Bio-stimulant	(El-Sawy et al., 2017a)
35	Silver nanoparticles (AgNPs)	*Tobacco mosaic virus (TMV)*	Tobacco	Treatment	(Wang et al., 2016)
36	Silver nanoparticles (AgNPs)	*Bean Yellow Mosaic Virus (BYMV)*	Fava bean crops	Treatment/ Bio-stimulant	(Elbeshehy et al., 2015)
37	Gold nanoparticles (AuNPs)	*Barley yellow mosaic virus (BaYMV)*	Barley	Treatment	(Aref et al., 2012)
38	Layered double hydroxide (LDH) clay nanosheets	*PMMoV and CMV strain 207.*	Extracted from *E. coli* HT115 expressing CMV2b-dsRNA	Treatment/ dsRNA Carrier	(Mitter et al., 2017)

Table 8.1 contd. ...

...Table 8.1 contd.

Sr. No.	Nanoparticles	Virus to be Treated/Detected	Experimental Plant/Host Plant	Applications	References
39	Layered double hydroxide nanoparticles (BioClay)	*Bean common mosaic virus (BCMVB)*	*Nicotiana benthamiana* & cowpea (*Vigna unguiculata*)	Treatment /dsRNA Carrier	(Worrall et al., 2019)
40	Cadmium telluride quantum dots (CdTe-QDs)	*Citrus Tristeza Virus (CTV)*	Citrus	Biosensor	(Safarnejad et al., 2017)
41	Lead sulfide (PbS) nanoparticles	*Cauliflower mosaic virus (CaMV) 35 S gene sequences*	CaMV 35 S gene of GMOs	Gene sensing	(Sun et al., 2008)
42	Gold nanoparticles	*Large cardamom chirke virus (LCCV)*	Cardamom leaf	Biosensor	(Maheshwari et al., 2018)
43	Polypyrrole (PPy) nanoribbon	*Cucumber mosaic virus (CMV)*	*Nicotiana clevelandii*	Biosensor	(Chartuprayoon et al., 2013)
44	Zinc oxide (ZnO) nanofilms	*Grapevine virus* A-type (GVA) proteins	Grapes	Biosensor	(Tereshchenko et al., 2017)

to complete suppression of the Sunhemp rosette virus (Jain and Kothari, 2014). In another study, AgNPs of size 77–92 nm synthesized by three different *Bacillus* sp. had shown antiviral activity against the bean yellow mosaic virus which had infected fava beans. It had been observed that AgNPs prevent all symptoms caused by the bean yellow mosaic virus when it was administered 24 hrs post-inoculation in concentration 0.1 $\mu g\ \mu L^{-1}$ (Elbeshehy et al., 2015). Similarly, AgNPs had shown antiviral properties against tomato mosaic virus and potato virus Y. At 50 ppm, AgNPs induce systemic acquired resistance and suppress the two viruses in the tomato plant (Noha et al., 2018).

4.1.2 Gold Nanoparticles (AuNPs)

Gold nanoparticles (AuNPs) were explored for the detection of phytoviruses but they have antiviral potential as well. AuNPs had shown antiviral activity against Barley yellow mosaic virus. From *in vitro* assessment, it was concluded that AuNPs caused complete viral particles dissociation (Aref et al., 2012). AuNPs of size 3.151 nm and 31.67 nm had also been reported to cause deterioration of barley yellow mosaic virus-PAV viral particles confirmed by TEM analysis (Alkubaisi and Aref, 2017).

4.1.3 Zinc Oxide Nanoparticles (ZnO NPs)

Zinc oxide NPs (ZnO NPs) had better antimicrobial activity as compared to their bulk counterpart (Duhan et al., 2017; Banerjee et al., 2021). ZnO NPs generate ROS which can cause cell death on the enhancement of the antioxidant capacity of the cell (Duhan et al., 2017; Hajong et al., 2019). Also, ZnO NPs consist of extraordinary magnetic, photocatalytic, optical, and electrical properties (Rajwade et al., 2020). In one of the studies, ZnO NPs of size 20 nm had shown an antiviral effect against tobacco mosaic virus (TMV) infection both in vitro and in vivo in the tobacco plant. ZnO NPs caused the deactivation of TMV and also up-regulated plant immunity (Cai et al., 2019).

4.1.4 Titanium Dioxide Nanoparticles (TiO_2 NPs)

Titanium dioxide NPs (TiO_2 NPs) are non-toxic and chemically stable due to which they are explored for various environmental applications. TiO_2 NPs shows long term antimicrobial activity mainly due to their strong oxidizing nature (Rajwade et al., 2020). TiO_2 NPs also exhibits photocatalytic property which helps in their role in plant protection (Ramezani et al., 2019). TiO_2 nanostructures were used to treat the broad bean stain virus (BBSV) infection in the faba bean plant. It has been observed that TiO_2 nanostructures treated plants showed a reduction in the severity of infection as compared to untreated plants. The antiviral effect of TiO_2 NPs was time-dependent means whether it was applied before or after viral infection (Elsharkawy and Derbalah, 2019). In one of the studies, TiO_2 NPs showed antiviral potential against turnip mosaic virus in the *Nicotiana benthamiana* plant (Hao et al., 2018).

4.1.5 Silica Nanoparticles

Silica NPs have been successfully explored for their activity against various agricultural pests and ectoparasites in animals. Silica NPs adsorb on cuticular lipids of insect pests which leads to their death by physical means. Although silica NPs have applications as carriers for various agrochemicals in agriculture (Ramezani et al., 2019), but they were also used for antiviral activity against phytoviruses. Silica NPs showed antiviral activity against tomato yellow leaf curl virus (TYLCV). It was found that silica NPs-treated tomato plants showed reduced virus titre and decreased the severity of the infection (El-Sawy et al., 2017a). Silica NPs showed a significant decrease in viral accumulation and severity of disease in PRSV infection in cucumber plants (Elsharkawy and Mousa, 2015).

4.1.6 Carbon-based Nanoparticles

Carbon-based NPs have the potential to eradicate plant pathogens but they are less studied for antiviral application (Banerjee et al., 2021). Recently it was reported that carbon nanotubes and C60 fullerenes at their application, that was before the viral infection, inhibited the tobacco mosaic virus (TMV) in *Nicotiana benthamiana* plants at a concentration of 200 mg L^{-1} (Adeel et al., 2021).

4.1.7 Chitosan Nanoparticles

Chitosan NPs are one of the most popular as they have advantages of biocompatibility, less toxicity, biodegradability and also have antimicrobial potential (Worrall et al., 2018). Chitosan had been reported to show antiviral properties against the TMV (Davydova et al., 2011). So, chitosan NPs can also be explored for their antiviral effect against plant viruses.

4.1.8 Other Nanoparticles

Nickel oxide nanostructures had shown antiviral activity against the cucumber mosaic virus (CMV) in the cucumber plant. It had reduced the viral load and severity of the infection (Derbalah and Elsharkawy, 2019). Iron oxide NPs have also been reported to show antiviral properties against the turnip mosaic virus (TuMV) in the tobacco plant (Hao et al., 2018). Cerium dioxide NPs were found to reduce viral necrotic lesions both in *D. stramonium* and *N. tabacum* plants infected with TMV. It has antiphytoviral activity against TMV (Eugene and Zholobak, 2016).

4.2 Nanomaterials as Delivery Systems

Well-functional delivery systems are required in agriculture for the delivery of pesticides, fertilizers, and genetic material for plant growth and disease management. NPs can play a good role in the preparation of highly efficient delivery systems due to their attributes like easy attachment, effective loading of desired chemical or genetic material due to their large surface area, speedy mass transfer, and as nanomaterials due to their slow release attribute provide the advantage of controlled release of active ingredients at the site (Manjunatha et al., 2016). Although there are some limitations associated with the delivery of genetic materials that can be overcome by research in the area. The nano-based delivery systems can be used for the delivery of potential antiphytoviral agents (pesticides), genetic materials, and fertilizers which, by enhancing plant growth and health, will indirectly help to combat various plant viral diseases.

4.2.1 Some Common Nanoparticles Used as Nanocarriers

To form efficient formulations for treating plant diseases various kinds of nanomaterials can be used. They can carry pesticides, fertilizers, other agrochemicals and genetic material by entrapping, encapsulating, attaching, or adsorbing it (Worrall et al., 2018). Some of the common nanocarriers are described below.

Silica Nanoparticles: These NPs are very advantageous for the delivery of various agricultural chemicals because they can be seamlessly synthesized and also their shape, size, and structure can be controlled easily (Worrall et al., 2018). They are mostly spherical structures with pores in them also known as mesoporous silica NPs (MSNs) or porous hollow silica NPs (PHSNs). These porous silica NPs generally load the pesticides in their inner core which help to protect the active molecules from UV light and helps in controlled sustained release of the pesticides. According to various reports, it was found that silicon has the ability to enhance plant tolerance for various biotic and abiotic stresses, so, silica NPs are the natural choice for the development of nanoformulations for plant diseases management (Barik et al., 2008).

Solid-Lipid Nanoparticles (SLPs): These NPs are formed of lipids that have the property of being solid at room temperature. SLPs are similar to emulsions. The main advantage of these NPs is that they can carry hydrophobic substances in their matrix without the use of organic solvents. Also, due to decreased mobility of hydrophobic active ingredients in the solid matrix, it provides the property of controlled release of the loaded substance. For dispersing SLPs in water, surfactants are used to stabilize them. SLPs have one disadvantage, the loaded active ingredients can leak from them while storage (Worrall et al., 2018).

Chitosan Nanoparticles (CS NPs): These NPs are also used to load agrochemicals or essential oil to manage pests (Tortella et al., 2021). CS NPs are less soluble in water due to their hydrophobic property. Therefore, CS NPs are commonly mixed with a copolymer which may be organic or inorganic to enhance their solubility. CS properties can be improved because it has reactive hydroxyl and amine groups which helps in their modification, ionic interactions, and graft reaction. CS strongly adheres to the epidermis of stems and leaves enhancing their contact time and expediating the uptake of bioactive substances (Worrall et al., 2018).

Layered double hydroxides (LDHs) Nanoparticles: These are clays that are arranged as hexagonal sheets in which active substances' layers are trapped into the interlayer spaces. Under acidic conditions like carbon dioxide from the environment and addendum of moisture leads to the breakdown of LDH NPs (Worrall et al., 2018). In one of the studies, it was found that positively charged LDH lactate NPs that were delaminated had facilitated biologically active material transport across the plant cell wall barrier (Bao et al., 2016).

These all are some commonly used NPs for delivery of pesticides and other agrochemicals to treat plant diseases, but not mainly for viruses. So, these NPs have great potential to be used for plant antiviral therapy. Other metallic, metal oxides, non-metallic, polymeric NPs can also be explored to combat various plant phytoviruses.

4.2.2 Delivery Systems for Pesticides

Both chemical and biological pesticides can be delivered with the help of nanocarriers or their nano form also termed as nanopesticides which means nanotization of pesticides themselves to treat plants (Manjunatha et al., 2016; Rajwade et al., 2020). Nanomaterials are used in pesticide formulations due to their exceptional properties like solubility, biodegradability, permeability, stiffness, thermal stability, and crystallinity (Rajwade et al., 2020; Shang et al., 2019). These properties help to increase the dispersion and wettability of agricultural nanoformulations which leads to a decrease in the organic solvent run-off and unnecessary movement of pesticides (Manjunatha et al., 2016; Shang et al., 2019; Rajwade et al., 2020). Also, nano-based formulations show systemic activity due to their small size (Manjunatha et al., 2016). Nanoformulation of pesticides decreases the amount of pesticides to be used that will reduce the use of hazardous pesticides and their harmful impact on the ecosystem, enhance their efficacy, prevent loss of pesticides due to leaching or evaporation and also serves as an efficient system for pathogen and plant growth management (Shang et al., 2019; Rajwade et al., 2020). Mainly natural polymers like chitosan, albumin, lignin, cellulose, etc., can be used to develop nanomaterial for delivery. These polymers are non-toxic, cost-effective, and biodegradable (Rajwade et al., 2020). Further, nanoformulation of biopesticides increases the stability of cells, enzymes, and other biological products. However, there are many challenges in on-field use of nanoformulations of pesticides like large spray area and environmental perturbations which can be overcome by applying target specific delivery (Manjunatha et al., 2016). PHSNs when loaded with pesticide validamycin had the ability to act as an efficient delivery system for controlled release of water-soluble pesticide (Liu et al., 2006).

Nanoformulations of pesticides are prepared by the nanoencapsulation process. This is the process in which active ingredients of pesticides are coated with nano-sized materials; the encapsulated material forms the inner phase while the outer coating is of nanomaterials. This encapsulation helps in the controlled release of active ingredients in the plant or in the root area without affecting its efficacy (Shang et al., 2019). The mechanisms by which the encapsulated substance is released are diffusion, biodegradation, dissolution, and osmotic pressure with a specific pH. It is the most promising technology in delivery systems for protecting plants from pests. Many leading companies in the areas are focusing on the nanoencapsulation technology to prepare nanoscale pesticides delivery systems (Prasad et al., 2014). For example, clay nanotubes were found to be used as a carrier for pesticides and they showed the property of long release of active ingredients and had low environmental contact (Shang et al., 2019). Chitosan-alginate-acetamiprid nanoformulation was used for the controlled release of the insecticide acetamiprid and was effective against insect pests (Kumar et al., 2015). Similarly, nanoformulations can be prepared which have certain agrochemicals that are effective against phytoviruses. This will help in the efficient treatment of plant viral diseases and will increase crop productivity. More research is required in this area, especially for combating plant viral infections.

4.2.3 Delivery Systems for Fertilizers

Fertilizers play a very vital role in enhancing the production of crops (Prasad et al., 2014). However, localized application of fertilizers like urea, ammonium salts, and nitrate compounds in large amounts are harmful to plants and the environment. Also, the fertilizers are lost and are not available to plants due to run-off leaching which causes pollution (Manjunatha et al., 2016). These issues can be overcome by using nanofertilizers. Nanofertlizers enhance nutrient efficiency and can solve the eutrophication issues too. Nanofertlizers have the ability to release nutrients in regulated amounts according to the plant's need; this property makes them more efficient than conventional chemical fertilizers. Nanoferltizers also have the ability to selectively release nutrients according to the time and environmental conditions. The attribute of the slow and regulated release of nutrients from nanofertilizers decreases the toxic effects on the soil as well as its overapplication is not required (Prasad et al., 2014). Nanocoating is the surface coating of nanomaterials on fertilizers particles that help them to adhere more firmly to plants on application due to the high surface tension of nanomaterials (Manjunatha et al., 2016). Sulfur nanocoated fertilizers were found to be beneficial for sulfur deficient soil and also effective in sustained and slow release of sulfur to plants. The rate of dissolution of nano-coated fertilizers was decreased due to the stability provided by nanocoating. Similarly, nanoencapsulation of other fertilizers like urea and phosphate helps in the controlled release of nutrients to plants and soil which also enhances plant and soil health. Other nanomaterials used for nanofertilizers were kaolin and polymeric NPs, like chitosan of size approximately 78 nm which had shown the ability of sustained release of NPK nutrients from various sources like potassium chloride, urea, and calcium phosphate (Manjunatha et al., 2016). Also, it was reported that nanofertilizers have the ability to decrease nitrogen loss due to extended incorporation of soil microbes, emission, and leaching (Prasad et al., 2014). By using nanoformulations of fertilizers, various vital nutrients sustained delivery to plants can be achieved which improves the health of plants and indirectly will help in combating various plant infections. The healthier plant will be able to generate effective and quick defence mechanisms against the invading pathogens. In this way, usage of nanofertilizers will be helpful in combating plant viral diseases.

4.2.4 Delivery Systems for Genetic Materials

The conventional delivery system for genetic material has various drawbacks like limited host range, size of genetic material to be used, difficulty in crossing the cell membrane and trafficking problem in the nucleus (Shang et al., 2019). Direct application of genetic material leads to their environmental degradation. The use of NPs as delivery systems to carry genetic materials is one of the promising fields to combat plant virus diseases. RNA interference (RNAi) method is generally used for treating phytoviruses. RNAi pathway offers resistance to various endogenous and exogenous pathogenic nucleic acids by using double-stranded (ds) RNA molecules via managing the mRNA stability and the translation process (Farooq et al., 2021). Biodegradable and safe layered double hydroxide bioclay nanosheets (BioClay) were efficiently used to deliver dsRNA to plants by Mitter et al. (2017). Plants have the ability to accumulate this externally provided dsRNA and, consequently, they initiate the RNAi pathway. By using this BioClay in genetic delivery, protection was achieved against two phytoviruses which are the pepper mild mottle virus (PMMoV) and the cucumber mosaic virus (CMV) and it lasted for 20 days (Mitter et al., 2017). In this study non-vectored inoculation was used, while in another study by the same group it was shown that BioClay had the ability to protect plant from virus which was transmitted by vector. Also in one study after five days of topical application of dsRNA, molecules nanoformulation prepared by using BioClay, protection against vectored transmission of bean common mosaic virus (BCMV) was obtained in cowpea plant (*Vigna unguiculata*) and tobacco plant (*N. benthamiana*) (Worrall et al., 2019). This BioClay dsRNA formulation should be tested in a plant-virus-vector combined system where the host plant transmission occurs through insect species. However, this area requires more research to acquire knowledge about pathogen resistance and resiliency. Further, more studies are needed to understand the dosage, fate of nanocarriers, and timing of application (Farooq et al., 2021).

Another interesting approach is the transfer of CRISPR/Cas9 single guide RNA (sg-RNA) with the help of NPs to treat plant diseases and this also enhances the efficiency and specificity of CRISPR/Cas9 system. This system helps in genome editing with the help of RNA. Chitosan NPs are also a good delivery system for genetic material because it has ability to bind with RNA and can easily cross the plasma membrane (Shang et al., 2019; Banerjee et al., 2021). Genetic materials can also be delivered by nanoencapsulation technology (Prasad et al., 2014). Further, more delivery systems can be explored to genetically treat plant viral diseases. Many NPs have the potential to be proved as an efficient delivery agent for genetic materials. Table 8.1 summarizes some nano delivery systems used to combat plant viral diseases.

4.3 Nanomaterials as Biostimulants

Plant biostimulants are substances that promote plant immunity and stimulate plant defence mechanisms (Vargas-Hernandez et al., 2020; Farooq et al., 2021). These are the substances other than pesticides and nutrients, which when applied to plants provide the potential advantage of improving plant growth, defences, and development (Vargas-Hernandez et al., 2020). The common innate immune responses which are observed in plants are the generation of ROS (Fu et al., 2020; Rajwade et al., 2020; Farooq et al., 2021), generation of various enzymes involved in defence mechanisms (Fu et al., 2020), systemic acquired resistance induction by activation of pathogenesis-related (PR) genes, callose deposition and induction of various hypersensitive responses (Rajwade et al., 2020). Also, biostimulants application leads to elevation of various phytohormone levels like salicylic acid (SA), abscisic acid (ABA), jasmonic acid (JA), zeatin riboside (ZR), and brassinosteroids (BR) (Farooq et al., 2021). The natural immune responses in plants are non-specific which helps to combat a broad range of plant pathogens. Nanomaterials along with antimicrobial potential have the ability to act as biostimulants when applied in small quantities and which helps to augment plant tolerance to various biotic stresses subsequently leading to disease suppression (Fu et al., 2020). NPs have the ability to provide both distress (negative) and eustress (positive) impact, but it depends upon the plant's state and the type of nanomaterial (Vargas-Hernandez et al., 2020).

Many metallic NPs have shown to generate plant immune responses against the invading phytopathogens (Fu et al., 2020). Silver NPs improved the quality parameter in PVY-infected tubers may be because of the generation of resistance or their impact on viral entry (Wang et al., 2016). AgNPs also acted as biostimulants in the fava beans plant which was infected with BYMV (Elbeshehy et al., 2015). Enhanced growth in healthy tobacco plants was observed when treated with SiO_2, ZnO, and Fe_2O_3 NPs (Cai et al., 2019; Cai et al., 2020). Tobacco plants infected with TuMV, when treated with 50 mg/L TiO_2 and Fe_2O_3 NPs, showed an increase in the plant's dry and fresh weight while at 200 mg/L concentration no effect was observed (Hao et al., 2018). This shows that a nanomaterial's biostimulant activity is concentration-dependent. Similar activity with NiO NPs had been observed in infected cucumber plants (Derbalah and Elsharkawy, 2019). Chitosan and selenium NPs also show immunomodulatory activity (Rajwade et al., 2020). In one study it was found that CS NPs induce the generation of ROS and enhance the peroxidase activity in finger millets leaves (Sathiyabama and Parthasarathy, 2016). CS NPs augment immune responses in plants (Rajwade et al., 2020).

Biostimulants can cause eustress in healthy as well as stressed plants like the virally infected plant; however, they counteract the harmful effect caused by them (Vargas-Hernandez et al., 2020). Biostimulation occurs at a specific concentration by nanomaterials although the phytotoxicity issue should be considered while using them in higher concentrations (Fu et al., 2020). Although more study is needed to understand the molecular mechanisms of biostimulation by NPs as well as to understand the stability and effects of plant resistance generated through their application (Fu et al., 2020).

Application of nitric oxide (NO) releasing nanomaterials help in combating plant viral infections as in NO, the gaseous free radical plays an important role in plant defence mechanisms. NO acts as a biological messenger for various growth and development processes in the plant. It is a tiny lipophilic molecule that can easily pass across cell membranes and is synthesized by an enzymatic pathway by the action of NO synthase enzymes. NO has a role to acclimatize plants during pathogen attacks by initiating defence responses against them. Various plant hormones, during defence against pathogens, increase the production of NO. These hormones are auxin, jasmonic acid, cytokinin, salicylic acid, brassinosteroids, and abscisic acid (Tortella et al., 2021). In one study salicylic acid and NO showed a synergistic effect against the tomato mottle mosaic virus (Nagai et al., 2020). In another study, it was reported that cucumber mosaic virus was lessened in *A. thaliana* due to the synthesis of NO which controlled salicylic acid to mediate the defence mechanism (Tortella et al., 2021). The external application of NO donating NPs will help in plant defence mechanism by generating ROS, activating various genes involved in it, or by developing systemic acquired resistance. Lu et al., 2019 confirmed the role of NO in plant defence mechanisms by administrating NO inhibitors on rice stripe virus-infected plants. It was found that on the administration of the inhibitors there was an enhancement in the rice stripe virus (Lu et al., 2019). NO-releasing nanomaterials have been applied to counter various abiotic stresses like salinity, drought, UV-light exposure in plants, but is not extensively used in biotic stress conditions. But after seeing the role of NO in plant defence response, NO-releasing nanomaterials should be explored for combating the biotic stresses (Tortella et al., 2021) as they have the ability to function as a biostimulant to control plant viral infections. Further, Table 8.1 gives the idea about nanomaterials used as biostimulants.

4.4 Nanomaterials in Detection of Plant Viral Diseases

Biosensors are analytic devices that comprise of a biological sensing element unified with the physiochemical transducer which measures the signals when analyte (pathogen) of interest comes in interaction with the sensing element. This interaction is measured in the readable digital format (Elmer and White, 2018; Hong and Lee, 2018). To overcome the drawbacks of conventional biosensors, nano-based biosensors came into the picture. Different types of NPs (metallic, carbon, polymeric, etc.) and nanostructures (nanowires, nanorods, nanotubes, etc.) have the ability to be used in the preparation of nanobiosensors (Elmer and White, 2018; Kumar and Arora, 2020). The conventional molecular and immunodiagnostic techniques for plant disease detection require a costly reagent, it is time-consuming, requires skilled individual, and their in-field application is difficult (Hong and Lee, 2018; Hajong et al., 2019; Rajwade et al., 2020).

Nanobiosensors are very selective and sensitive, that is, they can detect a very minute amount of analyte, they are accurate, reproducible, quantitative, small, portable, and robust. This will help to detect plant diseases before the appearance of the symptoms and enable efficient plant disease management (Kashyap et al., 2017). The unique tunable and properties of nanomaterials make them ideal for biosensing applications. The properties include optical properties, good electrical conductivity, good shock-bearing capacity that offers different types of ultra-sensitive responses/detection mechanisms by evaluating the thermal, electrical, optical, and piezoelectrical properties. These stirring properties of nanomaterials due to their size in nanometers enable their assembly with the biomolecules that will help to form specific nanobiosensors for desired applications (Kumar and Arora, 2020). Generally, it is very difficult to detect viruses and bacteria in less amounts or when their intensity is low; usage of nanosensors can be an effective strategy to solve this issue (Banerjee et al., 2021). The usage of nanomaterials in sensing and biosensing application have transformed conventional sensors and biosensors. Nanomaterials have multidimensional roles in biosensing applications that are that they act as a

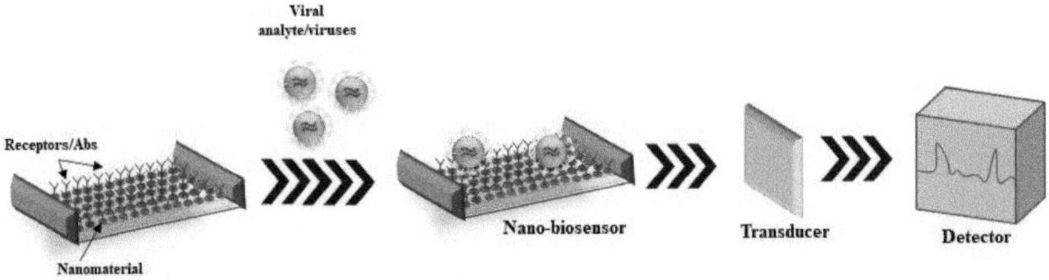

Fig. 8.2. Schematic representation for general working of nanobiosensors.

stable support for attachment of biological receptors and as a signal enhancer, labels, and sometimes as sensing receptors too. They have a significant capability to deal with different types of signals like magnetic, optical, thermal, electrochemical, etc. according to the type of biosensor and its detection principle (Kumar and Arora, 2020). Nanobiosensors are an ideal agent to detect plant viral infections. Figure 8.2 is the schematic representation of the general working of nanobiosensors. The various approaches and trends in nanobiosening to combat phytoviral diseases are explained below.

4.4.1 Different Approaches in Nanobiosensing of Plant Viral Diseases

The two receptors among all biosensing components which were explored in nanobiosensing plant viral disease are DNA and antibodies. These two have more advantageous attributes and prove to be efficient in the detection of plant pathogens (Dyussembayev et al., 2021). Also, quantum dots-based nanobiosensors are gaining immense importance to detect plant viral diseases.

4.4.1.1 Antibody-based Nanobiosensors

These type of nanobiosensors hold enormous importance for the detection of plant pathogens in agriculture. They have several advantages like efficient sensitivity and selectivity, rapid and real-time analysis. Antibodies are very suitable for various immunosensing applications. The basic principle of antibodies-based biosensors is that these are coupled with a transducer which on the binding of an antigen with antibodies helps in signal transduction and amplification. In the case of nanobiosensors, antibodies are immobilized on nanomaterials. Commonly explored antibody-based nanobiosensors for phytoviruses detection used electrochemical and optical transducers (Fang and Ramasamy, 2015; Dyussembayev et al., 2021). Antibody functionalized gold nanorods were used to detect the cymbidium mosaic virus (CymMV) and the odontoglossum ringspot virus (ORSV). The limits of detection (LODs) were found to be 48 pg/ml and 42 pg/ml for CymMV and ORSV, respectively, in leaf saps (Lin et al., 2014). Polypyrrole (PPy) nanoribbon based label-free electrochemical sensor was prepared to detect the cucumber mosaic virus. It was observed that the sensitivity increased by decreasing the thickness of nanoribbons from 100 nm to 25 nm. Further, the limit of detection was found to be 10 ng/ml and this nanobiosensor was more sensitive than the traditional ELISA (Chartuprayoon et al., 2013). A lateral flow immunoassay based on Au NPs was developed to detect the potato virus X (PVX). The time of assay was 15 minutes and a lower limit of detection was found to be 2 ng/ml in non-clarified leaf extract (Drygin et al., 2012).

4.4.1.2 DNA-based Nanobiosensors

This type of nanobiosensor uses nucleic acid fragments as a receptor to detect plant pathogens. The principle of DNA-based nanobiosensors is that there is a hybridization between the target *DNA* sequence and the DNA probe sequence which is immobilized on the sensing material. The DNA probe which acts as a receptor is formed of nucleotide sequence specific for the DNA sequence of interest (Fang and Ramasamy, 2015; Dyussembayev et al., 2021). This type of biosensor helps to detect the disease before the appearance of symptoms because detection is at the molecular level. This type of nanobiosensor is simple, economical, and rapid in its application (Fang and Ramasamy, 2015). Electrochemical and optical types of DNA-based nanobiosensors were explored to detect plant viral diseases. Citrus tristeza virus (CTV) was detected by using label-free impedimetric method consisting of Au NPs modified screen-printed carbon electrodes on which thiolated the ssDNA was immobilized to act as a probe for detection. The nanobiosensor was found to be selective for nucleic acid of and the limit of detection was 126–1.26 × 10^3 pg/ml (Khater et al., 2019a).

4.4.1.3 Quantum Dots (QDs) Based Nanobiosensors

Quantum dots (QDs) are nanomaterials of size 2–10 nm which have a semiconductor property. They are composed of elements of groups II–VI, III–V, and IV–VI for example CdTe, CdS, ZnSe, and ZnS (Hong and Lee, 2018; Rajwade et al., 2020). QDs are zero-dimensional nanomaterials (Hong and Lee, 2018). Owing to the phenomenon of quantum confinement, QDs consists of varied size-dependent optical properties. These properties are broad absorption spectra that lead to concurrent excitation of various fluorescent colors, emission wavelength which depends upon their size, excellent photostability, and narrow emission peak; (Duhan et al., 2017; Kashyap et al., 2017; Hong and Lee, 2018; Rajwade et al., 2020). Also, QDs are resistant to photobleaching and are brighter than other materials (Hong and Lee, 2018). Concerning the unique optical properties of QDs, they have been used for biosensing on the basis of the fluorescence dependent resonance energy transfer (FRET) mechanism (Rajwade et al., 2020). However, depending upon the molecular signal particle conjugated with QDs and transduction signals they are named as QD-based FRET immunosensors, QD-based bioluminescence resonance energy transfer (BRET) immunosensors, QD-based FRET genosensor, etc. (Hong and Lee, 2018). QDs-based nanobiosensors haves great potential in the detection of plant diseases. Tereshchenko et al. (2017) used ZnO nanofilms for the detection of the grapevine virus A-type (GVA) proteins (GVA-antigens). It was an optical biosensor with a detection limit of 1 pg/ml–10 ng/ml (Tereshchenko et al., 2017). Cadmium telluride QD (CdTe-QD)-based nanobiosensors were used to detect the citrus tristeza virus (CTV) infected plants. The limit of detection of this CdTe-QDs-based donor-acceptor complex biosensors ware found to be 198 ng/ml (Safarnejad et al., 2017). A rapid, specific, and sensitive AuNP and CdTe-QD-nanobiosensor was designed to detect the CTV. The sensor works on the FRET mechanism and the limit of detection was 0.13 μg/ml (Shojaei et al., 2016b).

Some other nanodiagnostic methods to detect plant viral diseases are nanobarcodes, nanopore sequencing, nano-based diagnostic kits, and nanofabrication. Nanobarcoading is an alternative method for PCR. This involves the detection and amplification of nucleic acid by tagging the pathogen. QD-based nanobarcodes were used for the detection of plant pathogens. Nanodiagnostic kits are simple, portable, and semiquantitative which can be used in fields for various plant-pathogen determination. Nanofabrication is an imaging tool used to visualize plant tissues and cells. It is very helpful in the primary detection of plant diseases. In nanofabrication, the signal strength, contrast, tissue specificity, and time of imaging can be enhanced just by tuning the various properties of NPs. Further, nanopore sequencing is the next-generation sequencing that has the functionality to determine the entire genome in minutes. This technique is very advantageous for the detection of plant viral infections in a limited time (Hajong et al., 2019). In this way, various NPs can be investigated for sensitive, early, and selective determination of plant viruses that ruin the plant's health and subsequently the agricultural productivity. Table 8.1 further gives the idea about various nanomaterials used in the detection of plant viral infections.

5. Uptake and Translocation of Nanoparticles in Plants

For uptake and translocation of NPs in the plant, it has to cross various physiological and chemical barriers (Hajong et al., 2019; Wang et al., 2016). Commonly, NPs can enter plant tissues either through either the root tissues or the shoot and leaves. There are various natural openings in plants such as the stomata, stigma, hydathode, wounds as well as root junctions from which NPs can pave their way inside plants (Wang et al., 2016; Hajong et al., 2019; Banerjee et al., 2021). NPs can enter plants through endocytosis (Hajong et al., 2019). The mucilage and exudates secreted in the rhizosphere region by the roots bind with NPs and help in their internalization which also depends upon the delivery method used (Banerjee et al., 2021). The vegetative parts of plants are covered with thick cuticles. Many NPs are inhibited by cuticle (Banerjee et al., 2021), but in one study it was found that TiO_2 NPs have the ability to make a hole in the cuticle layer for its internalization (Sanzari et al., 2019).

Further, the translocation of NPs in plants takes place by apoplastic or symplastic pathways (Wang et al., 2016; Hajong et al., 2019; Banerjee et al., 2021), in which NPs can transport through the xylem or phloem tissues and finally to the whole plant. In symplastic translocation, NPs have to be penetrated into the cell; however, due to the cellulosic and hemicellulosic composition of the cell wall, there is some restriction in the movement of NPs. In apoplastic translocation, NPs move radially in the vascular tissues of the root which help in their upward movement, that is, towards the aerial parts of the plants (Banerjee et al., 2021). MSNs (Sun et al., 2014) and SiO_2 NPs (Le et al., 2014) have been reported to be translocated by these pathways. In symplastic translocation, NPs has to pass through plasmodesmata so their size should be in range of 3–50 nm, while in the apoplastic pathway the NPs have to pass through the cell wall and casparian strips in the vascular system. The size exclusion limit in the cell wall is 5–20 nm while in the casparian strip it is less than 1 nm (Wang et al., 2016). NPs may also cross the membrane using embedded transport carrier proteins or through ion channels (Ramezani et al., 2019). However, reports are there in which large size NPs were also passed through plasmodesmata, cell wall, and Casparian strips. This may be due to the dynamic nature of size exclusion limits of plasmodesmata, cell wall and Casparian strips and also may be influenced by other substances like calcium, proteins, silica, environmental stress and viruses (Wang et al., 2016). NPs translocation in plants is also influenced by the type of plant species (Banerjee et al., 2021). It was found that Fe_3O_4 NPs have translocated successfully in *Cucurbita maxima* but not in *Phaseolus*

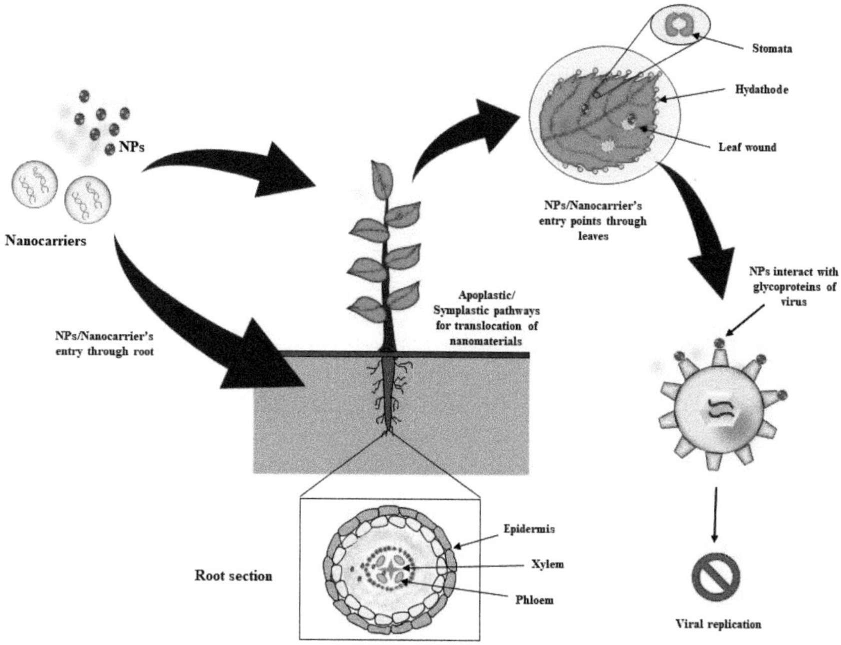

Fig. 8.3. General process of uptake and translocation of nanomaterials in plants to combat viral diseases.

limensis (Banerjee et al., 2021; Zhu et al., 2008). It is very important to understand effective translocation strategy of NPs in the plant; this will help in efficient disease management and plant growth. Further, in a detailed study of uptake and translocation of NPs in plants with understanding, the exact mechanisms will help the researchers to develop more specific, effective, and efficient nanomaterial-based methods to control plant viral diseases. Figure 8.3 depicts the general process of uptake and translocation of nanomaterials in plants for plant disease management.

6. Nanophytotoxicity

Plants are an integral part of our ecosystem, so, effective disease management strategies are required for their health. Nanomaterials have been proved to be a boon for plant disease management. However, in some reports, it was found that the application of nanomaterials in plant disease control causes harmful effects in plants. As seen earlier, mainly metal and metal oxide nanomaterials that are inorganic NPs were used for combating plant viral diseases and these NPs get internalized and translocated into the plants for effective results in disease control. This translocation of NPs can lead to the production of toxic substances which can cause plant phytotoxicity issues (Banerjee et al., 2021). In one of the studies, it was found that when various NPs like aluminium, carbon, zinc, alumina, and ZnO were applied to different plants, phytotoxicity was observed in them (Ling and Xing, 2007). In one report of a transcriptome study, it was found that few metal NPs suppressed expression of stress and pathogen-related genes which are required for acclimatization (Ruotolo et al., 2018). It was observed that when cerium oxide NPs were applied in soybean plants it hindered the nitrogen fixation potential of the plants and reduced the pod size as well as hampered plant growth. Further, it was observed that excessive use of metal and metal oxide NPs in the plants may result in an oxidative burst by intermingling with the electron transport chain (Sanzari et al., 2019; Banerjee et al., 2021).

Consideration of several intrinsic (size, shape, etc.) and extrinsic (surface charge, dissolution, etc.) properties of NPs are necessary to understand their toxicological effects in plants (Wang et al., 2016). The application of nanomaterials in agriculture should be according to the 'safety-by-design' principle which should be directed by plant physiology, nanomedicine stirred nano-based delivery systems and nanomaterials functionalization for proficient availability of pesticides and bioactive substances to plants; additionally, reducing the adverse effects of nanomaterials on plants (Wang et al., 2016). Although phytotoxicity occurs when nanomaterials are used above the threshold concentration, up to a certain range they are safe and efficient to control phytoviral diseases. In addition to this, more research in the direction to control phytotoxicity of nanomaterials will help to find out innovative strategies to control phytotoxicity issues in plants.

7. Conclusions and Future Prospects

Nanophytovirology is an evolving new-fangled field that bestows the use of multifunctional and exceptional nanomaterials to combat the phytoviral diseases. This chapter pours into the little information available about the new rising area, nanophytovirology. Due to the prompt augmentation of the human population, there is immense pressure on the agriculture sector to fulfil the food demands around the world. The various abiotic and biotic stresses on plants act as hurdles to accomplish the goal of food demands. While focussing on the biotic stresses, it was originated that viruses are the challenging moieties to control as compared to bacteria, fungi, and other plant pests. Nanomaterials have been proved as a promising solution to combat the viral diseases of plants. Nanomaterials chiefly can be used as antivirals directly, as delivery vehicles to carry antiviral components, and as biostimulants and also can be used in the detection and diagnosis of the phytoviral infections. However, various researches are there that report the use of nanomaterials as antivirals and efficient detection systems but initiation of research is required to scrutinize the usage of nanomaterials as a delivery system for plant viral diseases management. The chapter enlights the potential of nanomaterials as delivery vehicles due to which it can be proved as an efficient phytovirus management method. Further, investigation of more types of nanomaterials apart from metallic and metal-oxides are desired to be applied as antivirals. The biostimulation abilities of nanomaterials have great scope for reconnaissance which will help in crop improvement. Also, the detection and diagnosis application accounts for more studies to develop smart and portable nanobiosensors made up of varied types of NPs. In short, it can be concluded that all the chief applications of nanomaterials in plant viral disease management require more attention from researchers around the world, so that these nano-based techniques can be efficiently translated into the well versed commonly used methods to control plant viral infection and subsequently this will help in food security.

Nanobiosensors can be combined with GPS to real-time management of plant diseases. Also, nanotization of antiviral components itself could be implied as one of the innovative strategies in phytovirus management. Detailed understanding of the uptake and translocation of nanomaterials inside the plant will be good supportive knowledge to develop the in-field functional and competent disease management strategies. Nanophytotoxicity is the issue associated with the application of NPs which could be solved with little effort by the researchers as it is concentration-dependent. But, the residual effect of nanomaterials on the other components of the ecosystems needed to be considered in greater detail. The new strategies like multifunctional nanosystems which involve nanomaterials which help in diseases management as well as enhances crop production can be proved as a breakthrough for the agriculture field. Finally, we can conclude that nanopahytovirology has many loopholes which need to be closed so that the field can astonish the world with its over-the-top strategies.

References

Abubakar, A.L., Abarshi, M.M. and Maruthi, M.N. (2018). Testing the infectivity of a Begomovirus by particle bombardment method using a gene gun. *Nigerian Journal of Biotechnology*, 35(2) (December): 58–65. https://doi.org/10.4314/njb.v35i2.8.

Adeel, M., Farooq, T., White, J.C., Hao, Y., He, Z. and Rui, Y. (2021). Carbon-based nanomaterials suppress tobacco mosaic virus (TMV) infection and induce resistance in Nicotiana benthamiana. *Journal of Hazardous Materials*, 404, part A (February): 124167. https://doi.org/10.1016/j.jhazmat.2020.124167.

Ahsan, T. and Yuanhua, W. (2021). Plant virus disease management by two modern applications (dsRNA nano-clay sheet and CRISPR/Cas). *Archives of Phytopathology and Plant Protection*, 54(17–18) (November): 1292–1304. https://doi.org/10.1080/03235408.2021.1910415.

Akhtar, I., Iqbal, Z. and Saddiqe, Z. (2020). Nanotechnology in pest management. pp. 69–83. *In*: Javad, S. (ed.). *Nanoagronomy*. Springer. https://doi.org/10.1007/978-3-030-41275-3_5.

Alkubaisi, N.A. and Aref, N. (2017). Dispersed gold nanoparticles potentially ruin gold barley yellow dwarf virus and eliminate virus infectivity hazards. *Applied Nanoscience*, 7(1) (February): 31–40. https://doi.org/10.1007/s13204-016-0540-0.

Aref, N., Alkubaisi, N., Marraiki, N. and Hindi, A. (2012). Multi-functional effects of gold nano-particles inducing plant virus resistance crops. *Proceedings of the 5th Annual World Congress of Industrial Biotechnology-2012*.,Xi'an, China 25.

Ariffin, S.A.B., Adam, T., Hashim, U., Faridah Sfaridah, S., Zamri, I. and Uda, M.N.A. (2014). Plant diseases detection using nanowire as biosensor transducer. *Advanced Materials Research*, 832: 113–117. https://doi.org/10.4028/www.scientific.net/AMR.832.113.

Banerjee, A., Sarkar, A., Acharya, K. and Chakraborty, N. (2021). Nanotechnology: An emerging hope in crop improvement. *Letters in Applied NanoBioScience*, 10(4) (April): 2784–2803. https://doi.org/10.33263/lianbs104.27842803.

Bao, W., Wang, J., Wang, Q., O'Hare, D. and Wan, Y. (2016). Layered double hydroxide nanotransporter for molecule delivery to intact plant cells. *Scientific Reports*, 6(1) (May): 1–9. https://doi.org/10.1038/srep26738.

Barik, T.K., Sahu, B. and Swain, V. (2008). Nanosilica: From medicine to pest control. *Parasitology Research*, 103(2) (July): 253–258. https://doi.org/10.1007/s00436-008-0975-7.

Byzova, N.A., Vinogradova, S.V., Porotikova, E.V., Terekhova, U.D., Zherdev, A.V. and Dzantiev, B.B. (2018). Lateral flow immunoassay for rapid detection of grapevine leafroll-associated virus. *Biosensors*, 8(4) (November): 1–14. https://doi.org/10.3390/bios8040111.

Cai, L., Liu, C., Fan, G., Liu, C. and Sun, X. (2019). Preventing viral disease by ZnONPs through directly deactivating TMV and activating plant immunity in Nicotiana benthamiana. *Environmental Science. Nano*, 6(12) (October): 3653–3669. https://doi.org/10.1039/C9EN00850K.

Cai, L., Cai, L., Jia, H., Liu, C., Wang, D. and Sun, X. (2020). Foliar exposure of Fe_3O_4 nanoparticles on Nicotiana benthamiana: Evidence for nanoparticles uptake, plant growth promoter and defense response elicitor against plant virus. *Journal of Hazardous* Material,s 393 (July): 1–11. https://doi.org/10.1016/j.jhazmat.2020.122415.

Chartuprayoon, N., Rheem, Y., Ng, J.C.K., Nam, J., Chen, W. and Myung, N.V. (2013). Polypyrrole nanoribbon based chemiresistive immunosensors for viral plant pathogen detection. *Analytical Methods*, 5(14) (April): 3497–3502. https://doi.org/10.1039/c3ay40371h.

Davydova, V.N., Nagorskaya, V.P., Gorbach, V.I., Kalitnik, A.A., Reunov, A., Solov, T.F. and Ermak, I.M. (2011). Chitosan antiviral activity : Dependence on structure and depolymerization method. *Applied Biochemistry and Microbiology*, 47(1) (January): 103–108. https://doi.org/10.1134/S0003683811010042.

Derbalah, A.S.H. and Elsharkawy, M.M. (2019). A new strategy to control cucumber mosaic virus using fabricated NiO-nanostructures. *Journal of Biotechnology*, 306(December): 134–141. https://doi.org/https://doi.org/10.1016/j.jbiotec.2019.10.003.

Dharanivasan, G., Riyaz, S.U.M., Jesse, D.M.I., Muthuramalingam, T.R., Rajendran, G. and Kathiravan, K. (2016). DNA templated self-assembly of gold nanoparticle clusters in the colorimetric detection of plant viral DNA using a gold nanoparticle conjugated bifunctional oligonucleotide probe. *RSC Advances*, 6(14) (January): 11773–11785. https://doi.org/10.1039/C5RA25559G.

Dharanivasan, Gunasekaran, Jesse, D.M.I., Rajamuthuramalingam, T., Rajendran, G., Shanthi, S. and Kathiravan, K. (2019). Scanometric detection of tomato leaf curl new delhi viral DNA using mono-and bifunctional AuNP-conjugated oligonucleotide probes. *ACS Omega*, 4(6) (June): 10094–10107. https://doi.org/10.1021/acsomega.9b00340.

Drygin, Y.F., Blintsov, A.N., Grigorenko, V.G., Andreeva, I.P., Osipov, A.P., Varitzev, Y.A., Uskov, A.I., Kravchenko, D.V. and Atabekov, J.G. (2012). Highly sensitive field test lateral flow immunodiagnostics of PVX infection. *Applied Microbiology and Biotechnology*, 93(1) (January): 179–189. https://doi.org/10.1007/s00253-011-3522-x.

Duhan, J.S., Kumar, R., Kumar, N., Kaur, P., Nehra, K. and Duhan, S. (2017). Nanotechnology: The new perspective in precision agriculture. *Biotechnology Reports*, 15(September): 11–23. https://doi.org/10.1016/j.btre.2017.03.002.

Dyussembayev, K., Sambasivam, P., Bar, I., Brownlie, J.C., Shiddiky, M.J.A. and Ford, R. (2021). Biosensor technologies for early detection and quantification of plant pathogens. *Frontiers in Chemistry*, 9(June): 1–13. https://doi.org/10.3389/fchem.2021.636245.

El-Sawy, M.M., Elsharkawy, M.M., Abass, J.M. and Hagag, E.S. (2017a). Inhibition of tomato yellow leaf curl virus by zingiber off cinaleand mentha longifolia. *International Journal of Antivirals & Antiretrovirology*, 1(1) (December): 1–6.

El-Sawy, M.M., Elsharkawy, M.M., Abass, J.M. and Kasem, M.H. (2017b). Antiviral activity of 2-nitromethyl phenol, zinc nanoparticles, and seaweed extract against cucumber mosaic virus (CMV) in eggplant. *Journal of Virology and Antiviral Research,* 6(2) (August). https://doi.org/10.4172/2324-8955.1000173.

El-Shazly, M.A., Attia, Y.A., Kabil, F.F., Anis, E. and Hazman, M. (2017). Inhibitory effects of salicylic acid and silver nanoparticles on potato virus Y-infected potato plants in Egypt. *Middle East Journal of Agriculture Research*, 6(3) (September): 835–848.

Elazzazy, A.M., Elbeshehy, E.K.F. and Betiha, M.A. (2017). *In vitro* assessment of activity of graphene silver composite sheets against multidrug-resistant bacteria and tomato bushy stunt virus. *Tropical Journal of Pharmaceutical Research*, 16(11) (November): 2705–2711. https://doi.org/10.4314/tjpr.v16i11.19.

Elbeshehy, E.K.F., Elazzazy, A.M. and Aggelis, G. (2015a). Silver nanoparticles synthesis mediated by new isolates of *Bacillus* spp., nanoparticle characterization and their activity against bean yellow mosaic virus and human pathogens. *Frontiers in Microbiology*, 6(May): 1–13. https://doi.org/10.3389/fmicb.2015.00453.

Elmer, W. and White, J.C. (2018). The future of nanotechnology in plant pathology. *Annual Review of Phytopathology*, 56(August): 111–133. https://doi.org/10.1146/annurev-phyto-080417-050108.

Elsharkawy, M.M. and Mousa, K.M. (2015). Induction of systemic resistance against Papaya ring spot virus (PRSV) and its vector *Myzus persicae* by *Penicillium simplicissimum* GP17-2 and silica (Sio 2) nanopowder. *International Journal of Pest Management*, 61(4) (June): 353–358. https://doi.org/10.1080/09670874.2015.1070930.

Elsharkawy, M.M. and Derbalah, A. (2019). Antiviral activity of titanium dioxide nanostructures as a control strategy for broad bean strain virus in faba bean. *Pest Management Science*, 75(3) (March): 828–834. https://doi.org/10.1002/ps.5185.

Eugene, K. and Zholobak, N. (2016). Antiviral activity of cerium dioxide Nanoparticles on tobacco mosaic virus model. *Proceedings of the Topical Issues of New Drugs Development*. Kharkiv, Ukraine 1 (April).

Fang, Y. and Ramasamy, R.P. (2015). Current and prospective methods for plant disease detection. *Biosensors*, 5(3) (August): 537–561. https://doi.org/10.3390/bios5030537.

Farooq, T., Adeel, M., He, Z., Umar, M., Shakoor, N., Da Silva, W., Elmer, W., White, J.C. and Rui, Y. (2021). Nanotechnology and plant viruses: An emerging disease management approach for resistant pathogens. *ACS Nano*, 15(4) (March): 6030–6037. https://doi.org/10.1021/acsnano.0c10910.

Fu, L., Wang, Z., Dhankher, O.P. and Xing, B. (2020). Nanotechnology as a new sustainable approach for controlling crop diseases and increasing agricultural production. *Journal of Experimental Botany*, 71(2) (January): 507–519. https://doi.org/10.1093/jxb/erz314.

Hajong, M., Devi, N.O., Debbarma, M. and Majumder, D. (2019). Nanotechnology: An emerging tool for management of biotic stresses in plants. pp. 299–335. *In*: Prasad, R. (ed.). *Plant Nanobionics*, Springer. https://doi.org/10.1007/978-3-030-16379-2_11.

Hao, Y., Yuan, W., Ma, C., White, J.C., Zhang, Z., Adeel, M., Zhou, T., Rui, Y. and Xing, B. (2018). Engineered nanomaterials suppress turnip mosaic virus infection in tobacco (Nicotiana benthamiana). *Environmental Science: Nano*, 5(7) (May): 1685–1693. https://doi.org/10.1039/C8EN00014J.

Hong, S. and Lee, C. (2018). The current status and future outlook of quantum dot-based biosensors for plant virus detection. *Plant Pathology Journal*, 34(2) (April): 85–92. https://doi.org/10.5423/PPJ.RW.08.2017.0184.

Jain, D. and Kothari, S.L. (2014). Green synthesis of silver nanoparticles and their application in plant virus inhibition. *J. Mycol. Plant Pathol.*, 44(1) (February): 21–24.

Jarocka, U., Wkasowicz Michał, Radecka, H., Malinowski, T., Michalczuk, L. and Radecki, J. (2011). Impedimetric immunosensor for detection of plum pox virus in plant extracts. *Electroanalysis*, 23(9) (July): 2197–2204. https://doi.org/10.1002/elan.201100152.

Kashyap, P.L., Kumar, S. and Srivastava, A.K. (2017). Nanodiagnostics for plant pathogens. *Environmental Chemistry Letters*, 15(1) (March): 7–13. https://doi.org/10.1007/s10311-016-0580-4.

Khater, M., de la Escosura-Muñiz, A. and Merkoçi, A. (2017). Biosensors for plant pathogen detection. *Biosensors and Bioelectronics*, 93(July): 72–86. https://doi.org/10.1016/j.bios.2016.09.091.

Khater, M., de la Escosura-Muñiz, A., Quesada-González, D. and Merkoçi, A. (2019a). Electrochemical detection of plant virus using gold nanoparticle-modified electrodes. *Analytica Chimica Acta*, 1046(January): 123–131. https://doi.org/10.1016/j.aca.2018.09.031.

Khater, M., Escosura-Muñiz, A. de la, Altet, L. and Merkoçi, A. (2019b). *In situ* plant virus nucleic acid isothermal amplification detection on gold nanoparticle-modified electrodes. *Analytical Chemistry*, 91(7) (March): 4790–4796. https://doi.org/10.1021/acs.analchem.9b00340.

Kumar, S., Chauhan, N., Gopal, M., Kumar, R. and Dilbaghi, N. (2015). Development and evaluation of alginate--chitosan nanocapsules for controlled release of acetamiprid. *International Journal of Biological Macromolecules*, 81(November): 631–637. https://doi.org/10.1016/j.ijbiomac.2015.08.062.

Kumar, V. and Arora, K. (2020). Trends in nano-inspired biosensors for plants. *Materials Science for Energy Technologies*, 3(January): 255–273. https://doi.org/10.1016/j.mset.2019.10.004.

Le, V.N., Rui, Y., Gui, X., Li, X., Liu, S. and Han, Y. (2014). Uptake, transport, distribution and bio-effects of SiO_2 nanoparticles in Bt-transgenic cotton. *Journal of Nanobiotechnology*, 12(1) (December): 1–15. https://doi.org/10.1186/s12951-014-0050-8.

Lin, H.-Y., Huang, C.-H., Lu, S.-H., Kuo, I.-T. and Chau, L.-K. (2014). Direct detection of orchid viruses using nanorod-based fiber optic particle plasmon resonance immunosensor. *Biosensors and Bioelectronics*, 51 (January): 371–378. https://doi.org/10.1016/j.bios.2013.08.009.

Ling, D.H. and Xing, B.S. (2007). Phytotoxicity of nanoparticles: Inhibition of seed germination and root elongation. *Environmental Pollution*, 150(2) (November): 243–250. https://doi.org/10.1016/j.envpol.2007.01.016.

Liu, F., Wen, L.-X., Li, Z.-Z., Yu, W., Sun, H.-Y. and Chen, J.-F. (2006). Porous hollow silica nanoparticles as controlled delivery system for water-soluble pesticide. *Materials Research Bulletin*, 41(12) (December): 2268–2275. https://doi.org/10.1016/j.materresbull.2006.04.014.

Lu, R., Liu, Z., Shao, Y., Sun, F., Zhang, Y., Cui, J., Zhou, Y., Shen, W. and Zhou, T. (2019). Melatonin is responsible for rice resistance to rice stripe virus infection through a nitric oxide-dependent pathway. *Virology Journal*, 16(1) (November): 1–8. https://doi.org/10.1186/s12985-019-1228-3.

Manjunatha, S.B., Biradar, D.P. and Aladakatti, Y.R. (2016). Nanotechnology and its applications in agriculture: A review. *J. Farm Sci.*, 29(1) (March): 1–13.

Maheshwari, Y., Vijayanandraj, S., Jain, R.K. and Mandal, B. (2018). Field-usable lateral flow immunoassay for the rapid detection of a macluravirus, large cardamom chirke virus. *Journal of Virological Methods*, 253(March): 43–48. https://doi.org/10.1016/j.jviromet.2017.12.009.

Misra, A.N., Misra, M. and Singh, R. (2013). Nanotechnology in agriculture and food industry. *International Journal of Pure and Applied Sciences and Technology*, 16(2) (May): 1–9.

Mitter, N., Worrall, E.A., Robinson, K.E., Li, P., Jain, R.G., Taochy, C., Fletcher, S.J., Carroll, B.J., Lu, G.Q. and Xu, Z.P. (2017). Clay nanosheets for topical delivery of RNAi for sustained protection against plant viruses. *Nature Plants*, 3(2) (January): 1–10. https://doi.org/10.1038/nplants.2016.207.

Mokrini, F. and Bouharroud, R. (2019). Application of nanotechnology in plant protection by phytopathogens: Present and future prospects. pp. 261–279. *In*: Prasad R. (ed.). *Microbial Nanobionics*: *Nanotechnology in the Life Sciences*. Springer. https://doi.org/10.1007/978-3-030-16534-5_13.

Nagai, A., Torres, P.B., Duarte, L.M.L., Chaves, A.L.R., Macedo, A.F., Floh, E.I.S., de Oliveira, L.F., Zuccarelli, R. and Dos Santos, D.Y.A.C. (2020). Signaling pathway played by salicylic acid, gentisic acid, nitric oxide, polyamines, and non-enzymatic antioxidants in compatible and incompatible Solanum-tomato mottle mosaic virus interactions. *Plant Science*, 290(January). https://doi.org/10.1016/j.plantsci.2019.110274.

Noha, K., Bondok, A.M. and El-Dougdoug, K.A. (2018). Evaluation of silver nanoparticles as antiviral agent against ToMV and PVY in tomato plants. *Sciences*, 8(1) (January): 100–111.

Panferov, V.G., Safenkova, I.V., Zherdev, A.V. and Dzantiev, B.B. (2018). Post-assay growth of gold nanoparticles as a tool for highly sensitive lateral flow immunoassay. Application to the detection of potato virus X. *Microchimica Acta*, 185(11) (October): 1–8. https://doi.org/10.1007/s00604-018-3052-7.

Prasad, R., Kumar, V. and Prasad, K.S. (2014). Nanotechnology in sustainable agriculture: Present concerns and future aspects. *African Journal of Biotechnology*, 13(6) (October): 705–713. https://doi.org/10.5897/ajbx2013.13554.

Rai, M. and Ingle, A. (2012). Role of nanotechnology in agriculture with special reference to management of insect pests. *Applied Microbiology and Biotechnology*, 94(2) (April): 287–293. https://doi.org/10.1007/s00253-012-3969-4.

Rajwade, J.M., Chikte, R.G. and Paknikar, K.M. (2020). Nanomaterials: New weapons in a crusade against phytopathogens. *Applied Microbiology and Biotechnology*, 104(4) (February): 1437–1461. https://doi.org/10.1007/s00253-019-10334-y.

Ramezani, M., Ramezani, F. and Gerami, M. (2019). Nanoparticles in pest incidences and plant disease control. pp. 233–272. *In*: Panpatte, D. and Jhala, Y. (eds.). *Nanotechnology for Aagriculture: Crop Production & Protection*. Springer. https://doi.org/10.1007/978-981-32-9374-8_12.

Razmi, A., Golestanipour, A., Nikkhah, M., Bagheri, A., Shamsbakhsh, M. and Malekzadeh-Shafaroudi, S. (2019). Localized surface plasmon resonance biosensing of tomato yellow leaf curl virus. *Journal of Virological Methods*, 267(May): 1–7. https://doi.org/10.1016/j.jviromet.2019.02.004.

Razo, S.C., Panferov, V.G., Safenkova, I.V., Varitsev, Y.A., Zherdev, A.V., Pakina, E.N. and Dzantiev, B.B. (2018). How to improve sensitivity of sandwich lateral flow immunoassay for corpuscular antigens on the example of potato virus Y? *Sensors*, 18(11) (November): 1–16. https://doi.org/10.3390/s18113975.

Ruotolo, R., Maestri, E., Pagano, L., Marmiroli, M., White, J.C. and Marmiroli, N. (2018). Plant response to metal-containing engineered nanomaterials: an omics-based perspective. *Environmental Science & Technology*, 52(5) (January): 2451–2467. https://doi.org/10.1021/acs.est.7b04121.

Safarnejad, M.R., Samiee, F., Tabatabie, M. and Mohsenifar, A. (2017). Development of quantum dot-based nanobiosensors against citrus tristeza virus (CTV). *Sensors & Transducers*, 213(6) (June): 54–60.

Sanzari, I., Leone, A. and Ambrosone, A. (2019). Nanotechnology in plant science: to make a long story short. *Frontiers in Bioengineering and Biotechnology*, 7(May): 1–12. https://doi.org/10.3389/fbioe.2019.00120.

Sathiyabama, M. and Parthasarathy, R. (2016). Biological preparation of chitosan nanoparticles and its *in vitro* antifungal efficacy against some phytopathogenic fungi. *Carbohydrate Polymers*, 151(October): 321–325. https://doi.org/10.1016/j.carbpol.2016.05.033.

Shafie, R.M., Salama, A.M. and Farroh, K.Y. (2018). Silver nanoparticles activity against Tomato spotted wilt virus. *Middle East Journal of Agriculture Research*, 7(4) (October): 1251–1267.

Shang, Y., Kamrul Hasan, M., Ahammed, G.J., Li, M., Yin, H. and Zhou, J. (2019). Applications of nanotechnology in plant growth and crop protection: A review. *Molecules*, 24(14) (July): 1–23. https://doi.org/10.3390/molecules24142558.

Shojaei, T.R., Salleh, M.A.M., Sijam, K., Rahim, R.A., Mohsenifar, A., Safarnejad, R. and Tabatabaei, M. (2016a). Fluorometric immunoassay for detecting the plant virus Citrus tristeza using carbon nanoparticles acting as quenchers and antibodies labeled with CdTe quantum dots. *Microchimica Acta*, 183(7) (July): 2277–2287. https://doi.org/10.1007/s00604-016-1867-7.

Shojaei, T.R., Salleh, M.A.M., Sijam, K., Rahim, R.A., Mohsenifar, A., Safarnejad, R. and Tabatabaei, M. (2016b). Detection of Citrus tristeza virus by using fluorescence resonance energy transfer-based biosensor. *Spectrochimica Acta Part A: Molecular and Biomolecular Spectroscopy*, 169(December): 216–222. https://doi.org/10.1016/j.saa.2016.06.052.

Sun, D., Hussain, H.I., Yi, Z., Siegele, R., Cresswell, T., Kong, L. and Cahill, D.M. (2014). Uptake and cellular distribution, in four plant species, of fluorescently labeled mesoporous silica nanoparticles. *Plant Cell Reports*, 33(8) (August): 1389–1402. https://doi.org/10.1007/s00299-014-1624-5.

Sun, W., Zhong, J., Qin, P. and Jiao, K. (2008). Electrochemical biosensor for the detection of cauliflower mosaic virus 35 S gene sequences using lead sulfide nanoparticles as oligonucleotide labels. *Analytical Biochemistry*, 377(2) (June): 115–119. https://doi.org/10.1016/j.ab.2008.03.027.

Tahir, M.A., Hameed, S., Munawar, A., Amin, I., Mansoor, S., Khan, W.S. and Bajwa, S.Z. (2017). Investigating the potential of multiwalled carbon nanotubes based zinc nanocomposite as a recognition interface towards plant pathogen detection. *Journal of Virological Methods*, 249(November): 130–136. https://doi.org/10.1016/j.jviromet.2017.09.004.

Tahir, M.A., Bajwa, S.Z., Mansoor, S., Briddon, R.W., Khan, W.S., Scheffler, B.E. and Amin, I. (2018). Evaluation of carbon nanotube based copper nanoparticle composite for the efficient detection of agroviruses. *Journal of Hazardous Materials*, 346(March): 27–35. https://doi.org/10.1016/j.jhazmat.2017.12.007.

Tereshchenko, A., Fedorenko, V., Smyntyna, V., Konup, I., Konup, A., Eriksson, M., Yakimova, R., Ramanavicius, A., Balme, S. and Bechelany, M. (2017). Biosensors and Bioelectronics ZnO films formed by atomic layer deposition as an optical biosensor platform for the detection of grapevine virus A-type proteins. *Biosensors and Bioelectronic*, 92(June): 763–769. https://doi.org/10.1016/j.bios.2016.09.071.

Tortella, G.R., Rubilar, O., Diez, M.C., Padrão, J., Zille, A., Pieretti, J.C. and Seabra, A.B. (2021). Advanced material against human (including Covid-19) and plant viruses: Nanoparticles as a feasible strategy. *Global Challenges*, 5(3) (March): 1–13. https://doi.org/10.1002/gch2.202000049.

Vargas-Hernandez, M., Macias-Bobadilla, I., Guevara-Gonzalez, R.G., Rico-Garcia, E., Ocampo-Velazquez, R.V., Avila-Juarez, L. and Torres-Pacheco, I. (2020). Nanoparticles as potential antivirals in agriculture. *Agriculture (Switzerland)*, 10(10) (September): 1–18. https://doi.org/10.3390/agriculture10100444.

Wang, P., Lombi, E., Zhao, F.J. and Kopittke, P.M. (2016). Nanotechnology: A new opportunity in plant sciences. *Trends in Plant Science*, 21(8) (August): 699–712. https://doi.org/10.1016/j.tplants.2016.04.005.

Wang, Y., Sun, C., Xu, C., Wang, Z., Zhao, M., Wang, C., Liu, L. and Chen, F. (2016). Preliminary experiments on nano-silver against tobacco mosaic virus and its mechanism. *Tob. Sci. Technol.*, 49: 22–30. 10.16135/j.issn1002-0861.20160104.

Wei, J., Liu, H., Liu, F., Zhu, M., Zhou, X. and Xing, D. (2014). Miniaturized paper-based gene sensor for rapid and sensitive identification of contagious plant virus. *ACS Applied Materials & Interfaces*, 6(24) (November): 22577–22584. https://doi.org/10.1021/am506695g.

Worrall, E.A., Hamid, A., Mody, K.T., Mitter, N. and Pappu, H.R. (2018). Nanotechnology for plant disease management. *Agronomy*, 8(12) (November): 1–24. https://doi.org/10.3390/agronomy8120285.

Worrall, E.A., Bravo-Cazar, A., Nilon, A.T., Fletcher, S.J., Robinson, K E., Carr, J.P. and Mitter, N. (2019). Exogenous application of RNAi-inducing double-stranded RNA inhibits aphid-mediated transmission of a plant virus. *Frontiers in Plant Science*, 10(March): 1–9. https://doi.org/10.3389/fpls.2019.00265.

Zhu, H., Han, J., Xiao, J.Q. and Jin, Y. (2008). Uptake, translocation, and accumulation of manufactured iron oxide nanoparticles by pumpkin plants. *Journal of Environmental Monitoring*, 10(6) (May): 713–717. https://doi.org/10.1039/B805998E.

Miscellaneous Application of Nanoparticles

Chapter 9

Nanosilver and Smart Delivery in Agricultural System

Rythem Anand and *Madhulika Bhagat**

1. Introduction

Recent advances have been made to improve multiple agriculture-related problems by increasing the efficient inputs and minimizing the irrelevant losses through exhaustive research in nanotechnology (Fig. 9.1). The fabrication of nanomaterials of different shapes and sizes leading to a wide array of applications in agricultural development have gained attention of scientists around the world (Shang et al., 2019). These nanomaterials have helped in the reduction of toxicity issues and nutrient losses during the application of agrochemicals in the fields thereby increasing crop yields. To have proper management of crop production, the nanoparticles (NPs) are either used alone or as a carrier of the various agrochemicals such as insecticides, pesticides, fertilisers, or herbicides (Kaushik and Djiwanti, 2017). These NPs (nano-based agrochemicals) have a targeted and sustained delivery of active ingredients that decreases the cost and environmental impact (Kumar et al., 2019).

Despite these numerous advantages of nanotechnology and the growing trends in publications and patents, agricultural applications have not yet made it to the market. Therefore, there is an urgent need of integration of new tools and techniques in agriculture nanotechnology for better regulatory evaluation and environmental concerns of these nanomaterials.

It has been studied that the nanomaterials have distinct physical, chemical, and biological properties as compared to larger bulk materials. Therefore, nanoparticles with desired characteristics, like shape, pore size, and surface properties could be engineered to have precise and targeted delivery via adsorption, encapsulation, or conjugation (Khandelwal et al., 2016). NPs can be prepared in a several ways such as gamma irradiation, chemical reduction, photochemical methods, etc. These methods are costly having higher energy demands along with the liberation of

The School of Biotechnology, University of Jammu, Jammu-180006 (J&K), India.
* Corresponding author: madhulikasbt@gmail.com

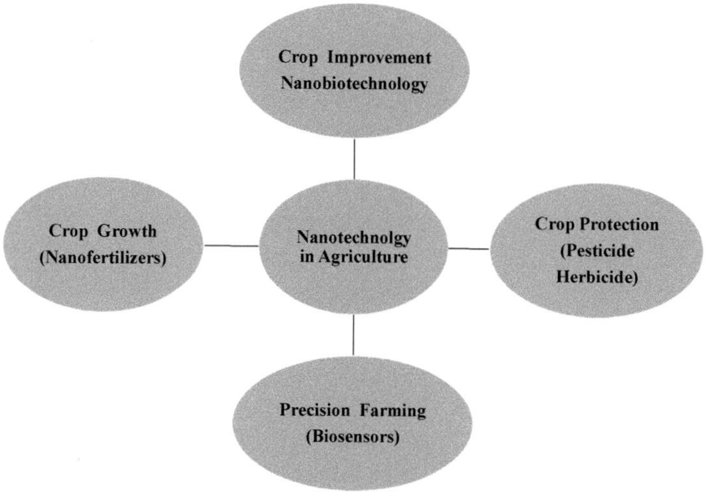

Fig. 9.1. Application of silver nanoparticles in agriculture.

hazardous chemicals into the environment (Clifford et al., 2017). Thus, application of biologically synthesized nanomaterials from various plants or microbial systems, often considered as green technology, could help in minimizing the toxic releases into the environment. Green nanotechnology, being a safe and energy efficient process, reduces waste and greenhouse gas emissions into the surroundings (Prasad et al., 2017). Another area of nanotechnology known as Phytonanotechnology involves development of eco-friendly NPs for their uptake and targeted delivery in plants for increased crop production, nutrient utilization as well as reduced disease susceptibility (Wang et al., 2016; Wang et al., 2017). Therefore, major emphasis should be given to the biogenic or green synthesis of NPs involving reduction by biological enzymes secreted by microorganisms, plant extract, or any other biological agent.

Over the years, scientists have been working on the synthesis of various nanoscale particles of organic or inorganic origins with properties superior to their bulk counterparts (Saratale et al., 2018). Organic NPs include liposomes, dendrimers, micelles whereas the inorganic ones consist of the metallic, magnetic and semiconductor nanoparticles (Elias and Saravanakumar, 2017). Amongst all the different types of nanomaterials mentioned above, the metallic ones are gaining attention of the researchers due to their specific properties (Khan et al., 2019).

Silver (Ag) NPs is the most popular nontoxic metallic NPs and has been in use since time immemorial due to their various inhibitory properties such as antifungal, antibacterial, anti-inflammatory, antiviral, antioxidant, etc., making them a suitable tool in agroecosystems (Firdhouse et al., 2015). In fact, silver ions and compounds related to them are known to have low toxicity toward animal cells but high toxicity towards microorganisms. AgNPs have a surface area/zone that provides them with striking biochemical reactivity, catalytic activity, and atomic behaviour (Khatoon et al., 2018). Therefore, AgNPs could be exploited for their use in various fields, especially in the agricultural sector due to their different physical and chemical properties. One of the important features of the AgNPs is high surface area to volume ratio that increases their efficiency in comparison to their bulk counterparts (Chouhan,

2018). Although bulk materials possess various physical and chemicals properties, the change in size and shape at nanoscale leads to change in these properties also. Lesser the size, more is the surface area to volume ratio (Beyene, 2017). The particles are also known to possess some phenomenon of arbitrary motion, dual nature of particles, and uncertainty of matter due to the prominence of Van der Waal forces and reduce the effect of gravity in nanostructures (Forbes, 2019). Moreover, the biological activity of AgNPs also depends on size distribution, shape, particle morphology, composition, capping, agglomeration, ion release efficiency, type of reducing agents, etc. (Rónavári et al., 2017). It was also investigated that the AgNPs showed catalytic properties when supported on the silica spheres (Jiang et al., 2005). These silica spheres prevent the flocculation of metal NPs which in turn helps in their proper catalytic activity. Surfactants are also used in the synthesis processes of the metal nanosized particles that are known to increase the catalytic activity of the particles which in turn could be used in various textile industries (Suvith et al., 2014).

2. Synthesis of Silver Nanoparticles

'Top–down' and 'Bottom–up' are the two approaches for synthesis of NPs. In the top–down approach, suitable bulk material splits into fine particles by size reduction with different techniques, i.e., pulse laser ablation, evaporation–condensation, ball milling, pulse wire discharge method, etc. In the bottom–up approach, NPs can be synthesized using chemical and biological methods by the self-assembly phenomenon of atoms to new nuclei which grow into a particle of nanoscale (Table 9.1). Chemical reduction is one of the bottom–up approaches being mostly in use for the synthesis of AgNPs (Rafique et al., 2017). Various organic and inorganic reducing agents in aqueous or non-aqueous solutions are used for the reduction of Ag ions, e.g., poly-ethylene glycol copolymers, sodium citrate, Ascorbate, N, N-dimethyl formamide (DMF), and sodium borohydride ($NaBH_4$) (Tran and Le, 2013; Rafique et al., 2017). Capping agents are additionally used for size stabilization of the NPs. One of the greatest advantages of the chemical reduction method is generation of substantial amounts of particles in a short period of time. However, these involve release of toxic and non-eco-friendly products which shifts the interest towards green methods over the chemical ones. The green methods are environment-friendly, cost-effective, and could easily be scaled up for large-scale production of NPs. Other benefits of the green method are non-involvement of high temperature, energy, pressure, and harmful chemicals wastes (Ahmed et al., 2016). In the green methods, there need to be a proper choice of solvent, reducing agent and stabiliser used for the synthesis (Shui et al., 2019). There is reduction of aqueous salt solution of silver by enzymes secreted by various biological agents such as microbes or plant extracts thereby yielding silver nanoparticles (Raveendran et al., 2003). Ambient temperature, pressure and neutral pH are the requirements for the efficient synthesis, and the different biological systems used in the process provide these suitable conditions. Therefore, it was reported that microorganisms are nano factories for producing of nanomaterials especially AgNPs (Burkett and Mann, 1996; Mann, 1996). Still the microbial sources for production of NPs are not entirely safe and need aseptic conditions. On the other hand, plants also act as a hub of the metabolites in reducing the metal ions as well as capping and stabilizing agents in the green methods.

Table 9.1. Various methodologies for synthesis of silver nanoparticles.

Methodology for the Synthesis of Silver Nanoparticles		
Top–down Approach	**Bottom–up Approach**	
Physical Methods	**Chemical Methods**	**Green Methods**
• Pulsed laser ablation • Evaporation–condensation • Arc discharge • Spray Pyrolysis • Ball milling • Vapor and gas phase • Pulse wire discharge • Lithography	• Chemical reduction • Sono-chemical • Microemulsion • Photochemical • Electrochemical • Pyrolysis • Microwave • Solvothermal • Co-precipitation	• Plant and their extracts • Enzymes and biomolecules • Microorganisms

Moreover, in the biological methods there are several parameters affecting the synthesis of AgNPs such as pH, temperature, reaction time, plant extract concentrations directly altering the shape and size of particles. pH determines the shape, size, production rate, and stability of the particles. Alkaline pH favours the synthesis process of the NPs while acidic pH does not favour the production (Sumi et al., 2017). Temperature plays an important role in the synthesis methods. It has a direct effect on the formation of the nanoparticles, Higher the temperature, higher is the synthesis rate of the particles. More the time given for the process to take place, more will be the synthesis of the particles via increased conversion rate of salt solution by reducing agents (Krishanraj et al., 2010). The concentration of plant extract directly affects the synthesis process of the AgNPs. Daniel et al. (2013) studied that the decrease in concentration of *Azadiracta indica* leaves extract leads to the formation of more stable AgNPs than the ones with decreased concentration of plant extract. It is studied that the lower quantity of the plant extract yields stable and uniform NPs (Song et al., 2009).

3. Characterisation of Silver Nanoparticles

To confirm the formation of AgNPs, the various aspects of NPs such as size, shape, functional groups, etc., need to be analysed by various techniques.

Ultraviolet visible (UV-Vis) spectral analysis of AgNPs is carried out using a spectrophotometer in the range of 350–470 nm. The silver ions, synthesized by biological and chemical methods exhibit a characteristic surface plasmon band, are monitored by the UV-Vis spectrum of the reaction mixture. Nano silver shows an absorbance peak around 420 nm–450 nm which is further supported by the change in colour of the NP suspension (Gudikandula et al., 2016).

Electron Microscopy consists of both Scanning Electron Microscopy (SEM) and Transmission Electron Microscopy (TEM) that are commonly used in the detection of the size and shape of the particles. SEM scans the surface of the sample with an electron beam that are usually the secondary electrons (Liu et al., 2016). TEM provides the details of the sample such as size, shape, and morphology by passing an electron beam through the sample (Jayaramudu et al., 2016). Both these imaging techniques are done in vacuum for solid and dehydrated liquid samples.

Energy dispersive spectroscopy (EDS) illustrates the elemental composition. The spectra show several peaks that correspond to elements present in the sample.

X-ray diffraction technique analyses the structural properties of the metal NPs. The different peaks formed on the XRD graph corresponds to different angles that help in identification of different phases present in the sample. AgNPs and the other metal NPs are found to have a crystalline structure.

Atomic force microscopy (AFM) gives the 3D images of the AgNPs by moving a tip on the surface of the particles that emits an electron beam.

Dynamic light scattering (DLS) analysis finds out the size of the particles that are suspended in the liquid solution showing the Brownian movement. There is scattering of light and the intensity of scattered light is measured by the instrument giving details about the size of synthesized NPs, i.e., Z size and net charge present on the particles known as Zeta potential of the particles. Higher the value, higher will be the electrostatic repulsion, less will be the aggregation and more will be the stability of particles.

Fourier transform infrared spectroscopy (FT-IR) gives the information about functional groups that are present in the AgNPs synthesized by both the biogenic and chemical methods. There is a conversion of silver nitrate to elemental silver by phytochemicals in green synthesis and proteins/chemicals in chemical reduction. These act as capping, reducing, and stabilizing agents. FT-IR spectra show different bands that correspond different bonds and stretching. The different bonds are due to the functional groups that confirms the formation of the NPs by the action of the protein or phytochemicals.

Thermogravimetric analysis (TGA) is used to measure the concentration of silver in the suspension. A known volume of the AgNP suspension is passed through nitrogen gas. The weight of the particles is measured as the temperature changes. When the temperature reaches beyond 100°C, the solvent evaporates and the particles remains in the pan (Kang et al., 2016).

4. Silver Nanoparticles in Agriculture

The potential uses and advantages of AgNPs in agriculture involve plant disease management (Liang et al., 2017), enhanced nutrient uptake efficiency (Zuverza Mena et al., 2016), improved plant growth (Sataraddi and Nandibewoor, 2012), prevention of fungal diseases (Mishra et al., 2014), enhancement of fruit ripening (Samrot et al., 2018), and sustainable release of agrochemicals (Duhan et al., 2017).

4.1 Nanosilver for Crop Protection

Usually, chemicals used in the fields for treatment of nutrient deficiencies and plant diseases, do not reach their target site and even if they reach, they are not sufficient to meet the requirement of the plant. Therefore, these are applied several times leading to harmful effects on the environment. To combat the problem, researchers are exploiting nano silver for the targeted delivery. These are designed for controlled, enhanced, and targeted delivery having high stability, solubility, effective concentration, and least toxicity towards the environment. AgNPs are reported to have a diverse role as antibacterial and antifungal agents controlling pathogens of the plants (Jo et al.,

2009). Flowering in *Gerbera* are known to be enhanced by the application of AgNPs that helps in preventing microbial invasion, increasing water uptake, and reducing vascular blockage (Solgi et al., 2009). Moreover, the well-dispersed and stabilized AgNPs are more adhesive on bacterial and fungal cell surface acting as bactericide and fungicide (Ocsoy et al., 2013) have also developed DNA-directed AgNPs grown on grapheme oxide and studied the antibacterial activity against Xanthomonas perforans, a causative agent of bacterial spot in tomatoes.

4.2 Nano Silver for Crop Growth

AgNPs have been used in agriculture to increase the crop yield by enhancing the seed germination and plant growth that is highly dose dependent and may have a positive or negative effect on growth. Exposure of plants to higher or lower concentrations of AgNPs may have an inhibitory effect on plant growth, therefore there is a requirement for proper concentration of nanomaterials. Large amounts of commercial fertilizers are used in the fields in the form of urea and nitrate, etc., but they possess toxic effects on plants and beneficial microflora, so nano silver provides an improved uptake of nutrients from the soil as compared to commercial ones. AgNPs were developed as plant-growth stimulators (Monica and Ceromonini, 2009), or agents for enhancement of fruit ripening (Vinkovic et al., 2017). Almutairi and Alharbi (2015) also examined the effect of Ag NPs dosage on the seed germination in species of corn, watermelon, and zucchini. It was found that germination rates get enhanced in response to AgNPs, keeping in view the concentration of the particles.

4.3 Smart Delivery of Nano Silver in Plants

NPs can serve as herbicides, pesticides, and fertilizers in agricultural systems due to their nanoscale properties making them the smart delivery systems having controlled and targeted delivery. It has been observed that plants prove to be the best transport pathway for NPs having better penetration through the cuticle of plants. The plant cell wall, a major barrier for the entry of most of external agents into the cell, have a pore size of mostly 5–20 nm (Yan et al., 2019). Therefore, NPs smaller than this diameter can easily pass through these pores further carrying out the activities. Moreover, it has also been studied that AgNPs can enlarge the pore size by interacting with the cell wall enhancing the uptake and internalization of large AgNPs (Tripathi et al., 2017). Once in the cells, the particles interact with the different organelles and effect their metabolic processes (Navarro et al., 2008). In other studies, foliar uptake of nanoparticles is also reported in the watermelon plant through aerosol route (Wang et al., 2013). Geisler-Lee (2014) found that the cotyledons of *Arabidopsis* when suspended in AgNPs containing the medium accumulated the particles in the stomatal guard cells that further penetrates leaf tissue and other organs. There has been more silver bioaccumulation in foliar exposure than root exposure (Li et al., 2017).

AgNPs are adsorbed easily by roots and their accumulation in roots depend upon the properties of particles and environmental conditions. Doolette et al. (2015) studied the plant uptake of Ag from the biosolids-amended soil containing Ag_2S-NPs, reported that ammonium thiosulfate and potassium chloride fertilization significantly increased the AgNPs concentrations of plant roots and shoots. Another such mechanism for increased bioavailability of metal NPs in rhizosphere is the presence of microbial siderophores and root exudates that help in nutrient limiting conditions (Patel, 2018). Thus, chelation between siderophores and metals promoted dissolution and bioavailability of metallic NPs. In addition, root exudates also enhance nutrients uptake from insoluble sources (Jones and Darrah, 1994). NPs accumulated in plant roots easily translocate to the shoot or other tissues including newly developing seeds (Rico et al., 2011).

5. Conclusions and Perspective

NPs present a platform for a varied range of biological applications in various fields in modern society where their dispersal and permeation into the ecosystem became inevitable. During the past decade, the research communities undertook the responsibility to increase our knowledge on the interaction between AgNPs and plants, keeping in view their impact on plants and the environment. In this article, the focus was on the biosynthesis of AgNPs giving direct or indirect input to the agricultural ecosystem. AgNPS prove to be an efficient tool in agriculture, providing green, eco-friendly alternatives, and efficient delivery systems. Due to the unique physical and chemical properties of metallic NPs, there is still a need to fully understand and predict their environmental behaviour. Fate and transport of metallic NPs largely depends on the conditions of surrounding environment such as pH, ionic strength, redox state, and the organic matter presented. Besides, the particles could also tend to undergo different physical, chemical, or biological transformations simultaneously (e.g., oxidation and aggregation). However, more information on the interaction of NPs with soil, plant, and soil microbial in the agriculture system need to be studied and analysed further. The development of methodologies, such as transcriptomics, proteomics, and metabolomics, could further be employed in this field soon.

Disclosure statement

No conflict of interest was reported by the authors.

References

Ahmed, S. (2016). A review on plant extract mediated synthesis of silver nanoparticles for antimicrobial applications: A green expertise. *J. Adv. Res.*, 7(1): 17–28.

Almutairi, Z.M. and Alharbi, A. (2015). Effect of silver nanoparticles on seed germination of crop plants. *J. Adv. Agric.*, 4(1): 283–288.

Beyene, H.D., Werkneh, A.A., Bezabh, H.K. and Ambaye, T.G. (2017). Synthesis paradigm and applications of silver nanoparticles (AgNPs): A review. *Sus. Mater. Tech.*, 13: 18–23.

Burkett, S.L. and Mann, S. (1996). Spatial organization and patterning of gold nanoparticles on self-assembled biolipid tubular templates. *Chemical Communications*, 1(3); 321–2.

Chouhan, N. (2018). Silver nanoparticles: Fabrication, characterisation, and applications. *Intechopen.*, 75611: 21–57. DOI: 10.5772/intechopen.75611.

Clifford, D.M., Castano, C.E. and Rojas, J.V. (2017). Supported transition metal nanomaterials: Nanocomposites synthesized by ionizing radiation. *Rad. Phy. Chem.*, 132: 52–64.

Daniel, S.K., Vinothini, G., Subramanian, N., Nehru, K. and Sivakumar, M. (2013). Biosynthesis of Cu, ZVI, and Ag nanoparticles using *Dodonaea viscosa* extract for antibacterial activity against human pathogens. *J. Nano. Res.*, 15(1): 1–10.

Doolette, C.L., McLaughlin, M.J. and Kirby, J.K. (2015). Bioavailability of silver and silver sulfide nanoparticles to lettuce (*Lactuca sativa*): Effect of agricultural amendments on plant uptake. *J. Hazard Mater.*, 300: 788–795.

Duhan, J.S., Kumar, R., Kumar, N., Kaur, P., Nehra, K. and Duhan, S. (2017). Nanotechnology: The new perspective in precision agriculture. *Biotech Rep.*, 15: 11–23.

Ealias, A.M. and Saravanakumar, M.P. (2017). A review on the classification, characterisation, synthesis of nanoparticles and their application. *In*: *IOP Conf. Ser. Mater. Sci. Eng.*, 263(3): 032019.

Firdhouse, M.J. and Lalitha, P. (2015). Biosynthesis of silver nanoparticles and its applications. *J. Nanotech.*, 2015: 1–18.

Forbes, K.A., Bradshaw, D.S. and Andrews, D.L. (2019). Optical binding of nanoparticles. *Nanophot.*, 9(1): 1–7.

Geisler-Lee, J., Brooks, M., Gerfen, J.R., Wang, Q., Fotis, C., Sparer, A., Ma, X., Berg, R.H. and Geisler, M. (2014). Reproductive toxicity and life history study of silver nanoparticle effect, uptake, and transport in *Arabidopsis thaliana*. *Nanomat.*, 4(2): 301–318.

Gudikandula, K. and Charya Maringanti, S. (2016). Synthesis of silver nanoparticles by chemical and biological methods and their antimicrobial properties. *J. Exp. Nanosc.*, 11(9): 714–721.

Jayaramudu, T., Raghavendra, G.M., Varaprasad, K., Reddy, G.V., Reddy, A.B., Sudhakar, K. and Sadiku, E.R. (2016). Preparation and characterization of poly (ethylene glycol) stabilized nano silver particles by a mechanochemical assisted ball mill process. *J. App. Poly Sc.*, 133(7): 1–8.

Jiang, Z.J., Liu, C.Y. and Sun, L.W. (2005). Catalytic properties of silver nanoparticles supported on silica spheres. *J. Phys Chem.*, 109(5): 1730–1735.

Jo, Y.K., Kim, B.H. and Jung, G. (2009). Antifungal activity of silver ions and nanoparticles on phytopathogenic fungi. *Plant Dis.*, 93(10): 1037–1043.

Jones, D.L. and Darrah, P.R. (1994). Role of root derived organic acids in the mobilization of nutrients from the rhizosphere. *Plant and Soil*, 166(2): 247–257.

Kang, C.K., Kim, S.S., Kim, S., Lee, J., Lee, J.H., Roh, C. and Lee, J. (2016). Antibacterial cotton fibres treated with silver nanoparticles and quaternary ammonium salts. *Carb. Poly.*, 151: 1012–1018.

Kaushik, S. and Djiwanti, S.R. (2017). Nanotechnology for enhancing crop productivity. pp. 249–262. *In*: *Nanotech*. Springer, Singapore.

Khan, I., Saeed, K. and Khan, I. (2019). Nanoparticles: Properties, applications, and toxicities. *Arab. J. Chem.*, 12(7): 908–931.

Khandelwal, N., Barbole, R.S., Banerjee, S.S. and Chate, G.P. (2016). Budding trends in integrated pest management using advanced micro-and nanomaterials: Challenges and perspectives. *J. Environ. Manag.*, 184: 157–169.

Khatoon, U.T., Rao, G.N., Mantravadi, K.M. and Oztekin, Y. (2018). Strategies to synthesize various nanostructures of silver and their applications: A review. *RSC Adv.*, 8(35): 19739–19753.

Krishnaraj, C., Jagan, E.G., Rajasekar, S., Selvakumar, P., Kalaichelvan, P.T. and Mohan, N.J. (2010). Synthesis of silver nanoparticles using Acalypha indica leaf extracts and its antibacterial activity against water borne pathogens. *Col. Surf. B: Bioint.*, 76(1): 50–56.

Kumar, S., Nehra, M., Dilbaghi, N., Marrazza, G., Hassan, A.A. and Kim, K.H. (2019). Nano-based smart pesticide formulations: Emerging opportunities for agriculture. *J. Cont. Rel.*, 294: 131–153.

Li, C.C., Dang, F., Li, M., Zhu, M., Zhong, H., Hintelmann, H. and Zhou, D.M. (2017). Effects of exposure pathways on the accumulation and phytotoxicity of silver nanoparticles in soybean and rice. *Nanotoxi.*, 11(5): 699–709.

Liang, Y., Yang, D. and Cui, J. (2017). A graphene oxide/silver nanoparticle composite as a novel agricultural antibacterial agent against *Xanthomonas oryzae pv. oryzae* for crop disease management. *New J. Chem.*, 41(22): 13692–13699.

Liu, T., Asiri, A.M. and Sun, X. (2016). Electrodeposited co-doped NiSe 2 nanoparticles film: A good electrocatalyst for efficient water splitting. *Nanoscale*, 8(7): 3911–3915.

Mann, S. ed. (1996). Biomimetic Materials Chemistry. John Wiley & Sons.

Mishra, S., Singh, B.R., Singh, A., Keswani, C., Naqvi, A.H. and Singh, H.B. (2014). Biofabricated silver nanoparticles act as a strong fungicide against *Bipolaris sorokiniana* causing spot blotch disease in wheat. *PLoS One*, 9(5): e97881.

Monica, R.C. and Cremonini, R. (2009). Nanoparticles and higher plants. *Caryolog.*, 62(2): 161–165.

Navarro, E., Baun, A., Behra, R., Hartmann, N.B., Filser, J., Miao, A.J., Quigg, A., Santschi, P.H. and Sigg, L. (2008). Environmental behavior and ecotoxicity of engineered nanoparticles to algae, plants, and fungi. *Ecotoxic.*, 17(5): 372–386.

Ocsoy, I., Paret, M.L., Ocsoy, M.A., Kunwar, S., Chen, T., You, M. and Tan, W. (2013). Nanotechnology in plant disease management: DNA-directed silver nanoparticles on graphene oxide as an antibacterial against Xanthomonas perforans. *ACS Nano.*, 7(10): 8972–8980.

Patel, P.R., Shaikh, S.S. and Sayyed, R.Z. (2018). Modified chrome azurol S method for detection and estimation of siderophores having affinity for metal ions other than iron. *Environ. Sust.*, 1(1): 81–87.

Prasad, R., Bhattacharyya, A. and Nguyen, Q.D. (2017). Nanotechnology in sustainable agriculture: recent developments, challenges, and perspectives. *Front. Microbio.*, 8: 1014.

Rafique, M., Sadaf, I., Rafique, M.S. and Tahir, M.B. (2017). A review on green synthesis of silver nanoparticles and their applications. *Art. Cel. Nanomed. Biotech.*, 45(7): 1272–1291.

Raveendran, P., Fu, J. and Wallen, S.L. (2003). Completely 'green' synthesis and stabilization of metal nanoparticles. *J. Amer. Chem. Soc.*, 125(46): 13940–13941.

Rico, C.M., Majumdar, S., Duarte-Gardea, M., Peralta-Videa, J.R. and Gardea-Torresdey, J.L. (2011). Interaction of nanoparticles with edible plants and their possible implications in the food chain. *J. Agri. Food Chem.*, 59(8): 3485–3498.

Rónavári, A., Kovács, D., Igaz, N., Vágvölgyi, C., Boros, I.M., Kónya, Z., Pfeiffer, I. and Kiricsi, M. (2017). Biological activity of green-synthesized silver nanoparticles depends on the applied natural extracts: A comprehensive study. *Int. J. Nanomed.*, 12: 871.

Samrot, A.V., Shobana, N. and Jenna, R. (2018). Antibacterial and antioxidant activity of different staged ripened fruit of Capsicum annuum and its green synthesized silver nanoparticles. *BioNanoSc.*, 8(2): 632–646.

Saratale, R.G., Saratale, G.D., Shin, H.S., Jacob, J.M., Pugazhendhi, A., Bhaisare, M. and Kumar, G. (2018). New insights on the green synthesis of metallic nanoparticles using plant and waste biomaterials: Current knowledge, their agricultural and environmental applications. *Env. Sc. Poll. Res.*, 25(11): 10164–10183.

Sataraddi, S.R. and Nandibewoor, S.T. (2012). Bio synthesis, characterization, and activity studies of Ag nanoparticles by *Costus ingneus* insulin plant extract. *Der. Pharma. Let.*, 4: 152–158.

Shang, Y., Hasan, M., Ahammed, G.J., Li, M., Yin, H. and Zhou, J. (2019). Applications of nanotechnology in plant growth and crop protection: A review. *Molecules*, 24(14): 2558.

Shui, L., Zhang, G., Hu, B., Chen, X., Jin, M., Zhou, G., Li, N., Muhler, M. and Peng, B. (2019). Photocatalytic one-step synthesis of Ag nanoparticles without reducing agent and their catalytic redox performance supported on carbon. *J. Energy Chem.*, 36: 37–46.

Solgi, M., Kafi, M., Taghavi, T.S. and Naderi, R. (2009). Essential oils and silver nanoparticles (SNP) as novel agents to extend vase-life of gerbera (*Gerbera jamesonii* cv.'Dune') flowers. *Posthar. BioTech.*, 53(3): 155–158.

Song, J.Y. and Kim, B.S. (2009). Rapid biological synthesis of silver nanoparticles using plant leaf extracts. *Biopro. Biosys. Eng.*, 32(1): 79.

Sumi, M.B., Devadiga, A., Shetty, K.V. and MB, S. (2017). Solar photo-catalytically active, engineered silver nanoparticle synthesis using aqueous extract of mesocarp of *Cocos nucifera* (Red Spicata Dwarf). *J. Exp. Nanosc.*, 12(1): 14–32.

Suvith, V.S. and Philip, D. (2014). Catalytic degradation of methylene blue using biosynthesized gold and silver nanoparticles. *Spec. Chim. Acta Part A: Mol. Biomol. Spec.*, 118: 526–532.

Tripathi, D.K., Tripathi, A., Singh, S., Singh, Y., Vishwakarma, K., Yadav, G., Sharma, S., Singh, V.K., Mishra, R.K. and Upadhyay, R.G. (2017). Uptake, accumulation, and toxicity of silver nanoparticle in autotrophic plants, and heterotrophic microbes: A concentric review. *Front. Microbiol.*, 8: 7.

Vinković, T., Novák, O., Strnad, M., Goessler, W., Jurašin, D.D., Parađiković, N. and Vrček, I.V. (2017). Cytokinin response in pepper plants (*Capsicum annuum* L.) exposed to silver nanoparticles. *Environ. Res.*, 156: 10–18.

Wang, W.N., Tarafdar, J.C. and Biswas, P. (2013). Nanoparticle synthesis and delivery by an aerosol route for watermelon plant foliar uptake. *J. Nanopart. Res.*, 15(1): 1417.

Wang, P., Lombi, E., Zhao, F.J. and Kopittke, P.M. (2016). Nanotechnology: A new opportunity in plant sciences. *Trends in Plant Science*, 21(8): 699–712.

Wang, P., Lombi, E., Sun, S., Scheckel, K.G., Malysheva, A., McKenna, B.A., Menzies, N.W., Zhao, F.J. and Kopittke, P.M. (2017). Characterizing the uptake, accumulation, and toxicity of silver sulfide nanoparticles in plants. *Env. Sc.: Nano.* 4(2): 448–460.

Yan, D., Yadav, S.R., Paterlini, A., Nicolas, W.J., Petit, J.D., Brocard, L., Belevich, I., Grison, M.S., Vaten, A., Karami, L. and El-Showk, S. (2019). Sphingolipid biosynthesis modulates plasmodesmal ultrastructure and phloem unloading. *Nat. Plants*, 5(6): 604–615.

Zuverza-Mena, N., Armendariz, R., Peralta-Videa, J.R. and Gardea-Torresdey, J.L. (2016). Effects of silver nanoparticles on radish sprouts: Root growth reduction and modifications in the nutritional value. *Front. Plant Sc.*, 7: 90.

Chapter 10

Recent Approaches in Nanobioformulation for Sustainable Agriculture

Ngangom Bidyarani,[1] *Gunjan Vyas,*[1] *Jyoti Jaiswal,*[1] *Sunil K Deshmukh*[2] and *Umesh Kumar*[1,*]

1. Introduction

Bioformulation could be a potent substitute for chemical formulation or chemical fertilizers for sustainable agriculture (Mishra et al., 2015). But there are certain limitations of bioformulations including poor shelf life, stability, efficiency, and their efficacy when applied in field in high amount. Recently nanotechnology has rapidly gained as a tool to promote plant growth and yield in the area of agriculture. PRNewswire (2015) has estimated that globally USD 75.8 billion would be noteworthy development by 2020 due to overall prominence and development in nanotechnology. Consequently, USD 1.08 billion could be sponsored in the agriculture-based business sector. Also (Sabourin, 2015) has evaluated approximately USD 3.4 trillion by 2020 would push the worldwide monetary development through nanotechnology in agribusiness. Thus agro-nanobiotechnology could provide the solution to global food security, sustainability, and environmental friendly with no other issues on biosafety regulatory (Mishra et al., 2014; Sangeetha et al., 2017a, b). Nanoparticles (NPs) are regarded as a vehicle to carry nutrients to the targeted site. Thus, these nanoformulations will deliver the nutrients to plants leading to the uptake and discharge of nutrients. Formulations prepared to infuse nanotechnology and using any biologically active substances from microorganisms-based products (green products) and their metabolites commercialized in the trade names of

[1] School of Nano Sciences, Central University of Gujarat, Gandhinagar-382030, Gujarat, India.
[2] Research and Development Division, Greenvention Biotech Pvt. Ltd., Uruli-Kanchan, Pune - 412202, Maharashtra, India.
* Corresponding author: umesh.kumar@cug.ac.in

biostimulants, bio inoculants, biofertilizers, biopesticides employed for plant growth promotion, nutrients acquisition, and disease control in an environmental free manner is known as nanobioformulation. Nanobioformulation has been reported widely for their application in delivering biofertilizers, controlling plant diseases, delivering pesticides, delivering plant stimulants, and crop productivity (Yang et al., 2009; Palmqvist et al., 2015; Naraghi et al., 2018; Patel et al., 2020). Some few commonly reported strains of microorganisms used for the preparation of bioformulations are strains of *Bacillus*, *Pseudomonas*, *Glomus*, *Actinobacteria*, *Lactobacillus*, *Acetobacter*, *Azospirillum*, *Paenibacillus*, *Serratia*, *Burkholderia*, *Herbaspirillum*, *Rhodococcus*, blue green algae (Backer et al., 2018), *Pseudomonas fluorescens*, *Paenibacillus elgii*, and *Bacillus subtilis* (Dikshit et al., 2013), *Pseudomonas putida*, *B. subtilis*, *Pseudomonas fluorescens*, and *Paenibacillus elgii* (Mishra and Kumar, 2009), *Bacillus amyloliquefaciens*, *Bacillus*, *Pseudomonas*, *Azospirillum*, *Azotobacter*, *Bradyrhizobium*, and *Rhizobium*, and fungus, viz., *Trichoderma* spp. and mycorrhizal fungi (Aamir et al., 2020). Various methods could be approached for the loading of nutrients in the prepared nanobioformulation like encapsulation/ entrapment, absorption, and attachment through ligands (Pirzadah et al., 2019).

2. Application of Nanobioformulations

Globally, agriculture is one of the most important and fruitful economic sectors. The agro-economic sector has a remarkable contribution to the world economy. In developing countries, more than 60% of the population is dependent on agriculture for their feed. Diverse challenges such as a fast-growing population, infrastructural development, urbanization, climate change, increased pollution have tremendously threatened agricultural production as well as productivity. Also, during the era of the green revolution, the agricultural sector has been compromised with soil fertility and its normal floral ecology. Due to this the agriculture sector is facing challenges with quality and quantity of products, stagnation in crop yield, low nutrition absorbance potency, declining soil quality, multi-nutrient deficiency, reduced water holding capacity, and shrinking of agricultural lands. As the world is growing dynamically, sources of food, feed, fibre, and agriculture requirement is increasing the quality product production by considering the importance of environmental safety and protection. For that, the adoption of more advanced technology and innovation in the area is the solution (Pirzadah et al., 2020). Nanotechnology and biotechnological approaches seem to be emerging tools in the field of agriculture with various outcomes, viz., increased production, productivity, disease-free, and disease-resistant crop varieties, stress-resistance products, high-nutritional products, improved soil quality, smart farming, etc. Nanotechnology contributes to major applications in crop improvement, protection, production as well as food processing. Nanotechnology fulfils agricultural demands such as seed management, nutrient management, disease control, pest control, precision farming, food harvest processing, and preservation in the safest, efficient, and cost-effective ways. Various nanotechnological tools as nanofertilizers, nanopesticides, nanoherbicides, metal nanoparticles, carbon-based nanocarriers for smart delivery systems, and nanosensors have widely started applicability in the recent era to combat agricultural challenges and fulfil the gaps shown in Fig. 10.1 (Mishra et al., 2017).

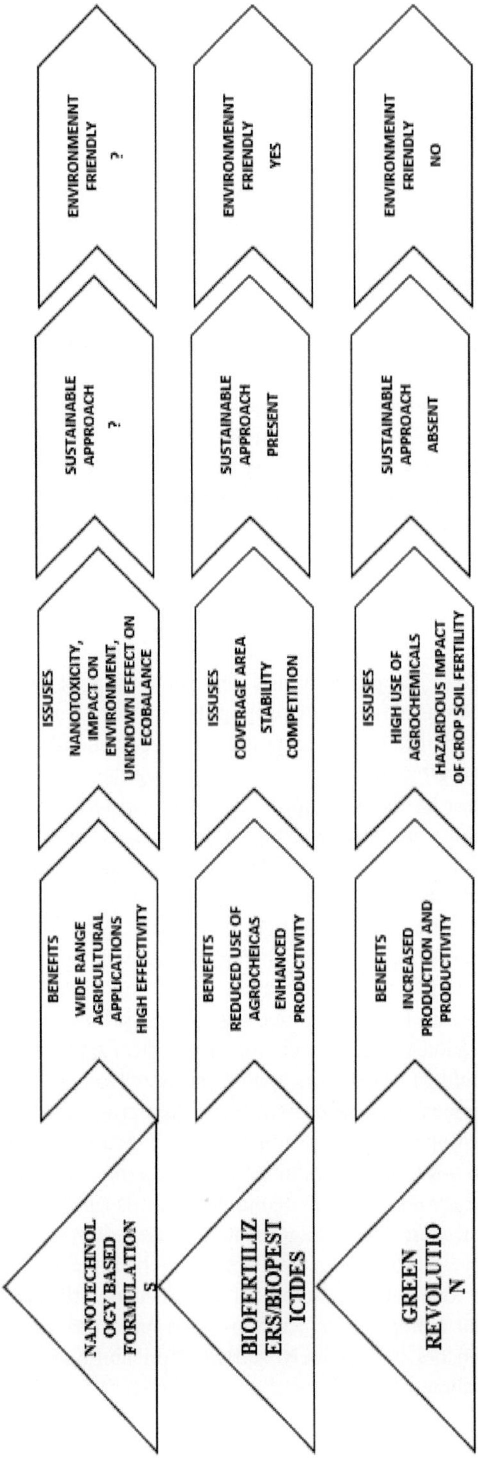

Fig. 10.1. An overview of revolution in the field of agriculture, its present scenario, benefits, and future with nanobioformulations application.

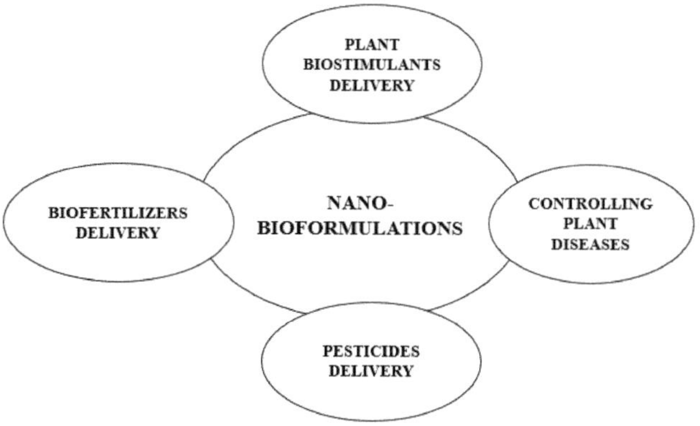

Fig. 10.2. Applications of nanobioformulations in agriculture.

Excessive use of synthetic fertilizers, pesticides, and chemicals is not an appropriate choice for agriculture in the long term as they cause irreparable damage to soil microflora, and the ecosystem to the environment. With enormous increment in crop production and productivity during the era of the green revolution, the world has compromised with the absence of a sustainable approach and environmental friendliness. With more awareness in agriculture, the world has moved towards the bio-based origin of biofertilizers and biopesticides which have faced issues regarding shelf life, on-field stability, active dose, and coverage area. Thus, one of the most visionary nanotechnology-based applications as nanobioformulations seems very promising in the long term as is depicted in Fig. 10.2.

Thousands to millions of bacterial and fungal species are found in soil. Such normal microflora of the soils is responsible for soil fertility and growth and promotion of plants while lowering biotic and abiotic stresses. These microbes are considered to be beneficial to both plants and soil (Blackwell, 2011). Being the latest line of innovation and technology, nanotechnology nanobioformulations can provide applicability of these normal florae and their products as nanoparticle-encapsulated formulation for slow and sustainable activity with least contamination of soil and precision farming.

2.1 Nanobioformulation for Delivering Biofertilizers

The use of synthetic fertilizers can be effectively minimized using biofertilizers where microbes convert organic matter into simpler compounds. These microbes of the rhizosphere play an important role in nutrient assimilation, soil texture improvement by modulating and secreting extracellular molecules, viz., secondary metabolites, hormones, antibiotics, various signalling molecules which altogether enhances plant growth. These biofertilizers also regulate plant stress responses against drought, salinity, and heat. Strains of *Bacillus*, *Pseudomonas*, *Glomus*, *Actinobacteria*, *Lactobacillus*, *Acetobacter*, *Azospirillum*, *Paenibacillus*, *Serratia*, *Burkholderia*, *Herbaspirillum*, *Rhodococcus*, blue-green algae can be applied as combinatorial in commercial agriculture. Drawbacks with biofertilizers are temperature sensitivity, storage desiccation, and lowered shelf-life. Thus, formulations can be developed using polymeric nanoparticles as a nanocoating of biofertilizers which reduces storage desiccation problems (Backer et al., 2018). The larvicide from *Lagenidium giganteum* was effectively encapsulated, stored, and distributed by preparing a water-in-oil emulsion using NaCl with effective cell viability. Within water-in-oil emulsion formulations, water had been trapped around the microorganism by the oil, hence reducing water evaporation (Vandergheynst et al., 2006). This system favours microorganisms to sustain against water desiccation. Thus, emulsion system improves cell viability and the storage material releases kinetics in water and/or oil phases. The emulsion phase is a challenging issue with sedimentation during storage. This could be reduced by addition of hydrophobic silica NPs which improves cell viability by thickening the layer of oil phase. For farming of horticulture crops, where limited land and water resources are available, metal oxide NPs could be effective as growth-promoting factors for microbial communities present in the rhizosphere as normal flora (Vandergheynst et al., 2007). Microbes such as *Pseudomonas fluorescens*, *Paenibacillus elgii,* and *Bacillus subtilis* are naturally existing biofertilizers that show increased efficiency of fertilization of crop in the presence of NPs such as silver and gold. Bacterial-based nano-bioformulations can widely play an important role in environmental clean-up activity by effectively reducing metal contaminants in the rhizosphere thus improving the plant growth and productivity (Dikshit et al., 2013). Reports show increased plant growth yield as well pathogen resistivity when plant-growth-promoting rhizosphere (PGPR) bacteria, such as *Pseudomonas putida, B. subtilis, Pseudomonas fluorescens,* and *Paenibacillus elgii* was given treatment with aluminium, gold, and silver NPs. Such formulations increase the adhesion of bacteria in the root rhizosphere. On the other hand, the wheat crop has also been experimented with nanobiofertilizer for its efficacy check (Mishra and Kumar, 2009). Nanofertilizers such as Fe, Zn, and Mn were used in combination with *Azotobacter* and *Pseudomonas* bacteria on wheat crop. Results show increased spike length; spike number, and seed numbers. Previously *Bacillus amyloliquefaciens* PGPR and titanium dioxide nanoparticles-based nanobio-formulations have been reported to provide protection to oilseed rape plants, i.e., *Brassica napus* against fungal pathogen *Alternaria brassicae* (Mardalipour et al., 2014). Thus, nanobiofertilizers enhance plant growth by making the soil environment more suitable for enriched rhizosphere (Palmqvist et al., 2015). The approach also confronted certain issues like performance under fluctuating climatic condition as high dosage requirement for maximal coverage area. The peculiar characteristics of high surface to volume ratio and physicochemical properties improve activity of such formulations. Thus agro-nanobiotechnology plays an important role in approachable and biosafety manner.

2.2 Nanobioformulation for Controlling Plant Diseases

Disease infected crops cause major loss to production. Various pathogenic fungus and bacteria are responsible for the infection. Fungi such as, e.g., *Bipolaris sorokiniana, Fusarium* sp., *Alternaria solani, Cylindrocarpon destructans, Penicillium digitatum, Klebsiella pneumonia, Aspergillus flavus, Aspergillus niger, Candida albicans, Fusarium oxysporum, Macrophomina phaseolina,* etc., are major infectious agents. Several antimicrobial chemicals are available in the market as fungicides and bactericides, but their widespread use in controlling plant infections compromise soil normal flora sustainability and environmental safety. To minimize the use of chemicals there are many metals, carbon, and metal oxide-based NPs that proved themselves as effective controllers at very low concentrations with high efficiency. Various NPs such as Ag, Cu, Zn, S, CNT, ZnO have been reported with higher effectivity against pathogenic fungus. On the other side, such NPs also placed challenges with bioaccumulation and genetic mutations in the biosphere (Duhan et al., 2017). Thus, advancement with nanobioformulation seems a better solution for all safety and concerning issues (Asadishad et al., 2018). One of the most fungal pathogens infecting cotton plants is *Verticillium dahliae.* This fungus is responsible for cotton wilt disease by bronzing of veins followed by interveinal chlorosis and yellowing of leaves. The leaves begin to dry and appear scorched and fall off, leaving the branches barren. *Talaromyces flavus* is an effective biocontrol agent against the pathogenic fungus. Reports are showing the development of nanobioformulation using any of the two combinations from polymeric monomers alginate, starch, and chitosan containing *Talaromyces flavus.* Researchers have developed nanobioformulations in nanocapsule, nanoemulsion, and nanopowder form. Different nanobioformulations have been compared for their efficiency and inhibitory check on *Verticillium dahliae*. All nanobioformulations showed effective control over pathogenic fungi. *Curvularia lunata* is the pathogenic fungus causing leaf spot disease in rice (Negahban, 2018). *Chaetomium cupreum* is the bioactive antifungal against the infectious agent. Nanobioformulation developed using *Chaetomium cupreum* obtained NPs and tebuconazole which showed higher effectivity against the pathogen than spore formulation and bioformulations of *Chaetomium cupreum* where leaf spot effectively reduced in field trials of rice crop (Song et al., 2020). *Fusarium oxysporum f.* sp. *Lycopercisi* is causing wilt disease in tomatoes and *Talaromyces flavus* is the effective biofungicide. *Talaromyces flavus* is also an antagonist against *Rhizoctonia solani*, *Verticillium dahliae*, and *Fusarium oxysporum*. The reported study formulated nanobioformulation in the form of nanocapsule, nanoemulsion, and nanopowder. The nanobioformulations developed using chitosan, starch, or alginate, any of the two polymeric monomers combinations, vegetable oil as an organic phase, calcium chloride as a crosslinker, and *T. flavus* culture. The nanobioformulations were effective for six months with a remarkable inhibitory effect on the pathogen. These results were as effective as metal oxide nanoparticle inhibition on the pathogenic fungus. The previously available report showed an inhibitory effect of nanobioformulation containing *Penicillium fallutanum* on *Candida albicans* (Naraghi et al., 2018). Ginger-based essential oils have been widely used in the form of nanoemulsion, as well as solid-lipid NPs to the enclosure of natural bioactive compounds to enhance antimicrobial activity. The surfactant used nanodroplets around or less than 100 nm size are widely used to control tropical plant diseases. Various garlic essential oil-based polymeric NPs such as poly (DL-lactide coglycolide), chitosan, zein are widely used to encapsulate active ingredients, viz., cinnamaldehyde, eugenol, and carvacrol. Various emulsion-based nanobioformulations have been widely reported to control plant pathogen *in vitro* as well some field trials. Targeted living entities are *Salmonella* sp., *Listeria* sp., *Escherichia coli*, *Staphylococcus aureus, Bacillus cereus, Pseudomonas aeruginosa, Fusarium oxysporum, F. vasinfectum, Listeria monocytogenes,* etc. (Abdullahi et al., 2020). Nanobio formulations are proving effective against plant pathogenic infection controlling agents. The future is aiming towards an environmentally friendly approach with nanobiotechnology in agriculture over regulation of plant disease and crop productivity.

2.3 Nanobioformulation for Delivering Pesticides

Agricultural production is facing a great challenge against biological disasters where pesticides, which directly impact higher jump of crop productivity, play an important role in defence. According to the UN Food and Agricultural Organization pesticides have restored 30% of agricultural production globally. Annually 4.6 million tons of pesticides input goes to agriculture among which 90% runoff into the environment which causes serious ecological imbalance, pollution to soil and water, bioaccumulation in the food chain, and loss of biodiversity. Generally, the important active ingredients are water-insoluble organic compounds which badly needed some carrier formulation for its spray application in the field. The uptake of input pesticide is lesser than 0.1% and the remaining wash off into the soil and water bodies. Conventionally wettable powders and emulsifier concentrates have been used. Nanotechnology provides a better alternative solution for all challenges (Zhao et al., 2017). Previously polymeric nanomaterials such as poly-caprolactone (PCL), polyglycolic acid (PGA), poly (lactic-co-glycolic) acid (PLGA), chitosan, and polylactic acid (PLA) have been widely used to deliver pesticides as precision farming. Here polymeric biocompatible shell degrades and delivers compound in the vicinity around the activity site. Though applicable concentration has reduced enormously, still the challenge with chemical pesticides is retained. In modern agriculture, recent nanobioformulation approaches have come up with a solution for pesticide replacement with a bioactive ingredient. Polyethylene glycol (PEG) NPs loaded with garlic essential oil showed effective insecticidal activity against *Tribolium castaneum*. The 80% loading of essential oil was achieved with 10% PEG. Nanobioformulation with particles of size less than 240 nm showed highly effective control against *Tribolium castaneum*, around 80% even after five months (Yang et al., 2009). Microbial biopesticides are designed based on their pathogenicity against the infectious organism. The antimicrobial metabolic agents of certain microorganisms are effectively used for nanobioformulations development. *Bacillus* sp., *Pseudomonas* sp., and *Trichoderma* sp. are examples of widely used as biopesticides. Chitosan NPs have been widely used as biocompatible nanocarriers for biopesticides delivery. Chitosan NPs are used to deliver etofenprox (Hwang et al., 2011), Lippia sidoides (Paula et al., 2010), oregano (Hosseini et al., 2013). Some of the organisms, on treating with NPs, show better effectiveness against infecting pests. *Escherichia coli*, *Bacillus subtilis*, *Streptococcus thermophiles*, and *Pseudomonas aeruginosa* treated with gold, aluminium, ZnO, silica, and silver NPs have been reported as effective biocontrol agents against well-known pests including *Fusarium* sp. (Kumari et al., 2019)—*P. syringae* and *F. oxysporum* are intense plant pathogenic microbes. Rotenone is a naturally occurring isoflavone with a broad insecticidal property. Reports show successful development of rotenone loaded zein-based nanobioformulations and their antimicrobial activity against mentioned phytopathogens (Bidyarani et al., 2019).

2.4 Nanobioformulation for Delivering Plant Stimulant

Recently, nanotechnology has gradually moved from experimental to successful field trials. The development of nanobioformulations can be prompt for slow/controlled release of plant stimulant which is critically important for upholding eco-friendly and viable agriculture. Various organic carbon-based nanomaterials as well as metal and metal oxide-based NPs have been widely used in agriculture as stimulants for seed germination, crop growth, and quality enrichment. Many crop species such as corn, tomato, soybean, wheat, barley garlic, maize, peanut have successfully benefited using multi-walled carbon nanotubes, nano silicon dioxide, and titanium dioxide for their effective seed germination. Despite significant research on nanomaterial-induced positive germination in plants, the actual mechanism of how nanomaterials could stimulate plant growth is unclear (Shang et al., 2019). Nano-bioformulations reduce unrequired losses in soil via direct internalization by crops such as chitosan, clay, and zeolites (Millán et al., 2008; Aziz et al., 2016). Various microbial bioformulations-based plant stimulants have been widely used for crop productivity under nutrient challenging conditions also for sustainable farming under biotic and abiotic stress. Various bacterial genus such as *Bacillus*, *Pseudomonas*, *Azospirillum*, *Azotobacter*, *Bradyrhizobium*, and *Rhizobium* and fungus, viz., *Trichoderma* spp. and mycorrhizal fungi are used as bioplant stimulants (Aamir et al., 2020). The research study explores the impact of ZnO NPs and a chitosan-ZnO-based nanobioformulation study on callus growth of *Nicotiana benthamiana*. At all various concentrations, tested results found remarkable growth of callus mass in CH-ZnO treated nanobioformulation while reduced callus mass was observed in ZnO treated samples. Also, a higher accumulation of proline was observed in CH-ZnO treated callus mass. Proline is a stress tolerant amino acid in plants. Thus, reports show that nanobioformulation treated plants are effective with increased stress tolerance than metal oxide-based NP treatment in the plant. Tannin and nicotine are the main active components of *Nicotiana benthamiana*. In the study, an increased level of tannin and nicotine was observed in callus treated with CH-ZnO-based nanobioformulation. Conclusively the nanobio-formulations show high effectivity as plant stimulants (Patel et al., 2020). Other reports with Cu-chitosan based nanobioformulation studied the effects of plant *Zea mays* L. The formulation studied against *Curvularia* leaf spot (CLS) disease and plant growth promoter activity. Cu-chitosan treated plants showed an effective response against disease with higher activity of antioxidants such as peroxidase and superoxide dismutase. Defence enzymes such as phenylalanine ammonia lyase and polyphenol oxidase were reported with significant increment. The treated plant showed significant control of disease in both pots as well as field trials. In addition to this, formulation treatment showed significant improvement in terms of growth enhancement as plant height, root length, and numbers, stem diameter, chlorophyll content, grain weight, grain yield. This nanobioformulation is significantly as effective as natural bioformulations (Prasad et al., 2017). Wood processing by-product lignin has been reported as a novel nanocapsule for delivering gibberellic acid in two model plants *Eruca vesicaria* and *Solanum lycopersicum*. Gibberellic acid (GA) is a crucial plant hormone responsible for plant growth and regulation. It stimulates the synthesis of hydrolases, specifically alpha-amylases, which makes the endosperm reserves available to the embryo (Miransari and Smith, 2014). Experimentation was carried out *in vitro* as well *in vivo* with GA loaded concentration of 0.5–1.5 mg/mL in lignin nanocapsule. Lignin-GA nanobioformulation was synthesized in size ranging from 200–500 nm where all seeds treated showed full germination and seedlings showed tolerance without toxicity. Reports suggest the presence of various hydrophilic active moieties on nanobioformulation positively increases water availability during the plant germination process. Thus, the water can pass the seed tegument and penetrate the rhizodermis of the seedling which eventually reaches the vascular system. Results show increased percentage of germination, the stem, root length as well primary dry and wet weight of seedlings as compared to non-treated sample (Falsini et al., 2019). Alginate/chitosan-based nanocarrier system has also been reported for delivering gibberellic acid 3 to study growth and productivity on *Solanum lycopersicum*. Results from reports showed a 4-fold increase of fruit production and productivity using nano Alginate/chitosan gibberellic acid nanobioformulation (Santo Pereira et al., 2019). Nanobioformulation based plant stimulants have successfully reported in field trials also which gives security about the environmentally friendly visionary future of agriculture.

3. Effects of Nanobioformulation on Crop Productivity

Protection of environment and agricultural products, diminishing culturable land resources, control regulation, and decrease of environmental pollutants are greatest challenges that agriculture is facing at present. Problems such as climate change, limited production area, and depletion in natural resources result in high loss of crop productivity (Norton, 2016; Duhan et al., 2017). Agriculture firm technologies such as the development of plant hybrids, chemical-based synthesis, and biotechnology-based products have successfully benefitted agriculture tremendously. Issues such as increasing concerns for sustainability, soil infertility, microflora survivability, productivity, environmental, and ecological disruption can be suitably addressed with nanotechnology. There is a wide range of nanotechnology-based products being used for raising productivity providing the solution for all found major challenges. It reduces chemical usage in the agricultural field, improving quality and productivity, better soil quality, reduced water contamination, less risk to consumers, and the agricultural worker (Fraceto et al., 2016). For these benefits, nanobioformulation seems to be the best fit for sustain release, target specific active ingredient delivery, reducing toxicity, enhanced bioavailability with ecosystem protection (Duhan et al., 2017). For the same benefits, plant growth promoting regulators (PGPR) and plant growth promoting microorganisms (PGPM) and their bioactive compound products play an important role. Recently some of the nanobio-formulations such as nanoAlginate/Chitosan-Gibberellic Acid 3 (Santo Pereira et al., 2019), Lignin nanocapsule-Gibberellic Acid (Falsini et al., 2019), nano Cu-Chitosan (Choudhary et al., 2017), nano ZnO-Chitosan (Patel et al., 2020) are reported and well established for successfully enhanced plant productivity field trials. Santo Pereira et al. (2019) studied the developed chitosan-based nanobiofromulation with GA3 and seeds of *Solanum lycopersicum* var. *cerasiforme* were treated with nanoAlginate/Chitosan-GA3 formulation and evaluated for effect on plant growth and fruit production under field conditions for completely eco-friendly and sustainable agriculture practice. NanoALG/CS based nanobioformulation developed with particles size of 450 nm, 0.3 polydispersity index (PDI) and –29 mv zeta potential. Seeds were given coating treatment of the formulation overnight and then sown for growth in a greenhouse for 30 days, after which the treated and control seedlings (treatments) were transferred to the soil and observed for growth for 90 days under greenhouse conditions with an average temperature of 25–30°C with relative humidity about 40%. After 90 days, measurements were made for shoot length, shoot fresh and dry masses, number of fruits per plant, and fruit weight and productivity calculation as fresh fruit weight produced per hectare in a measured cultivated area. The nanobioformulation % biological efficacy was measured by comparing the results with control treatments with GA3. Results after 30 days of initial germination showed increased shoot length and dry weight by 38% and 107%, respectively, when treated with a 100-fold dilution of nano-Alg/CS-GA3 compared to control. Also, nano-Alg/CS-GA3 treated seeds showed 113% increment of dry root which includes both main and lateral roots where only GA3 treated control seeds not resulted in effectivity. Developed nanobioformulation is completely non-toxic to plant and with no harm due to the absence of bioaccumulation as well as polymer can be utilized by microorganism as nutrient availability. Also, particles can be taken up due to electrostatic interaction with the cell wall followed by clathrin-dependent or clathrin-independent endocytosis. After 120 days of development, the plant showed 45%, 63%, and 61% enhancement in the shoot length, the shoot dry weight, and the number and weight of fruits than control treated plants. When nano-Alg/CS-GA3, Chitosan/Tripolyphosphate-GA3 nanoparticles, control GA3, and blank water-treated plants compared for productivity that is produced per hectare area which resulted in productivity 36.8 ± 5.6, 30.7 ± 3.2, 20.7 ± 1.1, and 10.1 ± 2.2 ton/ha, respectively. This shows nanobioformulation shows the highest effectivity in overall plant growth, development, production, and productivity in comparison to nanoformulation, the bare biological active ingredient, and regular water treatments. Other reports from, developed nanocapsule lignin-gibberellic acid nanobioformulation was checked for its effect on two model plant plantlets *Eruca vesicaria* and *Solanum lycopersicum*. These *in vivo* and *in vitro* conducted experiments investigated for a percentage of germination, the stem, the primary root lengths, fresh and dry weight of treated versus non-treated plants. Here the nanocapsules (NCs) were also loaded with dye fluorol yellow 088 to track their entrance and accumulation in plants. Seeds were treated with bare NCs, GA3-loaded NCs, and only GA3. *In vitro* and *in vivo* experiments were carried out using blank, control, and tests for all three samples for each plant. Sterilized seeds of plants *Eruca vesicaria* and *Solanum lycopersicum* were treated with sample formulations for 30 minutes before experiments. *In vitro* experiments were examined for 8/16 photoperiod where the germination percentage for arugula were 5 and 10 days, while growth parameters for tomato were evaluated for 7 and 14 days. For *in vivo* experiments, seeds grown up to 85% were evaluated for 3 days in case of arugula and 6 days for tomato. Particles having a size range of 200–250 nm developed with PDI between 0.17–0.38 reported. *In vivo* germination on arugula seed showed 97% percentage on treatment with lignin NCs-GA followed by 86% and 78% for the only GA and bare NPs-treated plant seeds. Also, lignin NCs-GA treated sample seedlings showed 28–50% increased total vegetative biomass concerning control. Also, seedling length, fresh and dry weight was higher than the control formulation. Here tomato seeds also showed significant germination of seedlings after completion of growth duration which was 95% for lignin NCs-GA than compared to 92% of a control system. *In vitro*, seed germination assay showed greater germinated seeds treated with loaded NC than to control and bare in both plants *Eruca vesicaria* and *Solanum lycopersicum*. Here nanobioformulation treated seeds showed enhanced effectivity for seed germination and seedling development without developing toxicity also hydrophilic moieties increased water retention capability thus positively affect seed germination and increase productivity used water stress conditions. As per fluorol yellow 088 dye observation, nanocapsules can pass seed tegument and penetrate the rhizodermis of the seedling and reach the vascular system. Being biocompatible to plant system thus, the further system is successful and worth for plant productivity improvement at the field level. As per this reported research, nanobioformulation increases plant productivity tremendously which needs to be explored further with more experimentation *in vitro* for some plant varieties and considering various geometric and environmental factors.

4. Current Status of Nanobioformulation

Nanobioformulation is the most emerging area in nanobiotechnology. There are very few scientific reports about nanobioformulation application in agriculture areas as *in vitro* seed germination and as a plant productivity enhancer. This area is very novel and newly born as far as agriculture is concerned. The approach is different in terms of eco-friendliness compared to conventional nanotechnology. As per the current scenario of development in agriculture and the status of nanotechnology, nanobioformulation is a highly innovative area that requires more focus from scientists across the globe.

4.1 Advantages of Nanobioformulations

The green methods for synthesizing nanoparticles with plant extract is advantageous because it is simple, convenient, eco-friendly, and has a short reaction time. Nanobioformulation prepared by green approach increases the agricultural potential to improve fertilization processes, plant growth regulators, to deliver pesticide of an active component at a target site, wastewater treatment, and nutrient absorption in plants. Nanotechnology is a rapidly developing field and has gained attention due to its wide range of applications in various fields such as agriculture, medicine, and the environment.

4.1.1 Agricultural Diagnostic and Drug Delivery

Certain nanoparticles are applied in the nanoforms like carbon, silver, silica, and aluminium silicate which help to control plant diseases. Carbon nanofiber helps to fortify natural fibers such as coconuts (*Cocus nucifera*) and sisal (*Agave sisalana*). It can also be used to encapsulate the pesticides in nanoparticles and control their release. Nanosilver possesses strong bactericidal and antimicrobial properties and is known to reduce various plant diseases that are produced by fungal pathogens (Nair et al., 2010). Furthermore, ZnO NPs inhibit the fungal growth of *Botrytis cinerea* and the growth of conidiophores of *Penicillium expansum*, eventually led to the death of fungal mats (Abd-Elsalam, 2012).

4.1.2 Nano Pesticides and Nano Herbicides

The pesticides encapsulated in nanoparticles are being developed in such a manner that they release slowly or have an environmental link release trigger (Nair et al., 2010). A pesticide like avermectin blocks insect neurotransmission by inhibiting the chloride channel. It is inactivated by UV with a half-life of 6 h in the field. Porous hollow silica NPs have an encapsulation capacity of avermectin 625 gm/kg, and its shell thickness is 15 nm with a pore diameter of 4–5 nm. This structure also protects the pesticides from UV degradation. It is reported that encapsulated avermectin shows its sustained release from the NPs for about 30 days (Ghormade et al., 2011).

4.1.3 Nanobiofarming

The latest research in the field of agriculture is focused on the fabrication of gold and silver NPs with various biomolecules of plants origin such as *Medicago sativa, Cyamopsis tetragonolobus, Zea mays, Pennisetum glaucum, Sorghum vulgare, Brassica junceaor* extracts from *B. juncea* and *M. sativa, Memecylon edule* or *Allium sativum*. Metal nanoparticles exhibit various shape and size dependent interesting properties. Thus, they possesses the advantage of being simple, cost-effective, and eco-friendly for agriculture (Groves and Titoria, 2009).

4.1.4 Nanobiosensors and Agriculture

Nanosensors are immobilized with bioreceptor probes for their selective analyte molecules known as nano-biosensors. These newly developed sensors have various applications such as detection of analytes like urea, glucose, pesticides, monitoring of metabolites, and detection of various microorganisms. Nanosensors possess the advantage of small size, portabilty, sensitive to real-time monitoring, precise, quantitative, reliability, accurate, reproducibility, and stability which can conquer the shortcoming of the current sensors. Controlled Environmental Agriculture (CEA) can be advanced with the help of implementing nanosensors in the field. These sensors help us to regulate the harvest time of crops, detect the health of the crop, and also help to determine microbial or chemical contamination of the crop (Rai et al., 2012).

5. Disadvantages of Nanobioformulations

5.1 Nanotechnology against Agriculture

Rapid advancement in the field of nanotechnology is now more a concern due to its accumulation of manufactured nanomaterial and its probablility to enter the food chain (Priester et al., 2012). The use of nanomaterials is not inherently dangerous. For example, traditional foods contain many nanoscale ingredients (such as proteins in milk, fat globules in mayonnaise, carbohydrates, DNA, etc.), but the use of such engineered nanoscale materials in agriculture, water, and food may have risks for human use and consumption, the environment or both (Gruère et al., 2011).

5.2 Nanomaterials Toxic to Human and Animals

Many toxicological studies reported that the use of high doses of fibrous and tubular nanostructured materials like single-walled carbon nanotube and multiple-walled carbon nanotubes are associated with fibrotic lung responses and thus result in inflammation and increased risk of carcinogenesis.

5.3 Phytotoxicity of Nanoparticles

To interact with plants, engineered NPs (ENPs) need to first penetrate the cell wall and plasma membrane of the root epidermal layer to enter the vascular tissue (xylem) and translocate through the stem to the leaves. Water molecules and other solutes must pass through the cell wall and enter the roots. An ENP of a smaller size is expected to easily enter and reach the plasma membrane while a larger particle aggregates and will not be able invade plant cells (Ma et al., 2010).

6. Prospects of Nanobioformulation

In the current scenario, there is massive use of fertilizers in the agricultural sector for increasing production which not only affects the soil quality but also has shown its harmful effects on the environment. Boosting agricultural production by the intervention of nanotechnology to meet the demands of people is essential, but its impact towards environmental safety should also be considered in the long run. Therefore, a novel approach needs to be developed. There are promising applications of nanotechnology in the agriculture sector such as nanofertilizers, nanobiosensors, nanopesticides, and environmental remediation agents.

Future opportunities where nanotechnology can be implemented in agriculture include genetic modification of plants (Eapen and D'Souza, 2005; Kuzma, 2007), delivery of genes and drug molecules to a specific site at the cellular level of plant and animals (Maysinger, 2007), and nano-array technology for gene expression in plants to overcomes stress along with the development of sensors (Ahmed et al., 2013), early detection of pathogens and contaminants in food products and smart delivery of agrochemicals such as fertilizers and pesticides (Chau et al., 2007). The advent of nanofertilizers (Derosa et al., 2010) as an alternative to conventional fertilizers, helps to increase the nutrients in the soil, and thus eutrophication and contamination of drinking water may be eliminated (Bhalla and Mukhopadhyay, 2009; Mukhopadhyay and Sharma, 2013). Excessive reliance on supplementary irrigation, climate vulnerability, inadequate inputs, and energy conversion are three major problems in the current agricultural production systems and nanotechnology might reduce their impacts. Nano remediation could be effective to reduce the overall cost of cleaning up large contaminated sites, and decrease its clean-up time by eliminating the need for treatment and disposal of contaminated soil (Karn et al., 2009). However, care must be taken to prevent potential adverse environmental impacts, especially in a full-scale ecosystem.

7. Recommendation for Future Research

Nature is a great teacher and successful implementation by simulating natural processes with more advanced scientific representation can lead to successful applications of nanotechnology in agriculture. For example, the goal may be to improve the soil's capability to use nutrients efficiently to increase productivity and environmental safety. We suggest some key points for futuristic research (Mishra et al., 2017):

- Future research must be emphasized to find many possible ways to avoid the risk aspects associated with NP usage. Research on NP synthesis provides some laboratory specific applications which cannot contribute to complete acceptance of nanotechnology in agriculture. Therefore, the scientific community must work together to improve future research on more realistic methods.
- The level of NP dose needs to be investigated, clarified, and validated within safety limits. This can only be achieved by trying various dose concentration-dependent studies with natural soil systems to better understand the dose of NPs.

- Physio-chemical attributes of agriculture fields where nanobioformulation can be applied to reduce the risks towards plants and soil biomes must be studied.
- Lastly, we recommend depth research for the biosynthesized NPs which play a vital role in agricultural fields. The green synthesis NPs are believed to have lesser or no toxicity thus, future exploration needs to be focused on their practical utility.

8. Conclusion

The outbreak and advancement of new pathogenic races is an ongoing problem, and to control pests we require chemicals which might be costly and not always effective. So, the use of nanoformulations is considered as an alternate solution to control plant pathogens. Opportunities for the application of nanotechnology in agriculture are tremendous. The research on various applications of nanotechnology in the field of agriculture has been a decade old. Traditional agricultural practices turn out to be inadequate and exceed the carrying capacity of the terrestrial ecosystem, leaving insufficient choices. For this reason, we need to explore nanotechnology in all sectors of agriculture. Nanotechnology helps to improve the nutrient utilization efficiency with its nanoformulation fertilizers, which can also be exploited for the removal of pollutants from soil and water bodies, vegetables, and flowers. Nanotechnologies require a detailed and in-depth understanding of science, manufacturing, and material technologies, and knowledge of agricultural production systems.

References

Aamir, M., Rai, K.K., Zehra, A., Dubey, M.K., Kumar, S., Shukla, V. and Upadhyay, R.S. (2020). Microbial bioformulation-based plant biostimulants: A plausible approach toward next generation of sustainable agriculture. pp. 195–225. *In*: Kumar, A. and Radhakrishnan, E.K. (eds.). *Microbial Endophytes*. U.K: Woodhead Publishing.

Abd-Elsalam, K.A. (2012). Nanoplatforms for plant pathogenic fungi management. *Fungal Genomics Biol.*, 2: e107. https://doi.org/10.4172/2165-8056.1000e107.

Abdullahi, A., Ahmad, K., Ismail, I.S., Asib, N., Ahmed, O.H., Abubakar, A.I., Siddiqui, Y. and Ismail, M.R. (2020). Potential of using ginger essential oils-based nanotechnology to control tropical plant diseases. *Plant Pathol. J.*, 36: 515.

Ahmed, F., Arshi, N., Kumar, S., Gill, S.S., Gill, R., Tuteja, N. and Koo, B.H. (2013). Crop improvement under adverse conditions. Nanobiotechnology: Scope and potential for crop improvement. pp. 245–269. *In*: Narendra Tuteja, N. and Gill, S.S. (eds.). *Crop Improvement under Adverse Conditions*. New York, NY.: Springer.

Asadishad, B., Chahal, S., Akbari, A., Cianciarelli, V., Azodi, M., Ghoshal, S. and Tufenkji, N. (2018). Amendment of agricultural soil with metal nanoparticles: Effects on soil enzyme activity and microbial community composition. *Environ. Sci. Technol.*, 52: 1908–1918.

Aziz, H.M.A., Hasaneen, M.N. and Omer, A.M. (2016). Nano chitosan-NPK fertilizer enhances the growth and productivity of wheat plants grown in sandy soil. *Spanish J. Agric. Res.*, 14: 17.

Backer, R., Rokem, J.S., Ilangumaran, G., Lamont, J., Praslickova, D., Ricci, E., Subramanian, S. and Smith, D.L. (2018). Plant growth-promoting rhizobacteria: Context, mechanisms of action, and roadmap to commercialization of biostimulants for sustainable agriculture. *Front. Plant. Sci.*, 9: 1473.

Bhalla. D. and Mukhopadhyay, S. (2009). Eutrophication: Can nanophosphorous control this menace? *Nat. Preced.*, 1–1.

Bidyarani, N. and Kumar, U. (2019). Synthesis of rotenone loaded zein nano-formulation for plant protection against pathogenic microbes. *RSC Adv.*, 9: 40819–40826.

Blackwell, M. (2011). The Fungi: 1, 2, 3… 5.1 million species? *Am. J. Bot.*, 98: 426–438.

Chau, C.F., Wu, S.H. and Yen, G.C. (2007). The development of regulations for food nanotechnology. *Trends Food Sci. Technol.*, 18: 269–280.

Choudhary, R.C., Kumaraswamy, R.V., Kumari, S., Sharma, S.S., Pal, A., Raliya, R., Biswas, P. and Saharan, V. (2017). Cu-chitosan nanoparticle boost defense responses and plant growth in maize (Zea mays L.). *Sci. Rep.*, 7: 1–11.

Derosa, M.C., Monreal, C., Schnitzer, M., Walsh, R. and Sultan, Y. (2010). Nanotechnology in fertilizers. *Nat. Nanotechnol.*, 5: 91. https://doi.org/10.1038/nnano.2010.2.

Dikshit, A., Shukla, S.K. and Mishra, R.K. (2013). *Exploring Nanomaterials with PGPR in Current Agricultural Scenario: PGPR with Special Reference to Nanomaterials.* CA, USA: LAP Lambert Academic Publishing.

Duhan, J.S., Kumar, R., Kumar, N., Kaur, P., Nehra, K. and Duhan, S. (2017). Nanotechnology: The new perspective in precision agriculture. *Biotechnol. Reports*, 15: 11–23.

Eapen, S. and D'souza, S.F. (2005). Prospects of genetic engineering of plants for phytoremediation of toxic metals. *Biotechnol. Adv.*, 23: 97–114.

Falsini, S., Clemente, I., Papini, A., Tani, C., Schiff, S., Salvatici, M.C., Petruccelli, R., Benelli, C., Giordano, C., Gonnelli, C. and Ristori, S. (2019). When sustainable nanochemistry meets agriculture: Lignin nanocapsules for bioactive compound delivery to plantlets. *ACS Sustain. Chem. Eng.*, 7: 19935–19942.

Fraceto, L.F., Grillo, R., de Medeiros, G.A., Scognamiglio, V., Rea, G. and Bartolucci, C. (2016). Nanotechnology in agriculture: Which innovation potential does it have? *Front. Environ. Sci.*, 4: 20.

Ghormade, V., Deshpande, M.V. and Paknikar, K.M. (2011). Perspectives for nano-biotechnology enabled protection and nutrition of plants. *Biotechnol. Adv.*, 29: 792–803.

Groves, K. and Titoria, P. (2009). Nanotechnology and the food industry. *Food Sci. Technol.*, 23: 22–24.

Gruère, G., Narrod, C. and Abbott, L. (2011). Agricultural, food, and water nanotechnologies for the poor opportunities, constraints, and role of the consultative group on International Agricultural Research. *Communications.*

Hosseini, S.F., Zandi, M., Rezaei, M. and Farahmandghavi, F. (2013). Two-step method for encapsulation of oregano essential oil in chitosan nanoparticles: Preparation, characterization and *in vitro* release study. *Carbohydr. Polym.*, 95: 50–56.

Hwang, I.C., Kim, T.H., Bang, S.H. and Kim, K.S. (2011). Insecticidal effect of controlled release formulations of etofenprox based on nano-bio technique. *J. Fac. Agric. Kyushu Univ.*, 56: 33–40.

Karn, B., Kuiken, T. and Otto, M. (2009). Nanotechnology and *in situ* remediation: A review of the benefits and potential risks. *Environ. Health Perspect.*, 117: 1823–1831.

Kumari, B., Mallick, M.A., Solanki, M.K., Hora, A. and Mani, M. (2019). Applying nanotechnology to bacteria: An emerging technology for sustainable agriculture. pp. 121–143. *In*: Kumar, A., Singh, A.K. and Choudhary, K.K. (eds.). *Role of Plant Growth Promoting Microorganisms in Sustainable Agriculture and Nanotechnology* Elsevier.

Kuzma, J. (2007). Moving forward responsibly: Oversight for the nanotechnology-biology interface. *J. Nanoparticle Res.*, 9: 165–182.

Ma, X., Geiser-Lee, J., Deng, Y. and Kolmakov, A. (2010). Interactions between engineered nanoparticles (ENPs) and plants: Phytotoxicity, uptake and accumulation. *Sci. Total Environ.*, 408: 3053–3061.

Mardalipour, M., Zahedi, H. and Sharghi, Y. (2014). Evaluation of nano biofertilizer efficiency on agronomic traits of spring wheat at different sowing date. p. 349. *In*: *Biological Forum.* New Delhi: Research Trend.

Maysinger, D. (2007). Nanoparticles and cells: Good companions and doomed partnerships. *Org. Biomol. Chem.*, 5: 2335–2342.

Millán, G., Agosto, F. and Vázquez, M. (2008). Use of clinoptilolite as a carrier for nitrogen fertilizers in soils of the Pampean regions of Argentina. *Int. J. Agric. Nat. Resour.*, 35: 293–302.

Miransari, M. and Smith, D.L. (2014). Plant hormones and seed germination. *Environ. Exp. Bot.*, 99: 110–121.

Mishra, S., Singh, A., Keswani, C. and Singh, H.B. (2014) Nanotechnology: Exploring potential application in agriculture and its opportunities and constraints. *Biotech. Today*, 4: 9–14.

Mishra, S., Singh, A., Keswani, C., Saxena, A., Sarma, B.K. and Singh, H.B. (2015). Harnessing plant-microbe interactions for enhanced protection against phytopathogens. pp. 111–125. *In*: *Plant Microbes Symbiosis: Applied Facets*. Springer.

Mishra, S., Keswani, C., Abhilash, P.C., Fraceto, L.F. and Singh, H.B. (2017). Integrated approach of agri-nanotechnology: Challenges and future trends. *Front. Plant. Sci.*, 8: 1–12. https://doi.org/10.3389/fpls.2017.00471.

Mishra, V.K. and Kumar, A. (2009). Impact of metal nanoparticles on the plant growth promoting rhizobacteria. *Dig. J. Nanomater Biostruct.*, 4: 587–592.

Mukhopadhyay, S.S. and Sharma, S. (2013). Nanoscience and nanotechnology: Cracking prodigal farming. *J. Bionanoscience*, 7: 497–502.

Nair, R., Varghese, S.H., Nair, B.G., Maekawa, T., Yoshida, Y. and Kumar, D.S. (2010). Nanoparticulate material delivery to plants. *Plant Sci.*, 179: 154–163.

Naraghi, L., Negahban, M., Heydari, A., Razavi, M. and Afshari-Azad, H. (2018). Growth inhibition of Fusarium oxysporum f. sp. lycopercisi, the causal agent of tomato fusarium wilt disease by nanoformulations containing *Talaromyces Flavus. Ekoloji*, 27: 103–112.

Negahban, L. (2018). The efficacy of nanoparticles containing Talaromyces Flavus in inhibiting the growth of *Verticillium Dahliae*, the causal agent of cotton wilt disease. *Int. J. Agric. Sci. Res.*, 8: 229–240.

Norton, L.R. (2016). Is it time for a socio-ecological revolution in agriculture? *Agric. Ecosyst. Environ.*, 235: 13–16.

Palmqvist, N.G.M., Bejai, S., Meijer, J., Seisenbaeva, G.A. and Kessler, V.G. (2015). Nano titania aided clustering and adhesion of beneficial bacteria to plant roots to enhance crop growth and stress management. *Sci. Rep.*, 5: 10146.

Patel, K.V., Nath, M., Bhatt, M.D., Dobriyal, A.K. and Bhatt, D. (2020). Nanofomulation of zinc oxide and chitosan zinc sustain oxidative stress and alter secondary metabolite profile in tobacco. *3 Biotech.*, 10: 1–15.

Paula, H.C.B, Sombra, F.M., Abreu, F.O.M.S. and Paul, R. (2010). Lippia sidoides essential oil encapsulation by angico gum/chitosan nanoparticles. *J. Braz. Chem. Soc.*, 21: 2359–2366.

Pirzadah, T.B., Malik, B., Maqbool, T. and Rehman, R.U. (2019). Development of nano-bioformulations of nutrients for sustainable agriculture. pp. 381–394. *In*: *Nanobiotechnology in Bioformulations*. Springer.

Pirzadah, T.B., Pirzadah, B., Jan, A., Dar, F.A., Hakeem, K.R., Rashid, S., Salam, S.T., Dar, P.A. and Fazili, M.A. (2020). Development of nano-formulations via green synthesis approach. pp. 171–183. *In*: Hakeem, K.R. and Pirzadah, T.B. (eds.). *Nanobiotechnology in Agriculture*. Springer.

Prasad, R., Bhattacharyya, A. and Nguyen, Q.D. (2017) Nanotechnology in sustainable agriculture: recent developments, challenges, and perspectives. *Front. Microbiol.*, 8: 1014.

Priester, J.H., Ge, Y., Mielke, R.E., Horst, A.M., Moritz, S.C., Espinosa, K., Gelb, J., Walker, S.L., Nisbet, R.M., An, Y.J., Schimel, J.P., Palmer, R.G., Viezcas, J.A.H., Zhao, L., Torresdey, J.L.G. and Holden, P.A. (2012). Soybean susceptibility to manufactured nanomaterials with evidence for food quality and soil fertility interruption. *Proc. Natl. Acad. Sci.*, U.S.A. 109(37): E2451–E2456. https://doi.org/10.1073/pnas.1205431109.

PRNewswire. (2015). FDAGrants 510 (k) Clearance to Life Technologies Stem Cell Growth Medium. (2011). Available at: http://www. prnewswire. com/news-releases/fda-grants-510k-clearance-to-life-technologies-stem-cell-growth-medium-117920569. html. Accessed: 25 August.

Rai, V., Acharya, S. and Dey, N. (2012). Implications of nanobiosensors in agriculture. *J. Biomater. Nanobiotechnol.*, 03: 315–324. https://doi.org/10.4236/jbnb.2012.322039.

Sabourin, V. (2015). Commercial opportunities and market demand for nanotechnologies in agribusiness sector. *J. Technol. Manag. Innov.*, 10: 40–51.

Sangeetha, J., Thangadurai, D., Hospet, R., Harish, E.R., Purushotham, P., Mujeeb, M.A., Shrinivas, J., David, M., Mundaragi, A.C., Thimmappa, S.C., Arakera, S.B. and Prasad, R. (2017). Nanoagrotechnology for soil quality, crop performance and environmental management. *In*: *Nanotechnology*. Singapore: Springer. https://doi.org/10.1007/978-981-10-4573-8_5.

Sangeetha, J., Thangadurai, D., Hospet, R., Purushotham, P., Karekalammanavar, G., Mundaragi, A.C., David, M., Shinge, M.R., Thimmappa, S.C., Prasad, R. and Harish, E.R. (2017). Agricultural Nanotechnology: Concepts, benefits, and risks. *In*: *Nanotechnology*. Singapore: Springer. https://doi.org/10.1007/978-981-10-4573-8_1.

Santo Pereira, A.D.E., Oliveira, H.C. and Fraceto, L.F. (2019). Polymeric nanoparticles as an alternative for application of gibberellic acid in sustainable agriculture: A field study. *Sci. Rep.*, 9: 1–10.

Shang, Y., Hasan, M., Ahammed, G.J., Li, M., Yin, H. and Zhou, J. (2019). Applications of nanotechnology in plant growth and crop protection: A review. *Molecules*, 24: 2558.

Song, J., Soytong, K., Kanokmedhakul, S., Kanokmedhakul, K. and Poeaim, S. (2020). Antifungal activity of microbial nanoparticles derived from *Chaetomium* spp. against *Magnaporthe oryzae* causing rice blast. *Plant Prot. Sci.*, 56(3): 180–190.

Umesha, S., Singh, P.K. and Singh, R.P. (2018). Microbial biotechnology and sustainable agriculture. pp. 185–205. *In*: *Biotechnology for Sustainable Agriculture*. Elsevier.

Vandergheynst, J., Scher, H., Guo, H.Y. and Schultz, D. (2007). Water-in-oil emulsions that improve the storage and delivery of the biolarvacide *Lagenidium giganteum*. *BioControl*, 52: 207–229.

VanderGheynst, J.S., Scher, H. and Guo, H. Y. (2006). Design of formulations for improved biological control agent viability and sequestration during storage. *Ind. Biotechnol.*, 2: 213–219.

Yang, F.L., Li, X.G., Zhu, F. and Lei, C.L. (2009). Structural characterization of nanoparticles loaded with garlic essential oil and their insecticidal activity against Tribolium castaneum (Herbst) (Coleoptera: Tenebrionidae). *J. Agric. Food Chem.*, 57: 10156–10162.

Zhao, X., Cui, H., Wang, Y., Sun, C., Cui, B. and Zeng, Z. (2017). Development strategies and prospects of nano-based smart pesticide formulation. *J. Agric Food Chem.*, 66: 6504–6512.

Chapter 11

Nanobiosensors for Monitoring Soil and Water Health

Archeka,[1] *Nidhi Chauhan*,[2] *Neelam*,[1] *Kusum*[1] and *Vinita Hooda*[1,*]

1. Introduction

The world population is increasing massively; currently it is about 7.6 billion and is estimated to reach about 8.6 billion in 2030 and about 9.8 billion in 2050 (Tamburino et al., 2020). For the ever-increasing population of the world, a demand for food is continuously increasing and to fulfill food demands farmers apply excessive fertilizers and pesticides into the field. Excessive and repetitive use of these agrochemicals is a matter of grave environmental concern as they adversely affect soil and aquatic ecosystems as well as human and animal health. In the wake of this situation, alternative promising approaches, such as precision farming, are becoming massively popular (Vecchio et al., 2020). Precision farming determines whether crops are growing at maximum efficiency, by identifying the nature and location of problems or any nutritional deficiency and applying targeted action with the goal to maximize output (i.e., crop yielding) and minimize input (i.e., fertilizers, pesticides, herbicides). It also helps in reducing agricultural waste and minimizes environmental pollution. At the heart of precision farming are sensing devices that measure soil conditions and determine the environmental variables (Kim et al., 2019; Griesche and Baeumner, 2020).

A biosensor is an analytical device that embeds communication between immobilized biological elements such as enzyme, substrate, complementary DNA, or antigen and a physiological transducer to bring forth an electronic signal in relation to the analyte concentration (Perumal and Hashim, 2014). Figure 11.1 depicts different components of a basic biosensor. Biosensors are generally preferred over traditional

[1] Department of Botany, Maharshi Dayanand University, Rohtak 124001, India.
[2] Amity Institute of Nanotechnology, Amity University, Noida 201303, India.
* Corresponding author: vinitahooda.botany@mdurohtak.ac.in

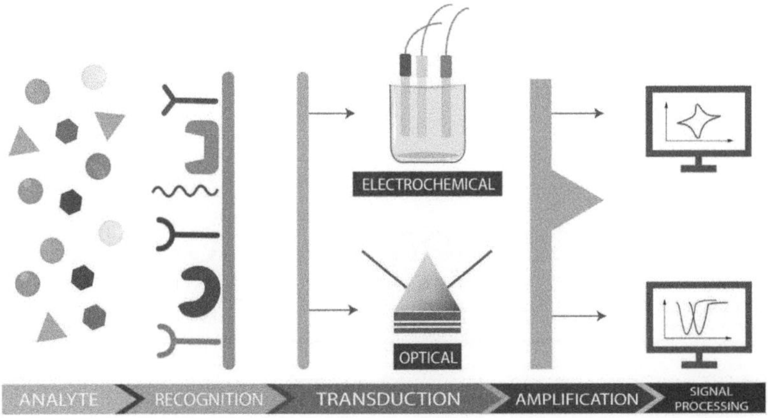

Fig. 11.1. Different components of the biosensor platform and amplification of the signal.

methods as they are sensitive, fast, and reliable and come with the possibility of onsite detection whereas conventional methods are time consuming, laborious, require expensive equipment and highly trained technicians. Predominantly, biosensors are disaggregated in pursuant to the transduction methods they utilize and hence, categorized into electrochemical, electrical, optical, piezoelectric (mass detection methods) and thermal detection. Electrochemical sensors are generally preferred over other sensors and are most widely used because they offer high sensitivity, lower cost, simple design, and small size (Baronas et al., 2016).

Nanoparticles are often employed in biosensors to enhance the performance of the sensing devices. As high surface area to volume ratio of nanoparticles is utilized to anchor the biological molecule of interest in greater quantities, excellent electrical properties help in better wiring of the biomolecule to the electrode surface and their biocompatible nature prolongs the working life of the biomolecule. The choice of nanoparticles depends upon the type of sensor developed and its application that may range from metallic to non-metallic to polymeric and carbon-based nanomaterials (Usman et al., 2020). The present chapter focuses on biosensing platforms developed for the determination of nitrates, phosphates, pesticides, and heavy metals, the four key contaminants that have already made food and water toxic in many parts of the world.

2. Nitrate Biosensors

Nitrate, an essential element for plant growth, is naturally present in the environment. Excessive use of manures and fertilizers has resulted in undesirable increase in its concentrations in the biota polluting agricultural soil, underground water reserves, and agricultural produce. Nitrate, when consumed in excess, either by drinking contaminated water or by eating nitrate loaded vegetables, is toxic to life. Inside the body, nitrate is converted to nitrite that causes oxidation of hemoglobin to methemoglobin impairing the ability of hemoglobin to carry oxygen. The condition is clinically known as methemoglobinemia or 'blue-baby' syndrome as it affects infants more and could even be fatal to them.

The condition is not prevalent in industrialized nations but exists in rural areas where people still use well water for drinking purposes. The World Health Organization (WHO) recommends that reducing nitrate level below 50 mg/l in drinking water can effectively prevent the occurrence of the disease (Hooda et al., 2016). This necessitates the development of biosensing devices that could effectively measure nitrate concentration in real-life samples, especially soil, to facilitate the need-based application of nitrogenous fertilizers.

In nitrate, biosensor detection is achieved with nitrate reductase (NR) and NAD(P)H or other suitable cofactors (Sohail and Adeloju, 2016). Madasamy et al. (2014) and Can et al. (2012) have successfully developed nitrate biosensors by covalently linking the NR to carboxylated carbon nanotubes (CNTs) with detection limits of 0.2 μM and 170 μM, respectively. The use of CNTs also prevented the denaturation and leaching of NR from the electrode surface. The authors also developed another biosensor with a detection limit of 0.5 μM, in which NR was linked to gold nanoparticles (GNPs)/polypyrrole via a self-assembled monolayer of cysteine (Madasamy et al., 2013). Recently, electro-polymerization of poly (3,4-ethylene dioxythiophene)/NR at 1.1 V for 300s was carried out to synthesize nanoscale wires and two-dimensional flat films. The structures were utilized for developing the biosensor, where two-dimensional flat films were deposited on a gold-coated electrode and the nanowires were grown inside the polycarbonate membrane. The performance of a flat two-dimensional poly (3,4-ethylene dioxythiophene)/enzyme sensor in terms of sensitivity and detection limit was found to have substantially improved compared to the performance of the nanoarray sensor (Gokhale et al., 2015). Since nitrate reductase is a multi-subunit enzyme with deep-seated reaction centers, electrical communication between the enzyme and the electrode surface often remains weak. However, the use of conducting polymers and metallic nanoparticles alone or in combination has improved the electrical signal transduction to a great extent. Zhang et al. (2019) developed a biosensor in which a combination of aminated graphene sheets and GNPs promoted the sensitivity of electrical response. On a glassy carbon electrode (GCE), aminated graphene sheets immobilized NR in correct orientation while GNPs acted as electronic wire to yield a detection limit of 7×10^{-7} mol L^{-1} with linearity from 1.0×10^{-6} mol L^{-1} to 2.0×10^{-3} mol L^{-1}.

A colorimetric strip for nitrate detection and determination was fabricated by immobilizing NR onto metal (Au and Ag) and metal oxide (Fe_2O_3 and ZnO) nanoparticles (Sachdeva and Hooda, 2014, 2015, and 2016). To facilitate easy handling and separation of nanoparticles/NR conjugates, the nanoparticles were attached to a multifunctional epoxy layer before being capped with NR. However, the minimum detection limit by the strip method was 0.05 mM, which was higher compared to other nanosensors, but adequate for analysis of real environmental samples. Recently, a paper strip sensor developed by Aukema et al. (2019) utilized *E. coli* impregnated paper strips to convert nitrates to nitrites. Nitrites were then detected colorimetrically based on their ability to produce azo dyes of different intensity. The method was free from interference by other co-contamination and detected water nitrates in the range 1–10 ppm. The developed strip-based methods were portable, offered ease of use, and eliminated the requirement of a technically trained person or a complicated instrument for analysis.

Despite the popularity of NR-based sensors, their commercialization is difficult to achieve mainly due to loss of NR activity over a period or after a few determinations. A gradual decrease in NR activity needs periodic calibrations to maintain the accuracy of measurements. Support vector machine learning methods trained to consider a decrease in NR activity while predicting nitrate concentrations were utilized for developing an intelligent nitrate biosensor. The sensor could analyse over 400 samples with acceptable accuracy integrated with the online sharing of results (Massah et al., 2019). Recently, Yi et al. (2020) developed an electrochemical biosensor for determination of both nitrates and biological oxygen demand employing *Shewanella loihica* PV-4 biofilms. The bacteria *S. loihica* is a model electricigen and produces electricity naturally while oxidizing the organic matter. The biofilms catalysed denitrification at the cathode and hence, the inward electron transfer was negatively correlated with nitrate concentrations in the range from 0 to 7.0 mg/l. The authors reported > 80% accuracy and < 1 h analysis time for the developed biosensor.

3. Phosphate Biosensor

Though phosphorus is a natural component of soil, which is found in organic and mineral forms, its bioavailability or solubility is a limiting factor affecting plant growth and yield. The fraction of soluble phosphorus is often increased by amending the soil with exogenous phosphorus fertilizers to meet crop requirements. However, a large proportion of this phosphorus ends up in streams, lakes, and other water bodies due to run-off, leaching, and soil erosion causing eutrophication along with nitrates. Eutrophication leads to the growth of algal plants resulting in depleting water of its oxygen and carbon resources. Natural water bodies such as lakes, reservoirs, etc., contain somewhere between 0.3–0.025 ppm of phosphate, whereas concentrations above 1 ppm can seriously offset the natural balance of the water reserves. Consumption of phosphate-rich water may induce aging and damage to the blood vessels in humans. Therefore, testing the soil for phosphates before fertilizer application will help meet the crop's requirement in a more cost-effective and environmentally friendly manner. A schematic representation of various biosensing platforms in Fig. 11.2.

In recent years, biosensors utilizing different nanomaterials such as GNPs, CuO nanodots, and ZnO nanoflakes have attracted the interest of researchers to their cost-effective and simple design. A promising field-effect transistor biosensor for phosphate detection based on immobilization of pyruvate oxidase onto high surface area platform of ZnO nanorods grown on seeded SiO_2/Si substrate. The biosensor offered high specificity, the sensitivity of 80.57 μA mM^{-1} cm^{-2} and linearity from 0.1 to 7.0 mM as well as good stability, reproducibility, and had little effect of electroactive species (Ahmad et al., 2017). Recently, an electrochemical biosensor developed using pyruvate oxidase immobilized onto the GNPs/CuO nanodots loaded poly (diallyl dimethylammonium chloride) functionalize graphene exhibited good biocompatibility and yielded a superior current response. The biosensor had acceptable analytical parameters such as linearity from 0.01 to 80 μM, a detection limit of 0.4 nM, good anti-interference, stability, and reproducibility (He et al., 2020).

Anew type of photoluminescence (PL) probe for highly selective detection of PO_4^{3-} based on the binding of Al^{3+} with either PO_4^{3-} or graphene quantum dots (QDs) has been proposed. In the absence of PO_4^{3-}, Al^{3+} causes aggregation of QDs resulting in enhanced PL, whereas, in the presence of PO_4^{3-}, Al^{3+} prefers to coordinate with PO_4^{3-} compared to QDs. As a result, PL intensity decreases. The method was found

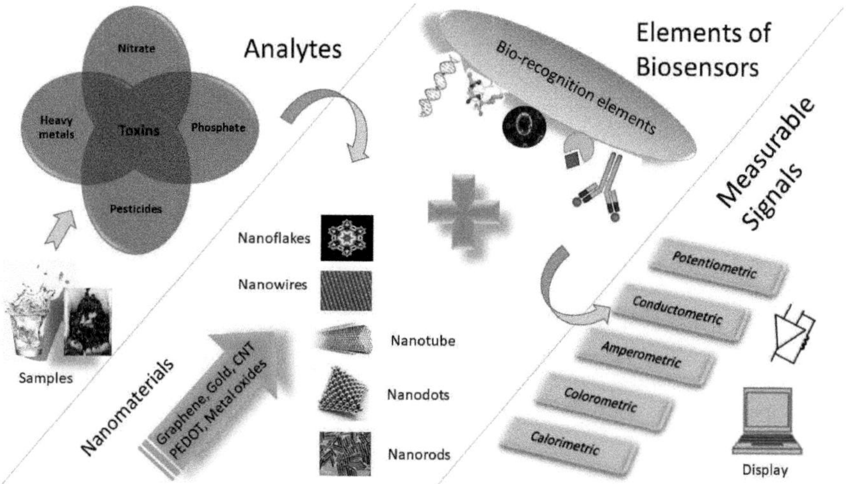

Fig. 11.2. Schematic representation of various biosensing platforms for determination of nitrates, phosphates, heavy metals, and pesticides in soil and water.

to be easy, rapid, low cost, label-free, selective, reproducible, and apt for phosphate determination in complex matrices (Chen et al., 2018).

4. Pesticide Biosensors

Pesticides are used in agriculture for preventing, destroying, repelling, or mitigating any pest. Undoubtedly, pesticides share the major credit for improving productivity in agriculture and controlling vector-borne diseases, yet they have brought along unwelcomed side effects on the environment and threat to the health and life of other non-target organisms. Serious health implications such as nerve disorders, infertility, immunological, and respiratory diseases in human beings have been correlated with harmful exposure to pesticides (Sassolas et al., 2012; Boudh and Singh, 2019). Hence, several electrochemical, enzymatic, whole-cell, immunochemical and DNA biosensors have been developed for determining pesticides in food, vegetables, water samples, soil, and agricultural products. Apparently, nano-sized particles are often employed in the sensors to increase the surface area for immobilization of biorecognition elements, and for achieving its better electrical wiring with electrode surface.

Enzyme-based pesticide biosensors are largely based on either inhibition of acetylcholinesterase (AChE) by the organophosphate pesticides (OPs) or catalysis of selective OPs and carbamate pesticides by organophosphate hydrolase (OPH) enzyme. In these sensors, different composite nanomaterials such as those of chitosan (CS)-TiO_2-graphene (Cui et al., 2018), Pd@Au nanorods (Lu et al., 2018), Pd@Au nanowires (Lu et al., 2019), TiO_2-nitrogen-doped carbon QDs (Cheng et al., 2019), CNTs-based polyaniline/polypyrrole polymer nanocomposite (Virutkar et al., 2019) and three-dimensional graphene CuO nanoflowers nanocomposites (Bao et al., 2019) have been successfully used for developing ultrasensitive and stable sensing platforms. Improvement in the electrochemical sensing of OPs have been achieved after incorporation of antimony tin oxide-CS, ordered mesoporous carbon-CS composite (Hou et al., 2019) and nitrogen-doped-ordered mesoporous carbon (Long et al., 2019). Wu et al. (2019) fabricated a fluorescence sensor based on aggregation-induced emission fluorogens-SiO_2–MnO_2 sandwich nanocomposites wherein an 'on–off' switch for fluorescence sensor was incorporated based on AChE inhibition of OPs. This fabrication paved the way for an effortless and appropriate test for visual onsite analysis for OPs.

Silva et al. (2020) have recently developed a low-cost electrochemical sensing platform based on alkaline phosphatase covalently linked to electropolymerized hydroxybenzoic acid isomers over carbon graphite electrode. The sensor detected 4-nitrophenol concentrations as low as 18.0 μM and, hence, was suggested to be an excellent platform for the determination of organophosphorus pesticides. Another chemiluminescent paper-based biosensor for acetylcholine inhibitors detected H_2O_2 generated via a sequential reaction of AChE, choline oxidase and horseradish peroxidase. The sensor has an assay time of 25 min and was used for the detection of organophosphate pesticides in food matrices (Montali et al., 2020). An ultrasensitive biosensor for the detection of chlorpyrifos in vegetables with a LOD of 0.02 nM was developed by immobilizing AChE onto CS-nano tin dioxide matrix over the GCE. The sensor showed good linearity (0.02 nM–20000 nM), reproducibility, selectivity, and acceptable storage stability (Liu et al., 2020). Table 11.1 describes important analytical parameters for some of the AChE-based biosensors.

In OPH-based biosensors, the analytical signal is the oxidation current of the *p*-nitrophenol, which is produced upon the OPH-catalysed hydrolysis of the selected OP substrates. Gothwal et al. (2014) developed an amperometric biosensor based on the glutaraldehyde coupling of OPH to the CNT electrode. The LOD detected for methyl parathion was 0.1 μM with a linearity range 0.1–200 μM and good operational stability of 50 days. The LOD of 20×10^{-3} μM and linearity of 20×10^{-3}–40.0 μM were acquired for detecting paraoxon by physisorption of OPH enzyme onto CS-CNTs/ZrO_2-modified GCE with nice storage stability of 30 days (Cabarcas et al., 2018). Du et al. (2010b) used OPH to form a nanocomposite biosensor based on the mixture of GNPs/MWCNTs/QDs settled on the surface of a GCE. The biosensor sensed methyl parathion out of all analytes or analogs with an LOD of 0.001 μM and linearity of 1.9×10^{-14}–7.6×10^{-13} μM. However, a potentiometric biosensor with OPH crosslinked onto Al/p–Si/SiO_2 showed detection limit of 2.0 μM for paraoxon with good storage stability of 60 days as reported by Schoning et al. (2003). The MWCNTs have strong electrocatalytic activity provided high sensitivity to the model analyte, without using any redox mediator. This ability of MWCNTs/GNPs/CdTeQDs-modified GCE was exploited by Du et al. (2008) for developing strong electron transfer and large immobilization sites for enzymes such asAChE or OPH. The authors inferred that the enzyme was sensitive towards pesticides with P–S bond, i.e., carbamates or OPs only. Also, they highlighted the advantage of using OPH over AChE in this device as using OPH made the device reusable and suitable for continuous monitoring. Amperometric biosensor with OPH/CNTs-coated screen-printed electrode (SPE) revealed higher sensitivity and stability through accelerated oxidation of hydrogen peroxide at the CNT transducer. The paraoxon detection limit for this biosensor was found to be 41.3 ppb within a 10-min detection time (Deo et al., 2005). A real-time analysis-based conductometric biosensor with OPH immobilized on a single-walled-carbon-nanotubes (SWCNTs) electrode served as a simple, reusable, and low-cost biosensing plan for detecting paraoxon with an LOD of 13.8 ppm (Liu et al., 2007).

Table 11.1. Attributes of AChE-based electrochemical biosensors for OP detection.

S.No.	Inhibitor/ analyte	Immobilization		Transduction/ Detection Method	(LOD) (μM)	Linearity (μM)	Time of Incubation (min)	Storage Stability (days)	Refs
		Method	**Matrix**						
1.	Trichlorfon	Crosslinking	Nylon and cellulose nitrate membrane/pH electrode	Potentiometric	0.038	50×10^3–2.5×10^3	15	30	Ivanov et al. (2000)
2.	Paraoxon	Crosslinking	Cellophane membrane/Au electrode	Amperometric	1.45	1.45–7.26	15	-	Rekha et al. (2008)
3.	DZN - oxon	Entrapment	Hybrid mesoporous SiO_2 membrane/Pt electrode	Amperometric	1.2×10^{-3}	1.0×10^{-3}–0.3	15	80	Shimomura et al. (2009)
4.	Dichlorvos	Encapsulation	Tetramethylorthosilicate sol–gel film/SPE	Amperometric	1.0×10^{-3}	1.0 and 3.0×10^{-3}	15	-	Sotiropoulou1 (2005)
5.	Chlorpyriphos-ethyl oxon	Encapsulation	Al_2O_3 sol-gel/sono gel–carbon electrode	Amperometric	2.5×10^{-4}	0.5	10	50	Zejli et al. (2008)
6.	Paraoxon	Electrostatic interactions	ZnO sol-gel/SPE	Amperometric	0.127	0.127–5.010	10	90	Sinha et al. (2010)
7.	Methyl parathion and Acephate	Encapsulation	Silica sol–gel film/CPE	Amperometric	3.0×10^{-4} and 0.47	3.7×10^{-4}–1.8×10^{-3}, 0.27–4.09	20 and 4	30	Raghu et al. (2011)
8.	Dichlorvos	Noncovalent	Polyethyleneimine/SPE	Amperometric	1.0×10^{-3}	-	2	-	Vakurov et al. (2004)
9.	Diazinon Fenthion	Adsorption	Mercaptobenzothiazole/ Polyaniline/gold electrode	Amperometric	0.48×10^{-3} 0.61×10^{-3}	-	20	-	Somers et al. (2007)

Table 11.1 contd. ...

...Table 11.1 contd.

S.No.	Inhibitor/ analyte	Immobilization		Transduction/ Detection Method	(LOD) (μM)	Linearity (μM)	Time of Incubation (min)	Storage Stability (days)	Refs
10.	Malathion	Electropolymeric entrapment	Polypyrrole (Ppy) and polyaniline (PAn)co-polymer doped with MWCNTs/GCE	Amperometric	1.0 ng/mL	0.01 to 0.5 g/mL and 1 to 25 g/mL	15	30	Du et al. (2010a)
11.	Malathion and chlorpyrifos	Covalent	Zn-sulphide and poly (indole-5-carboxylic acid)/ Au electrode	Amperometric	0.1×10^{-3} and 1.5×10^{-3}	$0.1–50 \times 10^{-3}$ and $1.5–40 \times 10^{-3}$	10	60	Chauhan et al. (2011)
12.	Dichlorvos	Adsorption	Al_2O_3 sol-gel matrix SPE	Amperometric	0.01	0.1–80	15	5	Shi et al. (2006)
13.	Chlorpyrifos-ethyl oxon	Adsorption	SPE	Amperometric	3.0×10^{-6}	$5 \times 10^{-2}–0.2$	10	50	Bonnet et al. (2003)
14.	Dichlorvos parathion, and azinphos	Entrapment	Polyaniline carbon/cobalt phthalocyanine/SPE	Amperometric	1×10^{-11}, 1×10^{-10} and 1×10^{-10}	$1 \times 10^{-11}–1 \times 10^{-2}$	10	92	Law et al. (2005)
15.	Paraoxon, Malaoxon	Covalent	SWCNTs–Co phtalocyanine /SPE	Amperometric	0.01 and 6.3×10^{-3}	0.018–0.181 and 6.36×10^{-3} –0.159	15	3	Ivanov et al. (2011)
16.	Monocrotophos	Covalent	CdTe QDs/GNPs/CS/GCE	Amperometric	1.34	$4.4 \times 10^{-3}–4.48$ and 8.96–67.20	8	30	Du et al. (2008)

17.	Paraoxon ethyl, sarin, and aldicarb	Crosslinking	Au–Pt nanoparticles/GCE	Amperometric	50×10^{-4}, 40×10^{-3} and 40	$50–200 \times 10^{-3}$ $1.40–50 \times 10^{-3}$, and 40–60	25	-	Upadhyay et al. (2009)
18.	Paraoxon and chlorpyrifos-ethyl oxon	Crosslinking	Prussian blue (PB) /CS/ GCE	Amperometric	0.113×10^{-4} 0.703×10^{-4} 0.194×10^{-4} and 0.33×10^{-4}	$0.45 \times 10^{-4}–0.045$, $0.234 \times 10^{-3}–0.046$, $0.116 \times 10^{-3}–0.0194$, and $0.167 \times 10^{-4}–0.0335$	10	-	Sun and Wang (2010)
19.	Paraoxon	Crosslinking	Poly (acrylonitrile-methylmethacrylate-sodium vinylsulfonate) -GNPs	Amperometric	7.3×10^{-11} g/L	$10^{-11}–10^{-7}$ g/L	-	20	Marinovet al. (2010)
20.	Malathion	Hydrophilic adhesion	MWCNTs-GNPs-CHIT/ GCE	Amperometric	0.6 ng/mL	1.0 to 1000 ng mL^{-1} and 2 to 15 µg mL^{-1}	8	30	Du et al. (2010b)
21.	Monocrotophos	Adsorption	GNPs/PB/GCE	Amperometric	3.5×10^{-9}	4.48×10^{-3} $–4.48 \times 10^{-2}$	10	30	Wu et al. (2011)
23.	Chlorpyrifos	Covalent	Graphite/CS/GCE	Voltametric	1.58×10^{-4}	$1 \times 10^{-4}–1.0$	10	10	Ion et al. (2010)
24.	Methyl Parathion, Monocrotophos, Chlorpyrifos and Endosulfan	Covalent	Nafion/cSWCNT/ MWCNT/GNPs-Au	Amperometric	1.9, 2.3, 2.2, 2.5 nM	0.1–130	10	60	Dhull (2020)

Table 11.1 contd. ...

...Table 11.1 contd.

S.No.	Inhibitor/ analyte	Immobilization		Transduction/ Detection Method	(LOD) (µM)	Linearity (µM)	Time of Incubation (min)	Storage Stability (days)	Refs
25.	Monocrotophos	Adsorption	GNPs/PB/GCE	Amperometric	3.5×10^{-9}	4.4×10^{-3} -4.48×10^{-2}	10	30	Chauhan et al. (2014)
26.	Malathion	Chemisorption	CS-GNPs/Au electrode	Amperometric	0.1×10^{-3}	0.3×10^{-3}– 60.5×10^{-3}	13	-	Cai et al. (2014)
27.	Chlorfenvinphos	Entrapment	MWCNTs/poly(4-(2,5-di(thiophen-2-yl)-1H-pyrrol-1-yl) benzenamine)	Amperometric	14×10^{-12} M	14–300 × 10^{-12} M, 30 × 10^{-12} M–35 × 10^{-9} M	-	-	Kesiket al. (2014)
28.	Dimethoate	Covalent coupling	CNTs/ZrO nanoparticles/ PB/Nafion/screen printed carbon electrode	Cyclic voltammetry and differential pulse voltammetry	2.4×10^{-12} M	4.4×10^{-12} M–44×10^{-9} M	-	-	Gan et al. (2010)
29.	Monocrotophos	Adsorption	GNPs/poly (dimethyldiallylammonium chloride) protected PB	Amperometric	3.6×10^{-12} M	4.5×10^{-12} M–4.5×10^{-9} M, 4.5–45 × 10^{-9} M	-	-	Wu et al. (2011)
30.	Phoxim	Covalent coupling	Reduced graphene oxide/ Ag nanoparticles/CS	Differential pulse voltammetry	81×10^{-12} M	0.2–250 × 10^{-9} M	-	-	Zhang et al. (2015)

Table 11.2. Different OPH-based electrochemical biosensors for OPs detection along with their characteristics.

S. No.	Analyte/ Inhibitor	Immobilization		Transduction Technique	Limit of Detection (μM)	Linearity (μM)	Storage Stability (days)	Refs
		Method	**Matrix**					
1.	Paraoxon	Entrapment	CNTs modified GCE	Amperometric	0.15	Up to 4	-	Deo et al. (2005)
2.	Paraoxon	Crosslinking	Al/p-Si/SiO_2	Potentiometric	2.0	NR	60	Schoning et al. (2003)
3.	Paraoxon	Entrapment	Mesoporous carbon/carbon black/GCE	Amperometric	0.12	0.2–8	-	Lee et al. (2010)
4.	Methyl Parathion	Covalent Binding	GNPs/MWCNTs/QDs/GCE	Amperometric	1×10^{-3}	1.9×10^{-14}–7.6×10^{-13}	-	Du et al. (2010c)
5.	Methyl parathion	Covalent Binding	Methyl parathion hydrolase/ GNPs/Si/MWCNTs	Amperometric	0.3×10^{-6}	Up to 1×10^{-6}	-	Chen et al. (2011)
6.	Paraoxon, Parathion	Covalent binding	Screen printed carbon electrode	Amperometric	1.8×10^{-5}	-	-	Mulyasuryani et al. (2014)
7.	Methyl parathion	Glutaraldehyde Coupling	CNTs electrode	Amperometric	0.1	$0.1–2 \times 10^{2}$	50	Gothwal et al. (2014)
8.	Methyl parathion, Parathion	Covalent Binding	Elastin-like-polypeptide-OPH/bovine serum albumin/ TiO_2nanofibers/MWCNTs	Amperometric	12×10^{-9} 10×10^{-9}	10×10^{-9}–12×10^{-9}	-	Bao et al. (2016)
9.	Paraoxon	Physical Adsorption	CS–CNT–ZrO_2-modified GCE	Amperometric	20.0×10^{-3}	20×10^{-3}–40	-	Cabarcas et al. (2018)

An amperometric biosensor proposed by Hossain et al. (2011) based on time-resolved differential pulse stripping voltammetry with a disposable proton-discriminative sensor allowed judicial detection of pesticides. The paraoxon was detected within a detection time of ~ 10 min with LOD of 145.6 ppb using this approach. A remote electrochemical biosensor was constructed by Wang et al. (1999) using polyvinyl chloride/nafion/carbon paste electrode (CPE) and obtained the reaction of the biosensor in the linearity range 4.6–46 μM for paraoxon and up to 5 μM for methyl parathion. The LODs of the biosensor for paraoxon and methyl parathion were 0.9 μM and 0.4 μM, respectively. Mulchandani et al. (1999) formed an amperometric biosensor with OPH entrapped on screen-printed carbon electrodes. The LOD of 9×10^{-14} μM and 7×10^{-14} μM and linearity of 40 μM and 5 μM were found for paraoxon and methyl parathion individually.

Bacterial electrochemical sensors employing *Moraxella* sp., *Pseudomonas putida*, and *Escherichia coli*, genetically engineered to express OPH on their surfaces offered direct OP determination. Such microbial sensors are attractive as they eliminate the need for enzyme extraction and purification, and are popular as low cost, simple, easy, and rapid analytical tools. In this regard, Mulchandani et al. (1998) developed a potentiometric bacterial sensor by immobilizing OPH expressing recombinant *E. coli* on the surface of the electrode. Hydrolysis of OPs such as paraoxon, methyl, and ethyl parathion, and diazinon by OPH causes a pH change, directly proportional to the concentration of the pesticide. The LOD for diazonium was found to be 5.0 μM. Another, amperometric bacterial sensor was designed by Mulchandani et al. in 2001, using a CPE as a transducer and OPH expressing *Moraxella* sp. as the biorecognition element with a detection limit of 1.0 and 0.2 μM and linearity up to 175 and 40 μM for methyl parathion and paraoxon, respectively. Lei et al. (2007) developed a microbial sensor, having a Clark-type electrode and recombinant OPH expressing bacteria *Pseudomonas putida*. The bacteria oxidize OPs to CO_2 with O_2 consumption, causing a shift in the O_2 reduction current proportional to the substrate concentration. The LOD for the fenitrothion detection with *P. putida* JS444 with surface-plane-expressed OPH O_2-electrode was found to be 277 ppb and linearity being 0.05 mM. Various OPH-based electrochemical biosensors and their important features have been listed in Table 11.2.

5. Heavy Metal Detection

Urbanization and industrialization have caused a substantial increase in heavy metal concentrations in the environment. Heavy metals can persist in the natural environment for a very long period and cause severe diseases in humans even at extremely low concentrations. Detection of heavy metals in the soil is one of the important parameters of measuring soil health and ensuring food safety. The dominant analytical method for heavy metals such as lead, arsenic, mercury, copper, and cadmium is based on DNAzymes and aptamers combined with nano-biosensing platforms. The principle for detection of nitrates, phosphates, pesticides, and heavy metals are shown in Fig. 11.3.

Wang et al. (2019) developed a biosensor for measuring Pb^{+2} in the soil from the forest. The authors utilized a specific Pbzyme to detect Pb^{2+} along with SWCNTs and field-effect transistor (Pbzyme/SWCNTs/FET). Change in conductivity of the Pbzyme/SWCNTs/FET upon binding of Pb^{2+} to the Pbzyme was used to evaluate the

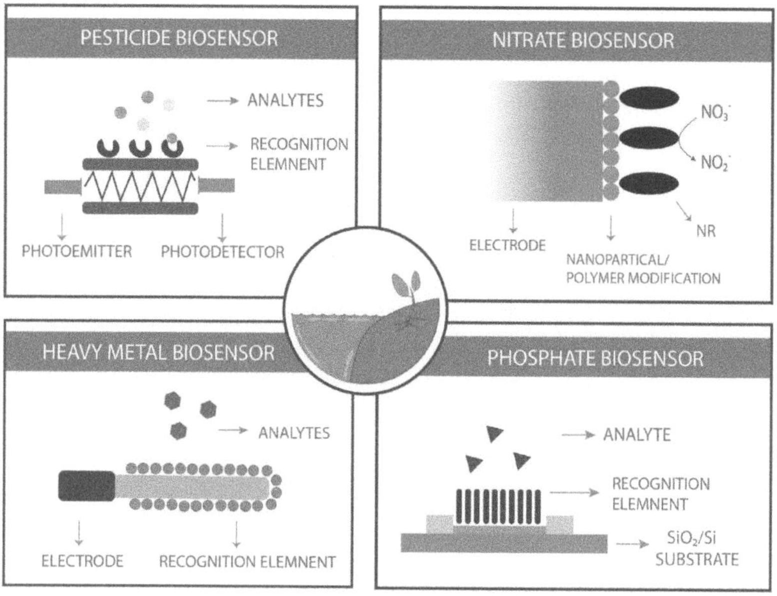

Fig. 11.3. Principles of biosensors for the detection of (1) pesticides, (2) nitrates, (3) heavy metals, and (4) phosphates.

Pb^{2+} concentration. Good linearity of 10 pM to 50 nM and a detection limit of 7.4 pM were achieved. Xu et al. (2018) presented a novel and simple strategy for the detection of Pb^{2+} based on G-quadruplex DNA and GNPs. The phosphate group of DNAs was functionalized with an electrochemical probe $[Ru (NH_3)_6]^{3+}$ (RuHex). In the presence of Pb^{2+}, the substrate DNA on the GNPs surface hybridized with complementary DNA to form a stable G-quadruplex, which fell off the electrode surface thereby decreasing the response proportional to the amount of RuHex lost. The biosensor had excellent selectivity for Pb^{2+}, showed linearity of 0.01 nM–200 nM and a low detection limit of 0.0042 nM under optimal conditions.

On a similar concept, a reusable biosensor for Pb (II) ions was designed employing a double-stranded DNA with a terminal amino group and containing a G-quadruplex (G4) aptamer covalently attached to SWCNTs/FET biosensor. Formation of a stable G4 aptamer/Pb(II) complex-induced structural changes in DNA duplex thereby affected the electrical conductivity of SWCNTs, which served as analytical signal. The obtained linearity range (1 $ng \cdot L^{-1}$–100 $\mu g \cdot L^{-1}$) and low LOD (0.39 $ng \cdot L^{-1}$) established the selective and sensitive nature of the biosensor. The sensor was successfully employed for the determination of Pb (II) in spiked water and soil samples with acceptable results (Wang et al., 2018). Ni et al. (2019) proposed a label-free electro-chemiluminescent biosensor for Pb^{2+} detection utilizing DNAzyme-based double-stranded DNA having an electrochemical indicator embedded in its groove. The sensor employed an electrically heated indium–tin-oxide electrode for temperature control, thus eliminating the bulk solution heating needed for electro-chemiluminescent detection by the traditional approach. This technique made the sensor simple, easy to operate, and eliminated tedious procedures and requirement of bulky instruments. The sensor produced satisfactory results for Pb^{2+} detection in soil samples with linearity from 0.25 to 500 pM with logarithm concentration of Pb^{2+} and a detection limit of 0.2 pM at 45°C.

An electrochemical sensor for Cu^{2+} detection based on gold nanoflower-modified ITO electrode was developed by Xu et al. (2019). The nanoflowers were used along with MIL-101(Fe) (MIL: Materials of Institute Lavoisier) for enhanced sensitivity and were functionalized with Cu^{2+}-dependent DNAzyme to impart selectivity to the sensor. The peroxidase-like catalytic activity of MIL-101(Fe) and high-affinity DNAzyme towards Cu^{2+} resulted in improved performance of the electrochemical biosensor with a detection range from 0.001 μM to 100 μM and a detection limit of 0.457 nM.

An attempt to improve arsenic sensing was made by Siddiqui et al. (2020), wherein the authors proposed aptamer-GNPs integrated smartphone-based optical sensing. The authors proposed a one-step acid extraction and different solid-phase extraction technique for high arsenic yield and removal of interfering ions from the soil samples. The detection limit of 14.44 ppb for water samples and 1.97 ppm for soil samples was obtained and the method correlated well with the inductively coupled plasma mass spectrometry method with a correlation coefficient of 0.997.

Kim et al. (2019) exploited a green fluorescent protein's ability to produce fluorescence in response to the presence of Cd and Hg. In the absence of metalloids Cd and Hg, the metal-binding loops remained inserted into a loop region of the fluorescent protein to render this protein inactive and the binding of metal ions to the metal-binding loops induced a positive conformational change in the fluorescent protein and restored its original activity. With this technique, the metal ions could be quantified in the concentration range 0 μM and its practical application was demonstrated for Cd analysis in artificially amended soil and water samples.

One other widely popular sensing platform is microbial biosensor due to its simplicity and ease of construction. A Cd-specific biosensing platform based on synthetic biology principles was constructed by Bereza-Malcolm et al. (2017). The multiplex biosensor consisted of a single-output Cd biosensor element, *cadRgfp*, a constitutively expressed *mrfp1* onto a broad-host-range vector (*Pseudomonas*, *Shewanella*, and *Enterobacter*) and green and red fluorescence proteins for detection of measurable signals. The constructed multiplex biosensors for Cd could detect Cd, As, Hg and Pb concentrations ranging from 0.01 to 10 μg ml^{-1}. Compared to other electrochemical sensing devices, the developed method was slow as it produced a response in 20–40 min when exposed to 3 μg ml^{-1} Cd concentrations.

6. Conclusion and Future Direction

Though biosensors are promising candidates for environmental monitoring owing to their exceptional performance capabilities such as high specificity, sensitivity, low cost, rapid feedback, relatively compact size, portability, user friendly and continuous real-time analysis, yet the usage and commercial application of biosensors is almost stagnant with respect to environmental monitoring. As most of the commercial biosensors are medical application-based so the challenge to develop improved and more reliable biosensors for fast and automatic environmental sample analysis is the need of the hour. So far, enzyme-based biosensors have been broadly used in the determination of a wide-range of contaminants such as nitrates, phosphates, pesticides (OPs and carbamates), and heavy metals. Since their inception, the main challenge faced by these biosensors is the loss of their recognition ability and stability (due to denaturation of the enzyme and loss of bioactivity). Moreover, analytical parameters such as sensitivity, reproducibility and interference from other analytes may not be fully reproduced when transferred from the research laboratories to the actual analysis of complex real-life samples. So, the researchers are compelled to explore novel biomolecules that may circumvent these limitations. From about a decade, the introduction and use of MIPs (i.e., molecular imprinting polymer sensors, with specific recognition site polymers) and Aptasensors (composed of aptamers that are nucleic acid sequences selected *in vitro* from oligonucleotide libraries) prove to be a good approach in this regard. These artificial elements possess many competitive benefits over other biorecognition elements such as remarkable target diversity, synthesis, specificity, chemical stability, convenient tight-binding capability, flexibility for favorable immobilization, easy modification, and low cost, all together owing to their improved stability and reusability. Moreover, the use of different types of nanomaterials (nanowires, nanotubes) and microfabrication or nanofabrication in these artificial biosensors is expected to pave the way for formulating low cost, more sensitive and user-friendly devices for field applications. Thus, the exploitation of the analytical potential offered by these elements can prove to be a solution to many biosensor-related problems in the future.

Furthermore, scientists have integrated microfluidic channels with paper-based sensors that might be very beneficial for small sample analysis for routine as well as random quality checking. In another direction, radio-frequency identification (RFID) tags proved extremely beneficial and futuristic for the detection of volatile compounds especially pesticides. Smartphone and portable aptamer-based analytical tools are still at the infancy stage, yet their integration with quartz crystal microbalance (QCM) and surface plasmon resonance (SPR)-based label-free and live monitoring sensors may improve sensitivity and stability of the platform. Overall, the development of next generation sensing platform for continuous monitoring of safe and healthy food and crops can be achieved by a combination of technological intervention, interdisciplinary research, and commercial interest.

Acknowledgements

Dr. Vinita Hooda is thankful to Department of Science and Technology (DST), Ministry of Science and Technology, Government of India grant under its "Fund for Improvement of Science & Technology Infrastructure (FIST)" scheme (No. SR/FST/LS1-529/2012(C)) (http://www.fist-dst.org/) and Haryana State Council for Science, Innovation and Technology (HSCSIT, Endst. No. HSCSIT/R&D/2020/474) for providing financial support to work in the field of Agriculture Nanotechnology.

References

Ahmad, R., Ahn, M.S. and Hahn, Y.B. (2017). ZnO nanorods array-based field-effect transistor biosensor for phosphate detection. *J. Colloid and Interface Science*, 498: 292–297. https://doi.org/10.1016/j.jcis.2017.03.069.

Aukema, K.G. and Wackett, L.P. (2019). Inexpensive microbial dipstick diagnostic for nitrate in water. *Environ. Sci. Water Res. Technol.*, 5(2): 406–416.

Bao, J., Hou, C., Dong, Q., Ma, X., Chen, J., Huo, D., Yang, M., Abd El Galil, K.H., Chen, W. and Lei, Y. (2016). ELP-OPH/BSA/TiO_2 nanofibers/c-MWCNTs-based biosensor for sensitive and selective determination of p-nitrophenyl substituted organophosphate pesticides in aqueous system. *Biosensors and Bioelectronics*, 100(85): 935–942.

Bao, J., Huang, T., Wang, Z., Yang, H., Geng, X., Xu, G., Samalo, M., Sakinati, M., Huo, D. and Hou, C. (2019). 3D graphene/copper oxide nano-flowers-based acetylcholinesterase biosensor for sensitive detection of organophosphate pesticides. *Sensors and Actuators B: Chemical*, 279: 95–101. https://doi.org/10.1016/j.snb.2018.09.118.

Baronas, R., Žilinskas, A. and Litvinas, L. (2016). Optimal design of amperometric biosensors applying multi-objective optimization and decision visualization. *Electrochimica Acta*, 100(211): 586–594.

Bereza-Malcolm, L., Aracic, S., Kannan, R., Mann, G. and Franks, A.E. (2017). Functional characterization of Gram-negative bacteria from different genera as multiplex cadmium biosensors. *Biosensors and Bioelectronics*, 100(94): 380–387.

Bonnet, C., Andreescu, S. and Marty, J.L. (2003). Adsorption: An easy and efficient immobilization of acetylcholinesterase on screen-printed electrodes. *Analytica Chimica Acta*, 481(2): 209–211.

Boudh, S. and Singh, J.S. (2019). Pesticide contamination: environmental problems and remediation strategies. *Emerging and Eco-Friendly Approaches for Waste Management*, 245–269. https://doi.org/10.1007/978-981-10-8669-4_12.

Cabarcas Beleno, M.T., Stoytcheva, M., Zlatev, R., Montero, G., Velkova, Z. and Gochev, V. (2018). Chitosan nanocomposite modified OPH-based Amperometric sensor for organophosphorus pesticides determination. *Current Analytical Chemistry*, 14(1): 75–82.

Cai, J.R., Zhou, L.N. and Han, E. (2014). A sensitive amperometric acetylcholine biosensor based on carbon nanosphere, and acetylcholinesterase modified electrode for detection of pesticide residues. *Analytical Sciences*, 30(6): 669–673.

Can, F., Ozoner, S.K., Ergenekon, P. and Erhan, E. (2012). Amperometric nitrate biosensor based on Carbon nanotube/Polypyrrole/Nitrate reductase biofilm electrode. *Materials Science and Engineering: C*, 32(1): 18–23.

Chauhan, N., Narang, J. and Pundir, C.S. (2011). Immobilization of rat brain acetylcholinesterase on ZnS and poly (indole-5-carboxylic acid) modified Au electrode for detection of organophosphorus insecticides. *Biosensors and Bioelectronics*, 29(1): 82–88.

Chauhan, N. and Pundir, C.S. (2014). Amperometric determination of acetylcholine–A neurotransmitter, by chitosan/gold-coated ferric oxide nanoparticles modified gold electrode. *Biosensors and Bioelectronics*, 61: 1–8. https://doi.org/10.1016/j.bios.2014.04.048.

Chen, B.B., Li, R.S., Liu, M.L., Zou, H.Y., Liu, H. and Huang, C.Z. (2018). Highly selective detection of phosphate ion based on a single-layered graphene quantum dots-Al^{3+} strategy. *Talanta*, 178: 172–177. https://doi.org/10.1016/j.talanta.2017.09.007.

Chen, J., Wang, W., Xu, Y. and Zhang, X. (2010). Slow-release formulation of a new biological pesticide, pyoluteorin, with mesoporous silica. *J. Agric Food Chem*., 59(1): 307–311.

Chen, S., Huang, J., Du, D., Li, J., Tu, H., Liu, D. and Zhang, A. (2011). Methyl parathion hydrolase-based nanocomposite biosensors for highly sensitive and selective determination of methyl parathion. *Biosensors and Bioelectronics*, 26(11): 4320–4325.

Cheng, W., Zheng, Z., Yang, J., Chen, M., Yao, Q., Chen, Y. and Gao, W. (2019). The visible light-driven and self-powered photoelectrochemical biosensor for organophosphate pesticides detection based on nitrogen doped carbon quantum dots for the signal amplification. *Electrochimica Acta*, 296: 627–636.

Cui, H.F., Wu, W.W., Li, M.M., Song, X., Lv, Y. and Zhang, T.T. (2018). A highly stable acetylcholinesterase biosensor based on chitosan-TiO_2-graphene nanocomposites for detection of organophosphate pesticides. *Biosensors and Bioelectronics*, 99: 223–229.

Deo, R.P., Wang, J., Block, I., Mulchandani, A., Joshi, K.A., Trojanowicz, M., Scholz, F., Chen, W. and Lin, Y. (2005). Determination of organophosphate pesticides at a carbon nanotube/organophosphorus hydrolase electrochemical biosensor. *Analytica Chimica Acta*, 530(2): 185–189.

Dhull, V. (2020). A Nafion/AChE-cSWCNT/MWCNT/Au-based amperometric biosensor for the determination of organophosphorous compounds. *Environmental Technology*, 41(5): 566–576.

Du, D., Chen, S., Song, D., Li, H. and Chen, X. (2008). Development of acetylcholinesterase biosensor based on CdTe quantum dots/gold nanoparticles modified chitosan microspheres interface. *Biosensors and Bioelectronics*, 24(3): 475–479.

Du, D., Ye, X., Cai, J., Liu, J. and Zhang, A. (2010a) Acetylcholinesterase biosensor design based on carbon nanotube-encapsulated polypyrrole and polyaniline copolymer for amperometric detection of organophosphates. *Biosensors and Bioelectronics*, 25(11): 2503–2508.

Du, D., Wang, M., Cai, J., Qin, Y. and Zhang, A. (2010b). One-step synthesis of multiwalled carbon nanotubes-gold nanocomposites for fabricating amperometric acetylcholinesterase biosensor. *Sensors and Actuators B: Chemical*, 143(2): 524–529.

Du, D., Chen, W., Zhang, W., Liu, D., Li, H. and Lin, Y. (2010c). Covalent coupling of organophosphorus hydrolase loaded quantum dots to carbon nanotube/Au nanocomposite for enhanced detection of methyl parathion. *Biosensors and Bioelectronics*, 25(6): 1370–1375.

Gan, N., Yang, X., Xie, D., Wu, Y. and Wen, W. (2010). A disposable organophosphorus pesticides enzyme biosensor based on magnetic composite nanoparticles modified screen printed carbon electrode. *Sensor*, 10(1): 625–638.

Gokhale, A.A., Lu, J., Weerasiri, R.R., Yu, J. and Lee, I. (2015). Amperometric detection and quantification of nitrate ions using a highly sensitive nanostructured membrane electrocodeposited biosensor array. *Electroanalysis*, 27(5): 1127–1137.

Gothwal, A., Beniwal, P., Dhull, V. and Hooda, V. (2014). Preparation of electrochemical biosensor for detection of organophosphorus pesticides. *I. J. of Analytical Chemistry*, 303641. https://doi.org/10.1155/2014/303641.

Griesche, C. and Baeumner, A.J. (2020). Biosensors to support sustainable agriculture and food safety. *TrAC Trends in Analytical Chemistry*, 115906. https://doi.org/10.1016/j.trac.2020.115906.

He, B. and Liu, H. (2020). Electrochemical biosensor based on pyruvate oxidase immobilized AuNRs@ Cu_2O-NDs as electroactive probes loaded poly (diallyldimethylammonium chloride) functionalized graphene for the detection of phosphate. *Sensors and Actuators B: Chemical*, 304: 127303. https://doi.org/10.1016/j.snb.2019.127303.

Hooda, V., Sachdeva, V. and Chauhan, N. (2016). Nitrate quantification: Recent insights into enzyme-based methods. *Reviews in Analytical Chemistry*, 35(3): 99–114.

Hossain, M.M., Faisal, S.N., Kim, C.S., Cha, H.J., Nam, S.C. and Lee, H.J. (2011). Amperometric proton selective strip-sensors with a microelliptic liquid/gel interface for organophosphate neurotoxins. *Electrochemistry Communications*, 13(6): 611–614.

Hou, W., Zhang, Q., Dong, H., Li, F., Zhang, Y., Guo, Y. and Sun, X. (2019). Acetylcholinesterase biosensor modified with ATO/OMC for detecting organophosphorus pesticides. *New Journal of Chemistry*, 43(2): 946–952.

Ion, A.C., Ion, I., Culetu, A., Gherase, D., Moldovan, C.A., Iosub, R. and Dinescu, A. (2010). Acetylcholinesterase voltammetric biosensors based on carbon nanostructure-chitosan composite material for organophosphate pesticides. *Materials Science and Engineering: C*, 30(6): 817–821.

Ivanov, A.N., Evtugyn, G.A., Gyurcsányi, R.E., Toth, K. and Budnikov, H.C. (2000). Comparative investigation of electrochemical cholinesterase biosensors for pesticide determination. *Analytica Chimica Acta*, 404(1): 55–65.

Ivanov, A.N., Younusov, R.R., Evtugyn, G.A., Arduini, F., Moscone, D. and Palleschi, G. (2011). Acetylcholinesterase biosensor based on single-walled carbon nanotubes-Co phtalocyanine for organophosphorus pesticides detection. *Talanta*, 85(1): 216–221.

Kesik, M., Kanik, F.E., Turan, J., Kolb, M., Timur, S., Bahadir, M. and Toppare, L. (2014). An acetylcholinesterase biosensor based on a conducting polymer using multiwalled carbon nanotubes for amperometric detection of organophosphorous pesticides. *Sensors and Actuators B: Chemical*, 205: 39–49. https://doi.org/10.1016/j.snb.2014.08.058.

Kim, H., Lee, W. and Yoon, Y. (2019). Heavy metal (loid) biosensor based on split-enhanced green fluorescent protein: development and characterization. *Applied Microbiology and Biotechnology*, 103(15): 6345–6352.

Kim, S.H., Thoa, T.T. and Gu, M.B. (2019). Aptasensors for environmental monitoring of contaminants in water and soil. *Current Opinion in Environmental Science & Health*, 10: 9–21. https://doi.org/10.1016/j.coesh.2019.09.003.

Kondawar, S.B., Virutkar, P.D., Mahajan, A.P. and Meshram, B.H. (2019). Conductive polymer nanocomposite enzyme immobilized biosensor for pesticide detection. *J. Materials NanoScience*, 6(1): 7–12.

Law, K.A. and Higson, S.P. (2005). Sonochemically fabricated acetylcholinesterase micro-electrode arrays within a flow injection analyser for the determination of organophosphate pesticides. *Biosensors and Bioelectronics*, 20(10): 1914–1924.

Lee, J.H., Park, J.Y., Min, K., Cha, H.J., Choi, S.S. and Yoo, Y.J. (2010). A novel organophosphorus hydrolase-based biosensor using mesoporous carbons and carbon black for the detection of organophosphate nerve agents. *Biosensors and Bioelectronics*, 25(7): 1566–1570.

Lei, Y., Mulchandani, P., Chen, W. and Mulchandani, A. (2007). Biosensor for direct determination of fenitrothion and EPN using recombinant *Pseudomonas putida* JS444 with surface-expressed organophosphorous hydrolase. 2. Modified carbon paste electrode. *Applied Biochemistry and Biotechnology*, 136(3): 243–250.

Liu, N., Cai, X., Lei, Y., Zhang, Q., Chan-Park, M.B., Li, C., Chen, W. and Mulchandani, A. (2007). Single-walled carbon nanotube based real-time organophosphate detector. *Electroanalysis: I. J. Devoted to Fundamental and Practical Aspects of Electroanalysis*, 19(5): 616–619.

Liu, X., Sakthivel, R., Liu, W.C., Huang, C.W., Li, J., Xu, C., Wu, Y., Song, L., He, W. and Chung, R.J. (2020). Ultra-highly sensitive organophosphorus biosensor based on chitosan/tin disulfide and British housefly acetylcholinesterase. *Food Chemistry*, 126889. https://doi.org/10.1016/j.foodchem.2020.126889.

Long, B., Tang, L., Peng, B., Zeng, G., Zhou, Y., Mo, D., Fang, S., Ouyang, X. and Yu, J. (2019). Voltametric biosensor based on nitrogen-doped ordered mesoporous carbon for detection of organophosphorus pesticides in vegetables. *Current Analytical Chemistry*, 15(1): 92–100.

Lu, X., Tao, L., Song, D., Li, Y. and Gao, F. (2018). Bimetallic Pd@ Au nanorods-based ultrasensitive acetylcholinesterase biosensor for determination of organophosphate pesticides. *Sensors and Actuators B: Chemical*, 255: 2575–2581. https://doi.org/10.1016/j.snb.2017.09.063.

Lu, X., Tao, L., Li, Y., Huang, H. and Gao, F. (2019). A highly sensitive electrochemical platform based on the bimetallic Pd@ Au nanowires network for organophosphorus pesticides detection. *Sensors and Actuators B: Chemical*, 284: 103–109. https://doi.org/10.1016/j.snb.2018.12.125.

Madasamy, T., Pandiaraj, M., Kanugula, A.K., Rajesh, S., Bhargava, K., Sethy, N.K., Kotamraju, S. and Karunakaran, C. (2013). Gold nanoparticles with self-assembled cysteine monolayer coupled to nitrate reductase in polypyrrole matrix enhanced nitrate biosensor. *Advanced Chemistry Letters*, 1(1): 2–9.

Madasamy, T., Pandiaraj, M., Balamurugan, M., Bhargava, K., Sethy, N.K. and Karunakaran, C. (2014). Copper, zinc superoxide dismutase and nitrate reductase coimmobilized bienzymatic biosensor for the simultaneous determination of nitrite and nitrate. *Biosensors and Bioelectronics*, 52: 209–215. https://doi.org/10.1016/j.bios.2013.08.036.

Marinov, I., Ivanov, Y., Gabrovska, K. and Godjevargova, T. (2010). Amperometric acetylthiocholine sensor based on acetylcholinesterase immobilized on nanostructured polymer membrane containing gold nanoparticles. *J. Molecular Catalysis B: Enzymatic*, 62(1): 66–74.

Massah, J. and Vakilian, K.A. (2019). An intelligent portable biosensor for fast and accurate nitrate determination using cyclic voltammetry. *Biosystems Engineering*, 177: 49–58.

Montali, L., Calabretta, M.M., Lopreside, A., D'Elia, M., Guardigli, M. and Michelini, E. (2020). Multienzyme chemiluminescent foldable biosensor for on-site detection of acetylcholinesterase inhibitors. *Biosensors and Bioelectronics*, 162: 112232. https://doi.org/10.1016/j.bios.2020.112232.

Mulchandani, A., Mulchandani, P., Kaneva, I. and Chen, W. (1998). Biosensor for direct determination of organophosphate nerve agents using recombinant *Escherichia coli* with surface-expressed organophosphorus hydrolase. 1. Potentiometric microbial electrode. *Analytical Chemistry*, 70(19): 4140–4145.

Mulchandani, A., Chen, W., Mulchandani, P., Wang, J. and Rogers, K.R. (2001). Biosensors for direct determination of organophosphate pesticides. *Biosensors and Bioelectronics*, 16(4–5): 225–230.

Mulchandani, P., Mulchandani, A., Kaneva, I. and Chen, W. (1999). Biosensor for direct determination of organophosphate nerve agents. 1. Potentiometric enzyme electrode. *Biosensors and Bioelectronics*, 14(1): 77–85.

Mulyasuryani, A. and Dofir, M. (2014). Enzyme biosensor for detection of organophosphate pesticide residues based on screen printed carbon electrode (SPCE)-bovine serum albumin (BSA). *Engineering*, 6(5): 230–235.

Ni, J., Zhang, H., Chen, Y., Luo, F., Wang, J., Guo, L., Qiu, B. and Lin, Z. (2019). DNAzyme-based Y-shaped label-free electro-chemiluminescent biosensor for lead using electrically heated indium-tin-oxide electrode for *in situ* temperature control. *Sensors and Actuators B: Chemical*, 289: 78–84. https://doi.org/10.1016/j.snb.2019.03.076.

Perumal, V. and Hashim, U. (2014). Advances in biosensors: Principle, architecture, and applications. *J. Applied Biomedicine*, 12(1): 1–5.

Raghu, P., Swamy, B.K., Reddy, T.M., Chandrashekar, B.N. and Reddaiah, K. (2012). Sol–gel immobilized biosensor for the detection of organophosphorous pesticides: A voltametric method. *Bioelectrochemistry*, 83: 19–24. https://doi.org/10.1016/j.bioelechem.2011.08.002.

Rekha, K. and Murthy, B.N. (2008). Studies on the immobilization of acetylcholine esterase enzyme for biosensor applications. *Food and Agricultural Immunology*, 19(4): 273–281.

Sachdeva, V. and Hooda, V. (2014). A new immobilization and sensing platform for nitrate quantification. *Talanta*, 124: 52–59. https://doi.org/10.1016/j.talanta.2014.02.014.

Sachdeva, V. and Hooda, V. (2015). Immobilization of nitrate reductase onto epoxy affixed silver nanoparticles for determination of soil nitrates. *I. J. Biological Macromolecules*, 79: 240–247. https://doi.org/10.1016/j.ijbiomac.2015.04.072.

Sachdeva, V. and Hooda, V. (2016). Effect of changing the nanoscale environment on activity and stability of nitrate reductase. *Enzyme and Microbial Technology*, 100(89): 52–62.

Sassolas, A., Prieto-Simón, B. and Marty, J.L. (2012). Biosensors for Pesticide Detection: New Trends. *American Journal of Analytical Chemistry*, 3(3): 210–232.

Schöning, M.J., Arzdorf, M., Mulchandani, P., Chen, W. and Mulchandani, A. (2003). Towards a capacitive enzyme sensor for direct determination of organophosphorus pesticides: fundamental studies and aspects of development. *Sensors*, 3(6): 119–127.

Shi, M., Xu, J., Zhang, S., Liu, B. and Kong, J. (2006). A mediator-free screen-printed amperometric biosensor for screening of organophosphorus pesticides with flow-injection analysis (FIA) system. *Talanta*, 68(4): 1089–1095.

Shimomura, T., Itoh, T., Sumiya, T., Mizukami, F. and Ono, M. (2009). Amperometric biosensor based on enzymes immobilized in hybrid mesoporous membranes for the determination of acetylcholine. *Enzyme and Microbial Technology*, 45(6–7): 443–448.

Siddiqui, M.F., Khan, Z.A., Jeon, H. and Park, S. (2020). SPE based soil processing and aptasensor integrated detection system for rapid on-site screening of arsenic contamination in soil. *Ecotoxicology and Environmental Safety*, 196: 110559. https://doi.org/10.1016/j.ecoenv.2020.110559.

Silva, T.S., Pereira, I.S., Lacerda, L.R., Cordeiro, T.A., Ferreira, L.F. and Franco, D.L. (2020). Electrochemical modification of electrodes with polymers derived from of hydroxybenzoic acid isomers: Optimized platforms for an alkaline phosphatase biosensor for pesticide detection. *Materials Chemistry and Physics*, 123221. https://doi.org/10.1016/j.matchemphys.2020.123221.

Sinha, R., Ganesana, M., Andreescu, S. and Stanciu, L. (2010). AChE biosensor based on zinc oxide sol–gel for the detection of pesticides. *Analytica Chimica Acta*, 661(2): 195–199.

Sohail, M. and Adeloju, S.B. (2016). Nitrate biosensors and biological methods for nitrate determination. *Talanta*, 100(153): 83–98.

Somerset, V.S., Klink, M.J., Baker, P.G. and Iwuoha, E.I. (2007). Acetylcholinesterase-polyaniline biosensor investigation of organophosphate pesticides in selected organic solvents. *J. Environmental Science and Health, Part B*, 42(3): 297–304.

Sotiropoulou, S. and Chaniotakis, N.A. (2005). Tuning the sol–gel microenvironment for acetylcholinesterase encapsulation. *Biomaterials*, 26(33): 6771–6779.

Sun, X. and Wang, X. (2010). Acetylcholinesterase biosensor based on Prussian blue-modified electrode for detecting organophosphorous pesticides. *Biosensors and Bioelectronics*, 25(12): 2611–2614.

Tamburino, L., Bravo, G., Clough, Y. and Nicholas, K.A. (2020). From population to production: 50 years of scientific literature on how to feed the world. *Global Food Security*, 24: 100346. https://doi.org/10.1016/j.gfs.2019.100346.

Upadhyay, S., Rao, G.R., Sharma, M.K., Bhattacharya, B.K., Rao, V.K. and Vijayaraghavan, R. (2009). Immobilization of acetylcholineesterase–choline oxidase on a gold–platinum bimetallic nanoparticles modified glassy carbon electrode for the sensitive detection of organophosphate pesticides, carbamates, and nerve agents. *Biosensors and Bioelectronics*, 25(4): 832–838.

Usman, M., Farooq, M., Wakeel, A., Nawaz, A., Cheema, S.A., ur Rehman, H., Ashraf, I. and Sanaullah, M. (2020). Nanotechnology in agriculture: Current status, challenges, and future opportunities. *Science of the Total Environment*, 137778. https://doi.org/10.1016/j.scitotenv.2020.137778.

Vakurov, A., Simpson, C.E., Daly, C.L., Gibson, T.D. and Millner, P.A. (2004). Acetylcholinesterase-based biosensor electrodes for organophosphate pesticide detection: I. Modification of carbon surface for immobilization of acetylcholinesterase. *Biosensors and Bioelectronics*, 20(6): 1118–1125.

Vecchio, Y., De Rosa, M., Adinolfi, F., Bartoli, L. and Masi, M. (2020). Adoption of precision farming tools: A context-related analysis. *Land Use Policy*, 94: 104481. https://doi.org/10.1016/j.landusepol.2020.104481.

Virutkar, P.D., Mahajan, A.P., Meshram, B.H. and Kondawar, S.B. (2019). Conductive polymer nanocomposite enzyme immobilized biosensor for pesticide detection. *Journal of Materials NanoScience*, 6(1): 7–12.

Wang, H., Liu, Y. and Liu, G. (2018). Reusable resistive aptasensor for Pb (II) based on the Pb (II)-induced despiralization of a DNA duplex and formation of a G-quadruplex. *Microchimica Acta*, 185(2): 142.

Wang, H., Yin, Y. and Gang, L. (2019). Single-gap microelectrode functionalized with single-walled carbon nanotubes and pbzyme for the determination of Pb^{2+}. *Electroanalysis*, 31(6): 1174–1181.

Wang, J., Chen, L., Mulchandani, A., Mulchandani, P. and Chen, W. (1999). Remote biosensor for in-situ monitoring of organophosphate nerve agents. *Electroanalysis: I. J. Devoted to Fundamental and Practical Aspects of Electroanalysis*, 11(12): 866–869.

Wu, S., Lan, X., Zhao, W., Li, Y., Zhang, L., Wang, H., Han, M. and Tao, S. (2011). Controlled immobilization of acetylcholinesterase on improved hydrophobic gold nanoparticle/Prussian blue modified surface for ultra-trace organophosphate pesticide detection. *Biosensors and Bioelectronics*, 27(1): 82–87.

Wu, X., Wang, P., Hou, S., Wu, P. and Xue, J. (2019). Fluorescence sensor for facile and visual detection of organophosphorus pesticides using AIE fluorogens-SiO2-MnO2 sandwich nanocomposites. *Talanta*, 198: 8–14.

Wu, Y., Ruan, X., Chen, C.H., Shin, Y.J., Lee, Y., Niu, J., Liu, J., Chen, Y., Yang, K.L., Zhang, X. and Ahn, J.H. (2013). Graphene/liquid crystal-based terahertz phase shifters. *Optics Express*, 21(18): 21395–21402.

Xu, S., Chen, X., Peng, G., Jiang, L. and Huang, H. (2018). An electrochemical biosensor for the detection of Pb^{2+} based on G-quadruplex DNA and gold nanoparticles. *Analytical and Bioanalytical Chemistry*, 410(23): 5879–5887.

Xu, S., Dai, B., Xu, J., Jiang, L. and Huang, H. (2019). An electrochemical sensor for the detection of Cu^{2+} based on gold nanoflowers-modifed electrode and dnazyme functionalized Au@ MIL-101 (Fe). *Electroanalysis*, 31(12): 2330–2338.

Yi, Y., Zhao, T., Xie, B., Zang, Y. and Liu, H. (2020). Dual detection of biochemical oxygen demand and nitrate in water based on bidirectional *Shewanellaloihica* electron transfer. *Bioresource Technology*, 123402. https://doi.org/10.1016/j.biortech.2020.123402.

Zejli, H., de Cisneros, J.L.H.H., Naranjo-Rodriguez, I., Liu, B., Temsamani, K.R. and Marty, J.L. (2008). Alumina sol–gel/sonogel-carbon electrode based on acetylcholinesterase for detection of organophosphorus pesticides. *Talanta*, 77(1): 217–221.

Zhang, K., Zhou, H., Hu, P. and Lu, Q. (2019). The direct electrochemistry and bioelectrocatalysis of nitrate reductase at a gold nanoparticles/aminated graphene sheets modified glassy carbon electrode. *RSC Advances*, 9(64): 37207–37213.

Zhang, Y., Liu, H., Yang, Z., Ji, S., Wang, J., Pang, P., Feng, L., Wang, H., Wu, Z. and Yang, W. (2015). An acetylcholinesterase inhibition biosensor based on a reduced graphene oxide/silver nanocluster/chitosan nanocomposite for detection of organophosphorus pesticides. *Anal. Methods*, 7: 6213–6219.

Chapter 12

Bacterial Small RNA and Nanotechnology

Vatsala Koul and *Mandira Kochar**

1. Introduction

The soil harbours different forms of organisms: fungi, algae, nematodes, protozoa, and essentially bacteria, though the existing isolation and cultivation methods have been able to reveal just 1% of the actual bacterial population (Handelsman, 2004; Glick, 2012). These organisms interact with each other as well as influence other life forms, such as plants. The plant-microbe interaction is an age-old process where both the partners have mutually selected each other over a long evolutionary journey (Nadeem et al., 2015). These interactions may be beneficial, harmful, or neutral (Whipps et al., 2011; Beneduzi et al., 2012). The commonly found bacterial genera in the soil are *Pseudomonas*, *Arthrobacter*, *Agrobacterium*, *Alcaligenes*, *Azotobacter*, *Klebsiella*, *Mycobacterium*, *Flavobacter*, *Cellulomonas*, and *Micrococcus* (Okon and Labandera-Gonzalez, 1994; Nadeem et al., 2015). The distribution of bacteria is usually uneven but the rhizosphere, i.e., the area around the plant roots is always rich due to the abundant presence of nutrients (sugars, amino-acids, organic acids) secreted by the plant roots as exudate (Lynch, 1990). In exchange of this high nutrition resource and protective habitat, the rhizobacteria influence the host plants thereby augmenting their productivity, pathogen resistance, and stress tolerance.

Under natural environments, the plants face numerous stresses which mitigate their growth (Cramer et al., 2011). Extreme variation in temperature and soil nutrient content are among the environmental stresses that most severely affect plant growth and production around the world (Haferkamp, 1988; Cramer et al., 2011; da Silva et al., 2011; Hatfield and Prueger, 2015; Jin et al., 2015). The plant growth and yield influence the productivity of food, fodder, feed, fibre, and medicines (Dutta and

TERI Deakin Nanobiotechnology Centre, Sustainable Agriculture Division, The Energy and Resources Institute, (TERI), Lodhi Road, New Delhi, 110003, India.
* Corresponding author: mandira.malhotra@gmail.com

Khurana, 2015). However, with the increase in the global demand of food production, it's important that the productivity is increased consistently, that too in an environment friendly manner, since the rampant use of chemical fertilizers and pesticides pose severe environmental threats (Shrivastava et al., 2015). Over the years, conventional methods such as cross breeding, crop rotation, trait selection, etc. have been used to improve plant yield but these methods are laborious, time-consuming, and not always successful.

Beneficial rhizosphere bacteria have gained world-wide attention due to their plant growth-promoting activities which also mitigate the adverse effects of environmental stresses thereby providing a promising future for sustainable agriculture (Beneduzi et al., 2012). Bacteria possess various regulatory mechanisms for stress adaptation and to exploit these beneficial bacteria for agricultural use, it is important to gather in-depth information of the underlying physiological and molecular mechanisms and unravel the participating intermediates especially during abiotic stress attacks.

2. Plant Growth Promoting Rhizobacteria (PGPR)

Based on the interaction of the bacteria with the plants, they may be categorized as beneficial or pathogenic (Beneduzi et al., 2012). The beneficial bacteria present in the soil which may or may not be associated with the plants are known as plant growth-promoting bacteria (PGPB) and are commonly present in many environments. The most widely studied group of PGPB is PGPR which colonize the host root surfaces and the closely adhering soil interface (Compant et al., 2005). The symbiotic relationship of the agronomically essential plants with the beneficial bacteria can be traced back for centuries (Bashan and de-Bashan, 2005). The beneficial bacteria such as *Azospirillum*, *Azotobacter*, *Bacillus*, *Pseudomonas*, *Rhizobium*, *Bradyrhizobium*, *Gluconacetobacter* and others are known as PGPR and usually share a similar mechanism to enhance the plant growth (Table 12.1). They act either directly by producing growth metabolites (phytohormones and other plant growth regulators) and facilitating nutrient uptake (nitrogen fixation, phosphate solubilization, and accelerating mineralization) or indirectly by exerting inhibitory effects on plant pathogenic agents (complex antibiotics and siderophores) (Glick, 2012; Kochar and Srivastava, 2012; Koul et al., 2015b).

The efficiency of the rhizosphere bacteria to exert beneficial effects on the host plant depends upon their ability to colonize the roots, exudates secreted and soil conditions (de Souza et al., 2015). Ability of the bacteria to colonize the plant rhizosphere is a key step since the microbe needs to be in proximity with the roots to exert maximum benefits to the plant.

The bacteria face constant competition from other microbes coexisting in the soil and is also influenced by different factors such as temperature, pH, nutrition pool, presence of chemicals, moisture, and toxic metals in the soil. Together, these factors may exert biotic and abiotic stresses on the bacteria (Nadeem et al., 2015).

Table 12.1. Some plant-microbe interactions and essential PGPR traits involved.

Bacteria	Host Plant	PGPR Trait	References
Azospirillum brasilense B510	Rice	N2-fixation, IAA-biosynthesis	Isawa et al., 2010; Bao et al., 2013
A. brasilense INTA Az-39	Wheat	N2-fixation, IAA-biosynthesis	Díaz-Zorita and Fernández-Canigia, 2009
A. brasilense Ab-V5	Wheat, Maize	N2-fixation, IAA-biosynthesis	Hungria et al., 2010
A. brasilense Ab-V6	Wheat, Maize	N2-fixation, IAA-biosynthesis	Hungria et al., 2010
A. brasilense SM	Sorghum	IAA-biosynthesis, NO production	Kochar and Srivastava, 2012; Koul et al., 2015a
A. brasilense Sp245	Wheat, Tomato, *Arabidopsis thaliana*	N2-fixation, IAA-biosynthesis	Pereg Gerk et al., 2000; Molina-Favero et al., 2008; Spaepen et al., 2014
A. brasilense Sp7	Wheat	N2-fixation, IAA-biosynthesis	Pereg Gerk et al., 2000
A. brasilense Cd	Tomato, Pepper, Cotton	IAA-biosynthesis	Bashan et al., 1991
Pseudomonas jessenii PS06	Chickpea	N2-fixation, P-solubilization	Valverde et al., 2006
P. alcaligenes PSA15	Maize	N2-fixation, IAA-biosynthesis, anti-fungal activity	Egamberdiyeva, 2007
P. fluorescens	Peanut, Pea, Rice, Wheat	ACC-deaminase, IAA-biosynthesis, anti-fungal activity, siderophore, phenazine	Dey et al., 2004; Zahir et al., 2008; Sirohi et al., 2015
P. putida Q-7	Pea	ACC-deaminase	Zahir et al., 2008
Bacillus subtilis SU47	Wheat	IAA-biosynthesis, P-solubilization	Upadhyay et al., 2012
B. amyloliquefaciens	Soyabean	ACC-deaminase, IAA-biosynthesis, anti-fungal activity, siderophore, phytases	Sharma et al., 2013
B. cereus UW85	Soyabean	ND[#]	Bullied et al., 2002
B. polymyxa BcP26	Maize	N2-fixation, anti-fungal activity	Egamberdiyeva, 2007
B. megaterium M-3	Chickpea	Biocontrol, P-solubilization	Elkoca et al., 2008
Rhizobium spp.	Chickpea	N2-fixation	Elkoca et al., 2008
R. leguminosarum	Pea, white clover	N2-fixation, IAA	Mathesius et al, 1998
Bradyrhizobium japonicum SEMIA 5079	Soyabean	N2-fixation, IAA-biosynthesis	Hungria et al., 2013

Table 12.1 contd. ...

...Table 12.1 contd.

Bacteria	Host Plant	PGPR Trait	References
B. japonicum SEMIA 5080	Soyabean	N2-fixation, IAA-biosynthesis	Hungria et al., 2013
Mesorhizobium ciceri C-2/2	Chickpea	N2-fixation	Valverde et al., 2006
Bukholderia ambifaria MC17	Maize	Anti-fungal activity, siderophore	Ciccillo et al., 2002
Gluconacetobacter diazotrophicus V127	Sugarcane	N2-fixation, IAA-biosynthesis, siderophore, P-solubilization	Beneduzi et al., 2013
G. diazotrophicus PAL	Sugarcane, *Phaseolus vulgaris* (common bean)	N2-fixation	Nogueira et al., 2001; Trujillo-Lopez et al., 2006
G. diazotrophicus LMG7603	Sugarcane	N2-fixation	Govindarajan et al., 2006
Herbaspirillum seropedicae LMG6513	Sugarcane	N2-fixation	Govindarajan et al., 2006
Herbaspirillum rubrisubalbicans	Sugarcane	N2-fixation	Nogueira et al., 2001
Mycobacterium phlei MbP18	Maize	N2-fixation, IAA-biosynthesis, anti-fungal activity	Egamberdiyeva, 2007

#ND: Not described

3. Regulation of Bacterial Stress Response

To survive in the rhizosphere, the microorganisms need to constantly adapt to the changing environment. These environmental conditions, for example, change in temperature, nutrition, pH, salt concentration, etc. may sometimes be favourable or otherwise pose a challenge and induce stress responses in the bacteria (Dimpka et al., 2009). The bacterial defence mechanism is usually a cascade of events with multilayer regulation. This response mechanism has been studied extensively but not understood completely (Picard et al., 2013). Tuning the expression of the stress responsive genes is the key to bacterial adaptation. Transcription is the vital step in gene expression and is, in turn, controlled by multiple factors and signals, though post-transcriptional regulation also exists (Balleza et al., 2009).

The process of gene transcription requires proteins known as sigma factors which serve as the master regulators (Bose et al., 2008; Seshasayee et al., 2011; Ron, 2013). The sigma-factor (σ-factor) is the dissociable subunit of the RNA-polymerase which is essential for the recognition of the proper promoter sequence upstream of the gene to be transcribed, followed by recruitment of the RNA-polymerase holoenzyme at the promoter to initiate the process of transcription. Two evolutionarily distinct σ-factor families exist: σ70 and σ54 (Merrick, 1993; Shingler, 1996, 2011; Buck et al., 2000; Guo et al., 2000). They play an essential role in modulating the gene expression of sets of genes when the cell encounters stress conditions or undergoes growth transformations (Chung et al., 2006; Seshasayee et al., 2011). Extensive studies have been carried out with model organisms, *Bacillus subtilis* and *Escherichia coli,* to elucidate the regulatory role of these factors. *E. coli* K12 is known to possess six members of the σ70 family (RpoD, RpoH, RpoS, RpoE, and FliA) and one σ54 protein (RpoN) (Seshasayee et al., 2011; Ron, 2013). The onset of stationary phase in *E. coli*, elicits drastic changes in its morphology, physiology, and membrane composition. In 1993, Hengge-Aronis, established the role of RpoS (σ^S) as the central regulator of gene expression in the early stationary phase in *E. coli.* RpoS in-turn is influenced by small molecules such as guanosine 5′-diphosphate 3′-diphosphate (ppGpp) and quorum sensing signalling molecules (homo serine lactones). In gram-negative bacteria, elevated temperature is mediated by heat-shock proteins which are influenced by σ32 and σE (Ron, 2013). The process of nitrogen-fixation is regulated by *nif* genes which are under the control of σ54 (Arsene et al., 1996). Chemotaxis and flagella synthesis is under the control of σ28 (Ding et al., 2009). Similarly, different σ factors control different cellular functions in all bacterial strains.

Transcription Factors (TFs) are another class of regulatory molecules which bind to the specific region on the DNA sequence called transcription factor-binding sites (TFBS) and either activate or repress the transcription of the genes (Latchman, 1997; Seshasayee et al., 2011; Shimizu, 2014; Yaryura et al., 2015). These specific regions may be up-stream of the gene promoters and facilitate the RNA polymerase-promoter interaction. On other occasions, some TFs inhibit transcription by blocking the RNA polymerase sites (Seshasayee et al., 2011). The TFs may be global in action which act under generic conditions such as stress, nutrition deficiency, utilization of alternate energy source, etc. The action of these global TFs is aided by various other local TFs whose action is restricted to a few genes or operons, thereby forming a transcriptional regulatory network (TRN) (Balleza et al., 2008). The bacterial TFs are present as a single unit possessing three domains: the DNA-binding domain, the trans-activation domain, and the signal sensing domain. However, in a two-component system, the signal sensing domain containing protein is localized in the cell periplasm to sense the change in surrounding conditions (Latchman, 1997). In the *E. coli* genome, more than 250 TFs are known which account for about 6% of its protein-coding genes and 10% of these TFs are a part of the two-component signalling cascade (Seshasayee et al., 2011).

Shimizu (2014) reviewed the stress regulation system in *E. coli* in response to nutrient limitation such as carbon, nitrogen, phosphate, and environmental stresses such as oxidative, heat, and acid shocks. Crp is a global TF in *E. coli* which senses carbon availability and is activated by cyclic-ATP (cAMP) under the conditions of glucose starvation. Similarly, NtrBC, CysB, and Fur are sensors for nitrogen, sulphur, and iron availability, respectively. Also, α-ketoglutarate (αKG) plays an important role in the coordinated regulation of carbon-nitrogen limitations (Seshasayee et al., 2011; Shimizu, 2014). One of the best studied local, transcriptional repressor protein in *E. coli* is tryptophan repressor (TrpR), which is functional during amino-acid biosynthesis pathway (Zubay et al., 1972; Grillo et al., 1999; Seshasayee et al., 2011).

The ensuing cellular cascade in response to environmental stresses is not always completely understood due to the lack of information about certain regulatory molecules which play an essential role at the post-transcriptional level but are difficult to be identified. Recently, with the advent of sophisticated molecular biology techniques, genome-wide searches have become possible which have led to the discovery of a plethora of small non-coding regulatory RNA molecules (Gottesman, 2004). These molecules have shown the capability to regulate the expression of multiple genes, transcriptional factors, sigma factors, and chaperones, leading to profound effect of the cell physiology. These small RNAs (sRNAs) have been now established as imperative regulatory elements in bacterial stress response (Gottesman, 2005; Grabowicz and Silhavy, 2017). The cumulative impact of all the members of the complex regulatory network enables the bacterial cells to adapt and survive under dynamic environment.

3.1 Bacterial Small Non-coding RNAs

RNA was initially considered to be important as an information carrying intermediate between DNA and proteins, and as transfer and ribosomal RNA involved in protein synthesis, but not viewed to have any regulatory function (Ghosh and Mallick, 2012). This opinion has changed dramatically in the last couple of decades because of genome-wide studies in different eukaryotic organisms that have led to the identification of RNA interference (RNAi), a phenomenon that revolutionized the concept of "central dogma" and established the role of non-coding RNA as regulatory molecules (Ghosh and Mallick, 2012). In eukaryotes, processes such as gene silencing, cell death, chromosome silencing, and developmental regulation all largely depend on such small regulatory RNAs, essentially microRNAs (miRNA) and short interfering RNAs (siRNA) (Gottesman, 2005; Ghildiyal and Zamore, 2009).

Today, we know that RNAs can carry out many functions in the cells, such as catalysis and gene regulation (Gottesman, 2005). Bacterial regulatory RNAs were identified in 1981, much earlier to their eukaryotic counterparts, with ColE1 and R1, the plasmid encoded antisense RNAs which regulated the plasmid replication and hence its copy number (Brantl, 2012). In 1984, the first chromosomal-encoded regulatory RNA, MicF (Fig. 12.1), was discovered in *E. coli* (Mizuno, 1984). This opened a whole new field; that of regulatory, non-coding sRNAs in bacteria but until

Fig. 12.1. Secondary structure of sRNA MicF (Source: BSRD).

1999, only a handful of these sRNAs had been identified in *E. coli* and their function remained largely unknown (Brantl, 2009).

In the year 2002, sRNAs gathered worldwide attention when a special issue of the journal *Science* was published, titled "Small RNAs – Breakthrough of the year". Since then, in-depth genome and transcriptome-based studies have been carried out to discover and attribute physiological function to these small molecules.

The bacterial sRNAs identified till now are heterogeneous in size and range from 40–550 bp, usually located in the intergenic regions, and demonstrate a characteristic stem-loop secondary structure (Storz et al., 2004, 2005; Altuvia, 2007; Gottesman and Storz, 2011; Khoo et al., 2012), though some exceptions exist (Wagner and Romby, 2015). A single bacterial genome is estimated to encode 200–300 sRNAs, possessing diverse functions (Brantl, 2012) and in *E. coli*, which has been the model organism for sRNA identification; the sRNAs identified so far account for about 0.3% of its genome (Khoo et al., 2012). The sRNAs can be categorized based on different parameters, as described by Storz et al. (2005) and Brantl (2012)

1. Base-pairing mechanism

   The sRNAs may be cis- or trans-encoded and are involved in the inhibition or activation of their target message. The cis-encoded antisense RNAs are perfectly complementary to their mRNA targets whereas the trans-encoded ones are partially complementary to their target. Due to partial complementarity, such sRNAs may have multiple targets.

   cis-encoded sRNAs

   The first identified bacterial sRNA, ColE1, and R1 were cis-encoded. Since then, the number of cis-encoded sRNAs identified in various bacteria have increased drastically. Silvaggi et al. (2005) reported the sRNA, RatA (for RNA antitoxinA) from *B. subtilis*, which regulated the accumulation of toxic peptide TxpA (for toxic peptide A). In *E. coli*, cis-encoded sRNAs include GadY (acid stress responsive sRNA; Opdyke et al., 2004), SymR (regulates expression of endonuclease SymE; Kawano et al., 2007), SibA-E (repress small toxic protein synthesis; Fozo et al., 2008), and various others.

   Certain sRNAs possess the dual feature of being a regulatory molecule as well as encode into protein. One of the first such cis-encoded sRNAs was identified while studying the mechanism of quorum-sensing in *Staphylococcus aureus*, RNAIII, also encodes a small protein (26aa) known to be involved in maintaining biofilm integrity (Novick and Geisinger, 2008).

   trans-encoded sRNAs

   MicF, the first chromosomally encoded sRNA to be discovered in *E. coli*, is trans-encoded and pairs imperfectly with its target *ompF* which encodes the outer membrane porin (Mizuno, 1984). Other than MicF, the trans-encoded sRNAs in *E. coli*, influencing outer membrane proteins are MicC (target: OmpC; Chen et al., 2004) and MicA (target OmpA; Guillier et al., 2006). A σ^E dependent cis-encoded sRNA, MicL was identified, which targets Lpp protein and is involved in outer membrane homeostasis (Guo et al., 2014). Unlike other sRNA, which exist in the intergenic region, MicL was transcribed from within the coding region of gene *cutC*.

The multi-target trans-encoded sRNA, RyhB was discovered to be involved in the downregulation of a set of iron-storage and utilizing genes under the conditions of iron-limitations (Massé and Gottesman, 2002). It pairs with the mRNAs of at least six operons encoding iron-binding proteins, including *bfr*, encoding bacterioferritin, *sdh*, encoding succinate dehydrogenase, and *sodB*, encoding superoxide dismutase, and leads to rapid degradation of their messages. The expression of RyhB is, in turn, regulated by the transcriptional repressor, Fur (Ferric uptake regulator) repressor and iron. A computational study was conducted to identify Fur-regulated sRNAs across diverse proteobacteria (e.g., *Enterobacter* sp., *Klebsiella pneumoniae*, *Photorhabdus luminescens*, *Salmonella typhimurium*, *Neisseria meningitidis*, and *Yersinia pestis*, among others) and found 38 novel sRNAs with high probability score (Sridhar et al., 2013b).

Just as a single sRNA can modulate the expression of multiple targets, studies have identified that a single mRNA could be regulated by more than one sRNA in response to variable environmental stimuli. In *E. coli*, the expression of the essential transcriptional regulator, RpoS is known to be influenced by at least three sRNAs: RprA (Majdalani et al., 2001), DsrA (Sledjeski et al., 1996), and OxyS (Zhang et al., 1998). Under stress conditions, the bacteria integrate multiple networks to develop proper response and ultimately survive. RprA and DsrA upregulate the translation of RpoS under the conditions of osmotic stress and low temperature, respectively. On the contrary, OxyS inhibits the RpoS translation under oxidative stress conditions (Wagner and Vogel, 2003; Gottesman, 2005).

2. Mode of action

Though the exact mechanism employed by most of the sRNAs is unknown, reports suggest that they modulate the bacterial gene expression by controlling mRNA stabilization and regulating translation (Fig. 12.2).

Translational inhibition

One of the most well-known mechanisms of sRNA action is binding to the ribosome binding site (RBS) of the target mRNA thus blocking the 30S subunit binding and in-turn inhibiting its translation (Fig. 12.2A) (Altuvia et al., 1998; Udekwu et al., 2005; Maki et al., 2008; Brantl, 2009). Some sRNAs inhibit translation by binding to the Shine-Dalgarno (SD) or AUG start codon of the mRNA coding region (Bouvier et al., 2008). Oxidative stress responsive sRNA, OxyS is known to modulate the expression of about 40 genes in *E. coli*, including *fhlA*-encoded transcriptional regulator. Altuvia et al. (1998) carried out mutation-based studies to decipher the OxyS regulation mechanism and found that it base-pairs with the sequence overlapping the SD-sequence of the *fhlA*, thus blocking its RBS, and inhibiting translation.

Translational activation

Apart from inhibiting translation, the sRNAs are equally capable of regulating the gene expression by positively influencing the process of translation (Fig. 12.2B). As discussed above, sigma factor σ^S encoding gene, *rpoS* is influenced by multiple sRNAs (Majdalani et al., 2001; Sledjeski et al., 1996; Zhang et al., 1998). Majdalani et al. (1998) carried out mutational studies to establish that sRNA DsrA modulates the expression of RpoS by activating its translation. The *rpoS* folds into a stable secondary structure such that its 5′ end

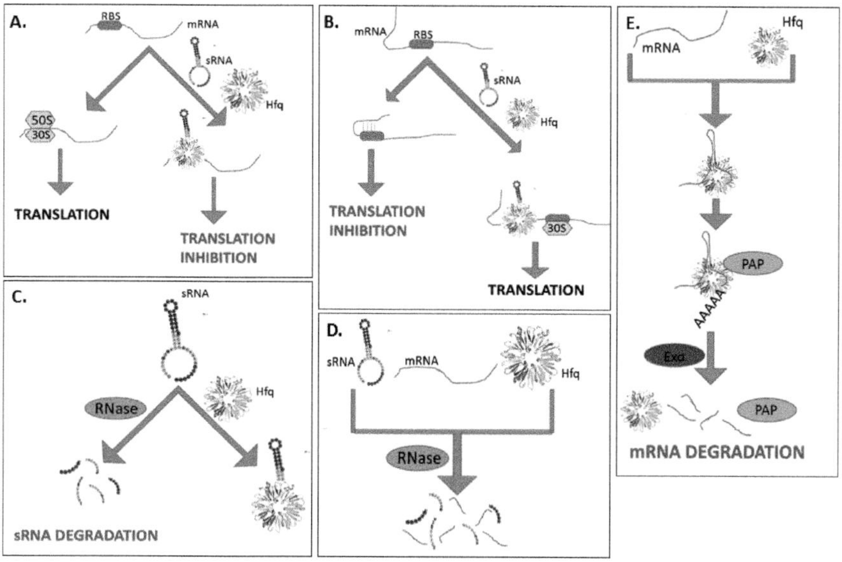

Fig. 12.2. Widely accepted modes of sRNA action: Translational inhibition, Translational activation, mRNA stabilization and mRNA degradation.

sequesters its SD sequence, rendering it unavailable to ribosomal subunit binding. DsrA binds to the sequence just above the rpoS translational initiation site, thus activating its translation.

mRNA stabilization

In the cellular environment, the mRNAs are prone to ribonuclease cleavage. The sRNA and associated proteins can bind and eventually protect and stabilize the mRNA (Fig. 12.2C). The *E. coli* sRNA GadY is known to regulate the expression of acid responsive transcriptional regulators, GadX and GadW. This sRNA exists in three forms, a long form (105 nt) and two processed forms (90 and 59 nt) and is induced during the stationary phase. The sRNA was found to overlap the 3'end (UTR) of the *gadX,* conferring it with increased stability leading to its accumulation and subsequent upregulation of acid-resistance genes (Opdyke et al., 2004).

mRNA degradation

Some sRNAs regulate gene expression by facilitating the degradation of the target mRNAs in response to suitable environmental signal (Figs. 12.2D, 12.2E). As described earlier, sRNA RyhB represses the expression of iron-utilizing genes in *E. coli* under the conditions of iron-limitations (Massé and Gottesman, 2002). Its mechanism was elucidated and described as coupled degradation of the sRNA-mRNA complex, dependent on ribonucleases (Massé et al., 2003). The stress signal induces the expression of RyhB, which binds to the RNA chaperone protein for protection against degradation till it binds to its target. The sRNA-mRNA complex is degraded by ribonuclease, RNaseE. An mRNA degradation mechanism was proposed for sRNA RatA which binds perfectly to its target mRNA, *txpA,* encoding the toxic peptide TxpA in *B. subtillis* (Silvaggi et al.,

2005). The presence of a *B. subtilis* RNaseE equivalent was also proposed which is possibly involved in the degradation of sRNA-mRNA duplex (Lehnik-Habrink et al., 2011).

3. Use of chaperones

   To accomplish their regulatory function successfully, certain sRNAs may require chaperone proteins (e.g., Host factor Hfq) which provide them stability, enhance rate of duplex formation, and participate in post-transcriptional regulation (Liu and Camilli, 2010). These sRNAs are known as Hfq-dependent sRNAs. However, other sRNAs may perform their functions independent of any chaperone proteins, despite their presence in the organism (Silvaggi et al., 2005). The presence of Hfq-independent sRNAs is mostly observed in gram-positive bacteria such as *Staphylococcus aureus*, and *B. subtilis* (Gottesman and Storz, 2011).

3.2 RNA-binding Chaperone: Hfq

With the increasing discovery of small RNAs, attention was drawn towards the RNA-binding chaperone, Hfq. These homo-hexameric, ring-shaped proteins were discovered in the 1960s in *E. coli* as the host factor required for the replication of Qβ-phage (Møller et al., 2002; Brennan and Link, 2007; Vogel and Luisi, 2011; Sauer, 2013). Soon they were discovered in other bacterial strains (both gram-negative and gram-positive) and were associated with plethora of different functions including stress tolerance, virulence, flagellar synthesis, exopolysaccharide synthesis, and motility (Fantappie et al., 2009; Chao and Vogel, 2010; Wang et al., 2014). The Hfq protein is highly abundant, and their length varies from 70–110 amino acids across different bacterial genomes (Brennan and Link, 2007). Later it was discovered that Hfq interacts with sRNAs and is mainly essential for their cellular stability and bringing the sRNA and the target mRNA in proximity.

Hfq belongs to the Sm(L) protein superfamily which is known to be involved in RNA metabolism. It has been established that the structural core of Hfq protein possesses α-β_{1-5} folds of which β_{1-3} forms the conserved Sm1 motif and β_{4-5} forms the variable Sm2 motif (Vogel and Luisi, 2011). They also described the two faces of the Hfq protein which interacts with the RNA molecules (sRNA or mRNA): "proximal face", the surface on which the aminoterminal αhelix is exposed and the opposite side as "distal face". Though, the proximal face has preferential binding towards the U-rich RNA strand and the distal face towards the RNA containing sequence motif ARN or ARNN, the two faces cannot be exclusively labelled as mRNA or sRNA binding faces (Vogel and Luisi, 2011; Sauer et al., 2012; De Lay et al., 2013).

Apart from the distal and proximal binding sites, Sauer et al. (2012) identified a third site-lateral surface as the independent RNA-binding site on the Hfq protein. They used the sRNA RybB to establish that the hexameric Hfq structure possesses RNA binding arginine patch on each monomer, thus a total of six patches, which act independently of the previously known two sites (proximal and distal; Fig. 12.3). This model could explain many of the Hfq/sRNA interactions which could not be explained earlier. Another study carried out by Murina et al. (2013) used x-ray crystallography to identify the Hfq-sRNA complex interactions in *Pseudomonas aeruginosa* and the presence of a third, lateral RNA-binding site was established.

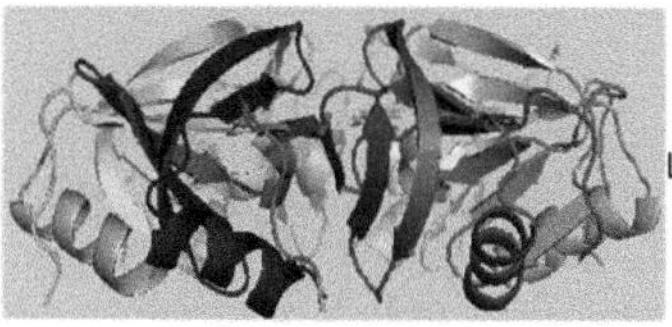

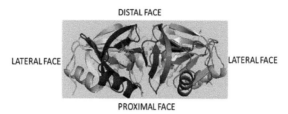

Fig. 12.3. The RNA-binding sites on the Hfq protein.

One of the initial studies that established the regulatory role of Hfq used mutational studies in *E. coli*, where the Hfq null mutant exhibited change in growth rate, cell length, and sensitivity to UV-light. Similar phenotypic characters were observed by the mutation in the gene encoding global stationary phase sigma factor, σ^S (Tsui et al., 1994).

It was later established in *E. coli* that Hfq improves the interaction of sRNA OxyS with its target mRNAs, fhlA, and rpoS (Zhang et al., 2002). In another study, Zhang et al. (2003) used a combinatorial study of co-immunoprecipitation and microarray to establish that at least 15 out of the 46 known sRNAs in *E. coli* base-pair with Hfq. The role of Hfq in sRNA-based regulation was demonstrated where it was shown to facilitate the action of *E. coli* sRNAs such as MicF, OxyR, DsrA, Spot42, and RyhB, which were involved in maintaining either the translational efficiency or mRNA stability of their respective targets (Storz et al., 2004). Torres-Quesada et al. (2010) established that Hfq influences central carbon metabolism and symbiotic interaction of alfalfa with *Sinorhizobium meliloti*. Also, seven Hfq-associated sRNAs were discovered in this study. Another study by Torres-Quesada et al. (2014) established that Hfq directly targets 18% of the predicted *S. meliloti* mRNAs, which encode functionally diverse proteins involved in transport and metabolism, σ^{E2}-dependent stress responses, quorum sensing, flagella biosynthesis, ribosome, and membrane assembly and symbiotic nitrogen fixation.

The number of sRNAs in a bacterial cell is high whereas the pool of required Hfq protein is saturated, hence the model for active RNA cycling on the Hfq protein was proposed (Fender et al., 2010; Wagner, 2013). This model suggested that the cycling of the RNA on the Hfq depends on the RNA concentration, instead of the half-life of the Hfq-RNA complex (passive-cycling). The competitor RNA binds to the Hfq-RNA complex, the RNAs exchange binding sites and the resident RNA molecule eventually dissociates (Wagner and Romby, 2015). This cycling of RNAs allows the multi-target regulation by the small RNAs which is an essential feature of sRNA particularly under stress conditions.

The most accepted mode of action of the Hfq chaperone has been described by Vogel and Luisi (2011) and includes:

1. Inhibition of translation

   The Hfq-sRNA complex blocks the mRNA RBS, rendering it unavailable to binding of the ribosomal subunits, thus repressing translation.

2. Unblocking translation

   The secondary structure of the mRNA may fold in a manner that the upstream sequence (UTR) of the RBS blocks it, thereby inhibiting translation. The Hfq-sRNA complex binds to the UTR, freeing the RBS and initiating the process of translation.

3. sRNA stabilization

   The sRNA molecules are susceptible to degradation by the ribonucleases (RNaseE) in the cellular environment. The formation of the Hfq-sRNA complex provides protection the sRNAs from such degradations, thereby improving their intracellular half-life.

4. Inducing cleavage

   The Hfq chaperone induces cleavage of some sRNAs and their mRNA targets by improving their proximity and employing the RNaseEs for degradation.

5. mRNA degradation

   The Hfq chaperone may stimulate the polyadenylation of the mRNA using poly (A) polymerases, which in turn triggers mRNA degradation (3–5′) by exonucleases (RNaseR, RNase II).

Scientists have reviewed various RNA binding proteins (RBP) which have been involved in bacterial post-transcriptional regulation. Some of these proteins assist sRNA-mRNA interaction, regulate the sRNA stability or the expression of mRNA targets, while others directly influence the protein synthesis. CsrA (carbon storage regulator) is the RNA-binding protein which is known to repress various metabolic pathways in *E. coli* (Gudapaty et al., 2001). The protein synthesis is repressed by the binding of CsrA to the 5′ region of the mRNA, and thus changing the RBS accessibility (Holmqvist and Vogel, 2013). In *B. subtilis*, three proteins, FbpA, B, and C were shown to be required for the control of iron metabolism by sRNA FsrA (Smaldone et al., 2012). However, the mechanism of a certain RBP is still unknown. Apart from Hfq, phosphorolytic RNases are important for certain sRNA-mRNA interactions and stabilization (Cameron and De Lay, 2016). Their work established that PNPase and RNasePH stabilize the sRNAs, RyhB, CyaR, and MicA, against premature degradation during exponential growth and additionally PNPase facilitates the interaction of RyhB with its targets. Recently, another RNA-binding protein, ProQ, which displayed domain similarities with Hfq was identified in *E. coli* (Chaulk et al., 2011; Gonzalez et al., 2017) and is now termed as the global RNA chaperone in Salmonella (Smirnov et al., 2016).

3.2.1 Identification of sRNAs

As the indispensable role of these small molecules is being identified in multiple regulatory pathways, various strategies are being applied for the global mining of sRNAs across the bacterial strains. The strategies can be either be dependent or independent of the sRNA expression (Podkaminski et al., 2014).

Next-generation transcriptome sequencing (RNA-Seq)

Next-generation transcriptome sequencing or RNA-Seq, is a common method to analyse gene expression, uncover novel RNA species, conduct analysis of unknown genes and novel transcript isoforms, thus carrying out the complete transcriptome profiling of the biological sample (Hrdlickova et al., 2017). It is one of the best methods to identify the presence of non-coding sRNAs in their genome. For better output and sRNA enrichment, the processed RNAs (rRNA and tRNA), which represent the majority in the total RNA, can be degraded and removed (Liu et al., 2009; Yoder-Himes et al., 2009) before proceeding to RNA-Seq. Fan et al. (2015) used the RNA-Seq approach to identify a large number of regulatory RNAs in gram-positive bacteria, *Bacillus amyloliquefaciens*, which included 53 cis-encoded riboswitches, 136 anti-sense RNAs, and 86 potential sRNA candidates, most of which were validated by northern-blotting. Seven potential novel, intergenic sRNAs have been identified using RNA-Seq in *P. aeruginosa* (Heera et al., 2015). In another study, RNA-Seq was used to uncover the antisense RNAs and new putative sRNAs in human pathogen, *Streptococcus pyogenes* (Le Rhun et al., 2015). Liu et al. (2009) improved the specificity of sequencing strategy by depleting the starting RNA sample of tRNA and rRNAs, and in addition to the 20 known sRNAs in *V. cholerae*, 500 new putative intergenic and 127 antisense sRNAs were revealed. An improved RNA sequencing strategy known as differential RNA sequencing (dRNA-seq) has been developed, where appropriate exonucleases [5′ monophosphate-dependent terminator exonuclease (TEX)] are used to degrade processed RNAs (Podkaminski et al., 2014). This approach has been used to map the primary transcriptome and identify sRNAs in diverse species. More than 60 sRNAs in gram-negative human pathogen, *Helicobacter pylori* were discovered using this strategy (Sharma et al., 2010). In the model cyanobacterium, *Synechocystis* sp. PCC6803, this approach was used to establish genome-wide map of 3,527 transcriptional start sites (TSS). Orphan TSS located in the intergenic regions subsequently led to the prediction 314 non-coding RNAs in this organism (Mitschke et al., 2011).

Microarray

Another strategy for transcriptome analysis is microarray, which has been successfully applied to discover potential bacterial sRNAs by using DNA probes for the sense and antisense strands of the genome, including the intergenic regions. One of the first microarray-based search was carried out in *E. coli* and uncovered more than 60 candidate intergenic regions which could encode potential sRNAs (Wassarman et al., 2001). From these regions, 17 potential sRNAs were identified, half of which were also found to interact with the RNA binding protein, Hfq. In another study, 27 novel sRNAs were identified in *Caulobacter crescentus* using microarray (Landt et al., 2008). A change in the expression of most of these sRNAs was observed in response to environmental challenges such as carbon starvation and entering the stationary phase.

Co-immunoprecipitation

The dependency of most of the sRNAs on RNA binding proteins has already been established across various bacterial strains (Wassarman et al., 2001; Zhang et al., 2002, 2003, 2013; Torres-Quesada et al., 2010). The most common protein to be used as a bait for this strategy is Hfq since it is present in many bacterial species. The combined use of co-immunoprecipitation (using FLAG epitope-tagged Hfq) and high-throughput pyrosequencing (HTPS) technology led to the identification of new Hfq-associated sRNAs in *Salmonella* (Sittka et al., 2008). In addition to Hfq, the use of such sRNA-binding proteins such as CsrA (Liu et al., 1997), Crc (Sonnleitner et al., 2009), and ProQ (Smirnov et al., 2016) to discover new sRNAs is now a well-exploited strategy.

Computational approach

One of the best strategies to identify the potential sRNAs in any sequenced bacterial genome is by conducting an *in silico* search (Podkaminski et al., 2014). Various sRNA features such as their location, size, secondary structure, termination sequences, transcriptional signals and others can form the basis of these searches. Argaman et al. (2001) exploited the genomic features of the known sRNA-encoding genes to search for additional sRNAs in *E. coli* and obtained 24 putative sRNA genes, two of which were the experimentally characterized sRNAs, RprA and GcvB. Computational approach was used to search for regions containing promoters close to the rho-independent terminators in *E. coli* (Chen et al., 2002). This study revealed more than 200 candidate sRNA genes. Different software and algorithms are now available for sRNA predictions (Rivas and Eddy, 2001; Livny et al., 2005; Sridhar et al., 2010). Khoo et al. (2012) developed a pipeline of successful sRNA gene-finding tools such as QRNA (Rivas and Eddy, 2001), RNAz (Gruber et al., 2010), sRNApredict (Livny et al., 2006), and sRNAscanner (Sridhar et al., 2010) to predict more than 1300 potential sRNA candidates in *B. pseudomallei*. Fifteen of these were found to be conserved in *Burkholderia* spp. or other *B. pseudomallei* strains and eight of these were subsequently validated by RT-PCR.

The best strategy for the most accurate bacterial sRNA identification is to incorporate both computational as well as experimental approaches. Rossi et al. (2016) carried out *in-silico* analysis to predict potential sRNAs in *Actinobacillus pleuropneumoniae*. Four computational algorithms (RNAz, INFERNAL, SIPHT, and BLASTn against Rfam) were used and only the candidates predicted by at least two of the algorithms were selected for further analysis. Experimental validation was subsequently carried out by RT-PCR and expression analysis using Northern blotting. Additionally, RNA-Seq was carried out to identify sRNAs in this organism (Rossi et al., 2016).

3.3 Identification of sRNA Targets

To identify the sRNA targets, first a computational search can be carried out to identify in which regulatory network the sRNA operates in. Over the years, different software, and algorithms such as TargetRNA2 (Kery et al., 2014), CopraRNA, IntaRNA (Wright et al., 2014), RNAup (Mückstein et al., 2006), and sTarPicker (Ying et al., 2011) have been developed to facilitate the process of putative target prediction in bacteria, based on features such as RNA-RNA binding thermodynamics, accessibility of sRNA and targets, conservation of sRNA, etc. (Wright et al., 2013; Kery et al., 2014).

Analysis of the phenotypic traits and protein profiles of sRNA knockout and over-expressing mutants is another way of identifying their targets. The genome-wide profiling can be carried out by microarray and the effect of sRNA overexpression or knockout on specific target mRNAs can be carried out using real-time PCR (qRT-PCR) and Northern blotting (Opdyke et al., 2004; Papenfort et al., 2006). Pyla et al. (2010) identified the target mRNAs of iron-responsive sRNA, ArrF in *A. vinelandii* which is negatively regulated by Fur protein. The targets were identified by the deletion of the sRNA followed by protein profiling and qRT-PCR study.

3.4 Role of sRNA in Stress Response in PGPRs

Different studies have been carried out to discover small RNAs in plant-associated bacteria (PAB) like *S. meliloti*, *Bacillus subtilis*, *Bradyrhizobium japonicum*, *Azotobacter vinelandii*, and many others (Valverde et al., 2008; Torres-Quesada et al., 2010, 2014; Madhugiri et al., 2012; Fan et al., 2015). One of the most important functions of relevance to the performance of PGPRs is that the expression of sRNAs helps cells cope up with environmental stress by redirecting cellular metabolism. Across various bacterial strains different sRNAs have been discovered which play essential roles under abiotic stress conditions like iron limitation (RyhB), oxidative and salt stress (OxyS), temperature shock (DsrA), and accumulation of glucose-phosphate (SgrS) (Gottesman, 2005; Gottesman and Storz, 2011). The oxidative stress in *E. coli*, causes the activation of transcriptional regulator, OxyR, leading to the synthesis of sRNA and OxyS, which subsequently regulates the synthesis of transcriptional regulators, RpoS and FhlA. This results in bacterial protection from oxidative stress caused due to hydrogen peroxide (Altuvia et al., 1997, 1998). Also, during low temperature stress, the promoter of the sRNA, i.e., DsrA is activated, causing the upregulation of RpoS translation and corresponding increase in RpoS-dependent genes which provide the cells the ability to endure low temperature stress (Sledjeski et al., 1996; Kandror et al., 2002).

There is a major loss to agronomically important plants and crops due to the abiotic and biotic stress conditions. PGPR strains with improved competence to endure environmental stress can be developed as biofertilizers and used successfully in soil to augment plant growth and development. PGPRs are an alternative to improve nitrogen use efficiency (NUE) in soil, since these PABs can increase the root-system development and improve acquisition of nutrients from the soil.

3.5 Secretion of Small RNAs as Extracellular Nanovesicles

Outer-membrane (OMV) vesicles are nano-sized (~ 20–300 nm) cargoes reported to be secreted by numerous bacterial species (Choi et al., 2017; Ahmadi Badi et al., 2020). These spherical molecules are derived by the vesiculation of the outer membrane of the bacteria during normal growth process (Jan et al., 2017; Cecil, 2019). The OMVs have been known to possess lipopolysaccharides (LPs), phospholipids, peptidoglycan, proteins (OMPs, periplasmic, cytoplasmic, and membrane-bound), and nucleic acids (DNA, RNA) (Lindmark et al., 2009; Koeppen et al., 2016; Vanaja et al., 2016) which allow these molecules to participate in various cellular pathways such as bacterial communication, quorum-sensing activity, transfer of virulence factors, among others (Jan et al., 2017). It is striking that studies have revealed that in addition to DNA and RNA, these vesicles also encompass bacterial small RNAs (Kuehn and Kesty, 2005; Mashburn-Warren et al., 2008; Lee et al., 2019) and the secretion of the OMVs depends upon the bacterial growth conditions, phases, and exposure of stress conditions.

Researchers have observed that both prokaryotes and eukaryotes are capable of RNA uptake from their surroundings, which can potentially cause silencing of the target genes in the host organism, a phenomenon described as Environmental RNA-interference (RNAi; Whanghbo and Hunter, 2008; Niu et al., 2021).

4. Future Perspective

Over the years, with global climate change, the environmental conditions are constantly changing, causing a drastic impact on agricultural productivity, thereby causing a constant threat to global food security. In addition to traditional methods of crop protection and enhanced productivity, innovative, scientifically viable, and environment friendly methods are required to fulfil the need for sufficient and nutritious food across the world.

The non-coding regulatory RNAs have been established as novel molecule existing across almost all the kingdoms of life. The once serendipitously discovered small RNAs are now known to be essential participant of most cellular pathways and enhance the ability of the bacteria to endure biotic as well as abiotic stresses, as well as capable of interspecies as well as interkingdom communications. It is known that the sRNAs can be released as nanovesicles and influence organisms in its vicinity. Such studies are in the infant stages but the field of host-microbe (plant-bacteria, plant-bacteria-mycorrhiza, mycorrhiza-bacteria among others) interactions and subsequent regulation of gene expression possess enormous scope and require detailed research. The detailed structure, mode of synthesis, required signal for synthesis, and uptake of the nanovesicles by the host organism and subsequent mode of action is largely unknown and requires investigation. Additionally, the controlled secretion of biologically synthesized nanoparticles (i.e., sRNA containing OMVs) by PGPRs and stabilized delivery of essential sRNAs to the host organisms, are additional areas where details studies are required and can lead to enhanced protection and stress endurance in the agriculturally relevant crops and can lead to the development of a new category of biofertilizers, bioinsecticides, and biopesticides.

References

Ahmadi Badi, S., Bruno, S.P., Moshiri, A., Tarashi, S., Siadat, S.D. and Masotti, A. (2020). Small RNAs in outer membrane vesicles and their function in host-microbe interactions. *Front Microbiol.*, 11: 1209.

Altuvia, S., Zhang, A., Argaman, L., Tiwari, A. and Storz, G. (1998). The *Escherichia coli* OxyS regulatory RNA represses *fhlA* translation by blocking ribosome binding. *EMBO J.*, 17: 6069–6075.

Altuvia, S. (2007). Identification of bacterial small non-coding RNAs: Experimental approaches. *Curr. Opin. Microbiol.*, 10: 257–261.

Argamann, L., Hershberg, R., Vogel, J., Bejerano, G., Wagner, E.G.H., Margalit, H. and Altuvia, S. (2001). Novel small RNA-encoding genes in the intergenic regions of *Escherichia coli*. *Curr. Biol.*, 11: 941–950.

Arsene, F., Kaminski, P.A. and Elmerich, C. (1996). Modulation of NifA activity by PII in *Azospirillum brasilense*: Evidence for a regulatory role of the NifA N-terminal domain. *J. Bacteriol.*, 178(16): 4830–4838.

Balleza, E., López-Bojorquez, L.N., Martínez-Antonio, A., Resendis-Antonio, O., Lozada-Chávez, I., Balderas-Martínez, Y.I., Encarnación, S. and Collado-Vides, J. (2009). Regulation by transcription factors in bacteria: Beyond description. *FEMS Microbiol. Rev.*, 33(1): 133–51.

Bao, Z., Sasaki, K., Okubo, T., Ikeda, S., Anda, M., Hanzawa, E., Kaori, K., Tadashi, S., Hisayuki, M. and Minamisaw,a K. (2013). Impact of *Azospirillum* sp. B510 inoculation on rice-associated bacterial communities in a paddy field. *Microbes Environ.*, 28: 487–490.

Bashan, Y., Levanony, H. and Whitmoyer, R.E. (1991). Root surface colonization of non-cereal crop plants by pleomorphic *Azospirillum brasilense*. *Can. J. Gen. Microbiol.*, 137: 187–196.

Bashan, Y. and de-Bashan, L.E. (2005) Bacteria/Plant growth-promotion. pp. 103–115. *In*: Hillei, D. (ed.). *Encyclopedia of Soils in the Environment*. Vol. 1. Elsevier, Oxford, U.K.

Beneduzi, A., Ambrosini, A. and Passaglia, L.M.P. (2012). Plant growth-promoting rhizobacteria (PGPR): Their potential as antagonists and biocontrol agents. *Genet. Mol. Biol.*, 35: 1044–1051.

Bose, D., Pape, T., Burrows, P.C., Rappas, M., Wigneshweraraj, S.R., Buck, M. and Zhang, X. (2008). Organization of an activator-bound RNA polymerase holoenzyme. *Mol. Cell*, 32(3): 337–346.

Brantl, S. (2009). Bacterial chromosome-encoded small regulatory RNAs. *Future Microbiol.*, 4: 85–103.

Brantl, S. and Mallick, B. (2012). Small regulatory RNAs (sRNAs): Key players in prokaryotic metabolism, stress response, and virulence. pp. 73–109. *In*: *Regulatory RNAs*. Berlin/Heidelberg: Springer.

Brennan, R.G. and Link, T.M. (2007). Hfq structure, function, and ligand binding. *Curr. Opin. Microbiol.*, 10(2): 125–133.

Bouvier, M., Sharma, C.M., Mika, F., Nierhaus, K.H. and Vogel, J. (2008). Small RNA binding to 5' mRNA coding region inhibits translational initiation. *Mol. Cell*, 32: 827–837.

Buck, M., Gallegos, M.T., Studholme, D.J., Guo, Y. and Gralla, J.D. (2000). The bacterial enhancer-dependent sigma (54) (sigma(N)) transcription factor. *J. Bacteriol.*, 182(15): 4129–4136.

Bullied, J.W., Buss, W.T.J. and Vessey, K.J. (2002). *Bacillus cereus* UW85 inoculation effects on growth, nodulation, and N accumulation in grain legumes: Field studies. *Can. J. Plant. Sci.*, 82: 291–298.

Cameron, T.A. and De Lay, N.R. (2016). The phosphorolytic exoribonucleases polynucleotide phosphorylase and RNase PH stabilize sRNAs and facilitate regulation of their mRNA targets. *J. Bacteriol.*, 198: 3309–3317.

Cecil, J.D., Sirisaengtaksin, N., O'Brien-Simpson, N.M. and Krachler, A.M. (2019). Outer membrane vesicle-host cell interactions. *Microbiol. Spectr.*, 7(1): 10.1128.

Chao, Y. and Vogel, J. (2010). The role of Hfq in bacterial pathogens. *Curr. Opin. Microbiol.*, 13(1): 24–33.

Chaulk, S.G., Smith Frieday, M.N., Arthur, D.C., Culham, D.E., Edwards, R.A., Soo, P., Frost, L.S., Keates, R.A., Glover, J.N. and Wood, J.M. (2011). ProQ is an RNA chaperone that controls ProP levels in *Escherichia coli*. *Biochemistry*, 50: 3095–3106.

Chen, S., Lesnik, E.A., Hall, T.A., Sampath, R., Griffey, R.H., Ecker, D.J. and Blyn, L.B. (2002). A bioinformatics-based approach to discover small RNA genes in the *Escherichia coli* genome. *Biosystems*, 65(2–3): 157–177.

Chen, S., Zhang, A., Blyn, L.B. and Storz, G. (2004). MicC, a second small-RNA regulator of Omp protein expression in *Escherichia coli*. *J. Bacteriol.*, 186(20): 6689–6697.

Choi, J.W., Um, J.H., Cho, J.H. and Lee, H.J. (2017). Tiny RNAs and their voyage via extracellular vesicles: Secretion of bacterial small RNA and eukaryotic microRNA. *Exp. Biol. Med. (Maywood)*, 242(15): 1475–1481.

Chung, H.J., Bang, W. and Drake, M.A. (2006). Stress Response of *Escherichia coli*. *Compr. Rev. Food Sci. Food Saf.*, 5: 52–64.

Ciccillo, F., Fiore, A., Bevivino, A., Dalmastri, C., Tabacchioni, S. and Chiarini, L. (2002). Effects of two different application methods of *Burkholderia ambifaria* MCI7 on plant growth and rhizospheric bacterial diversity. *Environ. Microbiol.*, 4: 238–245.

Compant, S., Duffy, B., Nowak, J., Clément, C. and Barka, E.A. (2005). Use of plant growth-promoting bacteria for biocontrol of plant diseases: Principles, mechanisms of action, and future prospects. *Appl. Environ. Microbiol.*, 71(9): 4951–4959.

Cramer, G.R., Urano, K., Delrot, S., Pezzotti, M. and Shinozaki, K. (2011). Effects of abiotic stress on plants: A systems biology perspective. *BMC Plant Biol.*, 11: 163.

da Silva, E.C., Nogueira, J.M.C., da Silva, M.A. and de Albuquerque, M.B. (2011). Drought stress and plant nutrition. *Plant Stress*, 1: 32–41.

De Lay, N., Schu, D.J. and Gottesman, S. (2013). Bacterial small RNA-based negative regulation: Hfq and its accomplices. *J. Biol. Chem.*, 288(12): 7996–8003.

de Souza, R., Ambrosini, A. and Passaglia, L.M.P. (2015). Plant growth-promoting bacteria as inoculants in agricultural soils. *Genet. Mol. Bio.*, 38(4): 401–419.

Dey, R., Pal, K.K, Bhatt, D.M. and Chauhan, S.M. (2004). Growth promotion and yield enhancement of peanut (*Arachis hypogaea* L.) by application of plant growth-promoting rhizobacteria. *Microbiol. Res.*, 159: 371–394.

Díaz-Zorita, M. and Fernández-Canigia, M.V. (2009). Field performance of a liquid formulation of *Azospirillum brasilense* on dryland wheat productivity. *Eur. J. Soil Biol.*, 45: 3–11.

Dimkpa, C., Weinand, T. and Asch, F. (2009). Plant–rhizobacteria interactions alleviate abiotic stress conditions. *Plant Cell Environ.*, 32: 1682–1694.

Ding, L., Wang, Y., Hu, Y., Atkinson, S., Williams, P. and Chen, S. (2009). Functional characterization of FlgM in the regulation of flagellar synthesis and motility in *Yersinia pseudotuberculosis. Microbiol.*, 155(Pt 6): 1890–1900.

Dutta, S. and Khurana, S.M.P. (2015). Plant growth promoting rhizobacteria (PGPR) for alleviating abiotic stresses in medicinal plants. pp. 167–200. *In*: Egamberdieva, D., Shrivastava, S. and Varma, A. (eds.). *Plant Growth Promoting rrhizobacteria (PGPR) and Medicinal Plants. Soil Biology*. Vol. 42. Heidelberg: Springer.

Egamberdiyeva, D. (2007). The effect of plant growth promoting bacteria on growth and nutrient uptake of maize in two different soils. *Appl. Soil Ecol.*, 36: 184–189.

Elkoca, E., Kantar, F. and Sahin, F. (2008). Influence of nitrogen fixing and phosphorus solubilizing bacteria on the nodulation, plant growth, and yield of chickpea. *J. Plant Nut.*, 31: 157–171.

Fan, B., Li, L., Chao, Y., Förstner, K., Vogel, J., Borriss, R. and Wu, X.Q. (2015). dRNA-Seq reveals genome-wide TSSs and noncoding RNAs of plant beneficial rhizobacterium *Bacillus amyloliquefaciens* FZB42. *PLoS One*, 10(11): e0142002.

Fantappie, L., Metruccio, M.M., Seib, K.L., Oriente, F., Cartocci, E., Ferlicca, F., Giuliani, M.M., Scarlato, V. and Delany, I. (2009). The RNA chaperone Hfq is involved in the stress response and virulence in *Neisseria meningitidis* and is a pleiotropic regulator of protein expression. *Infect. Immun.*, 77: 1842–1853.

Fender, A., Elf, J., Hampel, K., Zimmermann, B. and Wagner, E.G.H. (2010). RNAs actively cycle on the Smlike protein Hfq. *Genes Dev.*, 24: 2621–2626.

Fozo, E.M., Kawano, M., Fontaine, F., Kaya, Y., Mendieta, K.S., Jones, K.L., Ocampo, A., Rudd, K.E. and Storz, G. (2008). Repression of small toxic protein synthesis by the Sib and OhsC small RNAs. *Mol. Microbiol.*, 70(5): 1076–1093.

Ghildiyal, M. and Zamore, P.D. (2009). Small silencing RNAs: An expanding universe. *Nat. Rev. Genet.*, 10(2): 94–108.

Ghosh, Z. and Mallick, B. (2012). Renaissance of the regulatory RNAs. pp. 3–16. *In*: Mallick, B. (ed.). *Regulatory RNAs*. Berlin: Springer.

Glick, B.R. (2012). Plant growth-promoting bacteria: mechanisms and applications. *Scientifica (Cairo)*, 2012: 963401.

Gonzalez, G.M., Hardwick, S.W., Maslen, S.L. et al. (2017) Structure of the *Escherichia coli* ProQ RNA-binding protein. *RNA*, 23(5): 696–711.

Gottesman, S. (2004). The small RNA regulators of *Escherichia coli*: Roles and mechanisms. *Annu. Rev. Microbiol.*, 58: 273–301.

Gottesman, S. (2005) Micros for microbes: Non-coding regulatory RNAs in bacteria. *Trends Genet.*, 21: 399–404.

Gottesman, S. and Storz, G. (2011). Bacterial small RNA regulators: versatile roles and rapidly evolving variations. *Cold Spring Harb. Perspect. Biol.*, 3(12) pii: a003798.

Govindarajan, M., Balandreau, J., Muthukumarasamy, R., Revathi, G. and Lakshminarasimhan, C. (2006). Improved yield of micropropagated sugarcane following inoculation by endophytic *Burkholderia vietnamiensis*. *Plant Soil*, 280: 239–252.

Grabowicz, M. and Silhavy, T.J. (2017). Envelope stress responses: An interconnected safety net. *Trends Biochem. Sci.*, 42(3): 232–242.

Grillo, A.O., Brown, M.P. and Royer, C.A. (1999). Probing the physical basis for trp repressor-operator recognition. *J. Mol. Biol.*, 287(3): 539–554.

Gruber, A.R., Findeiss, S., Washietl, S., Hofacker, I.L. and Stadler, P.F. (2010). RNAZ 2.0: Improved noncoding RNA detection. *Pac. Symp. Biocomput.*, 15: 69–79.

Guillier, M. and Gottesman, S. (2006). Remodelling of the *Escherichia coli* outer membrane by two small regulatory RNAs. *Mol. Microbiol.*, 59: 231–247.

Guo, Y., Lew, C.M. and Gralla, J.D. (2000). Promoter opening by σ54 and σ70 RNA polymerases: σ factor-directed alterations in the mechanism and tightness of control. *Genes Dev.*, 14(17): 2242–2255.

Gudapaty, S., Suzuki, K., Wang, X., Babitzke, P. and Romeo, T. (2001). Regulatory interactions of Csr components: The RNA binding protein CsrA activates csrB transcription in *Escherichia coli. J. Bacteriol.*, 183(20): 6017–6027.

Haferkamp, M.R. (1988). Environmental factors affecting plant productivity. pp. 27–36. *In*: White, R.S. and Short, R.E. (eds.). *Achieving Efficient Use of Rangeland Resources.* Fort Keogh Research Symposium, Miles City.

Handelsman, J. (2004). Metagenomics: Application of genomics to uncultured microorganisms. *Microbiol. Mol. Biol. Rev.*, 68: 669–685.

Hatfield, J.L. and Prueger, J.H. (2015). Temperature extremes: Effect on plant growth and development. *Weather Clim. Extrem.*, 10: 4–10.

Hengge-Aronis, R. (1993). Survival of hunger and stress: The role of rpoS in early stationary phase gene regulation in *E. coli. Cell,* 72(2): 165–168.

Heera, R., Sivachandran, P., Chinni, S.V., Mason, J., Croft, L., Ravichandran, M. and Yin, L.S. (2015). Efficient extraction of small and large RNAs in bacteria for excellent total RNA sequencing and comprehensive transcriptome analysis. *BMC Res. Notes,* 8: 754.

Holmqvist, E. and Vogel, J. (2013). A small RNA serving both the Hfq and CsrA regulons. *Genes Dev.*, 27(10): 1073–1078.

Hrdlickova, R., Toloue, M. and Tian, B. (2017). RNA-Seq methods for transcriptome analysis. *WIREs RNA*, 8: e1364.

Hungria, M., Campo, R.J., Souza, E.M. and Pedrosa, F.O. (2010). Inoculation with selected strains of *Azospirillum brasilense* and *A. lipoferum* improves yields of maize and wheat in Brazil. *Plant Soil,* 331: 413–425.

Isawa, T., Yasuda, M., Awazaki, H., Minamisawa, K., Shinozaki, S. and Nakashita, H. (2010). *Azospirillum* sp. strain B510 enhances rice growth and yield. *Microbes Environ.*, 25: 58–61.

Jan, A.T. (2017). Outer membrane vesicles (OMVs) of gram-negative bacteria: A perspective update. *Front. Microbiol.*, 8: 1053.

Jin, X., Yang, G., Tan, C. and Zhao, C. (2015). Effects of nitrogen stress on the photosynthetic CO_2 assimilation, chlorophyll fluorescence, and sugar-nitrogen ratio in corn. *Sci. Rep.*, 5: 9311.

Kandror, O., DeLeon, A. and Goldberg, A.L. (2002). Trehalose synthesis is induced upon exposure of *Escherichia coli* to cold and is essential for viability at low temperatures. *Proc. Natl. Acad. Sci. USA*, 99: 9727–9732.

Kawano, M., Aravind, L. and Storz, G. (2007). An antisense RNA controls synthesis of an SOS-induced toxin evolved from an antitoxin. *Mol. Microbiol.* 64: 738–754.

Kery, M.B., Feldman, M., Livny, J. and Tjaden, B. (2014). TargetRNA2: Identifying targets of small regulatory RNAs in bacteria. *Nucleic Acids Res.*, 42: W124–W129.

Khoo, J.S., Chai, S.F., Mohamed, R., Nathan, S. and Firdaus-Raih, M. (2012). Computational discovery and RT-PCR validation of novel *Burkholderia* conserved and *Burkholderia pseudomallei* unique sRNAs. *BMC Genomics*, 13 Suppl 7: S13.

Kochar, M. and Srivastava, S. (2012). Surface colonization by *Azospirillum brasilense* SM in the indole-3-acetic acid dependent growth improvement of sorghum. *J. Basic Microbiol.*, 52: 123–131.

Koeppen, K., Hampton, T.H., Jarek, M., Scharfe, M., Gerber, S.A., Mielcarz, D.W. et al. (2016). A novel mechanism of host-pathogen interaction through sRNA in bacterial outer membrane vesicles. *PLoS Pathog.*, 12: e1005672.

Koul, V., Tripathi, C., Adholeya, A. and Kochar, M. (2015a). Nitric oxide metabolism and indole acetic acid biosynthesis crosstalk in *Azospirillum brasilense* SM. *Res. Microbiol.*, 166: 174–185.

Koul, V., Adholeya, A. and Kochar, M. (2015b). Sphere of influence of indole acetic acid and nitric oxide in bacteria. *J. Basic Microbiol.*, 55: 543–553.

Kuehn, M.J. and Kesty, N.C. (2005). Bacterial outer membrane vesicles and the host-pathogen interaction. *Genes Dev.*, 19: 2645–2655.

Landt, S.G., Abeliuk, E., McGrath, P.T., Lesley, J.A., McAdams, H.H. and Shapiro, L. (2008). Small non-coding RNAs in Caulobacter crescentus. *Mol. Microbiol.*, 68(3): 600–614.

Latchman, D.S. (1997). Transcription factors: An overview. *Int. J. Biochem. Cell Biol.*, 29(12): 1305–1312.

Le Rhun, A., Beer, Y.Y., Reimegård, J., Chylinski, K. and Charpentier, E. (2016). RNA sequencing uncovers antisense RNAs and novel small RNAs in Streptococcus pyogenes. *RNA Biol.*, 13(2): 177–195. doi:10.1080/15476286.2015.1110674.

Lee, H.J. (2019). Microbe-host communication by small RNAs in extracellular vesicles: Vehicles for Transkingdom RNA transportation. *Int. J. Mol. Sci.*, 20: E1487.

Lehnik-Habrink, M., Newman, J., Rothe, F.M., Solovyova, A.S., Rodrigues, C., Herzberg, C., Commichau, F.M., Lewis, R.J. and Stülke, J. (2011). RNase Y in *Bacillus subtilis*: A natively disordered protein that is the functional equivalent of RNase E from *Escherichia coli. J. Bacteriol.*, 193(19): 5431–5441.

Lindmark, B., Rompikuntal, P.K., Vaitkevicius, K., Song, T., Mizunoe, Y., Uhlin, B.E. et al. (2009). Outer membrane vesicle-mediated release of cytolethal distending toxin (CDT) from *Campylobacter jejuni. BMC Microbiol.*, 9: 220.

Liu, J.M., Livny, J., Lawrence, M.S., Kimball, M.D., Waldor, M.K. and Camilli, A. (2009). Experimental discovery of sRNAs in *Vibrio cholerae* by direct cloning, 5S/tRNA depletion and parallel sequencing. *Nucleic Acids, Res.*, 37: e46.

Liu, J.M. and Camilli, A. (2010). A broadening world of bacterial small RNAs. *Curr. Opin. Microbiol.*, 13: 18–23.

Liu, M.Y., Gui, G., Wei, B., Preston, J.F. III, Oakford, L., Yuksel, U., Giedroc, D.P. and Romeo, T. (1997). The RNA molecule CsrB binds to the global regulatory protein CsrA and antagonizes its activity in *Escherichia coli. J. Biol. Chem.*, 272(28): 17502–17510.

Livny, J., Fogel, M.A., Davis, B.M. and Waldor, M.K. (2005). sRNAPredict: An integrative computational approach to identify sRNAs in bacterial genomes. *Nucleic Acids Res.*, 33(13): 4096–4105.

Livny, J., Brencic, A., Lory, S. and Waldor, M.K. (2006). Identification of 17 *Pseudomonas aeruginosa* sRNAs and prediction of sRNA-encoding genes in 10 diverse pathogens using the bioinformatics tool sRNAPredict2. *Nucleic Acids Res.*, 34(12): 3484–3493.

Lynch, J.M. (1990). Introduction: Some consequences of microbial rhizosphere competence for plant and soil. pp. 1–10. *In*: Lynch, J.M. (ed.). *The Rhizosphere*. Chichester, UK: Wiley & Sons.

Madhugiri, R., Pessi, G., Voss, B., Hahn, J., Sharma, C.M., Reinhardt, R., Vogel, J., Hess, W.R., Fischer, H.M. and Evguenieva-Hackenberg, E. (2012). Small RNAs of the Bradyrhizobium/Rhodopseudomonas lineage and their analysis. *RNA Biol.*, 9: 47–58.

Majdalani, N., Chen, S., Murrow, J., St John, K. and Gottesman, S. (2001). Regulation of RpoS by a novel small RNA: The characterization of RprA. *Mol. Microbiol.*, 39: 1382–1394.

Maki, K., Uno, K., Morita, T. and Aiba, H. (2008). RNA, but not protein partners, is directly responsible for translational silencing by a bacterial Hfq-binding small RNA. *Proc. Natl. Acad. Sci. USA*, 105: 10332–10337.

Mashburn-Warren, L., Howe, J., Garidel, P., Richter, W., Steiniger, F., Roessle, M. et al. (2008). Interaction of quorum signals with outer membrane lipids: Insights into prokaryotic membrane vesicle formation. *Mol. Microbiol.*, 69: 491–502.

Massé, E. and Gottesman, S. (2002). A small RNA regulates the expression of genes involved in iron metabolism in *Escherichia coli. Proc. Natl. Acad. Sci. USA*, 99(7): 4620–4625.

Massé, E., Escorcia, F.E. and Gottesman, S. (2003). Coupled degradation of a small regulatory RNA and its mRNA targets in *Escherichia coli. Genes Dev.*, 17(19): 2374–2383.

Mathesius, U., Schlaman, H.R.M., Spaink, H.P., Sautter, C., Rolfe, B.G. and Djordjevic, M.A. (1998). Auxin transport inhibition precedes root nodule formation in white clover roots and is regulated by flavonoids and derivatives of chitin oligosaccharides. *Plant J.*, 14: 23–34.

Merrick, M.J. (1993). In a class of its own-the RNA polymerase sigma factor sigma 54 (sigma N). *Mol. Microbiol.*, 10(5): 903–909.

Mitschke, J., Georg, J., Scholz, I., Sharma, C.M., Dienst, D., Bantscheff, J., Voss, B., Steglich, C., Wilde, A., Vogel, J. and Hess, W.R. (2011). An experimentally anchored map of transcriptional start sites in the model cyanobacterium *Synechocystis* sp. PCC6803. *Proc. Natl. Acad. Sci. USA*, 108: 2124–2129.

Mizuno, T., Chou, M.Y. and Inouye, M. (1984). A unique mechanism regulating gene expression: Translational inhibition by a complementary RNA transcript (micRNA). *Proc. Natl. Acad. Sci. USA*, 81: 1966–1970.

Molina-Favero, C., Creus, C.M., Simontacchi, M., Puntarulo, S. and Lamattina, L. (2008). Aerobic nitric oxide production by *Azospirillum brasilense* Sp245 and its influence on root architecture in tomato. *Mol. Plant Microbe Int.*, 21: 1001–1009.

Møller, T., Franch, T., Højrup, P., Keene, D.R., Bächinger, H.P., Brennan, R.G. and Valentin-Hansen, P. (2002). Hfq: A bacterial Sm-like protein that mediates RNA–RNA interaction. *Mol. Cell*, 9(1): 23–30.

Mückstein, U., Tafer, H., Hackermuller, J., Bernhart, S.H., Stadler, P.F. and Hofacker, I.L. (2006). Thermodynamics of RNA–RNA binding. *Bioinformatics*, 22(10): 1177–1182.

Murina, V., Lekontseva, N. and Nikulin, A. (2013). Hfq binds ribonucleotides in three different RNA-binding sites. *Acta Crystallogr. D. Biol. Crystallogr.*, 69: 1504–1513.

Nadeem, S.M., Naveed, M., Ahmed, M. and Zaheer, Z.A. (2015). Rhizosphere bacteria for crop production and improvement of stress tolerance: Mechanisms of action, applications, and future prospects. pp. 1–36. *In*: *Plant Microbes Symbiosis: Applied Facets*. India: Springer.

Niu, D., Hamby, R., Sanchez, J.N., Cai, Q., Yan, Q. and Jin, H. (2021). RNAs: A new frontier in crop protection. *Current Opinion in Biotechnology*, 70: 204–212.

Nogueira, E.M., Vinagre, F., Masuda, H.P., Vargas, C., Muniz de Pádua, V.L., Da Silva, F.R., Dos Santos, R.V., Baldani, J.I., Ferreira, P.C.G. and Hemerly, A.S. (2001). Expression of sugarcane genes induced by inoculation with *Gluconacetobacter diazotrophicus* and *Herbaspirillum rubrisubalbicans*. *Gen. Mol. Biol.*, 24: 199–206.

Novick, R.P. and Geisinger, E. (2008). Quorum sensing in Staphylococci. *Annu. Rev. Genet.*, 42: 541–564.

Okon, Y. and Labandera-Gonzalez, C. (1994). Agronomic applications of *Azospirillum*: An evaluation of 20 years of worldwide field inoculation. *Soil Biol. Biochem.*, 26: 1591–1601.

Opdyke, J., Kang, J.G. and Storz, G. (2004). GadY, a small RNA regulator of acid response genes in *Escherichia coli*. *J. Bacteriol.*, 186(20): 6698–6705.

Papenfort, K., Pfeiffer, V., Mika, F., Lucchini, S., Hinton, J.C. and Vogel, J. (2006). SigmaE dependent small RNAs of *Salmonella* respond to membrane stress by accelerating global omp mRNA decay. *Mol. Microbiol.*, 62: 1674–1688.

Pereg Gerk, L., Gilchrist, K. and Kennedy, I.R. (2000). Mutants with enhanced nitrogenase activity in hydroponic *Azospirillum brasilense*-wheat associations. *Appl. Environ. Microbiol.*, 66: 2175–2184.

Picard, F., Loubière, P., Girbal, L. and Cocaign-Bousquet, M. (2013). The significance of translation regulation in the stress response. *BMC Genomics*, 14: 588.

Podkaminski, D., Bouvier, M. and Vogel, J. (2014). Identification and characterization of small non-coding RNAs in bacteria. pp. 719–786. *In*: *Handbook of RNA Biochemistry* (Second, Completely Revised and Enlarged Edition) Weinheim: Wiley-VCH.

Pyla, R., Kim, T.J., Silva, J.L. and Jung, Y.S. (2010). Proteome analysis of *Azotobacter vinelandii* ΔarrF mutant that overproduces poly-β-hydroxybutyrate polymer. *Appl. Microbiol. Biotechnol.*, 88(6): 1343–1354.

Rivas, E. and Eddy, S.R. (2001). Noncoding RNA gene detection using comparative sequence analysis. *BMC Bioinformatics*, 2(1): 8.

Ron, E.Z. (2013). Bacterial stress response. pp. 589–603. *In*: DeLong, E.F., Lory S., Stackebrandt, E. and Thompson, F. (eds.). *The Prokaryotes. Prokaryotic Physiology and Biochemistry* (4th Edn.). Berlin Heidelberg: Springer-Verlag.

Rossi, C.C., Bossé, J.T., Li, Y., Witney, A.A., Gould, K.A., Langford, P.R. and Bazzolli, D.M. (2016). A computational strategy for the search of regulatory small RNAs in *Actinobacillus pleuropneumoniae*. *RNA*, 22(9): 1373–1385.

Sauer, E., Schmidt, S. and Weichenrieder, O. (2012). Small RNA binding to the lateral surface of Hfq hexamers and structural rearrangements upon mRNA target recognition. *Proc. Natl. Acad. Sci. USA*, 109: 9396–9401.

Sauer, E. (2013). Structure and RNA-binding properties of the bacterial LSm protein Hfq. *RNA Biol.*, 10(4): 610–618.

Seshasayee, A.S., Sivaraman, K. and Luscombe, N.M. (2011). An overview of prokaryotic transcription factors: a summary of function and occurrence in bacterial genomes. *Sub. Cell Biochem.*, 52: 7–23.

Sharma, C., Hoffmann, S., Darfeuille, F. et al. (2010). The primary transcriptome of the major human pathogen *Helicobacter pylori*. *Nature*, 464: 250–255.

Sharma, S.K., Ramesh, A. and Johri, B.N. (2013). Isolation and characterization of plant growth promoting *Bacillus amyloliquefaciens* strain sks_bnj_1 and its influence on rhizosphere soil properties and nutrition of soybean (*Glycine max* L. Merrill). *J. Virol. Microbiol.*, 2013: 1–19.

Shimizu, K. (2014). Regulation systems of bacteria such as *Escherichia coli* in response to nutrient limitation and environmental stresses. *Metabolites*, 4(1): 1–35.

Shingler, V. (1996). Signal sensing by s54-dependent regulators: Derepression as a control mechanism. *Mol. Microbiol.*, 19: 409–416.

Shingler, V. (2011). Signal sensory systems that impact σ^{54}-dependent transcription. *FEMS Microbiol. Rev.*, 35(3): 425–440.

Shrivastava, S., Egamberdieva, D. and Varma, A. (2015). Plant growth-promoting rhizobacteria (PGPR) and medicinal plants: The state of the art. pp. 1–18. *In*: Egamberdieva, D., Shrivastava, S. and Varma, A. (eds.). *Plant Growth Promoting Rhizobacteria (PGPR) and Medicinal Plants. Soil Biology*, Vol. 42. Heidelberg: Springer.

Silvaggi, J.M., Perkins, J.B. and Losick, R. (2005). Small untranslated RNA antitoxin in *Bacillus subtilis*. *J. Bacteriol.*, 187: 6641–6650.

Sirohi, G., Upadhyay, A., Srivastava, P.S. and Srivastava, S. (2015). PGPR mediated Zinc biofertilization of soil and its impact on growth and productivity of wheat. *J. Soil Sci. Plant Nutr.*, 15(1): 202–216.

Sittka, A., Lucchini, S., Papenfort, K., Charma, C.M., Rolle, K., Binnewies, T.T., Hinton, J.C.D. and Vogel, J. (2008). Deep sequencing analysis of small noncoding RNA and mRNA targets of the global post-transcriptional regulator, Hfq. *PLoS Genet.*, 4(8): e1000163.

Sledjeski, D.D., Gupta, A. and Gottesman, S. (1996). The small RNA, DsrA, is essential for the low temperature expression of RpoS during exponential growth in *Escherichia coli*. *EMBO J.*, 15: 3993–4000.

Smaldone, G.T., Antelmann, H., Gaballa, A. and Helmann, J.D. (2012). The FsrA sRNA and FbpB protein mediate the iron-dependent induction of the *Bacillus subtilis* lutABC iron sulfur-containing oxidases. *J. Bacteriol.*, 194: 2586–2593.

Smirnov, A., Förstner, K.U., Holmqvist, E., Otto, A., Günster, R., Becher, D., Reinhardt, R. and Vogel, J. (2016). Grad-seq guides the discovery of ProQ as a major small RNA binding protein. *Proc. Natl. Acad. Sci. USA*, 113: 11591–11596.

Sridhar, J., Narmada, S.R., Sabarinathan, R., Ou, H-Y. and Deng, Z. (2010). sRNAscanner: A computational tool for intergenic small RNA detection in bacterial genomes. *PLoS ONE*, 5(8): e11970.

Sridhar, J., Sabarinathan, R., Gunasekaran, P. and Sekar, K. (2013). Comparative genomics reveals 'novel' Fur regulated sRNAs and coding genes in diverse proteobacteria. *Gene*, 516(2): 335–344.

Sonnleitner, E., Abdou, L. and Haas, D. (2009). Small RNA as global regulator of carbon catabolite repression in *Pseudomonas aeruginosa*. *Proc. Natl. Acad. Sci. USA*, 106: 21866–21871.

Spaepen, S., Bossuyt, S., Engelen, K., Marchal K. and Vanderleyden, J. (2014). Phenotypical and molecular responses of Arabidopsis thaliana roots as a result of inoculation with the auxin-producing bacterium *Azospirillum brasilense*. *New Phytol.*, 201(3): 850–861.

Storz, G., Opdyke, J.A. and Zhang, A. (2004). Controlling mRNA stability and translation with small, noncoding RNAs. *Curr. Opin. Microbiol.*, 7(2): 140–144.

Storz, G., Altuvia, S. and Wassarman, K.M. (2005). An abundance of RNA regulators. *Annu. Rev. Biochem.*, 74: 199–217.

Torres-Quesada, O., Oruezabal, R.I., Peregrina, A., Jofré, E., Lloret, J., Rivilla, R., Toro, N. and Jiménez-Zurdo, J.I. (2010). The *Sinorhizobium meliloti* RNA chaperone Hfq influences central carbon metabolism and the symbiotic interaction with alfalfa. *BMC Microbiol.*, 10: 71.

Torres-Quesada, O., Reinkensmeier, J., Schlüter, J.P., Robledo, M., Peregrina, A., Giegerich, R., Toro, N., Becker, A. and Jiménez-Zurdo, J.I. (2014). Genome-wide profiling of Hfq-binding

RNAs uncovers extensive post-transcriptional rewiring of major stress response and symbiotic regulons in *Sinorhizobium meliloti*. *RNA Biol.*, 11: 563–579.

Trujillo-Lopez, A., Camargo-Zendejas, O., Salgado-Garciglia, R., Cano-Camacho, H., Baizabal-Aguirre, V.M., Ochoa-Zarzosa, A., Lopez-Meza, J.E. and Valdez-Alarcon, J.J. (2006). Association of *Gluconacetobacter diazotrophicus* with roots of common bean (*Phaseolus vulgaris*) seedlings is promoted in vitro by UV light. *Can. J. Bot.*, 84: 321–327.

Tsui, H.C., Leung, H.C. and Winkler, M.E. (1994). Characterization of broadly pleiotropic phenotypes caused by an hfq insertion mutation in *Escherichia coli* K-12. *Mol. Microbiol.*, 13: 35–49.

Udekwu, K.I., Darfeuille, F., Vogel, J., Reimegard, J., Holmqvist, E. and Wagner, E.G. (2005). Hfq-dependent regulation of OmpA synthesis is mediated by an antisense RNA. *Genes Dev.*, 19: 2355–2366.

Upadhyay, S.K., Singh, J.S., Saxena, A.K. and Singh, D.P. (2012). Impact of PGPR inoculation on growth and antioxidant status of wheat under saline conditions. *Plant Biol.* (*Stuttg*), 14(4): 605–611.

Valverde, A., Burgos, A., Fiscella, T., Rivas, R., Velázquez, E., Rodríguez-Barrueco, C., Cervantes, E., Chamber, M. and Igual, J.M. (2006). Differential effects of coinoculations with *Pseudomonas jessenii* PS06 (a phosphate-solubilizing bacterium) and *Mesorhizobium ciceri* C-2/2 strains on the growth and seed yield of chickpea under greenhouse and field conditions. *Plant Soil*, 287: 43–50.

Valverde, C., Livny, J., Schlüter, J.P., Reinkensmeier, J., Becker, A. and Parisi, G. (2008). Prediction of *Sinorhizobium meliloti*s RNA genes and experimental detection in strain 2011. *BMC Genomics*, 9: 416.

Vanaja, S.K., Russo, A.J., Behl, B., Banerjee, I., Yankova, M., Deshmukh, S.D. et al. (2016). Bacterial outer membrane vesicles mediate cytosolic localization of LPS and caspase-11 activation. *Cell*, 165: 1106–1119.

Vogel, J. and Luisi, B.F. (2011). Hfq and its constellation of RNA. *Nat. Rev. Microbiol.*, 9: 578–589.

Wagner, E.G.H. and Vogel, J. (2003). Noncoding RNAs encoded by bacterial chromosomes. pp. 243–259. *In*: Barciszewski, J. and Erdmann, V. (eds.). Noncoding RNAs. Landes Bioscience, Georgetown, TX, USA.

Wagner, E.G.H. (2013). Cycling of RNAs on Hfq. *RNA Biol.*, 10: 619–626.

Wagner, E.G.H. and Romby, P. (2015). Small RNAs in bacteria and archaea: Who they are, what they do, and how they do it. *Adv. Genet.*, 90: 133–208.

Wang, M.C., Chien, H.F., Tsai, Y.L., Liu, M.C. and Liaw, S.J. (2014). The RNA chaperone Hfq is involved in stress tolerance and virulence in uropathogenic *Proteus mirabilis*. *PLoS One*, 9(1): e85626.

Wassarman, K.M., Repoila, F., Rosenow, C., Storz, G. and Gottesman, S. (2001). Identification of novel small RNAs using comparative genomics and microarrays. *Genes Dev.*, 15(13): 1637–1651.

Whangbo, J.S. and Hunter, C.P. (2008). Environmental RNA interference. *Trends Genet.*, 24(6): 297–305.

Whipps, J.M. (2001). Microbial interactions and biocontrol in the rhizosphere. *J. Exp. Bot.*, 52: 487–511.

Wright, P.R., Georg, J., Mann, M., Sorescu, D.A., Richter, A.S., Lott, S., Kleinkauf, R., Hess, W.R. and Backofen, R. (2014). CopraRNA and IntaRNA: Predicting small RNA targets, networks, and interaction domains. *Nucleic Acids Res.*, 42(Web Server issue): W119–123.

Yaryura, P.M., Conforte, V.P., Malamud, F., Roeschlin, R., de Pino, V., Castagnaro, A.P., McCarthy, Y., Dow, J.M., Marano, M.R. and Vojnov, A.A. (2015). XbmR, a new transcription factor involved in the regulation of chemotaxis, biofilm formation, and virulence in *Xanthomonas citri* subsp. citri. *Environ. Microbiol.*, 17(11): 4164–4176.

Ying, X., Cao, Y., Wu, J., Liu, Q., Cha, L. and Li, W. (2011). TarPicker: A method for efficient prediction of bacterial sRNA targets based on a two-step model for hybridization. *PLoS ONE*, 6(7): e22705.

Yoder-Himes, D.R., Chain, P.S., Zhu, Y., Wurtzel, O., Rubin, E.M., Tiedje, J.M. and Sorek, R. (2009). Mapping the *Burkholderia cenocepacia* niche response via high-throughput sequencing. *Proc. Natl. Acad. Sci. USA*, 106: 3976–3981.

Zahir, Z.A., Munir, A., Asghar, H.N., Shaharoona, B. and Arshad, M. (2008). Effectiveness of rhizobacteria containing ACC deaminase for growth promotion of peas (*Pisum sativum*) under drought conditions. *J. Microbiol. Biotechnol.*, 18: 958–963.

Zhang, A., Altuvia, S., Tiwari, A., Argaman, L., Hengge-Aronis, R. and Storz, G. (1998). The OxyS regulatory RNA represses *rpoS* translation and binds the Hfq (HF-I) protein. *EMBO J.*, 17: 6061–6068.

Zhang, A., Wassarman, K.M., Ortega, J., Steven, A.C. and Storz, G. (2002). The Sm-like Hfq protein increases OxyS RNA interaction with target mRNAs. *Mol. Cell*, 9: 11–22.

Zhang, A., Wassarman, K.M., Rosenow, C., Tjaden, B.C., Storz, G. and Gottesman, S. (2003). Global analysis of small RNA and mRNA targets of Hfq. *Mol. Microbiol.*, 50: 1111–1124.

Zhang, A., Schu, D.J., Tjaden, B.C., Storz, G. and Gottesman, S. (2013). Mutations in interaction surfaces differentially impact *E. coli* Hfq association with small RNAs and their mRNA targets. *J. Mol. Biol.*, 425(19): 3678–3697.

Zubay, G., Morse, D.E., Schrenk, W.J. and Miller, J.H. (1972). Detection and isolation of the repressor protein for the tryptophan operon of *Escherichia coli*. *Proc. Natl. Acad. Sci. USA*, 69(5): 1100–1103.

Chapter 13

Application of Nanoparticles for Quality and Safety Enhancement of Foods of Animal Origin

Kandeepan Gurunathan

1. Introduction

The potential application of nanotechnology in food is unlimited. Majority of the work in food application of nanoparticles (NPs) is being done on nanoscale biomaterials, packaging, sensors, enhancing the shelf-life of foods, and detection of contaminants present in food. Some applications on the horizon may include development of nanotechnology-based foods with lower calories, less fat, salt, and sugar while retaining flavor and texture; nanoscale vehicles for effective delivery of bioactive components; re-engineering of beneficial microbes at the genetic and cellular level; nanobiosensors for the detection of pathogens, toxins, and bacteria in foods; identification systems for tracking animal products from their origin to consumption; integrated systems for sensing, monitoring, and active response intervention for animal products; development of climate resilient animal products; and nanoscale films for food packaging and contact materials that extend shelf-life, retain quality, and reduce cooling requirements.

The application of nanomaterials in livestock products has tremendous potential. As predicted by the Institute for Health and Consumer Protection (IHPC), the NP-based market will touch USD 20 billion mark by 2020 (Montazer and Harifi, 2017). There are several nanomaterials which have great prospects for application in animal products due to their specific property and functionality. Carbon nanotubes (CNTs) exhibit antibacterial properties which are attributed to their direct penetration

ICAR-National Research Centre on Meat, Hyderabad-500092, India.
Email: drkandee@gmail.com

through microbial cells (Kuswandi, 2017). Quantum dots (QDs) have unique spectral properties compared with traditional organic dyes, thus recently, they have been applied as a new generation of fluorophores in bioimaging and biosensing (Bakalova et al., 2004). Addition of nanoclays improves the mechanical strength of biopolymers (Gutierrez et al., 2017). The assimilation of cellulose nanowhiskers and starch systems improve their thermo-mechanical characteristics, along with reduced water sensitivity and intact biodegradability (Lima and Borsali, 2004). Lu et al. (2004) observed that the combination of chitin whiskers with soyprotein isolate (SPI) thermoplastic significantly enhanced the tensile characteristics of the film matrix and their resistance to water. Silver (Ag) NPs inhibit the respiratory chain enzymes and can also stimulate the production of reactive oxygen species (ROS) (Emamifar et al., 2011). Copper NPs lead to multiple toxic effects such as generation of ROS, lipid peroxidation, protein oxidation and DNA degradation in food borne pathogens, which might be responsible for its antimicrobial activity (Chatterjee et al., 2014). Petchwattana et al. (2016) displayed significant antibacterial activity of polybutylenes succinate (PBS)/ zinc oxide (ZnO) composite films against *E. coli* and *S. aureus*.

Besides, the nanomaterials have application in various domains of animal product value chain. The NPs-based tiny edible capsules with the aim to improve delivery of vitamins or fragile micronutrients in the daily foods are being created to provide significant health benefits (Koo et al., 2005). In another study on smart nano-packaging, conductive inks for ink jet printing based on copper NPs have also been developed (Park et al., 2007). Zeng et al. (2010) developed triangular Ag nanoplates as colorimetric indicators for monitoring the time-temperature history, based on thermodynamic unsteadiness of Ag nanoplates. Lin et al. (2015) studied the antimicrobial effect of cellulose/chitosan Ag and cellulose/chitosan nanocomposite films and reported better activity in films with AgNPs. Many NPs such as Ag, copper, chitosan, and metal oxide NPs like titanium oxide or zinc oxide have been reported to have antibacterial properties (Singh et al., 2017). Nanomaterials are also applied in the traceability of livestock products. A nanobarcode detection system is being developed that fluoresces under ultraviolet light in a combination of colors that can be read by a computer scanner (Li et al., 2005). Nanotechnology can also assist in the detection of pesticides (Liu et al., 2008), pathogens (Inbaraj and Chen, 2015), and toxins (Palchetti and Mascini, 2008) in the food for tracking, tracing, and monitoring quality in the supply chain.

Apart from the prospective application of nanomaterials in livestock products, their safety to human health has considerable concerns. The studies on Ag and ZnO by Panea et al. (2013) have showed that particle migration from nano-packaging is within limits as set by the European Commission. McCracken et al. (2016) have proposed that the size, shape, material, surface charge, solubility, and surface chemistry of NPs are important in determining their toxicity. Hence, the selection of suitable nanomaterials for specific application is the foremost point in sprawling the usage of nanotechnology in foods of animal origin.

2. Food Products of Animal Origin in the Market

The demand for animal origin foods is rising day-by-day in the international market. Various foods of animal origin include milk, meat, egg, fish, and their variety of processed products. Animal products are a rich source of protein, essential amino acids, vitamins, and minerals. The biological value of these proteins is very high as compared to protein sources from plants. The satiety level is also high. Hence, consumers prefer animal origin foods. But the perishability of animal origin foods is a major concern leading to control measures on quality and safety.

Milk processing improves shelf-life by deterring microbial spoilage and results in nutritious products. Processed milk products provide convenience and variety to the milk to promote marketing and satisfy consumer demands. Some of the milk products include butter, ghee, cheese, candy, paneer, whey, etc. Ghee is used for frying flavored local dishes. Butter or cream is heated in an iron kettle at 700–800°C till the complete liquid form for ghee is obtained. Paneer is a well-known traditional dairy product prepared by acid coagulation of hot milk. Mostly paneer is prepared with full fat milk resulting in rich flavor, soft body, and texture (Gurunathan Kandeepan and Sudhir Sangma, 2011). Milk is often blamed for its high content of fat and lack of dietary fiber, a much-needed component for a healthy food. Value addition of low-fat paneer from milk with dietary fiber will promote paneer as a healthy food (Kandeepan and Sangma, 2010). Enrobing or coating is a method of value addition and preservation. Enrobed paneer fingers/bites are a crispy, snack food prepared from paneer. Enrobing preserves the nutritive value by inhibiting lipid peroxidation and microbial spoilage, thereby improving shelf-life of the product. *Dahi* (curd) is a most popular fermented dairy product in India. The flavor of dahi is due to the presence of lactic acid and diacetyl groups in the fermented milk. *Lassi* (buttermilk) is a beverage prepared from dahi by breaking the coagulum and adding sweetening and flavoring agents. The lassi is garnished with finely grated almonds and served or stored in the refrigerator. Whey, being rich in nutritional and functional components, needs to be utilized in beverages and food supplements to meet the increasing demand of protein and energy in human nutrition.

The different types of value-added meat products available in the market include cured meat products, comminuted/minced meat products, emulsion-based meat products, enrobed meat products, restructured meat products, smoked meat products, functional meat products, traditional/ethnic/heritage meat products, retort processed meat products, dehydrated meat products, extruded meat products, etc. The cured meat products are prepared through surface application/injection/dipping methods where the curing mix in the form of dry salt/curing solution consists of salt, polyphosphate, nitrite, sugar, and water. Curing is either followed by smoking or the cured products can also be directly consumed as slices in sandwiches. These types of products are quite popular in western countries with pork, ham, or bacon being the most preferred cuts. Nowadays, cured chicken products are gaining popularity in the market. The comminuted meat products are prepared from minced meat. Herein, the particle size of the meat is reduced by mincing the meat. The minced/comminuted meat can be used to make several products like patties (Gurunathan Kandeepan et al., 2009), balls, cutlets, *pakoda*s (a lentil-based fried delicacy), samosas, keema (Kandeepan et al., 2013), etc. by adding non-meat ingredients like spices, condiments, vegetables, salt, oil, flours, etc. and cooking them by deep fat frying/shallow frying/convection in oven/immersion in water. These products are mostly ready-to-eat type and products like patties and balls can be stored and served later. In the case of emulsion-based meat products, the comminuted meat is added with salt, polyphosphates, nitrite, sugar, oil, flour, spices, binders, fillers, and condiments to prepare a batter/emulsion in a bowl chopper. This meat emulsion can lead to at least a dozen variety of meat products yielding different shapes and flavors based on the mould and cooking methods. Some of the meat products prepared in this category include sausages, patties, slices, nuggets, bites, croquets, etc. The enrobed meat products are coated with a batter, breaded, and deep fat fried to obtain the enrobed meat products. Enrobing gives the product unique crunchy, crispy texture, and improves the shelf-life by conserving the nutrient contents and delaying the microbial growth in the product.

Restructured/reformed meat products are prepared by incorporating chunked meat in curing solution and binders, then extracting the protein from meat by tumbling/massaging to bind the meat pieces with each other. These tacky bound meat pieces can be filled into different moulds to give the desired shape and cooked to yield a unique product that would mimic the chewing of meat pieces when the product is tasted. Restructured meat products can be served as bites or slices. The smoked meat products are usually cured meat products that are smoked to impart a unique smoky flavor to the product and to extend its shelf-life. In this process, saw dust from desired wood is used for imparting the flavor in a smoke oven where smoking, cooking, and water spray on the meat product happens. In the case of functional meat products, apart from the usual nutrients present in meat, certain health-promoting ingredients are added to the formulation to produce functional meat products. These types of products are customized to serve the health-conscious consumers. Some of the meat products prepared in this category include low fat, high fiber, low salt, antioxidant rich meat products. Although many breakthroughs have happened through new equipment and technologies for the development of novel meat products, still traditional meat products have their own consumer niche. Some of the traditional meat products popular in India are kebab, biryani, meat curry (Gurunathan Kandeepan et al., 2011), paya, soup, haleem, yakini, momo, thukpa, etc. Retort processed meat products have a very long shelf life ranging from 6 months in retort pouches to 1–2 years in cans. Retort products are ready-to-eat meat products usually consisting of traditional/seasonal meat products. Dehydrated meat products are shelf stable at room temperature, and they can be rehydrated and used in the preparation of different culinary dishes. Dried meat powder and meat chunks are popular in this category. Apart from this, minced meat is made into a batter type with some non-meat ingredients and given different shapes/layers to yield dried meat products like chips, badi rings, fingers chips, etc. Extruded meat products are not popular in the market as the proportion of incorporation of meat into these types of products is limited to minimal level.

3. Emergence of Nanomaterials in Animal Origin Foods

Owing to high global interest, nanotechnology has been proposed to impact the global economy by around USD 3 trillion by 2020, generating a requirement of approximately 6 million professionals in different interrelated sectors (Duncan, 2011). It can be anticipated that nanotechnology will create a major thrust for the development of advanced packaging systems for the sake of consumers. Differently from the materials at macroscale, nanomaterials display specific and improved physicochemical properties. By virtue of their small size, NPs hold a huge surface-to-volume ratio and surface activity. When affixed to desirable polymers, nanomaterials result in improved mechanical strength, electrical conductivity, and thermal stability, etc. Thus, nanomaterials improve the mechanical and barrier properties of food packages, along with offering active and intelligent packaging systems. To meet effective food packaging requirements, advanced nanomaterial augmented polymers will help to amplify the benefits associated with the existing ones, i.e., enhanced safety, besides addressing environmental concerns. The developed packaging material will contribute to reducing any serious interaction between packaging and food matrices, impact over consumer's health, drop in quantity of waste material, improved biodegradability and barrier shielding to gases and light, and reduced CO_2 emissions.

Nanotechnology caters several areas of food sciences such as food safety, packaging, processing, bioavailability, fortification, encapsulation, pathogen detection, etc. Such progressions make it an ideal candidature for the development of nanomaterials in a wide array of food packaging applications such as processed meat products, cheese; in addition to this it also helps in extrusion-coating applications for dairy products. Numerous companies are already engaged in the production of packaging materials based on nanotechnology that are extending the shelf-life of food and drinks and improving the food safety. NanocorTM, a subsidiary of Illinois-based AMCOL International, offers an ample range of polymer nanocomposites for purchase in pellet form and packaging products developed with the montmorillonite minerals such as DurethanR KU2-2601 (blending of nylon 6 and nanoclay) has been used as food packaging material via improving several properties (gas and moisture barrier, strength, toughness, abrasion, and chemical resistance) of packaging (Bumbudsanpharoke and Ko, 2015). NanoTuffTM (nylon 6 based nanocomposite with 10% clay), commercialized by Nylon Corporation of America, exhibits improved barrier properties to H_2O, O_2, and CO_2, in comparison to neat Nylon 6 (Duncan, 2011). Nanosilvers are also engaged in packaging of different food items (cheese, soup, meat, etc.) due to their varied antimicrobial properties and globally it is marketed with different commercial names like Fresher LongerTM, Bags Fresher LongerTM (USA), e.WindowR Nano Silver Food Container (South Korea); Incense Nano Silver Food Container, Fresh Box NanoSilver Food Container (South Korea), Zeomic (Japan), Anson Nano, Nano Silver Food (China) (Bumbudsanpharoke and Ko, 2015). This trend is also reflected in the emergence of escalating numeral of national and European R & D (research and development) projects allied to active packaging.

4. Different Nanomaterials for Foods of Animal Origin

4.1 Carbon Nanotubes (CNTs)

It is a new form of carbon, equivalent to a two-dimensional graphene sheet rolled into a tube. CNTs are available as either a single wall nanotube (SWNT) or multiwalled nanotubes (MWNT). SWNT is generally one atom thick, whereas MWNT comprises of several concentric tubes with very high aspect ratios and elastic modulus. Its tensile strength is 200 GPa which is ideal for reinforced composites and nano electromechanical systems. Structurally, the nanotube systems consist of graphitic layers seamlessly wrapped into cylinders. The diameter of most SWNTs is about 1 nm and strongly correlated to synthesis techniques, mixing of s and a bonds and electron orbital rehybridization. Exploitation of these properties of CNTs will definitely open new possibilities to develop many types of nanodevices which confer unique conductive, optical, and thermal properties for various applications. Generally, the use of CNT nanosponges containing sulfur and iron increases efficiency in soaking up contaminants such as pesticides and pharmaceuticals.

The tensile strength/modulus of several polymers such as polyethylene naphtalate, polyvinyl alcohol, polyamide, and polypropylene have been improvised by incorporating carbon nanotubes and polyamides (Prashantha et al., 2009). CNTs possess elastic modulus of up to 1 TPa and tensile strength of 200 GPa (Lau and Hui, 2002). Dias et al. (2013) documented that CNTs, in combination with allyl isothiocyanate, could inhibit *Salmonella cholera suis* for over a period of 40 days of storage. Single-walled carbon nanotubes with cobalt mesoarylporphyrin complexes have also been explored for developing a chemiresistive detector which detects amines generated during spoilage of meat (Liu et al., 2015). However, desspite having a wide application, concerns associated with their processing and dispersion aspects, along with high cost, limits their incorporation in nanocomposites.

4.2 Quantum Dots

Generally, semiconductor QDs have high quantum yield and molar extinction coefficients, broad absorption spectra with narrow, symmetric fluorescence spectra spanning the ultraviolet to near-infrared, large effective excitation, high resistance to photobleaching, and exceptional resistance to photochemical degradation. Thus, these are excellent fluorescence, quantum confinement of charge carrier's materials and possess size tunable band energy. QDs also function as photocatalysts for the light driven chemical conversion of water into hydrogen as a pathway to solar fuel (Konstantatos and Sargent, 2009). Therefore, based on this transport approach, QDs can be utilized for live imaging of systems to verify known physiological processes (Das et al., 2015).

4.3 Nanocapsules

Nanocapsules are vesicular systems in which the substances are confined to a cavity consisting of an inner liquid core enclosed by a polymeric membrane. Recently, micro and NPs are getting significant attention for delivery of drugs, for protection and increase in bioavailability of food components or nutraceuticals, for food fortification and for the self-healing of several materials. Some drugs such as peptides or anti-inflammatory compounds are successfully nanoencapsulated (Haolong et al., 2011).

4.4 Nanoemulsions

Nanoemulsions are formed by very small emulsion nanoscale droplets (oil/water system) exhibiting sizes lower than 100 nm. Due to the size of droplets, the ratio of surface area to volume, Laplace pressure, and elastic modulus, nanoemulsions are significantly larger than that of ordinary emulsions. Moreover, unlike general emulsions, most of nanoemulsions appear optically transparent that, thus, technically can be incorporated into drinks. Unfortunately, the formulation of nanoemulsion needs very high energy, thus it requires some special devices that can generate extreme shear stress such as, high pressure homogenizer or ultrasonic generator.

4.5 Clay and Silicate Nanoplatelets

A wide number of NPs, including silica, silicate, clay, organomontmorillonite, and calcium carbonate, are used in nanocomposites for food packaging. These particles fall under the more general category of clay NPs, or 'nanoclays'. Clays exist in a structure held together in crystalline form. By breaking the crystal structure leaving only the platelets, a nanoclay is created. The high aspect ratio (width divided by height) and the large surface area create desirable barrier properties, reinforcing efficiency, and improving thermal stability. The nanoclays are then imbedded into a polymer film to create a nanocomposite. These nanocomposites decrease the diffusion of O_2 and CO_2 in and out of the packaging material, keeping food fresher for longer periods of time. They also help reduce the health risks associated with bacterial growth in food (Kuzma et al., 2008). Clay and silicates, owing to their availability, low cost, and relatively simple process have attracted focus of researchers as potential NPs. In addition to the typical tactoid structure of microcomposites, interaction among layered silicates and polymers may result in intercalated or exfoliated nanocomposites. The intercalated nanocomposites represent a multilayered structure with alternating polymer/inorganic layers lying apart by few nanometers. Such structures finally result through the penetration of polymers chains into the interlayer region of clay lead. The exfoliated nanocomposites comprise of extensive polymer penetration with random dispersion of clay layers (Luduena et al., 2007).

Application of montmorillonite (MMT) clay as a nanocomponent in an extensive variety of polymers such as, polyethylene, nylon, polyvinyl chloride, and starch dates to the 1990s (Montazer and Harifi, 2017). MMT [Mx (Al4-xMgx) Si8O20(OH)4] is the most common clay filler. It represents an octahedral sheet of $Al(OH)_3$ between silica tetrahedral bilayers, linked together by weak electrostatic forces. The imbalance between the surface negative charges is compensated by the presence of exchangeable cations, Na^+ and Ca^{2+} (Tan et al., 2008). The clay layers offer resistance to gases and water vapor permeability. The application of 5% (w/w) of clays in thermoplastic starch (TPS)/clay nanocomposites, improves the mechanical properties with decreased water vapor permeability of starch biopolymer (Muller et al., 2012). Clay NPs are further known to impact the glass transition and thermal degradation temperatures (Cyras et al., 2008). Nanoclays also have prospects in active and intelligent food nano-packaging. Recently, Gutierrez et al. (2017) proposed a nano-packaging system developed by inserting blueberry extract between the silicate interlayer spaces of clay. Blueberries possess anthocyanins which change color with pH, attributed to a shift between quinoidal and flavylium forms. Incorporation of blueberry extract could modify these clays into active and intelligent nanocomposites.

4.6 Cellulose-based Nanofibers or Nanowhiskers

Cellulose, the building material of long fibrous cells, is a highly strong natural polymer. It is ubiquitous, cost-effective, environmentally friendly, and easy to recycle. Cellulose is also explored as a supporting material for many nanomaterials. Application of cellulose increases the surface area of NPs associated with their enhanced activity. These additive features make cellulose nanofibers an attractive class of nanomaterials. Basically, two types of nano reinforcements, viz., microfibrils and whiskers, can be derived from cellulose. The cellulose chains appear as microfibrils (or nanofibers), which are bundles of molecules that are elongated and stabilized through hydrogen bonding (Wang and Sain, 2007). Each microfibril is further formed by aggregation of elementary fibrils, which contain crystalline and amorphous parts. The crystalline parts can be isolated by acid hydrolysis treatments and are referred to as whiskers, nanocrystals, nanorods, or rod-like cellulose microcrystals. The dimensions of whiskers, after hydrolysis mainly rely on the percentage of amorphous regions in the bulk fibrils, which differ between different organisms (Gardner et al., 2008). Microcrystalline cellulose (MCC) consists of a huge quantity of cellulose microcrystals associated with amorphous areas. Cellulose based nano reinforcements improve the strength, thermal characteristics, and modulus of polymers, with restricted elongation. Cellulose nano reinforcements also improvised the moisture barrier properties of polymer films. Earlier, Ghaderi et al. (2014) prepared an all-cellulose nanocomposite (ACNC) film consisting of sugarcane bagasse nanofiber and N, N-dimethylacetamide/lithium chloride as solvent. The study utilized a very low-value agricultural waste product for preparing a high-performance nanocomposite with a tensile strength of around 140 MPa.

4.7 Starch Nanocrystals

Starch has been extensively explored over decades as a choice material for food packaging applications. Several associated benefits, viz., abundance, biocompatibility, non-toxicity, low cost, biodegradability, easy availability, and stable in air further enhances its possible applications. The incorporation of inorganic materials and synthetic polymers further improves its water resistance properties. Native starch granules can be submitted to an extended-time hydrolysis at temperatures below the gelatinization temperature when the amorphous regions are hydrolyzed allowing separation of crystalline lamellae, which are more resistant to hydrolysis. The starch crystalline particles show platelet morphology with thicknesses of 6–8 nm, improve tensile strength and modulus of pullulan films, with decreased elongation property (Kristo and Biliaderis, 2007). It has been proposed that the positive charge present over any antimicrobial agent contributes to its antimicrobial activity. As a result, the antimicrobial spectrum of metals incorporated/adsorbed on to polysaccharides surface increases, owing to enhanced surface area.

4.8 Chitin/Chitosan Nanoparticles

Chitosan, a heteropolysaccharide is known for its biocompatibility, biodegradability, along with metal complexation. The polycationic nature of chitosan is mainly accountable for its wide antimicrobial activity. Chitosan NPs are formed through ionic gelation, where the positively charged amino groups of chitosan electrostatically interacts with the polyanions engaged as crosslinkers (Lopez-Leon et al., 2005). De Moura et al. (2008) added that the chitosan-based nanocomposites in hydroxypropyl methylcellulose (HPMC) helped in the improvement of the mechanical and barrier characteristics. Chitosan/Ag, chitosan/Au, and chitosan/cinnamaldehyde nanocomposite films have demonstrated antimicrobial activity against *E. coli*, *S. aureus*, *P. aeruginosa*, *Aspergillus niger*, and *Candida albicans* (Rieger et al., 2015).

4.9 Metal Nanoparticles

Various metals [Silver (Ag), Gold (Au), Zinc (Zn)] derived nanomaterials have been explored in diverse active packaging applications. These particles either function on direct contact or they can migrate slowly and react preferentially with organics present in the food. NPs antimicrobial activity might be due to one of these mechanisms: direct interaction with the microbial cells (interrupting transmembrane electron transfer, disrupting/penetrating the cell envelope); oxidizing cell components; and production of secondary products (e.g., ROS or dissolved heavy metal ions), leading to cell damage (Li et al., 2008).

4.9.1 Silver Nanoparticle

AgNPs are among the most explored, owing to their established antimicrobial potential against multiple commensal and pathogenic strains. Besides bacterial strains, they are known to be inhibitory against multiple fungi (Duncan, 2011). Ag targets bacterial metabolism by binding to its DNA, proteins, and enzymes, resulting into bacteriostatic effects (Cavaliere et al., 2015). AgNPs destabilize and disrupt both the outer and cytoplasmic membranes.

4.9.2 Copper Nanoparticle

Copper NPs were shown to inhibit the growth of *Saccharomyces cerevisiae*, *E. coli*, *S. aureus*, and *L. monocytogenes* on a polymer composite after 4 h exposure (Cioffi et al., 2005). Sheikh et al. (2011) also revealed good antibacterial effect of copper NPs against *E. coli* and *B. subtilis* in polyurethane nanofibers containing copper NPs.

4.9.3 Zinc Nanoparticle

Zinc nanocrystals have also been used as an antimicrobial and antifungal agent, when incorporated with plastic matrix (Vermeiren et al., 2002).

4.10 Metal Oxide Nanoparticles

Various metal oxide-derived nanomaterials [Titanium dioxide (TiO_2), zinc oxide (ZnO), silicon oxide (SiO_2), and magnesium oxide (MgO)] have been explored in diverse active packaging applications. Different NPs oxide(s) can act as UV blockers and photocatalytic disinfecting agents.

4.10.1 Titanium Dioxide (TiO_2)

Naturally, TiO_2 exists in three primary phases, i.e., anatase, rutile, and brookite, having varied crystal sizes. TiO_2 possesses photocatalytic abilities and at nanoscale, TiO_2 shows surface reactivity, which connects it with biological molecules (phosphorylated proteins and peptides) and DNA (Brown et al., 2008). The surface energy of TiO_2 NPs amplifies with size and is the significant factor in polymer/filler interaction. The surface energies of rutile particles are higher than those of anatase particles of similar size. TiO_2 is being highly explored in preparing several nanomaterials, viz., NPs, nanorods, nanowires, nanotubes, mesoporous, and nanoporous TiO_2 containing materials (Chen and Mao, 2006). The antibacterial properties of TiO_2 are well known, however, the antibacterial capacity of nano-TiO_2 particle is confined to the exposure of UV irradiation (Shi et al., 2008). Although, the exact mechanism of biocidal activity of TiO_2 is unclear, it may be attributed to its initial oxidative attack over the outer/inner bacterial cell membrane, alterations of Coenzyme A-dependent enzyme activity, and DNA damage through hydroxyl radicals (Kubacka et al., 2014). Cerrada et al. (2008) examined the photo-activated biocidal properties of TiO_2 NPs-based EVOH films against nine microorganisms (*B. stearothermophilus*, *S. aureus*, *E. coli*, *P. fluorescens*, *Bacillus* sp., *L. plantarum*, *E. caratovora*, *P. jadinii*, *Z. rouxii*). The TiO_2 NPs were uniformly dispersed ultrasonically. They reported over 5 log reduction for *B. stearothermophilus*, *Bacillus* sp., *L. plantarum*, and *P. jadinii* after 30 min of irradiation in the presence of the TiO_2/EVOH materials. Combination of TiO_2 NPs with Ag has been shown to enhance the antimicrobial properties to a significant level (Wu et al., 2010).

4.10.2 Silicon Oxide (SiO_2)

Silica nanoparticles (nSiO_2) also have the potential to improve the mechanical and/or barrier properties of various polymer matrices. Wu et al. (2002) examined the tensile characteristics (i.e., strength, modulus, and elongation) of nSiO_2 incorporated polypropylene (PP) matrix. Addition of nSiO_2 into starch matrix could improve the tensile properties along with decreased water absorption by starch (Xiong et al., 2008). Vladimiriov et al. (2006) integrated nSiO_2 in an isotactic polypropylene (iPP) matrix using maleic anhydride grafted polypropylene (PP-g-MA), and this nSiO_2 enhanced the storage modulus of iPP, making the material stiffer with improvised O_2 barrier capacity of matrix. Jia et al. (2007) generated polyvinyl alcohol and SiO_2 nanocomposites through radical copolymerization of vinyl silica NPs and vinyl acetate. The resultant nanocomposites exhibited enhanced thermal and mechanical properties in comparison to pure polyvinyl alcohol. This feature may be due to strong covalent interactions between nSiO_2 and the polymer matrix. In another study, Tang and Liu (2008) fabricated starch/PVOH/nSiO_2 biodegradable films and documented that the tensile and water resistance properties of films enhanced with an increase in their nSiO_2 content. Additionally, increased intermolecular H bonding and formation of C–O–Si groups between nSiO_2/starch and nSiO_2/PVOH were observed. Such bonding resulted in improvised miscibility and compatibility between the film components.

Food packaging applications of nSiO_2 has been explored as food contact surface materials. Bayer Polymers, Germany made silicate NPs enriched packaging film which retards the entrance of O_2 and other gases and loss of moisture, preventing food spoilage. Nanocor Inc, Chicago, IL, USA, developed a clay NP-based nanocomposite (Advantage Magazine, 2004). Salami-Kalajahi et al. (2012) reported that nanocomposites comprising of 5% silica NPs lead to the improvement of mechanical and physical properties. Recently, Farhoodi (2016) showed that the application of SiO_2 NPs as fillers in food packaging materials leaves a twisting pathway for gases. Chen et al. (2013) modified paper to form a lotus-like super hydrophobic surface by coating with R812S silica NPs and polydimethylsiloxane (PDMS) silicone oil. The coated paper displayed a strong water repellent property.

4.10.3 Zinc Oxide (ZnO)

Zinc oxide particles display good antibacterial activity, which further enhances with decrease in particle size. Its direct contact with microbial cell wall may result in destruction of bacterial cell integrity, liberation of antimicrobial ions, i.e., Zn^{2+} ions, and generation of ROS (Sirelkhatim et al., 2015). In another study, Jin et al. (2009) observed different approaches, i.e., powder, film, PVP capped and coating; for the application of nano-ZnO in food systems and concluded that nano-ZnO displayed antibacterial effects against *L. monocytogenes* and *Salmonella enteritidis* in liquid egg white and in culture media. Li et al. (2009) compared antimicrobial activity of ZnO powder and ZnO NPs against food borne pathogens (*Bacillus cereus*, *E. coli*, *S. aureus*, *S. enteridis*) and observed that ZnO NPs displayed better antibacterial activity than non-nano powder against all the tested bacteria.

5. Different Application of Nanomaterials in Foods of Animal Origin

5.1 Processing of Value-added Animal Products using Nanomaterials

The nanostructured food ingredients are being developed with the claims that they offer improved taste, texture, and consistency. Nanotechnology increases the shelf-life of different kinds of food materials and helps to bring down the extent of wastage of food due to microbial spoilage. Nowadays, nanocarriers are being utilized as delivery systems to carry food additives in food products without disturbing their basic morphology. Particle size may directly affect the delivery of any bioactive compound to various sites within the body as it was noticed that in some cell lines, only submicron NPs can be absorbed efficiently but not the larger size microparticles. An ideal delivery system is supposed to have following properties: (i) able to deliver the active compound precisely at the target place, (ii) ensure availability at a target time and specific rate, and (iii) efficient to maintain active compounds at suitable levels for long periods of time in storage condition. Nanotechnology is being applied in the formation of encapsulation, emulsions, and biopolymer matrices.

5.1.1 Nanoencapsulation

NPs have better properties for encapsulation and release efficiency than traditional encapsulation systems. Nanoencapsulations mask odors or tastes, control interactions of active ingredients with the food matrix, control the release of the active agents, ensure availability at a target time and specific rate, and protect them from moisture, heat, chemical, or biological degradation during processing, storage, and utilization, and exhibit compatibility with other compounds in the system (Weiss et al., 2006). Various synthetic and natural polymer based encapsulating delivery systems have been elaborated for the improved bioavailability and preservation of the active food components. Nanoencapsulation techniques have been used broadly to improve the flavor release and retention and to deliver culinary balance. NPs provide promising means of improving the bioavailability of nutraceutical compounds due to their subcellular size leading to a higher drug bioavailability. Many metallic oxides such as titanium dioxide and silicon dioxide (SiO_2) have conventionally been used as color or flow agents in food items.

Several bioactive compounds such as lipids, proteins, carbohydrates, and vitamins are sensitive to high acidic environment and enzyme activity of the stomach and duodenum. Encapsulation of these bioactive compounds enables them to resist such adverse conditions. Moreover, these delivery systems possess the ability to penetrate deeply into tissues due to their smaller size and thus allow efficient delivery of active compounds to target sites in the body. The nanocomposite, nano-emulsification, and nanostructuration are the different techniques which have been applied to encapsulate the substances in miniature forms to deliver nutrients more effectively like protein and antioxidants for precisely targeted nutritional and health benefits. Polymeric NPs are found to be suitable for the encapsulation of bioactive compounds (e.g., flavonoids and vitamins) to protect and transport bioactive compounds to target functions (Langer and Peppas, 2003). A real example: KD Pharma BEXBACH GMBH (Germany) provides nano-encapsulated Omega-3 fatty acids in two different forms—suspension and powder.

5.1.2 Nanoemulsions

NP emulsions are being used in ice cream and spreads as this can improve the texture and uniformity of the ice cream (Berekaa, 2015). The use of nanoemulsions to deliver lipid soluble bioactive compounds is much popular since they can be produced using natural food ingredients using easy production methods and may be designed to enhance water-dispersion and bioavailability (Ozturk et al., 2015). In functional foods, where a bioactive component often gets degraded, nanoencapsulation of these bioactive components extends the shelf-life of food products by slowing down the degradation processes or prevents degradation until the product is delivered at the target site. Encapsulating functional components within the droplets often enables a slowdown of chemical degradation processes by engineering the properties of the interfacial layer surrounding them. For example, curcumin, the most active and least stable bioactive component of turmeric (*Curcuma longa*), showed reduced antioxidant activity and found to be stable to pasteurization and at a different ionic strength upon encapsulation (Sari et al., 2015).

5.1.3 Nano-filtration

Nanofilter has a widespread application in eliminating microbial contaminants. The nanofilters are used in the processing of dairy products. Nanofilters provide a selective passing of particles. Also, nanofiltration is used to detect metabolites for quality control and pathogenic factors.

5.1.4 Nanocoatings

Oxygen is a problematic factor in food packaging because it can cause food spoilage and discoloration. NPs have been found to zigzag in the new plastic and prevent the penetration of oxygen as a barrier. Nanocoatings covering the food products completely, prevent weight loss and shrinkage. Anti-microbial NP coatings in the matrix of the packaging material can reduce the development of bacteria on or near the food product and maintaining the sterility of pasteurized foods by preventing post-processing contamination. Foods that are prone to spoilage on the surface, such as cheese, sliced meat, and bakery products, can be protected by contact packaging infused with antimicrobial NPs. CTC Nanotechnology GmbH, Merzig, Germany is selling a nanoscale dirt-repellent coating to create self-cleaning surfaces for use in food packages and meat-processing plants. This concept is based on a sol-gel process in which NPs are suspended in a fluid medium. By the action of nanohydrophobisation, the absorbency of the surfaces to be treated is eliminated so that they remain resistant to environmental factors after cleaning, with the added advantage that this product is biodegradable and approved and certified for use with food (Neethirajan and Jayas 2011). Nano-coated films are usually composed of layers of polymers that are designed as barriers to flavor, water, and/or gas.

5.1.5 Enzyme Immobilization

Enzyme immobilization at nanoscale increases the surface area and boosts the performance by enhancing the stability to pH, temperature, and resistance to proteases; besides it helps controlled release of enzymes into the food system (Brandelli et al., 2017). Rhim and Ng (2007) explored the approach of enzyme adsorption into nanoclays incorporated to polymers. It is further supported by Gopinath and Sugunan (2007) that nanoclay has a high affinity for protein adsorption and could be utilized as a competent enzyme carrier. Qhobosheane et al. (2001) modified SiO_2 NPs to immobilize glutamate dehydrogenase and lactate dehydrogenase, which displayed better enzyme activity after immobilization.

5.2 Packaging of Animal Products

A desirable packaging material must have gas and moisture permeability combined with strength and biodegradability. Nano-based 'smart' and 'active' food packaging confers several advantages over conventional packaging methods from providing better packaging material with improved mechanical strength, barrier properties, antimicrobial films to nanosensing for pathogen detection and alerting consumers to the safety status of food (Sharma et al., 2017).

5.2.1 Smart/Intelligent Packaging and Nanomaterials

Intelligent packaging offers superior functionality in terms of communication and marketing. It provides dynamic feedback on the actual quality of packaged food. In case of intelligent packaging systems, the nanotechnology-based indicator/sensor(s) are incorporated into the food package, where they can interact with internal (food components and headspace) and external environmental factors. Smart packaging system uses different innovative communication methods, i.e., nanosensors, time temperature indicators, oxygen sensors, freshness indicators, etc. Generally, nanosensors can be applied as labels or coatings to add an intelligent function to food packaging in terms of ensuring the integrity of the package through the detection of leaks (for foodstuffs packed in a vacuum or inert atmosphere), indications of time–temperature variations (e.g., freeze–thaw–refreeze), or microbial safety (the deterioration of foodstuffs). Gas sensors are basically used for revealing the gaseous analyte in the package. Optical O_2 sensors works on the principle of luminescence quenching or absorbance changes caused by direct contact with the analyte, whereas optochemical sensors are used to check the quality of products by sensing gas analytes such as, hydrogen sulfide, CO_2, and volatile amines. The unique chemical and electro-optical properties of nanoscale particles respond to environmental changes (e.g., temperature or humidity in storage rooms, levels of oxygen exposure), product degradation or microbial contamination. Therefore, such technology would apparently benefit consumers, industry stakeholders, and food regulators (Duncan, 2011).

5.2.1.1 O_2 Sensors

Food packaging systems with restricted oxygen availability are preferred over those giving free oxygen access. To achieve this, packaging needs to be carried out under vacuum or nitrogen gas along with incorporation of irreversible O_2 sensors. In case of modified atmosphere packaging (MAP), the headspace O_2 concentration is either deliberately reduced to an optimal level or eliminated as per food product requirements. Lee et al. (2005) developed a UV-based colorimetric O_2 indictor that utilizes TiO_2 NPs for photosensitization of triethanolamine induced reduction of methylene blue in a polymer encapsulation medium. UV irradiation leads to bleaching of sensor, which remains colorless until exposed to O_2. The pace of change in color is relative to the level of O_2 exposure. Mills and Hazafy (2009) used nanocrystalline SnO_2 as a photosensitizer and a colorimetric O_2 indicator consisting of glycerol (as electron donor), methylene blue (a redox dye), and a hydroxyethyl cellulose (as encapsulating polymer) was developed. Upon exposure to UV-B light, the indicator gets bleached and photoreduction of dye takes place by SnO_2 NPs. In another study done by Mihindukulasuriya and Lim (2013) generated UV based activated O_2 indicator membrane through the electrospinning method. Encapsulation of the active components, viz., TiO_2 NPs, glycerol, and methylene blue within electrospun polyfibers, enhanced the oxygen sensitivity of the membrane.

5.2.1.2 Detection of Spoilage and Pathogenic Microbes

Microbial sensors exploring NPs have unique optical and electrical characteristics in conjunction with other properties such as spacious and simply functionalized surfaces. Antibodies conjugated to nanomaterials such as QDs are mainly explored for detection of bacteria. QDs are mainly employed due to their characteristic high fluorescence efficiency, stability for photobleaching, high sensitivity, extended decay lifetime, and electronic properties such as wide and continuous absorption spectra and narrow emission spectra (Valizadeh et al., 2012). Yang and Li (2006) explored QDs for detection of *E. coli* O157: H7 and *Salmonella Typhimurium*. Highly fluorescent CdSe/ZnS QDs were conjugated to anti-*E. coli* O157 and anti-Salmonella antibodies. QDs for *E. coli* and *Salmonella* had different emission wavelengths but shared a common excitation wavelength, allowing simultaneous detection of the two test pathogens. The QD conjugate method has several benefits over typical fluorescence dyes. Multiplexed detection using different types of QDs is an additional advantage.

QDs and organic fluorescent compounds (OFCs) are classified into down-conversion phosphorus and up-conversion fluorescent NPs (UCNPs). The down-conversion phosphorus absorbs energy at low wavelength and emits radiation at higher wavelength, whereas UCNPs get excited by low energy radiations (near-infrared) and emit higher energy visible radiation. These characteristic features help them to rule out autofluorescence and photodamage related problems. UCNPs are extremely sensitive sensors that can also be used for bacteria, enzyme, protein, nucleic acid, pH, NH_3, CO_2, and other analytes (Gnach and Bednarkiewicz, 2012). Ong et al. (2014) conjugated *E. coli* specific antibodies to citrate modified oleic acid capped with NaYF4: Yb, Er UCNPs for selective detection of *E. coli*. Earlier, the up-conversion conjugates technique was used to design a quick lateral flow test, for *E. coli* detection (Niedbala et al., 2001). QDs and UCNPs have found application in labeling and food packaging, particularly for awaited advancement in intelligent label for pathogen and toxins detection in food matrix.

5.2.1.3 Freshness Indicators

Freshness indicators aim to provide the actual information regarding the quality of food product during storage, transit, and display. The reaction between microbial metabolites and incorporated indicators gives visual information about the microbial quality of product. The incorporated freshness indicators are sensitive to spoilage compounds or microbial metabolites generated during spoilage of food product, for example, volatile sulfides and amines. Smolander et al. (2004) detected the spoilage of meat products by depositing a transition metal (silver or copper) coating (1–10 nm thick) over plastic film or paper packaging structures. This coating turned dark upon reacting with sulfide volatiles produced from fresh meats undergoing spoilage. Food quality indicators have also been developed to provide visual indications to the consumer regarding when a packaged foodstuff starts to deteriorate. In another study in meat, a nanosilver layer is opaque light brown initially, but if the meat starts to deteriorate, silver sulphide is formed and the layer becomes transparent, indicating that the food may be unsafe to consume.

5.2.1.4 Time-temperature Indicators (TTIs)

The temperature exploitations met by food during transit and distribution are one of the main environmental factors responsible for the reduced shelf-life of foods. To counter different temperature abuses, TTIs are useful to monitor the thermal history during food storage, handling, and distribution. TTIs allow the retailers to ensure that the foods have been stored at the suitable temperatures; help consumers in determining the quality of food product they are purchasing; and help manufacturers in monitoring of supplied foodstuff throughout the supply chain.

Singh (2000) grouped TTIs into three basic types, viz., abuse indicators, partial temperature history indicators, and full temperature history indicators. Abuse indicators designate attainment of a particular temperature. Partial temperature history indicators present the time–temperature history only if the temperature surpasses a critical set limit. In contrast, the full temperature history indicators provide a constant monitoring of temperature changes with time. The communication regarding the change in food quality is generally linked with color development associated with a temperature dependent migration of dye through a porous material; or change in color of an indicator. Timestrip, an AuNP based iStrip for chilled foods appears red at temperatures beyond freezing. However, accidental freezing leads to irreparable agglomeration of the AuNP, resulting in loss of the red color (Robinson and Morrison, 2010). Zhang et al. (2013) formed a TTI working on a reaction of epitaxial overgrowth of Ag shell on Au nanorods dipped in cetyltrimethylammonium chloride (CTAC) solution appears red in color because of two absorbance bands that arise from transverse and plasmon resonances, respectively. With the introduction of $AgNO_3$ and a reducing agent (ascorbic acid), Ag atoms get deposited on Ag nanorods and form the shell of Au/Ag nanorods. The extinction bands of the longitudinal plasmon resonance transfer to lesser wavelengths as the Ag shell thickens which leads to changes in color from red to yellow, and lastly to green.

5.2.1.5 Humidity Indicators

Monitoring of humidity will provide an edge to know about the integrity of the package and actual status regarding the quality and safety of the food items in the package. Zhou (2013) developed a humidity indicator using iridescent technology, where a nanocrystalline cellulose film was made by casting to form a thick iridescent film and the color of these films was adjusted so that they can interact with the electromagnetic field and such films can be explored as a humidity indicator (HI). The dry film appears to be bluish-green in color whereas their color shifted to reddish-orange when exposed to high humidity.

5.2.2 Active Packaging and Nanomaterials

Active packaging assures preservation based on mechanisms activated by intrinsic and/or extrinsic factors. Unlike conventional food packaging, an active packaging is an intentionally designed packaging system that incorporates components that would release (antimicrobial or antioxidant agents) or absorb (oxygen or water vapor) material into or from the packaged food or the food environment. Amalgamation of active compounds, viz., antimicrobial agents, preservatives, O_2, and water vapor absorbers, ethylene removers, etc. with polymer renders it more effective for improving the shelf-life and quality of the food product (Prasad et al., 2017). Damm et al. (2008) evaluated the effectiveness of polyamide 6 and Ag micro and NPs incorporated films against *E. coli*. The developed NPs-based film completely inhibited the *E. coli* growth. However, films with Ag microparticles killed 80% of the bacteria. Rhim et al. (2014) reported noteworthy antimicrobial activity of AgNPs incorporated agar films against *Listeria monocytogenes* and *E. coli* O157: H7. Busolo et al. (2010) documented strong antibacterial activity of PLA/silver-OMMT nanocomposite against *Salmonella* spp. Antibacterial activity of nanocomposite systems such as PVA dispersed with cellulose nanocrystals and AgNPs has been displayed against *E. coli* and *Staphylococcus aureus* (Sadeghnejad et al., 2014).

Fayaz et al. (2009) studied biosynthesized AgNPs incorporated into sodium alginate films in food packaging and showed a noteworthy antibacterial effect against *E. coli* and *S. aureus*. Sanchez-Valdes et al. (2009) reported AgNPs deposited over multilayered linear low-density polyethylene (LLDPE) showed 70% reduction of *Aspergillus niger*. Hasim et al. (2015) observed the UV/ozone treated commercial low-density polyethylene (LDPE) films coated the layer-by-layer (LbL) method by alternating the deposition of polyethyleneimine (PEI) and poly (acrylic acid) (PAA) polymer solutions and AgNPs could be explored in antimicrobial packaging. De Moura et al. (2012) reported that the size of NPs also affects the antimicrobial efficacy as they evaluated the antibacterial properties of hydroxypropyl methylcellulose films containing AgNPs with diameters of 41 and 100 nm against *E. coli* and *S. aureus* and observed that AgNPs of a smaller size (i.e., 41 nm) had greater antibacterial properties than the larger one.

5.2.2.1 Nanocomposite and Antimicrobial Packaging

Nanocomposites, a fusion of traditional food packaging material with NPs are gaining active interest in the food packaging sector. Nanocomposites are a multiphase material resulting from the amalgamation of matrix (continuous phase) and a nano-dimensional material (discontinuous phase). Till date, many NPs have been identified as fillers for making polymer nanocomposites to improve their packaging performances. Among them, clays and silicates have attracted significant attention due to their layered structures. This is because they are abundant, inexpensive, easy to process, and provide considerable enhancements (Azeredo et al., 2009). There are three main polymer-clay morphologies, i.e., tactoid (or phase separated), intercalated, and exfoliated (McGlashan and Halley, 2003). In the tactoid structure, which usually occurs in microcomposites, the polymer chains and the clay gallery are immiscible because they have poor affinity for each other. Nanocomposite structures do not display this morphology (Alexandre and Dubois, 2000). In ideal polymer-clay nanocomposites, high affinity would exist between the polymer and clay, leading to exfoliated structures in which the polymer chains penetrate the interlayer space of the clay, making single sheets. If the clay shows a moderate affinity for the polymer, the results would be intercalated structures (Arora and Padua, 2010).

There are other particle fillers being used, including Ag, ZnO, TiO_2, carbon nanotubes, graphene nanoplates, copper, and copper oxides. It is reported that graphene nanoplates (GNPs) can form nanocomposites with improved heat resistance and barrier properties, making them a great option for food-packaging applications (Ramanathan et al., 2008). CNTs are another type of carbon-based NPs that have good electrical and mechanical characteristics. However, their use has been hindered, mainly due to their high cost and difficulty in processing dispersions (Arora and Padua, 2010). Copper, copper oxide, TiO_2, ZnO, and Ag have been used mainly for their antimicrobial properties (Duncan, 2011). In addition to its remarkable antimicrobial spectrum, it displays great mechanical performance and tough resistant characteristics (Montazer and Harifi, 2017). Based on the nanomaterial, the nanodimensional phase is generally characterized into nanospheres or NPs, nanowhiskers or nanorods, nanotubes and nanosheets, or nanoplatelets (Bratovcic et al., 2015). Nano-sized phases augment the mechanical properties of polymer, where the elastic strain is transferred to nano-reinforced material. Owing to this property, nanocomposite has been recognized as a gold standard for improvising the mechanical and barrier characteristics of polymers (Othman, 2014). Besides improving the mechanical and barrier characteristics, NPs also append active or smart properties to the packaging system.

The application of nanotechnology in polymer science can open new avenues for improving the characteristic features and cost-price-competence of packaging materials (de Azeredo et al., 2011). Polymer nanocomposites (PNCs), the mixtures of polymers with inorganic or organic fillers with geometries (fibers, flakes, spheres, particulates) have been recently introduced as novel packaging materials (Prateek Thakur and Gupta, 2016). The aspect ratio (ratio of the largest to the smallest dimension) of packaging filler material plays a significant role. Fillers having higher aspect ratios possess more specific surface area, with associated high reinforcing properties (Rafieian and Simonsen, 2014). Various nanomaterials such as silica (Bracho et al., 2012), clay, organo-clay (Ham et al., 2013), graphene (Lee et al., 2013), polysaccharide nanocrystals (Lin et al., 2012), carbon nanotubes (Swain et al., 2013), chitosan (Chang et al., 2010), cellulose-based (Sandquist, 2013), and other metal NPs, such as, ZnO_2 (Esthappan et al., 2013), colloidal Cu (Cardenas et al., 2009), or Ti (Li et al., 2011) are being extensively explored as fillers. Due to nanostructure, NPs are good barriers for diffusion gases such as oxygen and CO_2 and can thus be exploited in food packaging. The incorporation of NPs in packaging will slow down some biochemical processes such as oxidation, degradation, etc. It will help to extend the shelf-life of food products. In the food packaging industry, the most used materials are plastic polymers that can be incorporated or coated with nanomaterials for improved mechanical or functional properties. Moreover, nanocoatings on food contact surfaces act as barriers or antimicrobial properties.

Application of nanocomposites as an active material for packaging is used to improve food packaging. Many researchers have studied the antimicrobial properties of organic compounds like essential oils, organic acids, and bacteriocins (Schirmer et al., 2009) and their use in polymeric matrices as antimicrobial packaging. However, these compounds do not fit into many food-processing steps which require high temperature and pressure as they are highly sensitive to these physical conditions. Using inorganic NPs, a strong antibacterial activity can be achieved in low concentrations and more stability in extreme conditions. Therefore, in recent years, it has been a great interest for using these NPs in antimicrobial food packaging, which is a form of active packaging which is in contact with the food product or the headspace inside to inhibit or retard the microbial growth that may be present on food surfaces (Soares et al., 2009). The term "active nanocomposite" generally refers to a plastic composite (i.e., a polymer blend) that contains an active, nanostructured material that confers an activity on the plastic matrix (Yam and Lee, 2012). At least one of the dimensions of the active nano-structured material must be less than 100 nm in size. Nanofillers, such as Ag, ZnO, and MgO, have antimicrobial or antioxidant activities. Incorporation of these nanofillers in polymer or biopolymer matrices leads to an inhibiting or retarding effect on the growth of microorganisms, thereby reducing food spoilage (Huang et al., 2015). The main goal of active packaging systems is extending the product's shelf life. They can also be designed to improve food quality and safety and finally result in less food waste (Cushen et al., 2012). Montmorillonite, kaolinite clays, and graphene nanoplates are highly promising as nancomposites (Arora and Padua, 2010).

Silver zeolites are also used to create antibacterial polymer composites. AgNP-based nanocomposites are stable and offer slow release of Ag ion into stored foods, resulting in persistent antimicrobial activity. In a study where $AgNP/SiO_2$ nanocomposite material was compared with that of Ag zeolite and $AgNO_3/SiO_2$ composite, effective antimicrobial activity was displayed by both the materials; however, a longer period of activity was displayed by the nancomposite, while a better immediate effect was observed with zeolite-based material. Many anti-microbial nanocomposites used for food packaging are made from Ag, which has an intense toxicity to a large variety of microorganisms (Liau et al., 1997). Different mechanisms have been suggested for the anti-microbial activity of AgNPs, such as increasing cell permeability through the attachment to the cell surface and making pits in the membranes, damaging DNA followed by NP penetration inside the bacterial cell, and the release of Ag^+. Nanocomposite-based AgNPs may find application in food packages requiring longer transportation or storage. To explore the antimicrobial potential of AgNP on the shelf-life of food, various AgNP/polymer nanocomposite materials have been inspected within actual food systems. Fernandez et al. (2010) observed that AgNPs containing cellulose pads reduce the levels of microbial exudates of meat stored in modified atmosphere packaging. Further, cellulose pads containing AgNPs have also been successfully applied for beef coating, wherein significant reduction in the microbial load was reported (Smolkova et al., 2015).

NPs in precise have revealed broad-spectrum antibacterial properties against both Gram-positive and Gram-negative bacteria. ZnO NPs were found to inhibit *Staphylococcus aureus* (Liu et al., 2009) and AgNPs exhibit concentration-dependent antimicrobial activity against *Escherichia coli, Aeromonas hydrophila,* and *Klebsiella pneumoniae* (Aziz et al., 2016). The antimicrobial mechanism of NPs is typically considered as due to oxidative stress and cell damage, metal ion release, or non-oxidative mechanisms (Wang et al., 2017). These mechanisms can happen concurrently. Studies have suggested that AgNPs quickly neutralize the surface electric charge of the bacterial membrane and change its permeability, ultimately leading to apoptosis. Moreover, the generation of ROS prevents the antioxidant defense system and causes physiochemical damage to the intrinsic cell membrane. The major processes causing the antibacterial effects of NPs are as follows: disruption of the bacterial cell membrane, generation of ROS, penetration of bacterial cell membrane by passive or facilitated diffusion and induction of intracellular antibacterial effects, including interfaces with DNA replication, and inhibition of protein synthesis (Wang et al., 2017).

5.2.2.2 Antioxidant Packaging

Various polymers having nanostructured oxygen scavengers can be explored as active packaging material for sliced meat, poultry, fish, etc. Xiao et al. (2004) developed O_2 scavenger films and incorporated TiO_2 NPs to different polymers and this nanocomposite material was further explored as packaging films in O_2 sensitive food stuffs. Later, Busolo and Lagaron (2012) modified the high-density polyethylene (HDPE) films by integrating them with iron containing kaolinite to produce O_2 scavenging packaging films.

5.2.3 Biodegradable Packaging and Nanomaterials

Biopolymers can be derived from plant materials as well as animal and microbial products, such as polyhydroxybutyrate. Among these biopolymers, starch, cellulose, and their derivatives, as well as proteins, have been used extensively by many researchers to make biobased nanocomposites. The most reported biodegradable nanocomposite is starch clay, which has been investigated for several applications including food packaging. Plantic Technologies Ltd, Altona, Australia has manufactured and are selling biodegradable and fully compostable bioplastic packaging materials. This is constructed from organic corn starch using nanotechnology (Neethirajan and Jayas, 2011). Biodegradable bionanocomposites prepared from natural biopolymers such as starch and protein exhibit advantages as a food packaging material by providing enhanced organoleptic characteristics such as appearance, odour, and flavor.

Kriegel et al. (2009) have developed a methodology which uses an electrospinning technique to make biodegradable 'green' food packaging from chitin. Chitin is a natural polymer and one of the main components of lobster shells. The electrospinning technique used involves dissolving chitin in a solvent and drawing it through a tiny hole with applied electricity to produce nanoslim fibre spins. These strong and naturally antimicrobial nanofibers have been used for developing the 'green' food packaging. Many companies are creating a competitive advantage by producing food packaging bags and sachets from biodegradable polylactic acid and polycaprolactone obtained from the polymer nanocomposites of the corn plant (Bordes et al., 2009). Bionanocomposites open an opportunity for the use of new, high performance, light weight green nanocomposite materials making them to replace conventional non-biodegradable petroleum based plastic packaging materials. So far, the most studied bionanocomposites suitable for packaging applications are starch and cellulose derivatives, polylactic acid (PLA), polycaprolactone (PCL), poly (butylene succinate) (PBS) and polyhydroxybutyrate (PHB). The most promising nanoscale fillers are layered silicate nanoclays such as montmorillonite and kaolinite.

5.2.3.1 Nanofibers

Electro spinning technology has resulted in production of 100 nanometer diameter fibers that can be used as pesticide absorbents. Nanofiber-based fabrics are being used as a detection technology platform to capture and isolate pathogens. The nanofibers in this fabric are embedded with antibodies against specific pathogens. The fabric can be wiped across a surface and tested to determine whether the pathogens are present, perhaps indicating their presence by a change in color (Hager, 2011). Azeredo et al. (2010) described the use of cellulose nanofibers and glycerol as a plasticizer to improve the mechanical and water-vapor barrier properties of edible chitosan films. They reported that a nanocomposite film with 15% of cellulose nanofibers and plasticized with 18% glycerol was not only comparable in strength and stiffness to some synthetic polymers but was also extremely environmentally friendly.

5.3 Traceability of Animal Products through Labelling

NP based intelligent inks or reactive nanolayers provide smart recognition of relevant food product. Printed labels in the food package can indicate the following highlights: temperature, time, pathogens, freshness, humidity, etc. Nanobarcode particles with different patterns of gold template and silver stripes are available already. Nanobarcodes is also used to monitor the quality of animal produce. Scientists at Cornell University used the concept of grocery barcodes for cheap, efficient, rapid, and easy decoding and detection of diseases. They produced microscopic probes or nanobarcodes that could tag multiple pathogens in a farm which can easily be detected using any fluorescent-based equipment (Li et al., 2005). Oxonica in the United Kingdom offers solutions for food product identification and brand authenticity whereby the nanobarcodes become a biological fingerprint created by NPs which generate unique reading strips for every food item (Neethirajan and Jayas, 2011). To allow better information delivery in tracking and tracing, some nano-based products may be able to encrypt information technology in the form of nanodisks functionalized with dye molecules to emit a unique light spectrum when illuminated with a laser beam, so that they can be used as tags for tracking food products (Nam et al., 2003). Dip Pen Nanolithography involves using a scanning probe with a molecule-coated tip to deposit a chemically engineered ink material to create nanolithographic patterns on the food surface (Zhang et al., 2009).

5.4 Nanobiosensors of Quality

One of the major roles for nanotechnology-enabled devices will be the increased use of autonomous sensors linked into a GPS system for real-time monitoring. These nanosensors could be distributed throughout the field where they can monitor livestock growth. Ultimately, precision livestock farming, with the help of smart sensors, will allow enhanced productivity in animal husbandry by providing accurate information, thus helping farmers to make better decisions. Recently, some packaging materials incorporated with nanosensors to detect the oxidation process in food have been produced and used in the food industry. Working scheme is quite simple: when the oxidation occurs in the food package, NP-based sensors indicate the color change and information about the nature of the packed food is revealed. This technology has been successfully applied in package of milk and meat. Gold NPs functionalized with cyanuric acid groups selectively bind to melamine, an adulterant used to artificially inflate the measured protein content of pet foods and infant formulas (Ai et al., 2009). A promising photoactivated indicator ink for in-package oxygen detection based upon nanosized TiO_2 or SnO_2 particles and a redox-dye (methylene blue) has been developed (Mills, 2005). Nanosensors based on NPs have also been developed to detect the presence of moisture content inside a food packaging (Luechinger et al., 2007).

5.5 Nanobiosensors for Safety

Nanosensors can be used to prove the presence of contaminants, mycotoxins, and microorganisms in food. A biosensor is composed of a biological component, such as a cell, an enzyme, or an antibody, linked to a tiny transducer, a device powered by one system that then supplies power (usually in another form) to a second system. The biosensors detect changes in cells and molecules, measure and identify the test substance, even if there is a very low concentration of the tested material. When the substance binds with the biological component, the transducer produces a signal proportional to the quantity of the substance. So, if there is a large concentration of bacteria in a particular food, the biosensor will produce a strong signal indicating that the food is unsafe to eat. With this technology, mass amounts of food can be readily checked for their safety of consumption (Enisa Omanović-Mikličanin and Mirjana Maksimović, 2016). Smart sensors, which are obtained by nanotechnology, are the powerful tools to track, detect, and control animal pathogens. Detection of very small amounts of a chemical contaminant, virus, or bacteria in food systems is envisioned from the integration of chemical, physical, and biological devices working together as an integrated sensor at the nanoscale. The bioanalytical nano-sensors either use biology as a part of the sensor or are used for biological samples.

Nanomaterials for use in the construction of biosensors offer a high level of sensitivity and other novel attributes. In food microbiology, nanosensors or nanobiosensors are used for the detection of pathogens in food processing, quantification of food constituents, and alerting consumers and distributors on the safety status of food. The nanosensor works as an indicator that responds to changes in environmental conditions such as humidity or temperature in storage rooms, microbial contamination, or product degradation. Various nanostructures like thin films, nanorods, NPs and nanofibers have been examined to their possible application as biosensors (Jianrong et al., 2004). Thin film-based optical immunosensors for the screening of microbial substances or cells have led to the rapid and highly sensitive detection systems. In case of immunosensors, specific antibodies, antigens or protein molecules are immobilized on thin nanofilms or sensor chips which emit signals on detection of target molecules (Subramanian, 2006). A dimethylsiloxane microfluidic immunosensor integrated with specific antibody immobilized on an alumina nanoporous membrane was developed for rapid detection of foodborne pathogens *E. coli* O157: H7 and *S. aureus* with electrochemical impedance spectrum (Tan et al., 2011).

Biosensors based on carbon nanotubes have gained much attention due to their rapid detection, simplicity, and cost-effectiveness. The biosensors are successfully applied for the detection of microorganisms, toxins, and other degraded products in foods and beverages. Further, the use of electronic tongue or nose which consists of the array of nanosensors to monitor the food condition by giving signals on aroma or gases released from food has also been developed. The quartz crystal microbalance (QCM) based electric nose can detect the interaction between various odorants and chemicals that have been coated on its surface. Many studies on small molecule detection have used quartz crystal surfaces that have been modified with different functional groups or biological molecules, such as amines, enzymes, lipids, and various polymers (Kanazawa and Cho, 2009). In another development, a direct charge transfer (DCT) biosensor has been created that uses antibodies as sensing elements and polyaniline nanowire as a molecular electrical transducer (Pal et al., 2007). The resulting biosensor could be used for the detection of *Bacillus cereus*. Aptasensors are biosensors consisting of aptamers (the target-recognition element) and nanomaterial (the signal transducers and/or signal enhancers). Aptamers are single stranded nucleic acid or peptide molecules of size less than 25 kDa with natural or synthetic origin. They are highly specific and selective towards their target compound (ions, proteins, toxins, microbes, viruses) due to their precise and well-defined three-dimensional structures. Aptamers are named as synthetic antibodies. There are a wide variety of nanomaterials, which can be used in aptasensors (metal NPs, nanoclusters, semiconductor NPs, carbon NPs, magnetic NPs, etc.) (Sharma et al., 2015).

6. Safety of Nanomaterials for Food Application

The present century has observed a quick development of nanotechnology and its impact in every field; therefore, for the sake of consumers, it is mandatory to have comprehensive information regarding the interface between NPs, cells, tissues, and organisms, particularly in relation to possible hazards to human health (Zohreh Honarvar et al., 2016). NPs may enter the body through inhalation, ingestion, or cutaneous exposure (Maisanaba et al., 2015). Once they penetrate the biological environment, the NP will inevitably encounter a huge variety of biomolecules (proteins, sugars, and lipids) which are dissolved in body fluids such as the interstitial fluid between cells, lymph, or blood (Farhoodi, 2016). Studies on titania and AgNPs revealed that these materials may enter the blood circulation and their insolubility leads to accumulation in organs (Rhim et al., 2013). Liver and spleen are mainly responsible for distribution of NPs, mediating their passage from the intestine to the blood circulation (Dimitrijevica et al., 2015). NPs may unintentionally come in GIT contact via leaching/migration from nanopackaging to food commodities (He and Hwang, 2016). Migration leads to transfer of low molecular mass constituents of packaging material to packaged product.

This unintended transfer of undesirable packaging constituents raises the safety concerns for the consumers. The migration of NPs to food matrix mainly relies on the chemical and physical properties of food and the polymer involved. Other controlling parameters include the concentration, particle size, molecular weight, solubility, and diffusivity of a specific substance in the polymer, pH value, temperature, polymer structure, polymer viscosity, mechanical stress, contact time, and composition of food. The solubility of metallic NP in an aqueous solution is directly proportional to temperature and inversely proportional to the pH which ultimately augments the migration of metal into the food matrix (Huang et al., 2015). Besides the route of entrance into the body, concentration and duration of exposure, toxicity of NPs also depend upon the host susceptibility and state of organism (Sharma et al., 2009). Aschberger et al. (2011) studied the oral route of transmission and observed that signs of toxicity with relatively high doses of nano-Ag or nano-TiO_2.

Few reports point out toward the genotoxicity and carcinogenicity of NPs. ZnO NPs displayed genotoxicity in human epidermal cells, even if the bulk ZnO is non-toxic, implying the role of particle diameter (Sharma et al., 2009). Earlier, Chithrani et al. (2006) reported that the smaller NPs exhibit more toxicity than larger ones. High surface area comparative to total mass of smaller NPs enhances their prospects of interaction with the biological molecules, leading to adverse responses. Other factors, such as surface functionalization also plays vital role (Li Shang et al., 2014). Out of few reports, inhalation of very high doses (10 mg/m^3) of nano-TiO_2 has been associated with incidence of lung tumors. Inhalation and skin exposure routes are much more explored as compared to ingestion as a route for the entry of NPs. Inhaled MgO NPs can make their way to the olfactory bundle under the forebrain via the axons of olfactory nerve in the nose and can also travel to other parts of the brain through systemic inhalation. There are reports documenting the penetration of latex nanoparticles, smaller than 1 μm, through the outer layers of a skin, during constant flexing.

Furthermore, inhalation of nanomaterials and probability of their entrance through skin penetration is a matter of high concern, especially for workers and consumers in direct and regular contact (Youssef, 2013). Available data shows that the circulation time increases considerably in case the NPs are hydrophilic and positively charged. Love et al. (2012) also observed that cationic NP seems to be quite toxic than neutral or anionic ones, it might be due to their high affinity toward the negatively charged plasma membrane. Nel et al. (2009) also reported that a cationic NP leads to lysosomal damage and induced cytotoxicity. NPs that find their way to the blood stream may influence the blood vessel lining and their function, may lead to blood clotting, or may even contribute toward cardiovascular diseases. Hence, it becomes imperative to obtain data about influence/impact of NPs over blood vessels and chances of their crossing the blood brain barrier and migration to the fetus (Dimitrijevica et al., 2015).

There are ambiguous results leading to a vague situation regarding the toxicity of NPs to human beings. Some reports indicate total metal migration, over others showing particle migration. An imperative conclusion from the above discussion is that if NPs are completely covered or encapsulated by the host polymer matrix, then the probability of migration into food matrix is quite less. However, during an unintentional mechanical impact on the food contact surface, the smooth properties could get altered, else in case of cut edges or technically improperly manufactured polymer the nanocomposite may lead to its release. In lieu of the above, manufacturers should assure proper incorporation of NPs in films or molded articles (Stormer et al., 2017). Nevertheless, the available scientific data on toxicity or migration of NPs is still at an infant stage and additional meticulous analysis is required before their vast application. The fate and toxicity of nanomaterials in food packaging depend on the physiochemical characteristics and dosage. Avella et al. (2005) showed that tiny amounts of particle migration from nanocomposites to foods have been seen during packaging of food and this migration was within the limits prescribed by the European Commission (EC) for silica NPs in clay nanocomposites.

Besides these concerns, application of some nanocomposites triggered concerns regarding their environmental impacts, due to their non-biodegradable nature. Therefore, eco-toxicity studies on NPs are mandatory before their commercial applications. Overall, the existing reports regarding the toxicity of NPs are not in full agreement with each other and inconclusive. Safe application of nanotechnology to the food packaging requires systematic characterization and assessment *in silico*, *in vitro*, and *in vivo*. Altogether, taking into consideration, a varied number of physical, chemical, and biological factors, their absorption, distribution, metabolism, excretion and lastly their toxicity should be quantified and evaluated for risk assessment to consumers (He and Hwang, 2016). Overall data available concludes that more research is warranted before NPs may be tagged as toxic or safe. The forthcoming section briefly discusses the toxicity of some common NPs, viz., SiO_2, TiO_2, ZnO, and Ag.

6.1 Silicon Oxide Nanoparticles

Silica NPs may be toxic through oxidative stress generation, leading to DNA damage and induction of apoptosis. Different scientific reports support this hypothesis in intestinal epithelial cell line models. According to Tarantini et al. (2015), silica NPs induced oxidative stress is probably responsible for induction of apoptosis and DNA damage. They exposed Caco-2 cells (Colorectal adenocarcinoma cells) to silica NP (15 nm) and observed decreased cell viability and increased ROS production at 32 μg/ml dosage with overexpression of caspase-3 at 64 μg/ml. Silica NPs were genotoxic to cells and augmented the frequency of micronucleus formation (Tarantini et al., 2015). As reported by Yang et al. (2014), 10–50 nm silica NPs induced LDH release, indicating reduced cell viability, partial inhibition of cell proliferation and a slight S phase cell cycle arrest in human gastric epithelial cells (GES-1) cells, and S and G2/M phase arrest in Caco-2 cells with minor increase in ROS generation. Overall, they didn't report any apoptosis or necrosis, indicated that the ROS generation may accompany decreased viability and cell cycle arrest.

In contrast to the above reports, Moos et al. (2011) reported zero toxicity of silica NPs in intestinal epithelial cells. Caco-2 and colon carcinoma RKO cells showed minimal toxicity induction by silica NPs up to the dosage of 100 μg/cm^2. Cells exposed to 50 μg/cm^2 silica NPs for 4 hrs also showed minimal changes in gene expression as determined by whole genome microarray analysis. These results are further supported by Schubbe et al. (2012), who also observed no cellular toxicity in undifferentiated Caco-2 cells when co-cultured with 32 and 83 nm fluorescent silica NPs. No cytotoxicity or genotoxicity was observed in cells treated up to 200 μg/ml of NPs. It has also been found that the SiO_2 NPs can also induce allergen-specific Th2-type allergic immune responses, as observed in an *in vivo* study involving female BALB/c mice exposed to NPs. Intranasal exposure to ovalbumin (OVA) and SiO_2 NPs induces a relatively high level of OVA-specific immunoglobulin IgE, IgG, and IgG1 antibodies (Yoshida et al., 2011). From the above reports, it can be concluded that the harmful effects of silica NPs are associated with high dosage (i.e., 200 μg/ml) however, the same at a dose of 100 μg/ml are reported safe.

6.2 *Titanium Dioxide Nanoparticles*

TiO_2 NPs have been associated with cytotoxicity mediated through oxidative stress-dependent pathways leading to DNA damage, cell cycle arrest, or delay and mitochondrial dysfunction, particularly in pulmonary and inhalation models (Shi et al., 2013). However, experimental data observed in intestinal epithelial cells agree that TiO_2 NPs are nontoxic. Koeneman et al. (2010) observed that the treatment of Caco-2 cells with TiO_2 NPs (< 40 nm) displayed a decrease in epithelial monolayer integrity by decreased TEER (Trans Epithelial Electrical Resistance) measurements and a loss of localization of g-catenin to cell adherens junctions, beginning at 6 days after continuous TiO_2 NP treatment and long-lasting to 10 days at dose of 1,000 μg/ml. No decrease in TEER was examined after acute exposure and no induction of cell death was reported after acute or chronic exposure. Chalew and Schwab (2013) observed the toxicity of P25 TiO_2 (25% rutile and 75% anatase) treatment (100 μg/ml) on the intestinal epithelial cell lines, Caco-2 and SW480. No toxicity in Caco-2 cells was observed. Few studies had supported that TiO_2 NP can disrupt normal microvilli structure in intestinal epithelial cells, which affects the normal cellular functions, particularly nutrient absorption (McCracken et al., 2016).

6.3 *Zinc Oxide Nanoparticles*

As reported earlier, ZnO NPs may lead to toxicity due to the NP dissolution either in outside or within the cells, leading to enhanced availability of zinc ions, which interacts with enzymes and other cell components; oxidative stress and lysosomal destabilization; and mitochondrial dysfunction contributing to the cytotoxic response (Vandebriel and De Jong, 2012). Song et al. (2014) observed that 90 nm ZnO NPs at a concentration of 10 μg/ml decreases Caco-2 cell viability, inhibits cell proliferation, enhance ROS generation along with SOD (Super Oxide Dismutase) levels; signifying an oxidative stress response. In another study, MTT and LDH assays revealed the dose-dependent toxicity of ZnO NPs in Caco-2 cells (Kang et al., 2013).

6.4 *Silver Nanoparticles*

AgNPs are genotoxic, cytotoxic, and even carcinogenic. The nano size of NPs allows them to cross the cellular barrier, leading to the formation of free radicals in the tissues and eventually leading to oxidative damage to the cells and tissues (Pradhan et al.,

2015). In several *in vitro* studies, AgNPs displayed toxicity through an oxidative stress-dependent mechanism as well as through oxidative stress-independent intracellular effects. It has been observed that the exposure of human lung fibroblasts and glioblastoma cells to 6–20 nm AgNPs increased ROS production, induced mitochondrial injury, DNA damage, and induced G2/M phase cell cycle arrest (Rani et al., 2009). Similar toxicity has been observed in intestinal epithelial cells (Bohmert et al., 2012). In another study, Aueviriyavit et al. (2014) observed that < 100 nm AgNPs can be internalized by cells with a dose-dependent decrease in cell viability, starting at 10 μg/ml. Treatment with AgNPs also induced activation of the stress-responsive gene Nrf2 and heme oxygenase-1 (HO-1). Recently, McCracken et al. (2016) also demonstrated that AgNP-induced oxidative stress responses in intestinal epithelial cells. However, Song et al. (2014) reported that co-incubation of Caco-2 cells with 10 μg/ml of 90 nm AgNPs decreased the cell activity, but no induction of cell death was observed. In another study, Kumar (2015) reported that AgNPs caused depolarization of a-tubulin, a major component of microtubule, having adverse effect over the cellular structure and associated cytoskeleton of the cell. Other nanomaterials, such as carbon NPs are also known to cause allergic inflammation and it has been reported that the single and multiwalled carbon nanotubes increased lung inflammation and allergen specific IgE levels in mice sensitized to Ova egg allergen. In another study, multi-walled carbon nanotubes with preexisting inflammation increased airway fibrosis in mice with allergic asthma (He and Hwang, 2016).

7. Conclusion

The development in nanotechnology has a high potential to benefit the livestock products industry. The development of nano-packaging systems and nanobiosensors are the emerging field that focus on food safety and security which will grow exponentially in the years to come. The future of quality and safety of foods of animal origin depends largely on the technological advancement of nanosensors, its integration in food packaging, and generating pathbreaking innovative solutions for the livestock sector.

8. Future Perspectives

The future of nanomaterial for livestock products application is quite promising. Nanomaterial is expected to play a pivotal role in the quality and safety enhancement of animal products from farm to fork. The perspective areas of application include quality assurance, monitoring product safety, designer foods, preservation for enhanced shelf-life, and smart packaging of animal products.

References

Advantage Magazine. (2004). Nanotechnology and Food Packaging. Available online at: http://www.azonano.com/Details.asp?ArticleID=857.

Ai, K., Liu, Y. and Lu, L. (2009). Hydrogen bonding recognition-induced color change of gold nanoparticles for visual detection of melamine in raw milk and infant formula. *J. American Chem. Soc.*, 131(27): 9496–9497.

Alexandre, M. and Dubois, P. (2000). Polymer-layered silicate nanocomposites: preparation, properties, and uses of a new class of materials. *Materials Sci. Engineer.: R: Reports*, 28(1): 1–63.

Arora, A. and Padua, G.W. (2010). Review: Nanocomposites in food packaging. *J. Food Sci.*, 75(1): 43–9.

Aschberger, K., Micheletti, C., Sokull-Klüttgen, B. and Christensen, F.M. (2011). Analysis of currently available data for characterising the risk of engineered nanomaterials to the environment and human health lessons learned from four case studies. *Environ. Int.*, 37: 1143–1156.

Aueviriyavit, S., Phummiratch, D. and Maniratanachote, R. (2014). Mechanistic study on the biological effects of silver and gold nanoparticles in Caco-2 cells:I Induction of the Nrf2/HO-1 pathway by high concentrations of silver nanoparticles. *Toxicol. Lett.*, 224: 73–83.

Avella, M., De Vlieger, J.J., Errico, M.E., Fischer, S., Vacca, P. and Volpe, M.G. (2005). Biodegradable starch/clay nanocomposite films for food packaging applications. *Food Chem.*, 93: 467–474.

Azeredo, H., Mattoso, L.H.C., Wood, D., Williams, T.G., Avena–Bustillos, R.J. and McHugh, T.H. (2009). Nanocomposite edible films from mango puree reinforced with cellulose nanofibers. *J. Food Sci.*, 74(5): 31–35.

Azeredo, H.M., Mattoso, L.H., Avena-Bustillos, R.J., Filho, G.C., Munford, M.L., Wood, D. and McHugh, T.H. (2010). Nanocellulose reinforced chitosan composite films as affected by nanofiller loading and plasticizer content. *J. Food Sci.*, 75(1): N1–7.

Aziz, N., Faraz, M., Pandey, R., Sakir, M., Fatma, T. and Varma, A. (2015). Facile algae-derived route to biogenic silver nanoparticles: Synthesis, antibacterial and photocatalytic properties. *Langmuir*, 31: 11605–11612.

Bakalova, R., Zhelev, Z., Ohba, H., Ishikawa, M. and Baba, Y. (2004). Quantum dots as photosensitizers? *Nat. Biotechnol.*, 22: 1360–1361.

Berekaa, M.M. (2015). Nanotechnology in food industry: Advances in food processing, packaging, and food safety. *Int. J. Curr. Microbiol. App. Sci.*, 4: 345–357.

Bohmert, L., Niemann, B., Thunemann, A.F. and Lampen, A. (2012). Cytotoxicity of peptide-coated silver nanoparticles on the human intestinal cell line Caco-2. *Arch. Toxicol.*, 86: 1107–1115.

Bordes, P., Pollet, E. and Averou, L. (2009). Nano-biocomposites: Biodegradable polyester/nanoclay systems. *Progress in Polymer Sci.*, 34(2): 125–155.

Bracho, D., Dougnac, V.N., Palza, H. and Quijada, R. (2012). Fictionalization of silica nanoparticles for polypropylene nanocomposite applications. *J. Nanomater.*, 263915.

Brandelli, A., Brum, L.F.W. and dos Santos, J.H.Z. (2017). Nanostructured bioactive compounds for ecological food packaging. *Environ. Chem. Lett.*, 15: 193–204.

Bratovcic, A., Odobašić, A., Ćatić S. and Šestan, I. (2015). Application of polymer nanocomposite materials in food packaging. *Croat. J. Food Sci. Technol.*, 7: 86–94.

Brown, E.M., Paunesku, T., Wu, A., Thurn, K.T., Haley, B., Clark, J., Priester, T. et al. (2008). Methods for assessing DNA hybridization of peptide nucleic acid–titanium dioxide nanoconjugates. *Anal. Biochem.*, 383: 226–235.

Bumbudsanpharoke, N. and Ko, S. (2015). Nano-food packaging: An overview of market, migration research, and safety regulations. *J. Food Sci.*, 80: R910–R923.

Busolo, M.A., Fernandez, P., Ocio, M.J. and Lagaron, J.M. (2010). Novel silver-based nanoclay as an antimicrobial in polylactic acid food packaging coatings. *Food Addit. Contam.*, 27: 1617–1626.

Busolo, M.A. and Lagaron, J.M. (2012). Oxygen scavenging polyolefin nanocomposite films containing an iron modified kaolinite of interest in active food packaging applications. *Innov. Food Sci. Emerg. Technol.*, 16: 211–217.

Cardenas, G., Díaz, J., Meléndrez, M., Cruzat, C. and Cancino, A.G. (2009). Colloidal Cu nanoparticles/chitosan composite film obtained by microwave heating for food package applications. *Polym. Bull.*, 62: 511–524.

Cavaliere, E., De Cesari, S., Landini, G., Riccobono, E. and Pallecchi, L. (2015). Highly bactericidal Ag nanoparticle films obtained by cluster beam deposition. *Nanomedicine: Nanotechnolo. Biol. Med.*, 11: 1417–1423.

Cerrada, M.L., Serrano, C., Sánchez-Chaves, M., Fernández-García, M., Fernández-Martín F. and de Andrés, A. (2008). Self-sterilized EVOH-TiO_2 nanocomposites: effect of TiO_2 content on biocidal properties. *Adv. Funct. Mater.*, 18: 1949–1960.

Chalew, T.E.A. and Schwab, K.J. (2013). Toxicity of commercially available engineered nanoparticles to Caco-2 and SW480 human intestinal epithelial cells. *Cell Biol. Toxicol.*, 29: 101–116.

Chang, P.R., Jian, R., Yu, J. and Ma, X. (2010). Starch-based composites reinforced with novel chitin nanoparticles. *Carbohydr. Polym.*, 80: 420–425.

Chatterjee, A.K., Chakraborty, R. and Basu, T. (2014). Mechanism of antibacterial activity of copper nanoparticles. *Nanotechnol.*, 25: 135101.

Chen, X. and Mao, S.S. (2006). Synthesis of titanium dioxide (TiO_2) nanomaterials. *J. Nanosci. Nanotechnol.*, 6: 906–925.

Chen, W., Wang, X., Tao, Q., Wang, J., Zheng, Z. and Wang, X. (2013). Lotus-like paper/paperboard packaging prepared with nanomodified overprint varnish. *Appl. Surf. Sci.*, 266: 319–325.

Chithrani, B.D., Ghazani, A.A. and Chan, W.C. (2006). Determining the size and shape dependence of gold nanoparticle uptake into mammalian cells. *Nano Lett.*, 6: 662–668.

Cioffi, N., Torsi, L., Ditaranto, N., Tantillo, G., Ghibelli, L. and Sabbatini, L. et al. (2005). Copper nanoparticle/polymer composites with antifungal and bacteriostatic properties. *Chem. Mater.*, 17: 5255–5262.

Cushen, M., Kerry, J., Morris, M., Cruz-Romero, M. and Cummins, E. (2012). Nanotechnologies in the food industry: Recent developments, risks, and regulation. *Trends Food Sci. Technol.*, 24(1): 30–46.

Cyras, V.P., Manfredi, L.B., Ton-That, M.T. and Vazquez, A. (2008). Physical and mechanical properties of thermoplastic starch/montmorillonite nanocomposite films. *Carbohydr. Polym.*, 73: 55–63.

Damm, C., Munstedt, H. and Rosch, A. (2008). The antimicrobial efficacy of polyamide 6/silver-nano and microcomposites. *Mater. Chem. Phys.*, 108: 61–66.

Das, S., Wolfson, B.P., Tetard, L., Tharkur, J., Bazata, J. and Santra, S. (2015). Effect of N-acetyl cysteine coated CdS: Mn/ZnS quantum dots on seed germination and seedling growth of snow pea (*Pisum sativum* L.): Imaging and spectroscopic studies. *Environ. Sci.*, 2: 203–212.

de Azeredo, H.M.C., Mattoso, L.H.C. and McHugh, T.H. (2011). Nanocomposites in food packaging: A review. pp. 57–78. *In*: Reddy, B.S.R. (ed.). *Advances in Diverse Industrial Applications of Nanocomposites*. USA: InTech.

De Moura, M.R., Aouada, F.A., Avena-Bustillos, R.J., McHugh, T.H., Krochta, J.M. and Mattoso, L.H.C. (2008). Improved barrier and mechanical properties of novel hydroxypropyl methylcellulose edible films with chitosan/tripolyphosphate nanoparticlses. *J. Food Eng.*, 92: 448–453.

De Moura, M.R., Mattoso, L.H.C. and Zucolotto, V. (2012). Development of cellulose-based bactericidal nanocomposites containing silver nanoparticles and their use as active food packaging. *J. Food Eng.*, 109: 520–524.

Dias, V.M., Soares, N.F.F., Borges, S.V., de Sousa, M.M., Nunes, C.A., de Oliveira, I.R.N. and Medeiros, E.A.A. (2013). Use of allyl isothiocyanate and carbon nanotubes in an antimicrobial film to package shredded, cooked chicken meat. *Food Chem.*, 141(3): 3160–3166.

Dimitrijevica, M., Karabasila, N., Boskovica, M., Teodorovica, V., Vasileva, D. and Djordjevic, V. (2015). Safety aspects of nanotechnology applications in food packaging. *Procedia Food Sci.*, 5: 57–60.

Duncan, T.V. (2011). Applications of nanotechnology in food packaging and food safety: Barrier materials, antimicrobials, and sensors. *J. Colloid Interface Sci.*, 363: 1–24.

Emamifar, A., Kadivar, M., Shahedi, M. and Soleimanian-Zad, S. (2011). Evaluation of nanocomposites packaging containing Ag and ZnO on shelf-life of fresh orange juice. *Innov. Food Sci. Emerg. Technol.*, 11: 742–748.

Enisa Omanović-Mikličanin and Mirjana Maksimović. (2016). Nanosensors applications in agriculture and food industry. *Bulletin Chem. Technol. Bosnia and Herzegovina*, 47: 59–70.

Esthappan, S.K., Sinha, M.K., Katiyar, P., Srivastav, A. and Joseph, R. (2013). Polypropylene/zinc oxide nanocomposite fibers: morphology and thermal analysis. *J. Polym. Mater.*, 30: 79–89.

Farhoodi, M. (2016). Nanocomposite materials for food packaging applications: Characterization and safety evaluation. *Food Eng. Rev.*, 8: 35–51.

Fayaz, A.M., Balaji, K., Girilal, M., Kalaichelvan, P.T. and Venkatesan, R. (2009). Mycobased synthesis of silver nanoparticles and their incorporation into Sodium Alginate films for vegetable and fruit preservation. *J. Agric. Food Chem.*, 57: 6246–6252.

Fernandez, A., Picouet, P. and Lloret, E. (2010). Reduction of the spoilage-related microflora in absorbent pads by silver nanotechnology during modified atmosphere packaging of beef meat. *Food Prot.*, 73: 2263–2269.

Gardner, D.J., Oporto, G.S., Mills, R. and Azizi Samir, M.A.S. (2008). Adhesion and gel mineralisation of cellulose nanorod nematic suspensions. *J. Mater. Chem.*, 13: 696–699.

Ghaderi, M., Mousavi, M. and Labbafi, M. (2014). All-cellulose nanocomposite film made from bagasse cellulose nanofibers for food packaging application. *Carbohydr. Polym.*, 104: 59–65.

Gnach, A. and Bednarkiewicz, A. (2012). Lanthanide-doped upconverting nanoparticles: Merits and challenges. *Nano Tod.*, 7: 532–563.

Gopinath, S. and Sugunan, S. (2007). Enzymes immobilized on montmorillonite K 10: Effect of adsorption and grafting on the surface properties and the enzyme activity. *Appl. Clay Sci.*, 35: 67–75.

Gurunathan Kandeepan, Anne Seet Ram Anjaneyulu, Napa Kondaiah, Sanjod Kumar Mendiratta and Ramanathan Suresh. (2009). Comparison of quality and shelf life of buffalo meat patties stored at refrigeration temperature. *Int. J. Food Sci. Technol.*, 44(11): 2176–2182.

Gurunathan Kandeepan and Sudhir Sangma (2011). Comparison of quality characteristics of full fat and low-fat paneer developed from yak milk. *Int. J. Dairy Technol.* 64(1): 117–120.

Gurunathan Kandeepan, Anne Seet Ram Anjaneyulu, Napa Kondaiah and Sanjod Kumar Mendiratta. (2011). Comparison of quality attributes of buffalo meat curry at different storage temperature. *Acta Sci. Pol., Technol. Aliment.*, 10(1): 83–95.

Gutierrez, T.J., Ponce, A.G. and Alvarez, A.V. (2017). Nano-clays from natural and modified montmorillonite with and without added blueberry extract for active and intelligent food nanopackaging materials. *Mater. Chem. Phys.*, 194: 283–292.

Hager, H. (2011). Nanotechnology in agriculture. http://www.topcropmanager.com.

Ham, M., Kim, J.C. and Chang, J.H. (2013). Thermal property, morphology, optical transparency, and gas permeability of PVA/SPT nanocomposite films and equi-biaxial stretching films. *Polym. Korea*, 37: 579–586.

Haolong, L., Yang, Y., Yizhan, W., Chunyu, W., Wen, L. and Lixin, W. (2011). Self-assembly and ion-trapping properties of inorganic nanocapsule-surfactant hybrid spheres. *Soft Matter*, 7: 2668–2673.

Hasim, S., Cruz-Romero, M.C., Cummins, E., Kerry, J.P. and Morris, M.P. (2015). The potential use of a layer-by-layer strategy to develop LDPE antimicrobial films coated with silver nanoparticles for packaging applications. *J. Colloid Interface Sci.*, 461: 239–248.

He, X. and Hwang, H.M. (2016). Nanotechnology in food science: Functionality, applicability, and safety assessment. *J. Food Drug Anal.*, 24: 671–681.

Huang, J.Y., Li, X. and Zhou, W. (2015). Safety assessment of nanocomposite for food packaging application. *Trends Food Sci. Technol.*, 45: 187–199.

Inbaraj, B.S. and Chen, B.H. (2015). Nanomaterial-based sensors for detection of foodborne bacterial pathogens and toxins as well as pork adulteration in meat products. *J. Food Drug Anal.*, 24: 15–28.

Jia, X., Li, Y., Cheng, Q., Zhang, S. and Zhang, B. (2007). Preparation and properties of poly (vinyl alcohol)/silica nanocomposites derived from copolymerization of vinyl silica nanoparticles and vinyl acetate. *Eur. Polym. J.*, 43: 1123–1131.

Jianrong, C., Yuqing, M., Nongyue, H., Xiaohua, W. and Sijiao, L. (2004). Nanotechnology and biosensors. *Biotechnol. Adv.*, 22: 505–518.

Jin, T., Sun, D., Su, J.Y., Zhang, H. and Sue, H.J. (2009). Antimicrobial efficacy of zinc oxide quantum dots against *Listeria Monocytogenes*, *Salmonella Enteritidis*, and *Escherichia Coli* O157: H7. *J. Food Sci.*, 74(1): M46–52.

Kanazawa, K. and Cho, N.J. (2009). Quartz crystal microbalance as a sensor to characterize macromolecular assembly dynamics. *J. Sens.*, 6: 1–17.

Kandeepan, G. and Sangma, S. (2010). Optimization of the level of guar gum in low fat yak milk paneer. *J. Stored Products and Postharvest Res.*, 1(1): 9–12.

Kandeepan, G., Anjaneyulu, A.S.R., Kondaiah, N., Mendiratta, S.K. and Rajkumar, R.S. (2013). Evaluation of quality and shelf life of buffalo meat keema at refrigerated storage. *J. Food Sci. Technol.*, 50(6): 1069–1078.

Kang, T., Guan, R., Chen, X., Song, Y., Jiang, H. and Zhao, J. (2013). *In vitro* toxicity of different-sized ZnO nanoparticles in Caco-2 cells. *Nanoscale Res. Lett.*, 8: 496.

Koeneman, B.A., Zhang, Y., Westerhoff, P., Chen, Y., Crittenden, J.C. and Capco, D.J. (2010). Toxicity and cellular responses of intestinal cells exposed to titanium dioxide. *Cell Biol. Toxicol.*, 26: 225–238.

Konstantatos, G. and Sargent, E.H. (2009). Solution-processed quantum dot photodetectors. *Proc. IEEE.*, 97: 1666–1683.

Koo, O.M., Rubinstein, I. and Onyuksel, H. (2005). Role of nanotechnology in targeted drug delivery and imaging: a concise review. *Nanomed. Nanotechnol. Biol. Med.*, 1: 193–212.

Kriegel, C., Kit, K.M., McClements, D.J. and Weiss, J. (2009). Influence of surfactant type and concentration on electrospinning of chitosan–poly (Ethylene Oxide) blend nanofibers. *Food Biophysics*, 4(3): 213–228.

Kristo, E. and Biliaderis, C.G. (2007). Physical properties of starch nanocrystal reinforced pullulan films. *Carbohydr. Polym.*, 68: 146–158.

Kubacka, A., Diez, M.S., Rojo, D., Bargiela, R., Ciordia, S. and Zapico, I. (2014). Understanding the antimicrobial mechanism of TiO_2-based nanocomposite films in a pathogenic bacterium. *Sci. Rep.*, 4: 4134.

Kumar, L.Y. (2015). Role and adverse effects of nanomaterials in food technology. *J. Toxicol. Health*, 2: 1–11.

Kuswandi, B. (2017). Environmental-friendly food nano-packaging. *Environ. Chem. Lett.*, 15: 205–221.

Kuzma, J., Romanchek, J. and Kokotovich, A. (2008). Upstream oversight assessment for agrifood nanotechnology: A case studies approach. *Risk Analysis*, 28: 1081–1098.

Langer, R. and Peppas, N.A. (2003). Advances in biomaterials, drug delivery, and bionanotechnology. *AIChE J.*, 49: 2990–3006.

Lau, A.K.T. and Hui, D. (2002). The revolutionary creation of new advanced Materials–carbon nanotube composites. *Compos. B Eng.*, 33: 263–277.

Lee, S.K., Sheridan, M. and Mills, A. (2005). Novel UV-activated colorimetric oxygen indicator. *Chem. Mater.*, 17: 2744–2751.

Li, Q., Mahendra, S., Lyon, D.Y., Brunet, L., Liga, M.V., Li, D. et al. (2008). Antimicrobial nanomaterials for water disinfection and microbial control: Potential applications and implications. *Water Res.*, 42: 4591–4602.

Li, R., Liu, C.H., Ma, J., Yang, Y.J. and Wu, H.X. (2011). Effect of org-titanium phosphonate on the properties of chitosan films. *Polym. Bull.*, 67: 77–89.

Li, Shang., Karin Nienhaus and Gerd Ulrich Nienhaus. (2014). Engineered nanoparticles interacting with cells: Size matters. *J. Nanobiotechnol.*, 12: 5.

Li, X., Xing, Y., Jiang, Y., Ding, Y. and Li, W. (2009). Antimicrobial activities of ZnO powder-coated PVC film to inactivate food pathogens. *Int. J. Food Sci. Technol.*, 44: 2161–2168.

Li, Y., YHT, C.U. and Luo, D. (2005). Multiplexed detection of pathogen DNA with DNA-based fluorescence nanobarcodes. *Nature Biotechnol.*, 23: 885–889.

Liau, S., Read, D., Pugh, W., Furr, J. and Russell, A. (1997). Interaction of silver nitrate with readily identifiable groups: Relationship to the anti-bacterialaction of silver ions. *Letters Applied Microbial.*, 25(4): 279–83.

Lima, M.M.D. and Borsali, R. (2004). Rod like cellulose microcrystals: Structure, properties, and applications. *Macromol. Rapid Commun.*, 25: 771–787.

Lin, N., Huang, J. and Dufresne, A. (2012). Preparation, properties, and applications of polysaccharide nanocrystals in advanced functional nanomaterials: A review. *Nanoscale*, 4: 3274–3294.

Lin, S., Chen, L., Huang, L., Cao, S., Luo, X. and Liu, K. (2015). Novel antimicrobial chitosan–cellulose composite films bio-conjugated with silver nanoparticles. *Ind. Crops Prod.*, 70: 395–403.

Liu, S., Yuan, L., Yue, X., Zheng, Z. and Tang, Z. (2008). Recent advances in nanosensors for organophosphate pesticide detection. *Adv. Powder. Technol.*, 19: 419–441.

Liu, S.F., Petty, A.R., Sazama, G.T. and Swager, T.M. (2015). Single-walled carbon nanotube/metalloporphyrin composites for the chemiresistive detection of amines and meat spoilage. *Angew. Chem. Int. Ed. Engl.*, 54: 6554–6657.

Liu, Y., He, L., Mustapha, A., Li, H., Hu, Z.Q. and Lin, M. (2009). Antibacterial activities of zinc oxide nanoparticles against *Escherichia coli* O157: H7. *J. Appl. Microbiol.*, 107: 1193–1201.

Lopez-Leon, T., Carvalho, E.L.S., Seijo, B., Ortega-Vinuesa, J.L. and Bastos-Gonzalez, D. (2005). Physicochemical characterization of chitosan nanoparticles: Electrokinetic and stability behavior. *J. Colloid Interface Sci.*, 283: 344–351.

Love, S.A., Maurer-Jones, M.A., Thompson, J.W., Lin, Y.S. and Haynes, C.L. (2012). Assessing nanoparticle toxicity. *Annu. Rev. Anal. Chem.*, 5: 181–205.

Lu, Y., Weng, L. and Zhang, L. (2004). Morphology and properties of soy protein isolate thermoplastics reinforced with chitin whiskers. *Biomacromolecules*, 5: 1046–1051.

Luduena, L.N., Alvarez, V.A. and Vasquez, A. (2007). Processing and microstructure of PCL/clay nanocomposites. *Mater. Sci. Eng. A.*, 121–129.

Luechinger, N.A., Loher, S., Athanassiou, E.K., Grass, R.N. and Stark, W.J. (2007). Highly sensitive optical detection of humidity on polymer/metal nanoparticle hybrid films. *Langmuir*, 23(6): 3473–3477.

Maisanaba, S., Pichardo, S., Puerto, M., Gutiérrez-Praena, D., Cameán, A.M. and Jos, A. (2015). Toxicological evaluation of clay minerals and derived nanocomposites: A review. *Environ. Res.*, 138: 233–254.

McCracken, C., Dutta, P. and Waldman, W.J. (2016). Critical assessment of toxicological effects of ingested nanoparticles. *Environ. Sci. Nano.*, 3: 256–282.

McGlashan, S.A. and Halley, P.J. (2003). Preparation and characterisation of biodegradable starch-based nanocomposite materials. *Polymer International*, 52(11): 1767–1773.

Mihindukulasuriya, S.D.F. and Lim, L.T. (2013). Oxygen detection using UV-activated electrospun poly (ethylene oxide) fibers encapsulated with TiO_2 nanoparticles. *J. Mater. Sci.*, 48: 5489–5498.

Mills, A. (2005). Oxygen indicators and intelligent inks for packaging food. *Chemical Society Reviews*, 34(12): 1003–1011.

Mills, A. and Hazafy, D. (2009). Nanocrystalline SnO_2-based, UVB-activated, colorimetric oxygen indicator. *Sensor Actuat. B Chem.*, 36: 344–349.

Montazer, M. and Harifi, T. (2017). New approaches and future aspects of antibacterial food packaging: From nanoparticles coating to nanofibers and nanocomposites, with foresight to address the regulatory uncertainty. pp. 533–559. *In*: Grumezescu, A.M. (ed.). *Food Package.* Academic Press.

Moos, P.J., Olszewski, K., Honeggar, M., Cassidy, P., Leachman, S., Woessner, D. et al. (2011). Responses of human cells to ZnO nanoparticles: A gene transcription study. *Metallomics*, 3: 1199–1211.

Muller, C.M.O., Laurindo, B. and Yamashita, F. (2012). Composites of thermoplastic starch and nanoclays produced by extrusion and thermopressing. *Carbohydr. Polym.*, 89: 504–510.

Nam, J.M., Thaxton, C.S. and Mirkin, C.A. (2003). Nanoparticle-based bio-bar codes for the ultrasensitive detection of proteins. *Science*, 301(5641): 1884–1886.

Neethirajan, S. and Jayas, D.S. (2011). Nanotechnology for the food and bioprocessing industries. *Food Bioprocess Technol.*, 4: 39–47.

Nel, A.E., Mädler, L., Velegol, D., Xia, T., Hoek, E.M.V., Somasundaran, P. et al. (2009). Understanding biophysicochemical interactions at the nano–bio interface. *Nat. Mater.*, 8: 543–557.

Niedbala, R.S., Feindt, H., Kardos, K., Vail, T., Burton, J., Bielska, B. et al. (2001). Detection of analytes by immunoassay using upconverting phosphor technology. *Anal. Biochem.*, 293: 22–30.

Ong, L.C., Ang, L.Y., Alonso, S. and Zhang, Y. (2014). Bacterial imaging with photostable upconversion fluorescent nanoparticles. *Biomaterials*, 35: 2987–2998.

Othman, S.H. (2014). Bio-nanocomposite materials for food packaging applications: Types of biopolymers and nano-sized filler. *Agr. Sci. Procedia*, 2: 296–303.

Ozturk, A.B., Argin, S., Ozilgen, M. and McClements, D.J. (2015). Formation and stabilization of nanoemulsion-based vitamin E delivery systems using natural biopolymers: Whey protein isolate and gum. *Food Chem.*, 188: 256–263.

Pal, S., Alocilj, E.C. and Downes, F.P. (2007). Nanowire labeled direct-charge transfer biosensor for detecting Bacillus species. *Biosensors and Bioelectronics*, 22: 9(10): 2329–2336.

Palchetti, I. and Mascini, M. (2008). Electroanalytical biosensors and their potential for food pathogen and toxin detection. *Anal. Bioanal. Chem.*, 391: 455–471.

Panea, B.G., Ripoll González, J., Fernández-Cuello, A. and Albertí, P. (2013). Effect of nanocomposite packaging containing different proportions of ZnO and Ag on chicken breast meat quality. *J. Food Eng.*, 123: 104–112.

Park, B., Fu, J., Zhao, Y., Siragusa, G.R., Cho, Y.J., Lawrence, K.C. and Windham, W.R., (2007). Bio-functional Au/Si nanorods for pathogen detection. *In*: Islam, M.S. and Dutta, A.K. (eds.). *Nanosensing: Materials, Devices, and Systems III.* Proceedings of the Society of Photo-Optical Instrumentation Engineers (SPIE), 6769: O7690–O7690.

Petchwattana, N., Covavisaruch, S., Wibooranawong, S. and Naknaen, P. (2016). Antimicrobial food packaging prepared from poly (butylene succinate) and zinc oxide. *Measurement*, 93: 442–448.

Pradhan, N., Singh, S., Ojha, N., Shrivastava, A., Barla, A., Rai, V. et al. (2015). Facets of nanotechnology as seen in food processing, packaging, and preservation industry. *BioMed. Res. Int.*, 17.

Prasad, R., Bhattacharyya, A. and Nguyen, Q.D. (2017). Nanotechnology in sustainable agriculture: recent developments, challenges, and perspectives. *Front. Microbiol.*, 8: 1014.

Prashantha, K., Soulestin, J., Lacrampe, M.F., Krawczak, P., Dupin, G. and Claes, M. (2009). Masterbatch-based multi-walled carbon nanotube filled polypropylene nanocomposites: assessment of rheological and mechanical properties. *Compos. Sci. Technol.*, 69: 1756–1763.

Prateek Thakur, V.K. and Gupta, R.K. (2016). Recent progress on ferroelectric polymer-based nanocomposites for high energy density capacitors: Synthesis, dielectric properties, and future aspects. *Chem. Rev.*, 116: 4260–4317.

Qhobosheane, M., Santra, S., Zhang, P. and Tan, W.H. (2001). Biochemically functionalized silica nanoparticles. *Analyst*, 126: 1274–1278.

Rafieian, F. and Simonsen, J. (2014). Fabrication and characterization of carboxylated cellulose nanocrystals reinforced glutenin nanocomposite. *Cellulose*, 21: 4167–4180.

Ramanathan, T., Abdala, A., Stankovich, S., Dikin, D., Herrera-Alonso, M. and Piner R. (2008). Functionalized graphene sheets for polymer nanocomposites. *Nature Nanotechnol.*, 3(6): 327–331.

Rani, P.V.A., Mun, G.L.K., Hande, M.P. and Valiyaveettil, S. (2009). Cytotoxicity and genotoxicity of silver nanoparticles in human cells. *ACS Nano*, 3: 279–290.

Rhim, J.W. and Ng, P.K.W. (2007). Natural biopolymer-based nanocomposite films for packaging applications. *Crit. Rev. Food Sci. Nutr.*, 47: 411–433.

Rhim, J.W., Park, H.M. and Ha, C.S. (2013). Bio-nanocomposites for food packaging applications. *Prog. Polym. Sci.*, 38: 1629–1652.

Rhim, J.W., Wang, L.F., Lee, Y. and Hong, S.I. (2014). Preparation and characterization of bio-nanocomposite films of agar and silver nanoparticles: Laser ablation method. *Carbohydr. Polym.*, 103: 456–465.

Rieger, K.A., Eagan, N.M. and Schiffman, J.D. (2015). Encapsulation of cinnamaldehyde into nanostructured chitosan films. *J. Appl. Polym. Sci.*, 132: 41739.

Robinson, D.K.R. and Morrison, M.J. (2010). Nanotechnologies for food packaging: Reporting the science and technology research trends. *Observatory NANO*. Available onlineat: http://www.observatorynano.eu/project/filesystem/files/Food%20Packaging%20Report%202010%20DKR%20Robinson.pdf.

Sadeghnejad, A., Aroujalian, A., Raisi, A. and Fazel, S. (2014). Antibacterial nano silver coating on the surface of polyethylene films using corona discharge. *Surf. Coat. Tech.*, 245: 1–8.

Salami-Kalajahi, M., Haddadi-Asl, V. and Roghani-Mamaqani, H. (2012). Study of kinetics and properties of polystyrene/silica nanocomposites prepared via *in situ* free radical and reversible addition-fragmentation chain transfer polymerizations. *Sci. Iran.*, 19: 2004–2011.

Sanchez-Valdes, S., Ortega-Ortiz, H., Ramos-de Valle, L.F., Medellín-Rodríguez, F.J. and Guedea-Miranda, R. (2009). Mechanical and antimicrobial properties of multilayer films with a polyethylene/silver nanocomposite layer. *J. Appl. Polym. Sci.*, 111: 953–962.

Sandquist, D. (2013). New horizons for microfibrillated cellulose. *Appita J.*, 66: 156–162.

Sari, P., Mann, B., Kumar, R., Singh, R.R.B., Sharma, R., Bhardwaj, M. et al. (2015). Preparation and characterization of nanoemulsion encapsulating curcumin. *Food Hydrocol.*, 43: 540–546.

Schirmer, B.C., Heiberg, R., Eie, T., Møretrø, T., Maugesten, T. and Carlehøg, M. (2009). A novel packaging method with a dissolving CO_2 headspace combined with organic acids prolongs the shelf-life of fresh salmon. *Int. J. Food Microbiol.*, 133: 154–160.

Schubbe, S., Schumann, C., Cavelius, C., Koch, M., Muller, T. and Kraegeloh, A. (2012). Size-dependent localization and quantitative evaluation of the intracellular migration of silica nanoparticles in Caco-2 cells. *Chem. Mater.*, 24: 914–923.

Sharma, C., Dhiman, R., Rokana, N. and Panwar, H. (2017). Nanotechnology: An untapped resource for food packaging. *Front. Microbiol.*, 8: 1735.

Sharma, R., Ragavan, K.V., Thakur, M.S. and Raghavaro, K.S.M.S. (2015). Recent advances in nanoparticle based aptasensors for food contaminants. *Biosensors and Bioelectronics*, 74: 612–627.

Sharma, V., Shukla, R.K., Saxena, N., Parmar, D., Das, M. and Dhawan, A. (2009). DNA damaging potential of zinc oxide nanoparticles in human epidermal cells. *Toxicol. Lett.*, 185: 211–218.

Sheikh, F.A., Kanjawal, M.A., Saran, S., Chung, W.J. and Kim H. (2011). Polyurethane nanofibers containing copper nanoparticles as future materials. *Appl. Surf. Sci.*, 257: 3020–3026.

Shi, H., Magaye, R., Castranova, V. and Zhao, J. (2013). Titanium dioxide nanoparticles: A review of current toxicological data. *Particle Fibre Toxicol.*, 10: 15.

Shi, L., Zhao, Y., Zhang, X., Su, H. and Tan, T. (2008). Antibacterial and anti-mildew behavior of chitosan/nano-TiO_2 composite emulsion. *Korean J. Chem. Eng.*, 25: 1434–1438.

Singh, R.P. (2000). Scientific principles of shelf-life evaluation. pp. 2–22. *In*: Man, D. and Jones, A. (eds.). *Shelf-life Evaluation of Foods*. Aspen Publication.

Singh. T., Shukla, S., Kumar, P., Wahla, V., Bajpai, V.K. and Rather, I.A. (2017). Application of nanotechnology in food science: perception and overview. *Front. Microbiol.*, 8: 1501.

Sirelkhatim, A., Mahmud, S., Seeni, A., Kaus, N.H.M., Ann, L.C., Bakhori, S.K.M. et al. (2015). Review on zinc oxide nanoparticles: Antibacterial activity and toxicity mechanism. *Nano Micro Lett.*, 7: 219–242.

Smolander, M., Hurme, E., Koivisto, M. and Kivinen, S. (2004). Indicator. International Patent WO2004/102185 A1.

Smolkova, B., El Yamani, N., Collins, A.R., Gutleb, A.C. and Dusinska, M. (2015). Nanoparticles in food. Epigenetic changes induced by nanomaterials and possible impact on health. *Food Chem. Toxicol.*, 77: 64–73.

Soares, N.F.F., Silva C.A.S., Santiago-Silva, P., Espitia, P.J.P., Gonçalves, M.P.J.C., Lopez, M.J.G. et al. (2009). Active and intelligent packaging for milk and milk products. pp. 155–174. *In*: Coimbra, J.S.R. and Teixeira, J.A. (eds.). *Engineering Aspects of Milk and Dairy Products*, New York, NY: CRC Press.

Song, Y., Guan, R., Lyu, F., Kang, T., Wu, Y. and Chen, X. (2014). *In vitro* cytotoxicity of silver nanoparticles and zinc oxide nanoparticles to human epithelial colorectal adenocarcinoma (Caco-2) cells. *Mutat. Res.*, 769: 113–118.

Stormer, A., Bott, J., Kemmer, D. and Franz, R. (2017). Critical review of the migration potential of nanoparticles in food contact plastics. *Trends Food Sci. Technol.*, 63: 39–50.

Subramanian, A. (2006). A mixed self-assembled monolayer-based surface Plasmon immunosensor for detection of *E. coli* O157H7. *Biosens. Bioelectron.*, 7: 998–1006.

Swain, S.K., Pradhan, A.K. and Sahu, H.S. (2013). Synthesis of gas barrier starch by dispersion of functionalized multiwalled carbon nanotubes. *Carbohydr. Polym.*, 94: 663–668.

Tan, F., Leung, P.H.M., Liud, Z., Zhang, Y., Xiao, L., Ye, W. et al. (2011). Microfluidic impedance immunosensor for *E. coli* O157: H7 and *Staphylococcus aureus* detection via antibody-immobilized nanoporous membrane. Sensor. *Actuat. B. Chem.*, 159: 328–335.

Tan, W., Zhang, Y., Szeto, Y.S. and Liao, L. (2008). A novel method to prepare chitosan/montmorillonite nanocomposites in the presence of hydroxyl-aluminum olygomeric cations. *Compos. Sci. Technol.*, 68: 2917–2921.

Tang, C. and Liu, H. (2008). Cellulose nanofiber reinforced poly (vinyl alcohol) composite film with high visible light transmittance. *Compos. A Appl. Sci. Manuf.*, 39: 1638–1643.

Tarantini, A., Lanceleur, R., Mourot, A., Lavault, M.T., Casterou, G., Jarry, G. et al. (2015). Toxicity, genotoxicity and proinflammatory effects of amorphous nanosilica in the human intestinal Caco-2 cell line. *Toxicol. In Vitro.*, 29: 398–407.

Valizadeh, A., Mikaeili, H., Samiei, M., Farkhani, S.M., Zarghami, N., Kouhi M. et al. (2012). Quantum dots: Synthesis, bioapplications, and toxicity. *Nanoscale Res. Lett.*, 7: 480. Doi:10.1186/1556-276X-7-480.

Vandebriel, R.J. and De Jong, W.H. (2012). A review of mammalian toxicity of ZnO nanoparticles. *Nanotechnol. Sci. Appl.*, 5: 61–71.

Vermeiren, L., Devlieghere, F. and Debevere, J. (2002). Effectiveness of some recent antimicrobial packaging concepts. *Food Addit. Contam.*, 19: 163–171.

Vladimiriov, V., Betchev, C., Vassiliou, A., Papageorgiou, G. and Bikiaris, D. (2006). Dynamic mechanical and morphological studies of isotactic polypropylene/fumed silica nanocomposites with enhanced gas barrier properties. *Compos. Sci. Technol.*, 66: 2935–2944.

Wang, B. and Sain, M. (2007). Isolation of nanofibers from soybean source and their reinforcing capability on synthetic polymers. *Compos. Sci. Technol.*, 67: 2521–2527.

Wang, L., Hu, C. and Shao, L. (2017). The antimicrobial activity of nanoparticles: Present situation and prospects for the future. *Int. J. Nanomed.*, 12: 1227–1249.

Weiss, J., Takhistov, P. and Mc Clements, D.J. (2006). Functional materials in food nanotechnology. *J. Food Sci.*, 71: R107–R116.

Wu, C.L., Zhang, M.Q., Rong, M.Z. and Friedrich, K. (2002). Tensile performance improvement of low nanoparticles filled polypropylene composites. *Compos. Sci. Technol.*, 62: 1327–1340.

Wu, T.S., Wang, K.X., Li, G.D., Sun, S.Y., Sun, J. and Chen, J.S. (2010). Montmorillonite-supported Ag/TiO(2) nanoparticles: An efficient visible-light bacteria photodegradation material. *ACS Appl. Mater. Int.*, 2: 544–550.

Xiao, L., Green, A.N.M., Haque, S.A., Mills, A. and Durrant, J.R. (2004). Light-driven oxygen scavenging by titania/polymer nanocomposite films. *J. Photochem. Photobiol. A Chem.*, 162: 253–259.

Xiong, H.G., Tang, S.W., Tang, H.L. and Zou, P. (2008). The structure and properties of a starch-based biodegradable film. *Carbohydr. Polym.*, 71: 263–268.

Yam, K.L. and Lee, D.S. (2012). *Emerging Food Packaging Technologies, Principles and Practice*. Elsevier.

Yang, L. and Li, Y. (2006). Simultaneous detection of *Escherichia coli* O157: H7 and Salmonella Typhimurium using quantum dots as fluorescence labels. *Analyst*, 131: 394–401.

Yang, Y.X., Song, Z.M., Cheng, B., Xiang, K., Chen, X.X., Liu, J.H. et al. (2014). Evaluation of the toxicity of food additive silica nanoparticles on gastrointestinal cells. *J. Appl. Toxicol.*, 34: 424–435.

Yoshida, T., Yoshioka, Y., Fujimura, M., Yamashita, K., Higashisaka, K. and Morishita, Y. (2011). Promotion of allergic immune responses by intranasally-administrated nanosilica particles in mice. *Nanoscale Res. Lett.*, 6: 1–6.

Youssef, A.M. (2013). Polymer nanocomposites as a new trend for packaging applications. *Polym. Plast. Technol. Eng.*, 52: 635–660.

Zeng, J., Roberts, S. and Xia, Y. (2010). Nanocrystal-based time-temperature indicators. *Chem. A Eur. J.*, 16: 12559–12563.

Zhang, C., Yin, A.X., Jiang, R., Rong, J., Dong, L. and Zhao, T. (2013). Time-temperature indicator for perishable products based on kinetically programmable Ag overgrowth on Au nanorods. *ACS Nano*, 7: 4561–4568.

Zhang, H., Elghanian, R., Demers, L., Amro, N., Disawal, S. and Cruchon-dupeyrat, S. (2009). Direct-write nanolithography method of transporting ink with an elastomeric polymer coated nanoscopic tip to form a structure having internal hollows on a substrate. US Patent 7491422 (in English).

Zhou, C. (2013). Theoretical analysis of double-microfluidic-channels photonic crystal fiber sensor based on silver nanowires. *Opt. Commun.*, 288: 42–46.

Zohreh Honarvar, Zahra Hadian and Morteza Mashayekh. (2016). Nanocomposites in food packaging applications and their risk assessment for health. *Electronic Physician*, 8(6): 2531–2538.

Chapter 14

Exploring the Potential of Nanotechnology in Cotton Breeding:

Huge Possibilities Ahead

Sapna Grewal,[1,*] *Promila,*[1] *Santosh Kumari,*[1]
Sonia Goel[2] and *Shikha Yashveer*[3]

1. Introduction

Nanotechnology is a well-known, fast emerging field of science that deals with matter at nanoscale dimensions. The application of nanomaterials in some form or the other has been witnessed in diverse research areas, such as electronics, pharmaceutical sciences, energy, material sciences, chemistry, disease prevention, etc. (Carmen et al., 2003; Nair et al., 2010; Grewal et al., 2017; Goel et al., 2019; Sheorain et al., 2019). Within the paradigm of agriculture, most of the work in the field of nanotechnology has been done related to the production of nanofertilizers, nanopesticides and a few biosensor-based devices, that too restricted to some major cereal crops only. Other crops including cotton and jute, the two main fiber crops remain neglected even though they are the second most important economic crops in India. Out of the eight main cultivated fiber crops in India, cotton holds the most important place because of its extensive use in the textile industry (Pandey and Gupta, 2003). Though India is self-sufficient in cotton production, but the average productivity remains low mainly

[1] Department of Bio & Nanotechnology, Guru Jambheshwar University of Science & Technology, Hisar, 125011, Haryana, India.
[2] Faculty of Agricultural sciences, SGT University, Gurugram, 122505, Haryana, India.
[3] Department of Molecular Biology, Biotechnology & Bioinformatics, College of Basic Sciences, CCSHAU, Hisar, 125 004, Haryana, India.
* Corresponding author: sapnagrewal29@gmail.com

because of weather aberrations, pest incidence, and poor quality of soil among many others (Directorate of Cotton Development, GOI, 2017). There are many factors like pesticide resistance, decreasing per hectare yield, and abiotic stresses (drought, salinity, etc.) that pose a serious threat to the crops today and need immediate attention for solutions/alternate approaches. For a developing country like India, the advanced scientific technological interventions of nanotechnology can offer potential solution to these problems. The synthesis of nanoparticles (NPs) using green technologies, employing plant extracts, bacterial or fungal cultures instead of formerly used chemical-based methods has led to the formation of better and less toxic products. Such developments have helped reduce both the toxicity impact of nanomaterials and the environmental hazards, ultimately leading to better acceptability of nanoproducts for commercial applications. The major areas of research in cotton targets textile finishing and product improvement followed by methods of ensuring its better yield and productivity. This chapter focuses on some of the prominent applications of nanotechnology in the context of cotton, or the so-called 'white gold' (Fig. 14.1).

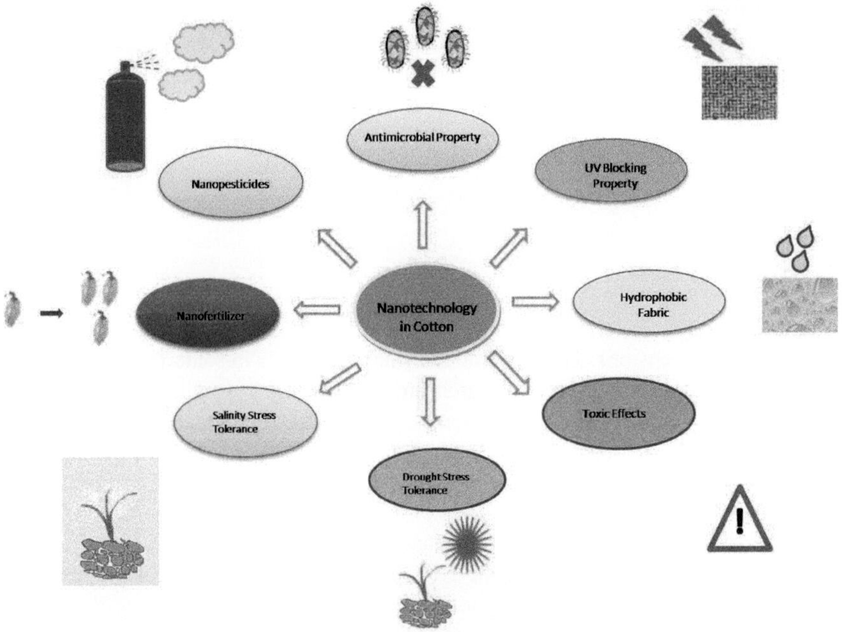

Fig. 14.1. Applications of nanotechnology in cotton.

2. Nanofertilizers in Cotton

When the agriculture sector faced serious limitations due to the continuous use of chemical fertilizers, the orientation of research studies shifted initially towards the possible use of biofertilizers and biopesticides (Mishra et al., 2015) and later to the latest entrant, i.e., nanofertilizers. In the last few decades, environmental pollution has become a grave concern due to the uncontrolled use of fertilizers and pesticides in crop fields, thereby demanding more sustainable alternatives. Nanofertilizers are nanoscale formulations of nutrients that are required in very less quantity and can be target specific too. As indicated by several studies, these are better alternatives than conventional fertilizers and biofertilizers in ensuring increased nutrient efficiency and overall plant nutrition (Auffan et al., 2009). Nanofertilizers can either be used as foliar spray or added in encapsulated form to assist their continuous and slow release over a longer period.

A study was conducted to analyse the impact of normal zinc (Zn), chelate of Zn, and nano-chelate of Zn on the morphological and physiological characteristics of cotton (Rezaei and Abbasi, 2014). In case of application of nano-chelate Zn, parameters such as plant height, dry as well as fresh weight, chlorophyll a and chlorophyll b were found to be positively affected though some other parameters did not show much variation. It was found that the nano-chelate of Zn improved the overall performance of cotton crop by directly increasing the number of bolls per plant. In India, after the invent of Bt cotton, 90% of area under cultivation is for Bt cotton (Shah, 2012). It seems important and interesting to compare transgenic Bt cotton with non-transgenic cotton varieties in their response towards NP application. Studies in the past have compared the two cotton varieties on various scales. When different concentrations of cerium oxide (CeO_2) NPs (0, 100, 500, and 2000 mg/L) were applied on transgenic Bt cotton and non-transgenic cotton, the former variety was found to be more susceptible to nanomaterials than the latter (Li et al., 2014). Though no significant change in plant height was observed, the root biomass of transgenic Bt cotton was reduced substantially in comparison to the control, clearly indicating that the transgenic Bt cotton variety was more vulnerable to CeO_2 NPs. Also, the effect of CeO_2 NPs was found to be more pronounced in roots than shoots. In the shoots of transgenic variety, the concentration of elements–calcium, potassium, and magnesium (Ca, K, and Mg) increased at 100 mg/L concentration of NPs but decreased thereafter, with further increase in concentration. In the non-transgenic variety, the concentration of zinc and potassium increased at a lower exposure of CeO_2 NPs. On similar lines, Le et al. (2014) studied the effects of SiO_2 NPs at concentrations of 0, 10, 100, 500, and 2000 mg/L in both transgenic and conventional cotton varieties for continuous three weeks. This nutrient element study showed SiO_2 NPs to enhance the transport of Mg in Bt-transgenic xylem sap and iron in xylem sap of both the varieties. Slightly different results were observed by Le Van Nhan et al. (2016) when they evaluated the effect of iron oxide (Fe_2O_3) NPs at concentrations of 100 and 1000 mg/L. They observed a decrease in plant height and root length and almost no impact on root and shoot biomass of transgenic Bt cotton. The study reported slightly enhanced levels of sodium and potassium (Na and K) in the roots of transgenic Bt cotton and reduced zinc concentration after the application of iron oxide NPs. All these studies have one thing in common that the transgenic cotton was found to be more sensitive to the exposure of NPs in comparison to the conventional cotton variety.

3. Nanopesticides for Pest Control

Globally, nearly 20,000 plant pathogen species, insects, mites, and weeds are causing harm to agriculture (Kagan, 2016; Zhang, 2018). USA and China are the highest consumers of pesticides with approximately 386 million and 1806 million kg of annual application of pesticides, respectively (Oerke and Dehne, 2004; FAO, 2015).

Due to the excessive and indiscriminate use of pesticides, alarmingly high level of soil and water pollution is caused every year. The problem is compounded further due to limited management options and inability of timely detection of pest related disease (Adisa et al., 2019). Conventional approaches like development of resistant crop varieties could be a reliable way to combat the problem related to pests' infestation, but since it takes considerable time to develop a resistant variety, the dependability on such an approach gets diluted (Servin et al., 2015). Several reports have shown the significant potential of nanoformulations in plant disease management for controlling weeds and pests. The encapsulated formulations of pesticides with slow-release property and enhanced solubility have the much-needed added advantages.

Cotton is infected by many pests and one of these is the cotton leaf worm, *Spodoptera littoralis,* that is known to cause serious damage to the crop. It is also known to attack tomatoes, peanuts, corn, cabbage, and few other crops as well and proves to be extremely damaging to all. The leaf worm, *Spodoptera littoralis*, has developed resistance towards the commonly used insecticides such as organophosphates and pyrethroids and therefore, research studies were carried out to evaluate the efficacy of various nanopesticides in controlling this pest. Zaki et al. (2017) studied the effectiveness of novel nanopesticides made of nanotubes of sodium titanate and its composite with the eco-friendly biopesticide, *Bacillus thuringiensis* bacteria to resist cotton leaf worm *Spodoptera littoralis.* The insects were reared in the laboratory and biological features were studied for second and fourth instars larvae after the treatment. Results showed 80% emergence under Bt-TNTs composite treatment for the second and fourth instars along with 20% total mortality for them and 11% increment in larval duration period for the second instar while there was 100% emergence in the control samples for both the stages. This composite treatment also led to 42% fecundity increment and 82% fecundity decline of the second and fourth instars, respectively, as compared to the control samples. The effectiveness of titanium dioxide (TiO_2) NPs as nanopesticides was also evaluated against the insect (Shaker et al., 2017). Six different concentrations of TiO_2 NPs ranging from 31.25 ppm to 1000 ppm were used against the second and fourth instar larvae and the mortality rate was checked two weeks after the application. The second instar larval stage was found to be more sensitive than the fourth instar at all concentrations of nanopesticides and LC50 value was found to be 62.5 ppm for the second instar stage, which indicated the effectiveness of the treatment. Ayoub et al. (2018) tested two different types of nanoparticles (CuO and CaO) by using a simple wet-chemical method and evaluated their pesticidal activity against the selected insect at different concentrations of NPs (150, 300, 450, and 600 mgl/L). The results showed that although both types of NPs exhibited significant entomotoxic effects against cotton leafworm but the effect of CuO NPs was very fast with LC50 = 232.75 mg/L following three days of post-treatments while CaO NPs showed a slow entomotoxic effect with LC50 = 129.03 mg/L following 11 days of post-treatments. The physical properties of NPs could be responsible for such a variation in pesticidal activity along with the cuticle layer present in the insect's body wall and interfacial surfaces on the insect mid-gut.

These studies indicated the possible application of nanopesticides in controlling pest infection that affects not just cotton, but almost all the crops around the world. As the studies were restricted to controlled laboratory conditions and in most cases, no clear-cut comparison was made with the currently used chemical pesticides, so conducting a large-scale field level study could be more useful to get a clear picture about how conveniently these nanopesticides can replace the conventional pesticides.

4. Role of Nanoparticles in Mitigating Abiotic Stress

Abiotic stresses are major factors that limit crop productivity due to drastic climatic changes that are happening in recent years. Abiotic factors such as salinity, drought, and temperature change are affecting the overall growth of plants by hampering their physiological and biochemical parameters. Such abiotic stress conditions are responsible for huge losses in crop production (to the tune of 70%) throughout the world (Li et al., 2017). The alarming fact is that such threats are predicted to intensify in the coming years due to climatic change risks. Therefore, it becomes crucial to work towards finding sustainable solutions to these problems. Use of NPs to mitigate such abiotic stresses has been supported by researchers through their studies (Hatami et al., 2016; Reddy et al., 2016). Nanomaterials in low concentrations can enhance the oxidative stress defence mechanism of the plant and some of these such as nanosilica, nanosilver, nanotitanium (SiO_2, $AgNO_3$, TiO_2) have shown promising results in ameliorating the stress levels as well (Wang et al., 2011; Akbari et al., 2014; Almutairi et al., 2016). The fact that SiO_2 can promote germination and reactive oxygen species (ROS) cascade under salt stress has been proved in various crops such as tomato, squash, and rice, among others (Haghighi et al., 2012; Siddiqui et al., 2014; Wang et al., 2014). Nano-Zn fertilizers influenced the overall yield of cotton plants grown under salinity stress conditions and this was studied by Hussein and Baker in 2018. The foliar application of 200 ppm of nano-Zn was found to mitigate most of the negative effects of salt, though the researchers suggested the use of combination fertilizers to prevent any imbalance.

Drought, another abiotic stress, is a complex phenomenon and changes the physiology and metabolism of the whole plant. It leads to the overall reduction of plant growth, decreased chlorophyll content, and increased oxidative stress (Anjum et al., 2011). To check if NPs can reduce the effects of drought on cotton plants, Shallon et al. (2016) sprayed nano TiO_2 and nano SiO_2 on cotton plants under drought stress conditions. They found that an optimum concentration of 50 ppm of nano TiO_2 and 3200 ppm of nano SiO_2 were sufficient to reverse the effects of drought stress by increasing total phenolics, proline content, antioxidant capacity, and enhancing yield traits.

5. Influence of Nanoparticles on Plant Hormones

Phytohormones or plant hormones are present in very low amounts in plants but have an important role to play in their developmental processes such as cell expansion, elongation, and growth. Some of these are auxin, gibberellic acid (GA), cytokinin, abscisic acid (ABA), brassinosteroids, ethylene, salicylic acid, jasmonic acid, and strigolactone. These are small endogenous compounds in plants and several reports have revealed their pivotal role in plant growth and regulation of cotton fibre development (Daviere and Achard, 2016; Xiao et al., 2019). While performing their studies on the impact of Fe_2O_3 NPs on cotton plants, Nhan et al. (2016a) also studied their effect on plant hormones. They observed an increment in the concentration of indole-3-acetic acid (IAA) in both transgenic and non-transgenic cotton after the application of 100 mg/L of Fe_2O_3 NPs. But the amount of ABA decreased in the conventional cotton variety and remained almost unchanged in the transgenic variety. The treatment enhanced the concentration of the plant hormones (IAA, ABA, GA) in the roots of Bt cotton and decreased the amount of all except GA (which was not impacted) in the case of conventional cotton. As far as the hormonal level is concerned, this indicated different behaviours of the two types of cotton plants towards the same type of treatment.

Le Van Nhan et al. (2016b) evaluated the toxic nature along with other biological effects of copper (Cu) NPs on transgenic Ipt cotton. Their work showed the consequences of nanoparticles on the hormonal levels of transgenic cotton with Ipt gene. They tracked the values of ABA, trans-zeatinriboside (t-ZR), IAA, GA, and isopentenyl adenosine (iPA). The outcome indicated that in Ipt-cotton, CuO NPs can increase GA levels in roots, but their formation is stopped in leaves. On the one hand, these results indicated the possible role of CuO NPs in affecting the concentration of phytohormones in plants though the effect was very different in the case of leaves and roots. On the other hand, the concentration of iPA increased, which suggests that CuO NPs could be used to delay senescence. Nhan et al. (2015) showed that plant hormone IAA content was higher in transgenic Bt cotton in comparison to traditional cotton upon CuO NPs treatment, while these nanoparticles had no effect on GA and t-ZR in cotton roots. These studies illustrate variable effects of nanoparticles on different parts of the plants. Overall, it can be concluded that when applied at very high concentrations, NPs can adversely affect the production of phytohormones. They can induce beneficial effects only when applied in lower amounts and that too would depend on the type of nanoparticles and cotton varieties.

6. Toxicity Studies

Studies done so far indicate both positive and negative effects of various NPs on plant systems and therefore concerns over toxicity have been raised (Cifuentes et al., 2010; Lee et al., 2010; Zhang et al., 2012). It is more or less evident that the particle size and concentration influence toxicity. Smaller size particles are shown to be toxic at even lesser concentrations because they can pass through cellular membranes. A higher dose of several NPs such as ZnO, TiO_2, and CeO_2 among others have shown deleterious effects in various crops. Le et al. (2014), investigated the toxicity effect of SiO_2 NPs in Bt-transgenic and conventional cotton. The plant height and root biomass decreased in both varieties with the increasing concentration of SiO_2 NPs. Additionally, SiO_2 NPs were found to bioaccumulate in cotton plant, which could have potential impact on the overall plant system. Nhan et al. (2015, 2016) demonstrated the adverse consequences of CeO_2 and Fe_2O_3 on Bt cotton. These nanoparticles significantly diminished the absorption of minerals such as Fe, Mg, P, and Zn in the xylem sap of non-transgenic cotton but they enhanced manganese (Mn) absorption. The CeO_2 NPs accumulated on the chloroplasts, which then ruptured specially in transgenic cotton resulting into reduced levels of some minerals in the xylem sap. Besides this, the structure of vascular bundles got destroyed because several fragments emerged in them. The exposure of Fe_2O_3 NPs increased Bt-toxin levels in both the roots and leaves of the cotton plant. In cotton, Le Van et al. (2016) showed that CuO NPs, in a concentration greater than 10 mg/L, inhibited the overall growth of the plants in both transgenic

Table 14.1. Application of various nanoparticles in reference to cotton (*Gossypium hirsutum*).

S. No.	Nanoparticle	Use Type/ Application	Details	References
1	Nano-chelate of Zn	Nanofertilizer (Foliar spray)	Increased plant height, dry as well as fresh weight, chlorophyll a & chlorophyll b, and improved overall performance of cotton crop	Rezaei and Abbasi (2014)
2	CeO_2	Nanofertilizer (Nutrient solution)	No significant change in plant height but the root biomass of transgenic cotton was reduced	Li et al. (2014)
3	SiO_2 NPs	Nanofertilizer (Nutrient solution)	Stimulated the transport of Magnesium in Bt-transgenic xylem sap	Le et al. (2014)
4	Fe_2O_3 NPs	Nanofertilizer (Nutrient solution)	Decrease in plant height and root length, and almost no impact on root and shoot biomass of Bt-transgenic cotton.	Nhan et al. (2016)
5	Zn NPs	Stress alleviation (Foliar spray)	Mitigate most of the negative effects of salinity	Hussein and Baker (2018)
6	nano-TiO_2 and nano-SiO_2	Nanofertilizer in stress alleviation (Foliar spray)	Increased total phenolics, proline content, total antioxidant capacity, and enhanced yield traits along with reversing effects of drought stress	Slallon et al. (2016)
7	Na_2TiO_3	Nanopesticides (Feeding solution)	80% emergence under Bt-TNTs composite treatment for both 2nd and 4th instars alongwith 20% total mortality for them	Zaki et al. (2017)
8	TiO_2	Nanopesticides (Feeding solution)	High level of toxic effect of TiO_2NPs	Shaker et al. (2017)
9	CuO and CaO NPs	Nanopesticides (Feeding solution)	CuO NPs showed toxic effects after 3 days whereas CaO NPs showed results after 11 days post treatment	Ayoub et al. (2018)
10	SiO_2 NPs	Toxicity (Nutrient solution)	Bioaccumulation of SiO_2 NPs in cotton plants can affect food crops and human health	Le et al. (2014)
11	CeO_2 NPs	Toxicity (Nutrient solution)	CeO_2 NPs accumulated in chloroplasts which then rupture, especially in transgenic variety	Nhan et al. (2015)
12	Fe_2O_3 NPs	Toxicity (Nutrient solution)	Higher level of Bt-toxin in leaf and root parts of the plant	Nhan et al. (2016)
13	CuO NPs	Toxicity (Nutrient solution)	Inhibited overall growth of the plants in both transgenic and non-transgenic varieties	Le et al. (2016a)
14	CuO NPs	Toxicity (Nutrient solution)	Resulted in decrease of root length and plant height by 42.80% and 26.91%, respectively, after a 10-day exposure	Le Van et al. (2016b)

and non-transgenic varieties. They also studied the expression of exogenous gene in response to nanoparticle application. The expression level of Bt-toxin protein in plant leaves and roots of Bt-transgenic cotton was found to have increased when CuO NPs were used in lower concentrations, which can be very useful in improving the resistance against pests of transgenic insecticidal crops. The phytotoxicity impact of CuO NPs on Ipt-transgenic cotton, harbouring the *Ipt* gene, was assessed in another study conducted by Le Van et al. (2016). When CuO NPs were applied at a high concentration of 1000 mg/L, the root length and plant height decreased by 42.80% and 26.91%, respectively, after a treatment for 10 days (Table 14.1). A higher concentration of NPs also hindered the formation of phytohormones in cotton. The harmful effects of CuO NPs at higher concentrations were also visible and noted.

7. Nanoparticles in Cotton Textile Industry

The textile industry is one of the major beneficiaries of nanotechnology due to diverse end-use possibilities. Presently, incorporation of nanoparticles into the fabrics ensures better breathability, fire retardancy, increased bacterial resistance, and even makes the fabric stronger and more elastic. Such fabrics are referred to as 'nanofabrics' and are developed by applying coatings of NPs onto the fabric through electro-spinning and these NPs form bonds with cotton fibrils. The two main goals of nanotechnological research in the textile industry are: (i) to target function and enhance performance of fabrics, and (ii) to develop unique products (Anita et al., 2011). These nanoparticles have a high surface area-to-volume ratio, which renders them very reactive. They easily form a layer over the cotton fabric in the form of a nanofibre coating, which in turn strengthens the fabric's special properties, such as extreme hydrophobicity and hence quick drying, self-cleaning, UV blocking, being stain-proof and microbial resistant (Yadav et al., 2006) (Fig. 14.2).

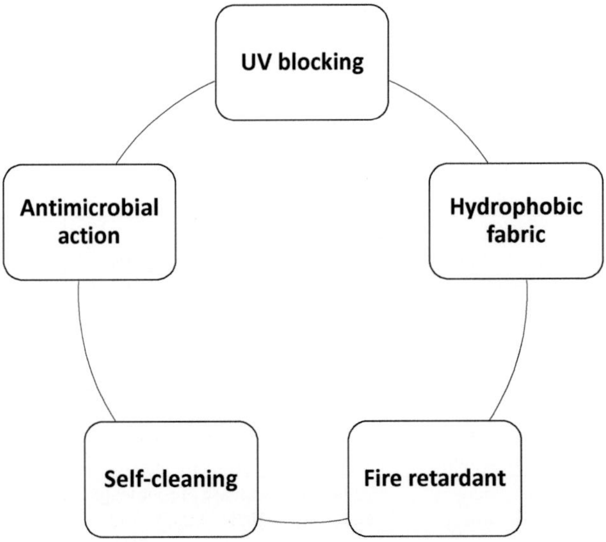

Fig. 14.2. Various commercial applications of nanotechnology in cotton textile industry.

7.1 UV Blocking Property

The flip side to using conventional methods for imparting different properties to the fabric is that after washing or wearing, the fabric loses its strength and durability. Contrarily, NPs-treated fabrics (NTFs) showed improved properties that stay intact even after the fabric has undergone repeated washings (Becheri et al., 2008). ZnO NPs have been greatly exploited for their several unique properties, such as these being magnetic, catalytic, antimicrobial, and anticancer, and useful in UV blocking and semiconducting (Akhtar et al., 2012; Sirelkhatim et al., 2015). ZnO can naturally absorb UV radiation and has unique photocatalytic and photo-oxidizing ability too (Sato and Ikeya, 2004). For being capable of absorbing the UV radiations efficiently, the band gap of the material should ideally be in the range of 3 eV–3.5 eV, which can be achieved by using some metal oxide NPs such as ZnO and TiO_2. According to the Rayleigh Scattering principle, scattering is inversely dependent on the wavelength and therefore to scatter UV rays (200 nm–400 nm), the particles' size should be in the range of 20–40 nm (Burniston, et al., 2004). Studies by Hafizah et al. (2009) and Arputharaj et al. (2017) confirmed that application of ZnO NPs onto cotton fabrics can impart UV protective traits. ZnO NPs of average size of 40 nm were developed by Yadav et al. (2006) using a chemical method and coated on bleached cotton fabrics using an acrylic binder for studying functional properties of coated fabrics. Results showed enhanced breathability, lower friction, and UV blocking ability of cotton fabrics. Both cotton and wool fabrics were targeted by Becheri et al. (2008) for applying a coat of ZnO NPs and in both cases significant enhancement in UV-absorbing activity was observed in the treated fabrics. Whether shape of the NPs can influence their properties was studied by Sricharussin et al. (2011). They prepared and analysed the effect of differently shaped ZnO NPs (rod-shaped, petal-shaped, and round-shaped) for UV blocking property after application on cotton fabric. They found that both round-shaped and petal-shaped ZnO NPs provide good UV blocking property to the cotton fabrics while the rod-shaped ZnO NP coated fabrics were least efficient in UV blocking. A completely green approach was used by El-Naggar et al. (2018), employing date seed waste for stabilizing ZnO NPs and applied them on cotton fabric for anti-bacterial and anti-UV properties.

7.2 Imparting Hydrophobicity

Cotton fibres are best known for their water-absorbing quality, which is a desirable property in making apparels suitable for hot and humid areas but there are situations where the property of water repellence is needed, that too without altering other characteristics of the cotton fabric like air permeability. Nano-coated fabrics are being developed that have a lower surface tension than water and hence do not interact much with water droplets. Water repellence can be achieved by adding nano-whiskers onto the fabric that shake off water droplets or creating 3D structures by adding gel-forming compounds (El-Khatib et al., 2012). Si NPs are particularly used in combination with some alkyl silanes for imparting hydrophobic character to the cotton fibres (Gao et al., 2009; Przybylak et al., 2016). Bae et al. (2009) carried out a detailed study to analyse and understand the suitability of Si NPs for this trait. They developed water-repellent super-hydrophobic cotton fabrics, which were treated with Si NPs as well as water-repelling components and concluded that super-hydrophobic cotton fabric can be obtained only after a combined treatment of both Si NPs and water-repelling components. Ramaratnam et al. (2007) had developed the ultra-hydrophobic textile with a permanent coating of non-fluorinated hydrophobic polymer and Si NPs with self-cleaning and water-repellent effects.

7.3 Antimicrobial Property

There is a need for fabrics with antimicrobial properties in various areas such as hospitals (doctor gowns), home furnishings, hotels (beddings), etc. The commonly used NPs with antimicrobial properties are Ag, ZnO, and TiO_2 NPs. They catalyse the conversion of atmospheric oxygen into active form, thereby creating a sterilizing impact (Saito, 1993). Ag is known to possess antimicrobial properties and is non-toxic too (Jeong et al., 2005). Bactericidal and strong inhibitory effects of Ag or Ag ions are due to their high specific surface area (Lee et al., 2003). The NPs can lead to cell death by causing inactivation of certain essential enzymes, disrupting electron transport mechanism, causing DNA damage or destruction of cellular membranes. Various studies have shown that Ag NPs can be used in clothes which make them sterile and can be used to control bacterial infection. Topical dressings, which are silver based are now widely used to treat burns, chronic ulcers, and open wounds (Lansdown, 2002). The Ag NPs can lead to the formation of ROS in the presence of oxygen, which leads to microbial cell death. Duran et al. (2007) synthesized Ag NPs of very small size (1.6 nm) from fungus *Fusarium oxysporum* and applied them on the cotton fabric. The antibacterial potential of these fabrics was found to be significant when tested against bacteria *S. aureus*. Zhang et al. (2009) impregnated the colloidal solution of Ag NPs on the cotton cloth to strengthen it with antibacterial properties. They observed almost 99% reduction against *S. aureus* and *E. coli*. This activity was maintained at 98.77% reduction even after 20 home laundering conditions. After re-confirming the results obtained by previous studies by El-Rafie et al. (2010), a stable bacterial reduction of 97% and 91% for *S. aureus* and *E. coli*, respectively, using Ag NPs was reported. On using a binder with the finishing formulation, coated fabrics showed bacterial reduction of as high as 94% and 85%, respectively, even after 20 washing cycles. Plant extracts of *E. citriodora* and *F. bengalensis* were used to synthesize Ag NPs by Ravindra et al. (2010). Their results also showed that the cotton fabric possessed excellent antibacterial activity on being coated with Ag NPs. The antibacterial properties against both gram-positive and gram-negative bacteria as well as UV-protection properties were reported by AbdElhady in his 2012 study on chitosan/nano-ZnO-coated cotton fabrics.

Presently, Cu NPs are being increasingly used in the textile industry. According to various studies, copper-based NPs are better at promoting catalytic activity and are good antimicrobial agents in comparison to bulk Cu (Nel et al., 2006). The use of ultrasound radiations to deposit CuO NPs on cotton fabric was tried by Perelshtein et al. (2009). The fabrics showed excellent antibacterial activity and had potential applications in bed linings, bandages, and wound dressing. CuO NPs below 100 nm were developed, and they showed effectiveness in killing several bacterial strains at very minimal concentrations (Ravishankar and Jamuna, 2011; El-Nahhal et al., 2012). Recently, Hasan (2018) carried out similar work by synthesizing CuO NPs using a chemical approach and application of coatings on cotton-based fabric. The antibacterial activity of the finished fabric was found to be significant against *S. aureus* bacteria and that too up to 25 washings.

7.4 Self-cleaning Property

The concept of self-cleaning fabrics was taken from lotus leaves which have nanoscale-like hair bumps that repel water and dirt. There are different approaches that can be used for this, such as photocatalysis, heat-based cleaning, etc. Researchers are now working on combining two or more NPs, such as TiO_2 and Ag to have UV protection, self-cleaning, and anti-bacterial properties. When the nano TiO_2-integrated cotton fabrics were investigated, the self-cleaning abilities were clearly visible in samples coated with higher concentrations of TiO_2 NPs (Pisitsak et al., 2013). Sivakumar et al. (2013) studied the UV-protection property and self-cleaning ability of nano-ZnO- and nano-TiO_2-coated cotton fabrics and found that the coating imparts not only UV protection and self-cleaning but also an antimicrobial and soil-repelling property to the fabric. Self-cleaning clothes are a reality now as it is possible to control the surface interaction as well as wetting ability of the fabrics with the incorporation of nanocrystals. The dirt from the clothes can be cleaned by simply spraying water on them. Such technologies are already being used and self-cleaning clothes are expected to hit the market soon. Since fabrics incorporated with NPs are more reliable and durable, there is scope in the arena of self-cleaning fabrics. Silicon is also being explored for its abrasion-resistance and water-repellent properties. Such unique properties of these fabrics make them promising agents for use in the form of medical clothes and protective garments, particularly in hospitals (Table 14.2).

8. Conclusion

There is no doubt that some reasonably fruitful work has been done with nanotechnology in agriculture, but much remains to be done at the ground level, especially with reference to addressing safety concerns and dealing with regulatory constraints. Public awareness and detailed risk and toxicity studies can give a fillip to this new and promising technology at a wider spectrum. There are intermittent knowledge gaps that must be filled for reaping expected benefits. NPs have significantly influenced yield and other morphological and biochemical parameters of cotton and studies support their use as nanofertilizers. The innovative and emerging use of nanoparticles on the cotton fabrics to impart them several unique properties such as UV-blocking, self-cleaning, enhanced tensile strength, and antimicrobial feature must be promoted at commercial scale. The textile industry is aggressively exploiting nanotechnology for fabric improvement, and we hope to see extensive involvement of nanotechnology in other areas of research too as far as cotton is concerned.

Table 14.2. Application of nanoparticles in cotton textile sector.

S. No.	Nanoparticle	Applications	Details	References
1.	ZnO NPs	UV-blocking	Enhanced breathability, lower friction, and UV-blocking ability of cotton fabrics	Yadav et al. (2006)
2.	ZnO NPs	UV-blocking	Significant enhancement in UV-absorbing activity was observed in the treated fabrics	Becheri et al. (2008)
3.	ZnO NPs	UV-blocking	Fabric coated with ZnO NPs exhibited promising blocking property against harmful UV rays	Sricharussin and associates (2011)
4.	Silica NPs	Hydrophobic fabric	Developed water-repellent super-hydrophobic cotton fabrics	Bae et al. (2009)
5.	Silica NPs	Hydrophobic fabric	Developed the ultra-hydrophobic textile with very good self-cleaning and water-repellent effects	Ramaratnam et al. (2007)
6.	Silver NPs	Antimicrobial property	Antibacterial potential of these fabrics was found to be significant when tested against bacteria *S. aureus*	Duran et al. (2007)
7.	Silver NPs	Antimicrobial property	Study reported more than 99% reduction against both *S. aureus* and *E. coli*	Zhang et al. (2009)
8.	Silver NPs	Antimicrobial property	Reported more than 90% bacterial reduction for both *S. aureus* and *E. coli*	El-Rafie et al. (2010)
9.	Silver NPs	Antimicrobial property	Cotton fabric possessed excellent antibacterial activity	Ravindra et al. (2010)
10.	Chitosan/ZnO NPs	Antimicrobial property	Cotton fabrics exhibited antibacterial properties against both gram-positive and gram-negative bacteria	AbdElhady (2012)
11.	CuO NPs	Antimicrobial property	Fabrics showed excellent antibacterial activity with potential applications in bed lining, bandages, wound dressing	Perelshtein et al. (2009)
12.	CuO NPs	Antimicrobial property	Antibacterial activity of finished fabric was found to be significant for *Staphylococcus aureus* bacteria. Even after more than 20 washes, the fabric could retain its durability	Hasan (2018)

References

AbdElhady, M.M. (2012). Preparation and characterization of chitosan/zinc oxide nanoparticles for imparting antimicrobial and UV protection to cotton fabric. *Int. J. Carbohydr. Chem.*, https://doi.org/10.1155/2012/840591.

Adisa, Ishaq Olarewaju, Venkata Laxma Reddy Pullagurala, Jose R. Peralta-Videa, Christian O. Dimkpa, Wade H. Elmer, Jorge Gardea-Torresdey and Jason White. (2019). Recent advances in nano-enabled fertilizers and pesticides: A critical review of mechanisms of action. *Environmental Science: Nano*, 6: 2002–2030. https://doi.org/10.1039/C9EN00265K.

Akbari, Gholam-Ali, Elham Morteza, Payam Moaveni, Iraj Alahdadi, Mohammad-Reza Bihamta, and Tahereh Hasanloo. (2014). Pigments apparatus and anthocyanins reactions of borage to irrigation, methylalchol, and titanium dioxide. *Int. J. Biosci.*, 4(7): 192–208.

Akhtar, Mohd. Javed, Maqusood Ahamed, Sudhir Kumar, M.A., Majeed Khan, Javed Ahmad and Salman A. Alrokayan. (2012). Zinc oxide nanoparticles selectively induce apoptosis in human cancer cells through reactive oxygen species. *Int. J. Nanomedicine*, 7: 845–857.

Almutairi, Zainab M. (2016). Influence of silver nanoparticles on the salt Resistance of tomato (*Solanum lycopersicum*) during germination. *Int. J. Agric. Biol.*, 18: 2.

Anita, S., Ramachandran, T., Rajendran, R., Koushik, C.V. and Mahalakshmi M. (2011). A study of the antimicrobial property of encapsulated copper oxide nanoparticles on cotton fabric. *Text. Res. J.*, 81(10): 1081–1088.

Anjum, Shakeel Ahmad, Xiao-yuXie, Long-chang Wang, Muhammad Farrukh Saleem, Chen Man and Wang Lei. (2011). Morphological, physiological, and biochemical responses of plants to drought stress. *Afr. J. Agric. Res.*, 6(9): 2026–2032.

Arputharaj, A., Vigneshwaran, N. and Shukla, Sanjeev R. (2017). A simple and efficient protocol to develop durable multifunctional property to cellulosic materials using in situ generated nano-ZnO. *Cellulose*, 24(8): 3399–3410.

Auffan, Mélanie, Jérôme Rose, Jean-Yves Bottero, Gregory V. Lowry, Jean-Pierre Jolivet and Mark R. Wiesner. (2009). Towards a definition of inorganic nanoparticles from an environmental, health and safety perspective. *Nat. Nanotechnol.*, 4(10): 634–641.

Ayoub, Haytham A., Mohamed Khairy, Salaheldeen Elsaid, Farouk A. Rashwan and Hanan F. Abdel-Hafez. (2018). Pesticidal activity of nanostructured metal oxides for generation of alternative pesticide formulations. *J. Agric. Food Chem.*, 66(22): 5491–5498.

Bae, GeunYeol, Byung Gil Min, Young GyuJeong, Sang Cheol Lee, Jin Ho Jang and Gwang Hoe Koo. (2009). Super hydrophobicity of cotton fabrics treated with silica nanoparticles and water-repellent agent. *J. Colloid Interface Sci.*, 337(1): 170–175.

Becheri, Alessio, Maximilian Dürr, Pierandrea Lo Nostro and Piero Baglioni. (2008). Synthesis and characterization of zinc oxide nanoparticles: Application to textiles as UV-absorbers. *J. Nanopart Res.*, 10(4): 679–689.

Burniston, N., Bygott,C. and Stratton, J. (2004). Nano technology meets titanium dioxide. *Surf. Coat. Int. Part A, Coatings Journal*, 87(4): 179–184.

Carmen, I.U., Chithra, P., Huang, Q., Takhistov, P., Liu, S. and Kokini, J.L. (2003). Nanotechnology: A new frontier in food science. *Food Technol.*, 57: 24–29.

Cifuentes, Zuny, Laura Custardoy, Jesús M. de la Fuente, Clara Marquina, M. Ricardo Ibarra, Diego Rubiales and Alejandro Pérez-de-Luque. (2010). Absorption and translocation to the aerial part of magnetic carbon-coated nanoparticles through the root of different crop plants. *J. Nanobiotechnol.*, 8: 26–33.

Davière, J.M. and Achard P.A. (2016). A pivotal role of DELLAs in regulating multiple hormone signals. *Mol. Plant*, 9(1): 10–20.

Directorate of Cotton Development Government of India. (2017). https://www.nfsm.gov.in/StatusPaper/Cotton2016.pdf.

Durán Nelson, Priscyla D., Marcato, Gabriel I.H. De Souza, Oswaldo L. Alves and Elisa Esposito. (2007). Antibacterial effect of silver nanoparticles produced by fungal process on textile fabrics and their effluent treatment. *J. Biomed. Nanotech.*, 3(2): 203–208.

El-Khatib, E.M. (2012). Antimicrobial and self-cleaning textiles using nanotechnology. *Res. J. Text. Appar.*, 16(3): 156–174.

El-Naggar, Mehrez, E., Shaarawy, S. and Hebeish, A.A. (2018). Multifunctional properties of cotton fabrics coated with *in situ* synthesis of zinc oxide nanoparticles capped with date seed extract. *Carbohydr. Polym.*, 181: 307–316.

El-Nahhal, I.M., Shehata M. Zourab, Fawzi S. Kodeh, Mohamed Selmane, Isabelle Genois and Florence Babonneau. (2012). Nanostructured copper oxide-cotton fibers: Synthesis, characterization, and applications. *Int. Nano Lett.*, 2(1): 14.

El-Rafie, M.H., Mohamed, A.A., Shaheen, Th I. and Hebeish, A. (2010). Antimicrobial effect of silver nanoparticles produced by fungal process on cotton fabrics. *Carbohydr. Polym.*, 80(3): 779–782.

FAO News, FAO report: Keeping plant pests and diseases at bay: Experts focus on global measures at annual meeting of the Commission on Phytosanitary Measures (CPM). http://www.fao.org/news/story/en/item/280489/icode/ (Accessed 20 October 2019).

Gao, Qinwen, Quan Zhu, Yuliang Guo and Charles Q. Yang. (2009). Formation of highly hydrophobic surfaces on cotton and polyester fabrics using silica sol nanoparticles and nonfluorinated alkylsilane. *Ind. Eng. Chem. Res.*, 48(22): 9797–9803.

Goel, S., Grewal, S., Singh, K. and Dwivedi, N. (2019). Impact of biotechnology and nanotechnology on future bread improvement: An overview. *Indian J. Agric. Sci.*, 89(9): 39–41.

Grewal, S., Goel, S., Kaushik, S. and Gill, A. (2017). NANO science having GIANT potential in Indian agriculture. *Trends in Biosciences*, 10(11): 1967–1969.

Hafizah, Nor and IisSopyan. (2009). Nanosized TiO_2 photocatalyst powder via sol-gel method: Effect of hydrolysis degree on powder properties. *International Journal of Photoenergy*, Article ID 962783. https://doi.org/10.1155/2009/962783.

Haghighi, Maryam, Zahra Afifipour and Maryam Mozafarian. (2012). The effect of N-Si on tomato seed germination under salinity levels. *J. Biol. Environ. Sci.*, 6(16): 87–90.

Hasan, Redwanul. (2018). Production of antimicrobial textiles by using copper oxide nanoparticles. *Int. J. Contemp. Res. Rev.*, 9(8): 20195–20202.

Hatami, Mehrnaz, Khalil Kariman and Mansour Ghorbanpour. (2016). Engineered nanomaterial-mediated changes in the metabolism of terrestrial plants. *Sci. Total Environ.*, 571: 275–291.

Hussein, M.M. and Abou-Baker, N.H. (2018). The contribution of nano-zinc to alleviate salinity stress on cotton plants. *R. Soc. Open Sci.*, 5(8): 171809.

Jeong, Sung Hoon, Yun Hwan Hwang and Sung Chul Yi. (2005). Antibacterial properties of padded PP/PE nonwovens incorporating nano-sized silver colloids. *J. Mater. Sci.*, 40(20) : 5413–5418.

Kagan, Cherie R. (2016). At the nexus of food security and safety: Opportunities for nanoscience and nanotechnology. *ACS Nano*, 10(3): 2985–2986.

Lansdown, A.B.G. (2002). Silver 2: Toxicity in mammals and how its products aid wound repair. *J. Wound Care*, 11(5): 173–177.

Le, Van Nhan, Chuanxin Ma, Yukui Rui, Weidong Cao, Yingqing Deng, Liming Liu and Baoshan Xing. (2016). The effects of Fe_2O_3 nanoparticles on physiology and insecticide activity in non-transgenic and Bt-transgenic cotton. *Front. Plant Sci.*, 6: 1263.

Le Van, Nhan, Chuanxin Ma, Jianying Shang, Yukui Rui, Shutong Liu and Baoshan Xing. (2016a). Effects of CuO nanoparticles on insecticidal activity and phytotoxicity in conventional and transgenic cotton. *Chemosphere*, 144: 661–670.

Le Van, Nhan, Yukui Rui, Weidong Cao, Jianying Shang, Shutong Liu, Trung Nguyen Quang and Liming Liu. (2016b). Toxicity and bio-effects of CuO nanoparticles on transgenic Ipt-cotton. *J. Plant Interact.*, 11(1): 108–116.

Le, V.N., Rui, Y., Gui, X., Li, X., Liu, S. and Han, Y. (2014). Uptake, transport, distribution, and bio-effects of SiO_2 nanoparticles in Bt-transgenic cotton. *J. Nanobiotechnol.*, 12(50): 5 Dec. doi:10.1186/s12951-014-0050-8.

Lee, Chang Woo, Shaily Mahendra, Katherine Zodrow, Dong Li, Yu-Chang Tsai, Janet Braam and Pedro J.J. Alvarez. (2010). Developmental phytotoxicity of metal oxide nanoparticles to *Arabidopsis thaliana*. *Environ. Toxicol. Chem.*, 29(3): 669–675.

Lee, H.J., Sang Young Yeo and Sung Hoon Jeong. (2003). Antibacterial effect of nanosized silver colloidal solution on textile fabrics. *J. Mater. Sci.*, 38(10): 2199–2204.

Li, Na-na, Wen-jun Qian, Lu Wang, Hong-li Cao, Xin-yuan Hao, Ya-jun Yang and Xin-chao Wang. (2017). Isolation and expression features of hexose kinase genes under various abiotic stresses in the tea plant (*Camellia sinensis*). *J. Plant Physiol.*, 209: 95–104.

Li, Xuguang, Xin Gui, Yukui Rui, Weikang Ji, Zihan Yu and Shengnan Peng. (2014). Bt-transgenic cotton is more sensitive to CeO_2 nanoparticles than its parental non-transgenic cotton. *J. Hazard. Mater.*, 274: 173–180.

Mishra, Sandhya, Akanksha Singh, Chetan Keswani, Amrita Saxena, Sarma, B.K. and Singh, H.B. (2015). Harnessing plant-microbe interactions for enhanced protection against phytopathogens. pp. 111–125. *In*: *Plant Microbes Symbiosis*: *Applied Facets*. New Delhi: Springer.

Nair, Remya, Saino Hanna Varghese, Baiju G. Nair, Maekawa, T., Yoshida, Y. and Sakthi Kumar, D. (2010). Nanoparticulate material delivery to plants. *Plant Sci.*, 179(3): 154–163.

Nel, Andre, Tian Xia, Lutz Mädler and Ning Li. (2006). Toxic potential of materials at the nano level. *Science*, 311(5761): 622–627.

Nhan, L.V., Ma, C., Rui, Y., Liu, S., Li, X., Xing, B. and Liu, L. (2015). Phytotoxic mechanism of nanoparticles: Destruction of chloroplasts and vascular bundles and alteration of nutrient absorption. *Sci. Rep.*, UK 5: 1–13.

Oerke, E-C. and Dehne, H-W. (2004). Safeguarding production: Losses in major crops and the role of crop protection. *Crop Prot.*, 23(4): 275–285.

Pandey, Anjula and Rita Gupta. (2003). Fibre yielding plants of India Genetic resources, perspective for collection and utilisation. *Nat. Prod. Radiance*, 2 (4): 194–204.

Perelshtein, I., Guy Applerot, Nina Perkas, Eva Wehrschuetz-Sigl, Andrea Hasmann, Georg Gübitz and Aharon Gedanken. (2009). CuO–cotton nanocomposite: Formation, morphology, and antibacterial activity. *Surf. Coat. Technol.*, 204(1): 54–57.

Pisitsak P., Samootsoot, A. and Chokpanich, N. (2013). Investigation of self-cleaning properties of cotton fabrics finished with nano-TiO_2 and nano-TiO_2 mixed with fumed silica. *KKU Res J.*, 18(2): 200–211.

Przybylak, Marcin, Hieronim Maciejewski, Agnieszka Dutkiewicz, Izabela Dąbek and Marek Nowicki. (2016). Fabrication of super-hydrophobic cotton fabrics by a simple chemical modification. *Cellulose*, 23(3): 2185–2197.

Ramaratnam, Karthik, Volodymyr Tsyalkovsky, Viktor Klep and Igor Luzinov. (2007). Ultrahydrophobic textile surface via decorating fibers with monolayer of reactive nanoparticles and non-fluorinated polymer. *Chemical Communications*, 43: 4510–4512.

Ravindra, Sakey, Murali Mohan, Y., Narayana Reddy, N. and Mohana Raju, K. (2010): Fabrication of antibacterial cotton fibres loaded with silver nanoparticles via Green Approach. *Colloids and Surfaces A: Physicochemical and Engineering Aspects*, 367, no. 1(3): 31–40.

Ravishankar Rai, V. and Jamuna Bai, A. (2011). *Nanoparticles and Their Potential Application as Antimicrobials*. Méndez-Vilas, A. (ed.). Mysore, India: Formatex, pp. 197–209.

Reddy, P., Venkata Laxma, J.A., Hernandez-Viezcas, J.R., Peralta-Videa and Gardea-Torresdey, J.L. (2016). Lessons learned: Are engineered nanomaterials toxic to terrestrial plants? *Sci. Total Environ.*, 568: 470–479.

Rezaei, Mohammadali and Hossein, Abbasi. (2014). Foliar application of nanochelate and non-nanochelate of zinc on plant resistance physiological processes in cotton (*Gossipium hirsutum* L.). *Iran J. Plant Physiol.*, 4: 1137–1144.

Saito, Mitumasa. (1993). Antibacterial, deodorizing, and UV absorbing materials obtained with zinc oxide (ZnO) coated fabrics. *J. Coat. Fabr.*, 23(2): 150–164.

Sato, H. and Ikeya, M. (2004). Organic molecules and nanoparticles in inorganic crystals: Vitamin C in $CaCO_3$ as an ultraviolet absorber. *Int. J. Appl. Phys.*, 95(6): 3031–3036.

Servin, Alia, Wade Elmer, Arnab Mukherjee, Roberto De la Torre-Roche, Helmi Hamdi, Jason C. White, Prem Bindraban and Christian Dimkpa. (2015). A review of the use of engineered nanomaterials to suppress plant disease and enhance crop yield. *J. Nanopart Res.*, 17(2): 1–21.

Shaker, A.M., Zaki, A.H., Abdel-Rahim, E.F.M. and Khedr, M.H. (2017). TiO_2 nanoparticles as an effective nanopesticide for cotton leaf worm. *Agricultural Engineering International: CIGR Journal*, Special: 61–68.

Shah, D.K. (2012). Bt cotton in India: A review of adoption, government interventions and investment initiatives. *Indian J. Agric. Econ.*, 67: (902-2016-67854).

Shallan, Magdy A., Hazem M.M. Hassan, Alia A.M. Namich and Alshaimaa A. Ibrahim. (2016). Biochemical and physiological effects of TiO_2 and SiO_2 nanoparticles on cotton plant under drought stress. *Res. J. Pharm. Biol. Chem. Sci.*, 7(4): 1540–1551.

Sheorain, Jyoti, Meenakshi Mehra, Rajesh Thakur, Sapna Grewal and Santosh Kumari. (2019). *In vitro* anti-inflammatory and antioxidant potential of thymol loaded bipolymeric (tragacanth gum/chitosan) nanocarrier. *Int. J. Biol. Macromol.*, 125: 1069–1074.

Siddiqui, Manzer H. and Mohamed H. Al-Whaibi. (2014). Role of nano-SiO_2 in germination of tomato (*Lycopersicum esculentum* seeds Mill.). *Saudi J. Biol. Sci.*, 21(1): 13–17.

Sirelkhatim, Amna, Shahrom Mahmud, Azman Seeni, Noor Haida Mohamad Kaus, Ling Chuo Ann, Siti Khadijah Mohd Bakhori, Habsah Hasan and Dasmawati Mohamad. (2013). Review

on zinc oxide nanoparticles: Antibacterial activity and toxicity mechanism. *Nanomicro Lett.*, 7(3): 219–242.
Sivakumar, A., Murugan, R., Sunderesan, K. and Periyasamy, S. (2013). UV protection and self-cleaning finish for cotton fabric using metal oxide nanoparticles. *IJFTR*, 38: 285–292.
Sricharussin, W., Threepopnatkul, P. and Neamjan, N. (2011). Effect of various shapes of zinc oxide nanoparticles on cotton fabric for UV-blocking and anti-bacterial properties. *Fibers Polym.*, 12(8): 1037–1041.
Wang, Aiwu, Yuhong Zheng and Feng Peng. (2014). Thickness-controllable silica coating of CdTeQDs by reverse microemulsion method for the application in the growth of rice. *J. Spectrosc*, 1–5.
Wang, Xiaoshan, Zhenwu Wei, Dalin Liu and Guoqi Zhao. (2011). Effects of NaCl and silicon on activities of antioxidative enzymes in roots, shoots, and leaves of alfalfa. *Afr. J. Biotechnol.*, 10(4): 545–549.
Xiao, Guanghui, Peng Zhao and Yu Zhang. (2019). A pivotal role of hormones in regulating cotton fiber development. *Front. Plant Sci.*, 10: 3389.
Yadav, A., Virendra Prasad, A.A. Kathe, Sheela Raj, Deepti Yadav, Sundaramoorthy, C. and Vigneshwaran, N. (2006). Functional finishing in cotton fabrics using zinc oxide nanoparticles. *Bull. Mater. Sci.*, 29(6): 641–645.
Zaki, A.M., Zaki, A.Z., Farghali, A.A. and Elham F. Abdel-Rahim. (2017). Sodium titanate-bacillus as a new nanopesticide for cotton leaf-worm. *J. Pure Appl. Microbiol.*, 11: 725–732.
Zhang, Feng, Xiaolan Wu, Yuyue Chen and Hong Lin. (2009). Application of silver nanoparticles to cotton fabric as an antibacterial textile finish. *Fibers Polym.*, 10(4): 496–501.
Zhang, Peng, Yuhui Ma, Zhiyong Zhang, Xiao He, Zhi Guo, Renzhong Tai, Yayun Ding, Yuliang Zhao and Zhifang Chai. (2012). Comparative toxicity of nanoparticulate/bulk Yb2O3 and YbCl3 to cucumber (*Cucumis sativus*). *Environ. Sci. Technol.*, 46(3): 1834–1841.
Zhang, WenJun. (2018). Global pesticide use: Profile, trend, cost/benefit and more. *Proc. Int. Acad. Ecol. Environ. Sci.*, 8(1): 1–27.

Section B
Environmental Nanotechnology

Recycling of Agricultural Waste

Chapter 15

Synthesis of Agro-waste-mediated Silica Nanoparticles: An Approach Towards Sustainable Agriculture

Rita Choudhary, Pawan Kaur and *Alok Adholeya**

1. Introduction

Agriculture is a vital segment of the Indian economy. The versatile use of crop residue includes, though not constrained to, animal feeding, soil compost, roofing for rural homes, and fuel for industrial use nationwide. Despite these advantages, growers burn a significant portion of the crop residue on-farm so that the next crop is planted on a clear field (Devi et al., 2016). Burning of agricultural residue has been identified as a major health hazard in many countries. In India, especially in the northern states of Punjab, Haryana, and Uttar Pradesh, biomass burning after wheat, rice, or sugarcane harvesting is a cyclical problem. The issues turn grave during winters when major parts of northern India stifle on exhaust cloud and fog activated by large-scale crop choke on smog residue burning. This leads to significant levels of particulate matter (PM), thereby affecting the well-being of the individuals and ecosystem (Yadav et al., 2017). In north-western part of India, more than 75% of rice area is harvested mechanically using a combine harvester, which cuts the crop in a way that a substantial amount of stubble remains after harvesting, which is hard to pick by farmers from the fields. Another problem arising from this stubble burning

TERI-Deakin Nanobiotechnology Centre, The Energy and Resources Institute (TERI), Lodhi Road, New Delhi, 110003, India.

* Corresponding author: adholeya1@gmail.com

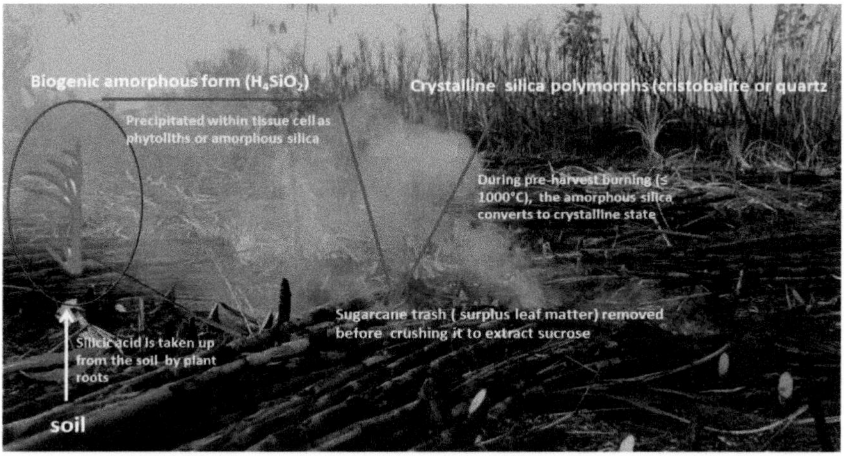

Fig. 15.1. Burning of sugar cane crop and conversion of biogenic amorphous silica into crystalline silica due to open field burning.

is that most of the crop residue has a high content of silica, which is unpalatable for animals. Since agriculture mostly generates unpalatable biomass, instead of incurring a high cost on collecting crop residue (90–140 Mt annually), farmers are left with the only alternative to burn it (Smil, 1999). Such crop burning practices have a damaging effect on the agro-ecosystems and cause global warming by producing tropospheric methane, a greenhouse gas, which is 60 times more hazardous than carbon monoxide (Smil, 1999). Hence, finding sustainable, technological solutions can help farmers and simultaneously allow a better management of crop residue (Singh and Kumar, 2018). In this chapter, the discussion points include crop residue potential, management options related to green approaches to develop biosilica from crop residue, and conversion of biosilica into valuable delivery systems by replacing potentially harmful and expensive organosilicates as a silica source among other things (Fig. 15.1).

1.1 Types of Agricultural Residue

India is an agrarian country and generates a large quantity of agricultural residue. Given the growing population and the consequent necessity to increase agricultural production, this quantity of crop residue is projected to rise in the future. Agriculture residue is distinguished based on availability and characteristics, and can be categorized into two types, i.e., "field residue" that remains on the field and "process residue", which is generated when crops are processed into alternate valuable resources (Sadh et al., 2018; Asakereh et al., 2014). These residues consist of leaves, husks, straws, molasses, bagasse, seeds, and roots as shown in Fig. 15.2. In sugarcane fields, leaves are the major residue in fields and farmers burn leaves of sugarcane in the fields, which produce fly ash that severely affect soil microbial flora and the ecosystem. Therefore, sugarcane leaves, because of their high silicon content (Kamwilaisak, 2016), being an abundant, inexpensive (Krishna et al., 1998; Singh et al., 2008; Krishnan et al., 2010), untouched, and readily available source of silicon are useful raw materials for

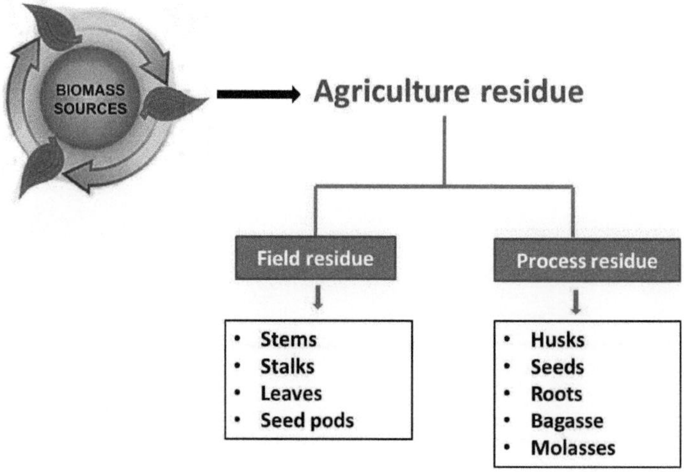

Fig. 15.2. Types of agricultural residue and their components.

synthesizing nanosilica (Kaupp, 1984; Kapur, 1985; James and Rao, 1986; Jenkins et al., 1996; Natarajan et al., 1998; Stephens et al., 2002; Noor-ul-Amin, 2014; Vaibhav et al., 2015).

Despite being the second most plenteous element in the Earth's crust, silicon is never considered important; however, it is advantageous for crop growth. Silica strengthens the plant, protects it from pests, increases crop production and quality, enhances plant nutrition, and neutralizes heavy metal toxicity in acid soils. Plants vary in their capacity to take up silicon (Rao and Susmitha, 2017). The amount of biosilica present in crop residue is directly proportional to the amount and accessibility of silicon in soil. In plants, transportation of silica occurs from the roots to the other parts via absorption of silicic acid from soil (Tubana et al., 2016). With reference to crop residue in India, sugarcane leave ash (SLA), having 80.14% of silica, is most widely explored for its application as a raw material in different sectors such as ethanol industry, building material, energy production, etc. (Arumugam and Ponnusami, 2013). The percentage of silica content in sugarcane leaves generally varies from crop to crop and depends on climatic conditions and soil type (Norsuraya et al., 2016).

1.2 Use of Crop Residue for Developing Value-added Applications in Agriculture

Nearly 50% of crop residue is used by canneries as a source of energy and the remainder is either piled up or burned, which further generates 1–4% of ash. Considering the number of crops being produced across India, each year almost 3–12 million tonnes of ash is generated by crop residue burning. Generally, biowaste generated from sugar industries is either disposed in landfills or used as fertilizers for plants. Such practices have adversely affected not only soil and water but public health too (Sales and Lima, 2010).

Owing to the emerging interest in the perspectives on ecological contamination and the need to moderate vitality and material assets, the use of crop residue and derived silica in various applications has been energized. On the one hand, several experts have concluded that crop residue is an excellent source of high-grade amorphous silica. On the other hand, the silica/carbon composition obtained from the thermal decomposition of crop residue under inert atmosphere has demonstrated to be a good crude material for the synthesis of silicon nitride (Si_3N_4). However, what is missing is a study concerning the use of crop residue as a raw material for obtaining silica with small particle sizes and high specific surface area that, when sufficient temperature and porosity are maintained, can be used as a carrier or delivery system.

Many proposals based on the torrefaction method have resulted in gathering these agriculture wastes to combust and generate electricity. Though, generation of high ash content from crop residue is one of the major shortcomings to energy (electricity) (Karmakar et al., 2013) and during combustion, undesirable waste is also formed (Lim et al., 2012), which leads to functioning problems, agglomeration, discolouring, and decay of heat transfer surfaces (Armesto et al., 2002; Bakker et al., 2002; Van Caneghem et al., 2012). Similarly, various pre-treatment techniques appear to be available for rice straw in principle but then again, the commercial use of rice straw for energy is still not found in numerous rice-producing countries due to the associated costs and lack of incentives or paybacks for farmers to gather the straw rather than burning it in the field, which increases greenhouse gas emissions and environmental pollution (Kargbo et al., 2010). Silicon-based material and its nano form have been the key research territory in the recent past on account of their broad capacities (Beall, 1994). In 2005, Siriluk and Yuttapong (2005) used biosilica, extracted from rice husk ash (RHA), to synthesize MCM-41 type silica material. The results demonstrated that the crystallinity and porosity of silica synthesized from biosilica were like those synthesized from commercial silica. Similarly, various research groups have shown that the synthesis of mesoporous silica nanoparticles and their surface functionalization using biosilica, extracted from RHA (Grisdanurak et al., 2003; Bhagiyalakshmi et al., 2010), had the same efficiency for absorption of chlorinated volatile organic molecules as depicted by conventional silica material. Further, Rahman et al. (2015) also reported the synthesis of porous silica with controlled pore structure from sugarcane waste known as bagasse. But, as per our knowledge, there are no reports about the use of field residue, i.e., sugarcane leaves as a raw material to produce amorphous silica nanoparticles. Such untouched waste biomass from sugarcane is opening opportunities for developing smart delivery systems for agri-inputs (Rovani et al., 2018). But a modest quantity of metal elements is difficult to expel, and the cost of silica is higher. The application extent of silica is constrained due to the presence of metal elements. Economically too, it continues to pose a challenge to remove metals from the agriculture waste biomass.

1.3 Approaches of Extracting Silicon from Agricultural Waste

Silica is a porous material with high chemical, thermal, and mechanical stability. But this approach is costly, unsafe, and requires meticulous synthesis conditions (Shen, 2017). Synthetic silica precursors, such as tetraethoxysilane (TEOS), tetramethoxysilane (TMOS), and sodium silicate [$Na_{2x}Si_yO_{2y+x}$ or $(Na_2O)_x(SiO_2)_y$] are commonly used for the synthesis of silica particles. However, these are not promising sources for synthesizing silica commercially as they are flammable, costly, and pose difficulties in handling transportation and storage (Kalapathy et al., 2000; Rida and

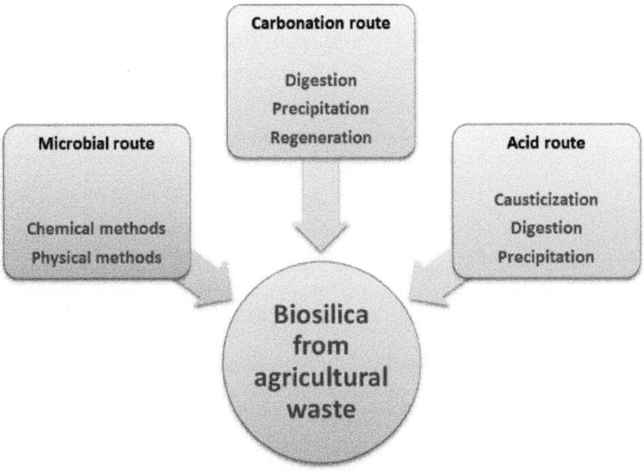

Fig. 15.3. Different routes for extracting biosilica from agricultural waste.

Harb, 2014). Thus, the extraction of biosilica from various types of crop residue (Bageru and Srivastava, 2017) is considered cost-effective and the best alternative to synthesized silica as compared to conventional synthetic precursors. There are two conventional methods for extracting reactive silica from biomass, namely, the carbonation route or the acid route, and the microbial route, which, according to some studies, has also been used for the synthesis of silica (Fig. 15.3).

1.4 Synthesis of Nanosilica from Crop Residue

Efforts have been made by researchers to produce nano-silica using crop residue. Scientists have been working on the synthesis of silica from RHA within a range of 5–10 nm, using the precipitation method (Thuadaij and Nuntiya, 2008), with 80 nm particle size using the sol-gel method (Amutha et al., 2010; Adam et al., 2011) within a range of 5–30 nm using the dissolution-precipitation process (Liou and Yang, 2011). Likewise, Witoon et al. (2008) also used RHA as a raw material and chitosan as a template to produce porous nano-silica. A mesoporous product, which was developed, appeared like a worm, and when the chitosan template was removed, macroporous silica was obtained. Further, Alves et al. (2017) demonstrated that sugarcane waste is also a good source to extract biosilica using sodium hydroxide (NaOH) and hydrochloric acid (HCl) precipitation and characterized by chemical and spectroscopic techniques. Various factors were assessed that could impact the tuning of size and shape of silica particles during synthesis. The experimental data proposed that sugarcane waste ash could be an eco-friendly source transforming biosilica into nano-silica as a value-added product. Likewise, Norsuraya et al. (2016) also showed in their study that silica nanoparticles were synthesized using sodium silicate $Na_{2x}Si_yO_{2y+x}$ or $(Na_2O)_x{\cdot}(SiO_2)_y$, extracted from sugarcane bagasse ash. To increase its purity, HCl was used after the treatment and the results showed that the quantity of SiO_2 was enhanced up to 88.13% as compared to the quantity in raw sugarcane bagasse ash. Extensive literature is available on the fabrication of the silica nanoparticles and the reasons of it gaining popularity is its abundant occurrence and low cost. The commonly reported methods for the fabrication of both pure and amorphous biogenetic silica include both chemical and thermal approaches.

2. Application of SILICA for Potential Agronomic Benefits

2.1 Silicon Fertilizers

Silica has the potential to enhance the immunity of plants under biotic and abiotic stresses (Ma, 2004) and it reduces the toxicity due to metal content elevation, increases water-use efficiency due to its swelling property, and acts as a potent bioprotectant against microbial attack (Datnoff et al., 1997) from pests, root reinforcement, and upregulation of antioxidative defense responses (Ma and Takahashi, 2002). Several studies have demonstrated and explained the role and mechanism of silica in boosting phosphorus availability. Therefore, silica can be an additional element to support plant growth. As compared to conventional silica, nanosilica has way more advantages, namely, it is more reactive because of its larger surface area, directly reaches the target, and is only required in small amounts (Suciaty et al., 2018). Yuvakkumar et al. (2011) reported that the application of nanosilica in soil can enhance the growth of maize plants. Also, nanosilica has been efficiently used to control a variety of agricultural insect including parasites, which are ectodermal, in animals (Ulrichs et al., 2005). With these advantages, the use of nanosilica as a fertilizer could be a catalyst for a breakthrough technology to intensify agricultural productivity that is both continuous and environmentally friendly.

2.2 Smart Delivery Systems

Since the role of agriculture is to ensure adequate supply of micronutrients and macronutrients to the plants, today, the major concern is that most soils found across the world are nutrient deficient (Reetz, 2016). On the one hand, this problem triggered manufacturers the world over to produce synthetic agrochemicals and market these to farmers to be used as fertilizers and pesticides so that crop loss could be minimized, and crop yield enhanced. On the other hand, the non-judicial use of these synthetic agrochemicals raised serious concerns over the environmental repercussions. The accumulation of these chemicals in the food chain and the consequent contamination in the neighbouring ecosystems (Puoci et al., 2008) had detrimental effects.

Using smart delivery systems for the sustained release of synthetic agrochemicals has been an alternative to the problem discussed earlier. Through these delivery systems, agrochemicals could be delivered directly to the soil and plants, thereby fulfilling the nutritional needs without contamination (Puoci et al., 2008). The implementation of delivering agrochemicals to the targeted area is based on either physical or chemical processes. For example, release kinetics of the agrochemicals from the delivery vehicle to the soil or to the plants is completely based on the physical mechanism, whereas the loading and stabilization of the agrochemicals within the delivery system is a chemical process. The criteria of selecting the best material to develop the delivery system are dependent on the biological and chemical properties of the agrochemical that must be released from the cargo including its physiochemical interactions (Puoci et al., 2008). In addition, the release profile of the loaded agrochemicals from the cargo is categorized into slow- and controlled-release processes. The release mechanism of agrochemicals in a slow-release fertilizer (SRF) is dependent on factors, such as soil temperature and microbial activity in which the release is delayed, whereas the controlled-release process depends on the mode of mechanism to release the agrochemicals into the environment like diffusion mode (Trenkel, 2010; Pereira et al., 2015). The controlled-release fertilizer (CRF) has a greater degree of control on rate, pattern, and duration (Trenkel, 1997; Shaviv, 2001; Trenkel, 2010). There are a lot of reports which are based on slow- and controlled-release patterns of agrochemicals in water (Shaviv, 2001; Du et al., 2006; Melaj and Daraio, 2013; Noppakundilograt et al., 2015).

The development of slow- and controlled-release agrochemicals in the agriculture sector is not a recent innovation. In 1962, Oertli and Lunt (1962) developed a controlled-release system for fertilizer release, using membrane encapsulation strategy (Al-Zahrani, 1999). In 1971, Allan et al. (1971) published their research data on the development of controlled-release pesticides. With the emergence of nanotechnology and its application in agriculture, researchers are now focused in reshaping the conventional technologies using nanotechnology to make them much smarter to provide solutions to a broader range of problems under one umbrella. One such example is extracting biosilica from agricultural waste and converting it into mesoporous silica nanoparticles using sol-gel methodology. There are several reports that have confirmed the use of mesoporous silica nanoparticles as a delivery vehicle for the controlled release of agrochemicals. Wen et al. (2005) was the first research group that used porous hollow silica nanoparticles (PHSNs) as cargo for controlled release. Their findings highlighted that PHSNs successfully delayed the release of avermectin and indicated that theycould be explored for delivery of agrochemicals. All these studies highlighted the possible opportunities to develop biosilica-based porous silica materials, abundant in the environment, particularly those involving the utilization of waste material. This would improve the worth of the wastes and proffer an answer to the menace caused by their improper disposal.

Further, the use of silica nanoparticles in agriculture has gained much interest in the areas of soil amendments and pesticide delivery systems (Lin et al., 2009; Kurepa et al., 2010). The reason for research groups' interest in porous silica materials is chiefly because of the materials' thermal stability and tuneable structures to develop nanomaterials of different sizes and shapes, including those having pores where the active agent can be loaded (Torney et al., 2007; Shi et al., 2010; Jang et al., 2013). Another reason is the natural existence of silica in soil and plants. As per available literature, there are two kinds of engineered silica: non-porous and porous (Slomberg and Schoenfisch, 2012; Wanyika et al., 2012) (Table 15.1 gives studies showing silica nanoparticles as delivery systems).

All the research outcomes clearly indicated that MSNs have potential to be the most innovative system to deliver agrochemicals in plants. Figure 15.4 depicts the possible applications of biosilica in agriculture.

Table 15.1. Description of interesting studies related to mesoporous silica nanoparticles as delivery systems.

Application of MSNs in Plants	**References**
Use of MSNs coated with gold particles as smart delivery system to deliver DNA into tobacco protoplasts	Torney et al., 2007
Uptake and phytotoxicity of nonporous silica nanoparticles on rice seedlings and on the roots of *Arabidopsis*	Nair et al., 2011; Slomberg and Schoenfisch, 2012
Translocation of nanocomposites of MSNs with metal oxides has also been studied in *Arabidopsis*, ryegrass, and pumpkin	Wang et al., 2012; Kurepa et al., 2010; Zhu et al., 2008
Surface functionalized MSNs used for transporting key molecules of cell nucleus and chemicals to plant tissues and plant cells using the biolistic method	Martin-Ortigosa et al., 2012
Hollow MSNs as smart cargo sustained release of avermectin pesticide	Wen et al., 2005
Developed a formulation of a bio-pesticide (pyoluteorin) with slow-release function through mesoporous silica as delivery system	Chen et al., 2011
MSNs for controlled release of urea in soil and water through pores of MSNs	Wanyika et al., 2012
MSNs (20 nm) with surface functionalization, were successfully moved into the roots of lupin and wheat plants without impacting the growth of these model plants	Sun et al., 2016
Diquat dibromide (DQ), a quaternary ammonium herbicide used globally, was loaded into negatively charged MSN-SO_3 nanoparticles	Shan et al., 2019

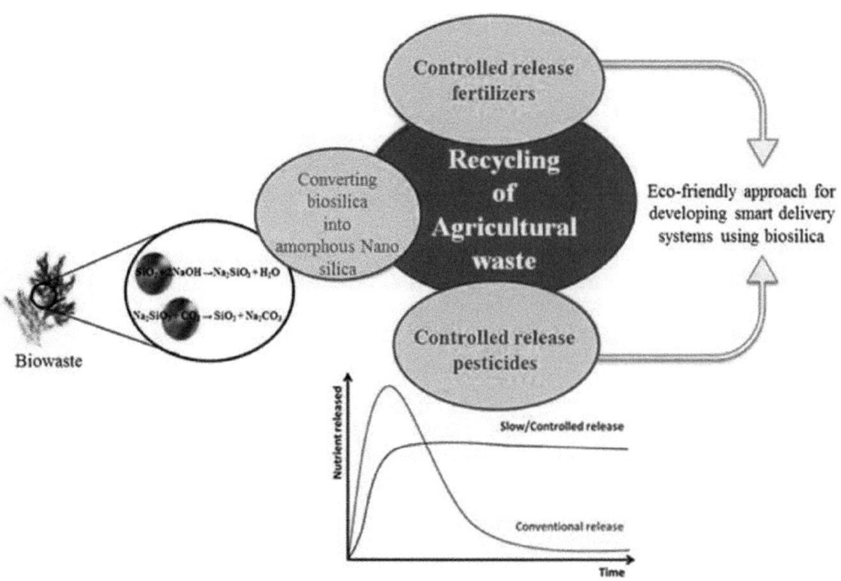

Fig. 15.4. Schematic diagram of the possible applications of biosilica in agriculture.

3. Global Market of Silicon Fertilizer

Considering the world market, silicon fertilizer is rapidly growing in the agriculture sector of the global economy. Naturally, silicon occurs as oxides (silica) and silicates, but it is normally locked up in a crystalline form and plants only take up amorphous silica from the soil. As silica exists in the earth's covering, plants use it as macronutrients. It has the property of nourishing crops and is resistant to several plant diseases. Yet it is not considered an essential plant nutrient. Due to lack of awareness, silicon fertilizers have not been used as fertilizers over the past years. However, present-day agricultural practices are reforming, and the focus is on the use of nutrients that already exist in soil. Silicon displayed great effect on the development of plant roots, thus permitting better root resistance and its faster growth in drought conditions. These are used in varied crops such as corn, barley, sugarcane, wheat, citrus, cucumber, tomato, and others for improved productivity along with sustainable production. Due to its wide range of uses, by 2025, the global silicon fertilizer market is likely to exceed USD 125.84 million. The market is expected to grow at a CAGR of around 4.5% during the forecasted period of 2020–2025, specifically in Asia-Pacific region. The key players of the global silicon fertilizer market are Agripower, Aries Agro Ltd., BASF, BAYER International, CRODA, Compass Minerals International, Inc., Denka Co. Ltd., MaxSil, Plant Tu Inc., Redox Pty Ltd., The Mosaic Company, and Yara International ASA.[1]

With the implementation of modern agriculture practices, the call for silicon fertilizers is likely to grow manifold. Besides, the rise in population and growth in the economy of the developing countries are predicted to fuel the silicon fertilizer market. Limited accessibility of arable land is probable to generate untapped opportunity for the market participants. Similarly, modern agriculture practices offer new prospects for the growth of silicon fertilizers in the market.[2] An increased demand for most essential crops such as rice and wheat in different regions of the world can foster the growth of silicon-based fertilizers. High demand and consumption are the two major factors which contribute to an escalation in demand for silicon fertilizers by growers, providing new prospects for the silicon fertilizer.

4. Conclusions and Future Aspects

The extraction of biosilica from ash and its further conversion into valuable products are potential steps to ensure the solid waste management of crop residue. Syntheses of silica from agriwaste using inexpensive as well as eco-friendly strategies have been summarized in this chapter. Superfine silica powder of high purity can be prepared from biosilica under controlled calcination settings. For these reasons, the conversion of biosilica from waste into commercial grade nanosilica is a promising approach. Silica can be used in agriculture as a plant stimulator to protect plants from insects as well as phytopathogens, biotic and abiotic stresses. This chapter proposes that amorphous silica nanoparticles generated from agricultural waste ash or material could become an exceptional carrier material for loading agrochemicals. Such biowaste occurring as

[1] Details are available at https://www.alliedmarketresearch.com; last accessed on 12 December 2021.
[2] Details are available at https://www.alliedmarketresearch.com; last accessed on 10 December 2021.

crop-harvesting residue after harvesting or during processes in agricultural systems should be used to develop fertilizers from natural materials. These kinds of initiatives, for example, "take-make dispose" approach would eventually help in strengthening agricultural systems in a sustainable manner. Future investigation should continue to discover and assess the composition, manufacture, and agronomic and environmental performance of MSNs with various fertilizers and agrochemicals.

Acknowledgement

The research activities of the authors are supported by TERI-Deakin Nanobiotechnology Centre, Gurugram, India and by the Centre of Excellence for Advanced Research in Agricultural Nanotechnology supported by the Department of Biotechnology, Govt. of India (Grant No. BT/NNT/28/SP30280/2019) and TERI-Deakin Nano Biotechnology Centre (TDNBC), India.

References

Abou Rida, M. and Harb, F. (2014). Synthesis and characterization of amorphous silica nanoparticles from aqueous silicates using cationic surfactants. *J. Met. Mater. Miner.*, 24(1). doi: 10. 14456/jmmm.2014.7.

Adam, F., Chew, T.S. and Andas, J. (2011). A simple template-free sol–gel synthesis of spherical nanosilica from agricultural biomass. *J. Sol-Gel. Sci. Technol.*, 59: 580–3. doi: 10.1007/s10971-011-2531-7.

Allan, G.G., Chopra, C.S., Neogi A.N. and Wilkins, R.M. (1971). Design and synthesis of controlled release pesticide-polymer combinations. *Nature*, 234: 349–51. doi: 10.1038/234349a0.

Alves, R.H., Reis, T.V., Rovani, S. and Fungaro, D.A. (2017). Green synthesis and characterization of biosilica produced from sugarcane waste ash. *J. Chem.1*. doi: 10.1155/2017/6129035.

Al-Zahrani, S.M. (1999). Controlled-release of fertilizers: modelling and simulation. *Int. J. Eng. Sci.*, 37: 1299–307. doi: 10.1016/j.proeng.2016.06.444.

Amutha, K., Ravibaskar, R. and Sivakumar, G. (2010). Extraction, synthesis, and characterization of nanosilica from rice husk ash. *Int. J. Nanotechnol.*, 4: 61–66.

Armesto, L., Bahillo, A., Veijonen, K., Cabanillas, A. and Otero, J. (2002). Combustion behaviour of rice husk in a bubbling fluidised bed. *Biomass Bioenergy*, 23: 171–179. doi: 10.1016/S0961-9534(02)00046-6.

Arumugam, A. and Ponnusami. (2013). Modified SBA-15 synthesized using sugarcane leaf ash for nickel adsorption. India: NISCAIR-CSIR, pp. 101–106.

Asakereh, A., Omid, M., Alimardani, R. and Sarmadian, F. (2014). Spatial analysis the potential for energy generation from crop residues in Shodirwan, Iran. *Int. J. u- and e- Service Sci. Technol.*, 7(1): 275–284.

Bageru, A.B. and Srivastava V.C. (2017). Preparation and characterisation of biosilica from teff (*Eragrostis tef*) straw by thermal method. *Mater. Lett.*, 206: 13–17. doi: 10.1016/j.matlet.2017.06.100.

Bakker, R.R., Jenkins, B.M. and Williams R.B. (2002). Fluidized bed combustion of leached rice straw. *Energy Fuels*, 16(2): 356–365. doi: 10.1016/S0961-9534(03)00053-9.

Beall, G.H. (1994). Industrial applications of silica. *Rev. Mineral. Geochem.*, 469–505. doi: 10.1515/9781501509698-019.

Bhagiyalakshmi, M., Lee, J.Y. and Jang, H.T. (2010). Synthesis of mesoporous magnesium oxide: Its application to CO_2 chemisorption. *Int. J. Greenh. Gas.*, 4(1): 51–56. doi: 10.1016/j.ijggc.2009.08.001.

Chen, J., Wang, W., Xu, Y. and Zhang, X. (2011). Slow-release formulation of a new biological pesticide, pyoluteorin, with mesoporous silica. *J. Agric. Food Chem.*, 59(1): 307–11. doi: 10.1021/jf103640t.

Datnoff, L.E., Deren, C.W. and Snyder, G.H. (1997). Silicon fertilization for disease management of rice in Florida. *Crop Prot.*, 16(6): 525–531. doi: 10.1016/S0261-2194(97)00033-1.

Devi, S., Gupta, C., Jat, S.L. and Parmar, M.S. (2016). Crop residue recycling for economic and environmental sustainability: The case of India. *Open Agric.*, 2(1): 486–494.

Du, C.W., Zhou, J.M. and Shaviv, A. (2006). Release characteristics of nutrients from polymer-coated compound-controlled release fertilizers. *J. Polym. Environ.*, 14(3): 223–230. doi: 10.3390/plants10020238.

Grisdanurak, N., Chiarakorn, S. and Wittayakun, J. (2003). Utilization of mesoporous molecular sieves synthesized from natural source rice husk silica to chlorinated volatile organic compounds (CVOCs) adsorption. *Korean J. Chem Eng.*, 20(5): 950–955. doi: 10.1007/BF02697304.

James, J. and Rao, M.S. (1986). Silica from rice husk through thermal decomposition. *Thermochim. Acta.,* 97(1986): 329–336. doi: 10.1016/0040-6031(86)87035-6.

Jang, H.R., Oh, H.J., Kim, J.H. and Jung, K.Y. (2013). Synthesis of mesoporous spherical silica via spray pyrolysis: Pore size control and evaluation of performance in paclitaxel pre-purification. *Microporous Mesoporous Mater.*, 165(2013): 219–227. doi: 10.1016/J.MICROMESO.2012.08.010.

Jenkins, B.M., Bakker, R.R. and Wei, J.B. (1996). On the properties of washed straw. *Biomass Bioenergy*, 10(4): 177–200. doi: 10.1016/0961-9534(95)00058-5.

Kalapathy, U., Proctor, A. and Shultz, J. (2000). A simple method for production of pure silica from rice hull ash. *Bioresour. Technol.*, 73(3): 257–262. doi: 10.1016/S0960-8524(99)00127-3.

Kamwilaisak, K. (2016). Comparison study of sugarcane leaves and corn stover as a potential energy source in pyrolysis process. *Energy Procedia.*, 100: 26–29. doi: 10.1016/j.egypro.2016.10.142.

Kapur, P.C. (1985). Production of reactive bio-silica from the combustion of rice husk in a tube-in-basket (TiB) burner. *Powder Technol.*, 44(1): 63–67.

Kargbo, F.R., Xing, J. and Zhang, Y. (2010). Property analysis and pre-treatment of rice straw for energy use in grain drying: A review. *Agric. Biol. J. N. Am.*, 1(3): 195–200. doi: 10.5251/abjna.2010.1.3.195.200.

Karmakar, M.K., Mandal, J., Haldar, S. and Chatterjee, P.K. (2013). Investigation of fuel gas generation in a pilot scale fluidized bed autothermal gasifier using rice husk. *Fuel*, 1(111): 584–591. doi: 10.1016/j.fuel.2013.03.045.

Kaupp, A. (1984). *Gasification of Rice Hulls: Theory and Practices.* Eschborn: Deutsches Zentrum Fuer Entwicklungs Technologien (GATE).

Krishnan, C., Sousa, L.D., Jin, M., Chang, L., Dale, B.E. and Balan, V. (2010). Alkali-based AFEX pretreatment for the conversion of sugarcane bagasse and cane leaf residues to ethanol. *Biotechnol. Bioeng.*, 107(3): 441–450. doi: 10.1002/bit.22824.

Krishna, S.H., Prasanthi, K., Chowdary, G.V. and Ayyanna, C. (1998). Simultaneous saccharification and fermentation of pretreated sugar cane leaves to ethanol. *Process Biochem.*, 33(8): 825–830. doi: 10.1016/s0960-8524(00)00151-6.

Kurepa, J., Paunesku, T., Vogt, S., Arora H., Rabatic, B.M., Lu, J., Wanzer, M.B., Woloschak, G.E. and Smalle, J.A. (2010). Uptake and distribution of ultrasmall anatase TiO_2 Alizarin red S nanoconjugates in *Arabidopsis thaliana. Nano Lett.*, 10(7): 2296–2302. doi: 10.1021/nl903518f.

Lim, J.S., Manan, Z.A., Alwi, S.R. and Hashim, H. (2012). A review on utilisation of biomass from rice industry as a source of renewable energy. *Renew. Sust. Energ. Rev.*, 16(5): 3084–3094. doi: 10.1016/j.rser.2012.02.051.

Lin, S., Reppert, J., Hu, Q., Hudson, J.S., Reid, M.L., Ratnikova, T.A., Rao, A.M., Luo, H. and Ke, P.C. (2009). Uptake, translocation, and transmission of carbon nanomaterials in rice plants. *Small.*, 5(10): 1128–1132. doi: 10.1002/smll.200801556.

Liou, T.H. and Yang, C.C. (2011). Synthesis and surface characteristics of nanosilica produced from alkali-extracted rice husk ash. *Mater. Sci. Eng. B.*, 176(7): 521–529. doi: 10.1016/j.mseb.2011.01.007.

Ma, J.F. and Takahashi, E. (2002). *Soil, Fertilizer, and Plant Silicon Research in Japan.* Elsevier.

Ma, J.F. (2004). Role of silicon in enhancing the resistance of plants to biotic and abiotic stresses. P.C., 50(1): 11–18. doi: 10.1080/00380768.2004.10408447.

Martin-Ortigosa, S., Valenstein, J.S. Lin, V.S. Trewyn, B.G. and Wang, K. (2012). Gold functionalized mesoporous silica nanoparticle mediated protein and DNA codelivery to plant cells via the biolistic method. *Adv. Funct. Mater.*, 22(17): 3576–3582. doi: 10.1002/adfm.201200359.

Melaj, M.A. and Daraio, M.E. (2013). Preparation and characterization of potassium nitrate controlled-release fertilizers based on chitosan and xanthan layered tablets. *J. Appl. Polym. Sci.*, 130(4): 2422–2428. doi: 10.1021/acs.jafc.5b00518.

Nair, R., Poulose, A.C. Nagaoka, Y., Yoshida, Y., Maekawa, T. and Kumar, D.S. (2011). Uptake of FITC labeled silica nanoparticles and quantum dots by rice seedlings: effects on seed germination and their potential as biolabels for plants. *J. Fluoresc.*. 21(6): 2057. doi: 10.1007/s10895-011-0904-5.

Natarajan, E., Öhman, M., Gabra, M., Nordin, A., Liliedahl, T. and Rao, A.N. (1998). Experimental determination of bed agglomeration tendencies of some common agricultural residues in fluidized bed combustion and gasification, *Biomass Bioenergy*, 15(2): 163–169. doi: 10.1504/IJRET.2011.042727.

Noppakundilograt, S., Pheatcharat, N. and Kiatkamjornwong, S. (2015). Multilayer coated NPK compound fertilizer hydrogel with controlled nutrient release and water absorbency. *J. Appl. Polym. Sci.*, 132(2). doi: 10.1002/app.41249.

Noor-ul-Amin. (2014). A multi-directional utilization of different ashes. *RSC Adv.*, 4(107): 62769–62788. doi: 10.1039/C4RA06568A.

Norsuraya, S., Fazlena, H. and Norhasyimi, R. (2016). Sugarcane bagasse as a renewable source of silica to synthesize santa barbara amorphous-15 (SBA-15). *Procedia Eng.*, 148: 839–846. doi: 10.1016/j.proeng.2016.06.627.

Oertli, J.J. and Lunt, O.R. (1962). Controlled release of fertilizer minerals by incapsulating membranes: I. Factors influencing the ate of elease. *Soil Sci. Soc. Am. J.*, 26(6): 579–583. doi: 10.2136/sssaj1962.

Pereira, E.I., Giroto, A.S., Bortolin, A., Yamamoto, C.F., Marconcini, J.M., de Campos Bernardi, A.C. et al. (2015). Perspectives in nanocomposites for the slow and controlled release of agrochemicals: Fertilizers and pesticides. pp. 241–265. *In*: *Nanotechnologies in Food and Agriculture*. Cham: Springer.

Puoci, F., Iemma, F. and Picci. (2008). Stimuli-responsive molecularly imprinted polymers for drug delivery: A review. *Curr. Drug Deliv.*, 5(2): 85–96. doi: 10.2174/156720108783954888.

Rahman, N.A., Widhiana, I., Juliastuti, S.R. and Setyawan, H. (2015). Synthesis of mesoporous silica with controlled pore structure from bagasse ash as a silica source. *Colloids Surf. A Physicochem. Eng.*, 476: 1–7. doi: 10.1016/j.colsurfa.2015.03.018.

Rao, G.B. and Susmitha, P. (2017). Silicon uptake, transportation, and accumulation in rice. *J. Pharmacogn. Phytochem.*, 6: 290–293.

Reetz, H.F. (2016). Fertilizers and Their Efficient Use. International Fertilizer industry Association, IFA.

Rovani, S., Santos, J.J., Corio, P. and Fungaro, D.A. (2018). Highly pure silica nanoparticles with high adsorption capacity obtained from sugarcane waste ash. *ACS Omega*, 3(3): 2618–27. doi: 10.1021/acsomega.8b00092.

Sadh, P.K., Duhan, S. and Duhan, J.S. (2018). Agro-industrial wastes and their utilization using solid state fermentation: A review. *Bioresour. Bioprocess.*, 5(1): 1. doi: 10.1186/s40643-017-0187-z.

Sales, A. and Lima, S.A. (2010). Use of Brazilian sugarcane bagasse ash in concrete as sand replacement. *Waste Manage.*, 30(6): 1114–1122. doi: 10.1016/j.wasman.2010.01.026.

Shaviv, A. (2001). Advances in controlled-release fertilizers. *Adv. Agron.*, 71(1): 1–49. doi: 10.1016/S0065-2113(01)71011-5.

Shan, Y., Cao, L., Xu, C., Zhao, P., Cao, C., Li, F., Xu, B. and Huang, Q. (2019). Sulfonate-functionalized mesoporous silica nanoparticles as carriers for controlled herbicide diquat dibromide release through electrostatic interaction. *Int. J. Mol. Sci.*, 20(6): 1330. doi: 10.3390/ijms20061330.

Shen, Y. (2017). Rice husk silica derived nanomaterials for sustainable applications. *Renew. Sust. Energ. Rev.*, 80: 453–66. doi: 10.1016/j.rser.2017.05.115.

Shi, Y.T., Cheng, H.Y., Geng, Y., Nan, H.M., Chen, W., Cai, Q., Chen, B.H., Sun, X.D. and Yao, W.Y. (2010). The size-controllable synthesis of nanometer-sized mesoporous silica in extremely dilute surfactant solution. *Mater. Chem. Phys.*, 120(1): 193–198.

Singh, B. and Kumar, D. (2018). Crop residue management through options. *Int. J. Agric. Environ. Biotechnol.*, 11: 427–432. doi: 10.30954/0974-1712.06.2018.2.

Singh, P., Suman, A., Tiwari, P., Arya, N., Gaur, A. and Shrivastava, A.K. (2008). Biological pre-treatment of sugarcane trash for its conversion to fermentable sugars. *World J. Microbiol. Biotechnol.*, 24(5): 667–673. doi: 10.1016/j.rser.2013.06.033.

Siriluk, C. and Yuttapong, S. (2005). Structure of mesoporous MCM-41 prepared from rice husk ash. *In*: *Asian Symposium on Visualization*. Chaingmai, Thailand.

Slomberg, D.L. and Schoenfisch, M.H. (2012). Silica nanoparticle phytotoxicity to *Arabidopsis thaliana*. *Environ. Sci. Technol.*, 46(18): 10247–10254. doi: 10.1021/es300949f.

Smil, V. (1999). Crop Residues: Agriculture's Largest Harvest: Crop residues incorporate more than half of the world's agricultural phytomass. *Bioscience*, 49(4): 299–308.

Stephens, D.K., Wellen, C.W., Smith, J.B. and Kubiak, K.F. (2002). Agritec Co., Occidental Chemical Corp., assignee. Precipitated silicas, silica gels with and free of deposited carbon from caustic biomass ash solutions and processes. United States patent US 6,375,735.

Suciaty, T., Purnomo, D. and Sakya, A.T. (2018). The effect of nano-silica fertilizer concentration and rice hull ash doses on soybean (*Glycine max* (L.) Merrill) growth and yield. *In*: *IOP Conference Series: Earth and Environmental Science*. 129(1): 012009, IOP Publishing.

Sun, D., Hussain, H.I., Yi, Z., Rookes, J.E., Kong, L. and Cahill, D.M. (2016). Mesoporous silica nanoparticles enhance seedling growth and photosynthesis in wheat and lupin. *Chemosphere*, 152: 81–91. doi: 10.1016/j.chemosphere.2016.02.096.

Thuadaij, N. and Nuntiya, A. (2008). Preparation of nanosilica powder from rice husk ash by precipitation method. *Chiang Mai J. Sci.*, 35(1): 206–211.

Trenkel, M.E. (1997). *Controlled-release and Stabilized Fertilizers in Agriculture*. Paris: International Fertilizer Industry Association.

Trenkel, M.E. (2010). *Slow- and Controlled-release and Stabilized Fertilizers: An Option for Enhancing Nutrient Use Efficiency in Agriculture*. IFA, International Fertilizer Industry Association.

Torney, F., Trewyn, B.G., Lin, V.S. and Wang, K. (2007). Mesoporous silica nanoparticles deliver DNA and chemicals into plants. *Nat. Nanotechnol.* 2(5): 295–300. doi: 10.1038/nnano.2007.108.

Tubana, B.S., Babu, T. and Datnoff, L.E. (2016). A review of silicon in soils and plants and its role in US agriculture: history and future perspectives. *Soil Sci.*, 181(9/10): 393–411. doi: 10.1097/SS.0000000000000179.

Ulrichs, C., Mewis, I. and Goswami, A. (2005). Crop diversification aiming nutritional security in West Bengal: biotechnology of stinging capsules in nature's water-blooms. *Ann. Tech Issue of State Agri Technologists Service Assoc.*, 1–8.

Van Caneghem, J., Brems, A., Lievens, P., Block, C., Billen, P. et al. (2012). Fluidized bed waste incinerators: Design, operational and environmental issues. *Prog. Energy Combust.*, 38(4): 551–582. doi: 10.1016/j.pecs.2012.03.001.

Vaibhav, V., Vijayalakshmi, U. and Roopan, S.M. (2015). Agricultural waste as a source for the production of silica nanoparticles. *Spectrochim. Acta A.*, 139: 515–20. doi: 10.1016/j.saa.2014.12.083.

Wang, Z., Xie, X., Zhao, J., Liu, X., Feng, W., White, J.C. et al. (2012). Xylem-and phloem-based transport of CuO nanoparticles in maize (*Zea mays* L.). *Environ. Sci. Technol.*, 46(8): 4434–4441. doi: 10.1021/es204212z.

Wanyika, H., Gatebe, E., Kioni, P., Tang, Z. et al. (2012). Mesoporous silica nanoparticles carrier for urea: Potential applications in agrochemical delivery systems. *J. Nanosci. Nanotechnol.*, 12(3): 2221–2228. doi: 10.1166/jnn.2012.5801.

Wen, L.X., Li, Z.Z., Zou, H.K., Liu, A.Q. et al. (2005). Controlled release of avermectin from porous hollow silica nanoparticles. *Pest Manag. Sci.*, 61(6): 583–590. doi: 10.1002/ps.1032.

Witoon, T., Chareonpanich, M. and Limtrakul, J. (2008). Synthesis of bimodal porous silica from rice husk ash via sol–gel process using chitosan as template. *Mater. Lett.*, 62(10–11): 1476–1479. doi: 10.1016/j.matlet.2007.09.004.

Yadav, I.C., Devi, N.L., Li, J., Syed, J.H., Zhang, G. et al. (2017). Biomass burning in Indo-China peninsula and its impacts on regional air quality and global climate change—a review. *Environ. Pollut.*, 227: 414–427. doi: 10.1016/j.envpol.2017.04.085.

Yuvakkumar, R., Elango, V., Rajendran, V., Kannan, N.S. and Prabu P. (2011). Influence of nanosilica powder on the growth of maize crop (*Zea mays* L.). *Int. J. Green Nanotechnol.*, 3(3): 180–190. doi: 10.1080/19430892.2011.628581.

Zhu, H., Han, J., Xiao, J.Q. and Jin, Y. (2008). Uptake, translocation, and accumulation of manufactured iron oxide nanoparticles by pumpkin plants. *J. Environ. Monit.*, 10(6): 713–717. doi: 10.1039/b805998e.

Chapter 16

Recent Advances in Heavy Metal Removal:

Using Nanocellulose Synthesized from Agricultural Waste

Mandeep Kaur,[1] *Parveen Sharma*[1] and *Santosh Kumari*[2,*]

1. Introduction

Agriculture is fortitude for developing countries like India. Exhaustion of fossil fuels and escalated environmental problems generated by exploiting non-renewable resources are unfitting. Therefore, agriculture based green materials have captured the path of innovations in environment. However, waste generated from the agricultural practices need to be addressed. Nanotechnology can unite with agriculture to curb the pollution generated by its waste. Agricultural waste being biodegradable, and renewable accommodates huge potential to contribute its utilization in the production of nanomaterials. One such entity is nanocellulose which can be produced from agricultural waste of various crops, viz., rice husk, rice stem, rice straw, wheat straw, coconut husk fibers, sugarcane bagasse, cassava bagasse, corncob, banana rachis, soy hulls, arecanut, bamboo, etc., and can also be derived efficiently from plants, animals, algae, bacteria, fungi, etc. (Thakur et al., 2021) presented in Table 16.1. Nanocellulose can be categorized into nanocellulose crystals, nanocellulose fibers and bacterial cellulose depending upon the methods of isolation (Lasrado et al., 2020). In the past

[1] Department of Environmental Sciences and Engineering, Guru Jambheshwar University of Science & Technology, Hisar-125001, Haryana, India.

[2] Department of Bio & Nano Technology, Guru Jambheshwar University of Science & Technology, Hisar-125001, Haryana, India.

* Corresponding author: kaushdbs@gmail.com

Table 16.1. Sources for nanocellulose production.

Group	Source
Agricultural Waste	Rice husk, rice stem, rice straw, wheat straw, coconut husk fibres, sugarcane bagasse, cassava bagasse, corncob, banana rachis, soy hulls, arecanut, bamboo, peanut shells, kneaf, garlic straw residues
Bacteria	Azotobacter, Rhizobium, Acetobacter, Agrobacterium, Salmonella, Gluconaacetobacter
Algae	Cladophora, Posidonia
Plants	Hemp, Jute, Oil palm, Agave, Sisal, Mulberry
Animals	Tunicates, Chordata

decade, use of nanocellulose has been boosted up in various research zones due to its brilliant properties and a wide spectrum of potential applications (Kaur et al., 2018). Owing to increased requisite of sustainability, the use of nanocellulose is anticipated to grow further (Kaur et al., 2017). Nanocellulose not only possess extraordinary surface area, but it also has size in the nanoscale range and is highly crystalline in nature (Kaur et al., 2021). It also has the advantage of being biodegradable and renewable (Siro and Placket, 2010; Abraham et al., 2011; Trache et al., 2017). It is considered as a potential moiety due to its demand in various application domains like bioethanol production (Duran et al., 2011), adhesion behavior of cellulose (Gardner et al., 2008), foams and low density aerogels based on cellulose biocomposites (Berguland, 2010), reinforcing phase in nanocomposite applications (Azizi et al., 2005; Peng et al., 2011), chemically modified nanocellulose particles (Habibi, 2014), packaging applications (Johansson et al., 2012; Li et al., 2015; Ferrer et al., 2017) and many more.

Recently, nanocellulose has shown high potential in heavy metal removal (Kaur et al., 2019a; Kaur et al., 2019b) as shown in Fig. 16.1. Heavy metal contamination is a serious threat to the environment. Highly toxic heavy metal ions released from various industrial processes become cumulative, create discomfort to aquatic life and deteriorate the aesthetic quality of water bodies, and have direct influence on other components of the environment. Therefore, these heavy metals need special

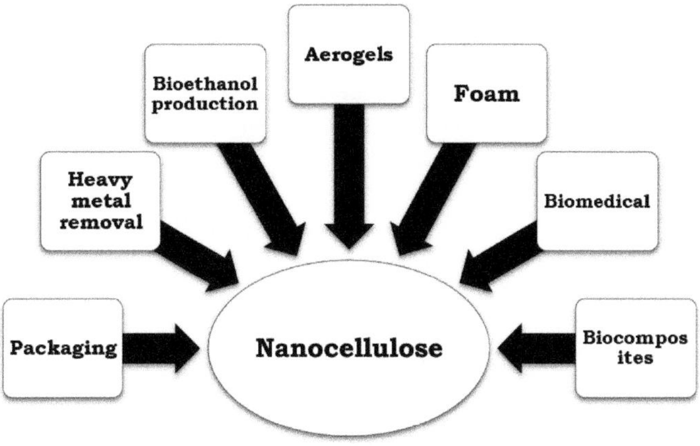

Fig. 16.1. Applications of nanocellulose.

technologies for their removal. Various treatment technologies are chemical precipitation, ion exchange, reverse osmosis, ultrafiltration, electrodialysis, and adsorption (Fu and Wang, 2011), but the adsorption process is the most competent and economically beneficial (Sayago et al., 2020). Heavy metal removal from wastewater using nanocellulose as nanoadsorbent is a promising area as it addresses the minimization of toxic effects of pollutants (Kaur et al., 2019a).

2. Nanotechnology and Nanocellulose

Nanotechnology is an interdisciplinary new science with superfluous innovations (Chirayil et al., 2014). The term nanotechnology (meaning *dwarf* in Greek) was given in 1959 by a physicist Richard Feynman, the father of nanotechnology. A nanometer is one part in a billion equal parts of a meter or 80,000 times thinner than the hair of human (Jonoobi et al., 2015). So, nano dimension is within 1 and 100 nanometers approximately and the nanometer domain covers sizes bigger than several atoms (Kamel, 2007). CelluForce in Canada started the first manufacturing plant to produce cellulose nanocystals (CNC) with the capacity of producing one tonne per day. One of the first pilot plant for nanocelluose production was started in the year 2011 by Inventia in Sweden and in the year 2012, United States started another pilot plant. In the year 2015, MoRe Research and in the year 2016, Ornskoldsvik in north Sweden set up a CNC production unit (MoRe Research, 2015). The prime target was upscaling the production of nanocellulose.

Nanocellulose is the most promising green material in terms of strength, flexibility, and nanostructured architecture in current era (de Amorim et al., 2020). It can be synthesized by employing different treatments such as mechanical, enzymatic, and chemical or a combination of these (Khalil et al., 2014; Kaur et al., 2017) as illustrated in Table 16.2. Nanocellulose is derived from cellulose, one of the major constituents of the plant cell wall (Wegner and John, 2006). Cellulose is biodegradable, renewable, non-toxic, and inexhaustible (Kargarzadeh et al., 2017). It is made up of D-glucopyranose units interlinked by β-1,4-linkages (Kumar et al., 2014). Intermolecular hydrogen bonds between the fibers assemble the individual polymer chains and bring into conformity the physical properties of cellulose (Dufresne and Castano, 2016). Cellulose fibres comprise of crystalline and amorphous regions with the latter being preferably dispelled to chemically liberate nanoscale components after hydrolysis (Rajnipriya et al., 2018). Acid hydrolysis individualizes the fibrils and results in dissolution of amorphous domains which promote the hydrolytic cleavage of glycosidic bonds and ultimately the release of individual crystallites (Chen et al., 2016) as presented in Fig. 16.2.

Table 16.2. Different treatment methods for synthesis of nanocellulose.

Chemical Treatments	Mechanical Treatments	Biological Treatments
Acid hydrolysis Alkaline Solvent extraction Ionic Oxidation	Steam explosion Disintegration Homogenization Ultrasonication Ball milling Electrospinning	Bacteria Fungi Enzymatic hydrolysis

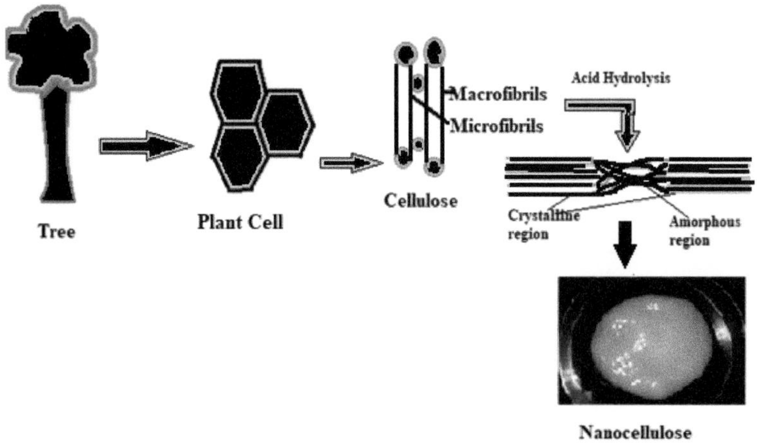

Fig. 16.2. Acid hydrolysis route to synthesis of nanocellulose.

3. Heavy Metals

Among various pollutants, grievous and unsafe heavy metals are toxic, non-biodegradable, and their bioaccumulation in the food chain can drastically affect human health and other living organisms (Chowdhury et al., 2016). Although heavy metals are naturally occurring compounds, yet they pierce into different environmental matrices through natural and anthropogenic activities (Kaur et al., 2019b). A heavy metal is an element with a density usually more than 5.0 g/cm^3 (Zvinowanda et al., 2009; Malik et al., 2016) (six times heavier than water) and specific gravity that surpasses the specific gravity of water by five or more times at 4°C (Kara et al., 2017). Heavy metals enter our body via food, water, and air to a small extent (Shah et al., 2018). Some of the heavy metals are treacherous because they tend to bioaccumulate (Yang et al., 2019). Bioaccumulation is the elevation in substance concentration in an organism than the concentration in the surroundings of organism (Harvey et al., 2015). Toxic heavy metals, viz., iron, chromium, mercury, arsenic, cobalt, cadmium, nickel, lead, copper, and zinc slowly metabolize, concentrate in the body above acceptable levels, and affect different organs (Zhan et al., 2018). Heavy metals are divided into two groups depending upon their importance for living beings: (i) essential or less toxic including Cu, Zn, Co, Mn, Fe, etc. for the growth of living things, but may be toxic if present in the system in higher concentrations, (ii) non-essential or highly toxic, viz., As, Pb, Cr, Cd, Hg, etc. are considered potentially toxic for living systems (Fig. 16.3). Toxic heavy metals may cause acute poisoning, attributing to metabolic and hormonal alterations in females, disablement of physical and mental development and central nervous system impairment in infants and children (Barakat, 2011). Gastrointestinal tract disorders, kidney damage, nausea, vomiting, muscle and joint pain, behavioral changes, leukemia, high blood pressure, etc. are also the risks associated with heavy metal toxicity summarized in Table 16.3. Different natural adsorbents like fungi, bacteria, algae, mud, fly ash, clay, and zeolites, and agricultural waste are utilized in heavy metal removal from wastewater (Azizi and Ameri, 2016). Cellulose and

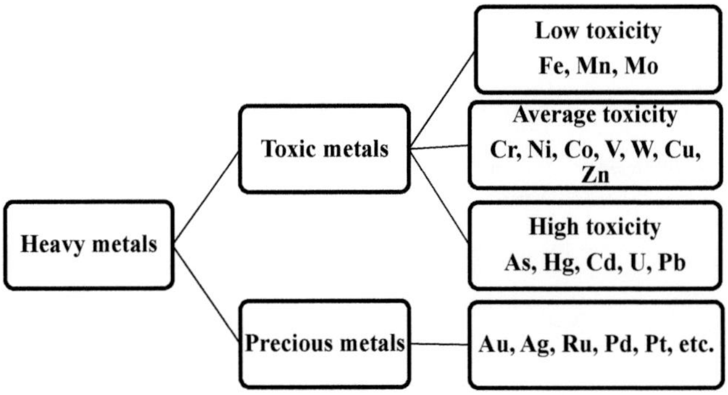

Fig. 16.3. Categorization of heavy metals (Volesky, 1990; Bishop, 2002; Thakur, 2006).

Table 16.3. MCL standards and toxic effects of heavy metals (Babel and Kurniawan, 2003).

Toxic Heavy Metal	Harmful Effects	MCL (mg/L)
Copper	Wilson disease, liver damage, insomnia	0.25
Zinc	Lethargy, depression, increased thirst, neurological signs	0.80
Lead	Cognitive dysfunction, CNS damage, fetal brain impairment, kidney dysfunction	0.006
Chromium	Kidney damage, nausea, headache, vomiting, diarrhea, carcinogen	0.05
Cadmium	Renal disorder, kidney damage, anemia, hypertension, carcinogen	0.01
Arsenic	Skin lamentation, visceral cancer, vascular disease	0.05
Nickel	Chronic asthma, dermatitis, nausea, myocardial infarction, carcinogen	0.20
Mercury	Depression, confusion, kidney disease, rheumatoid arthritis, nervous system	0.00003

Note: MCL: Maximum concentration limit.

chemical moieties like esters, phosphate, etc. present in agricultural wastes enhance sequestration of heavy metals by complexation (Volesky and Holan, 1995; Hossain et al., 2012). Adsorption is a surface phenomenon in which a molecular species gets deposited onto the surface as demonstrated in Fig. 16.4. The molecular species which get adsorbed on the surface is adsorbent and the surface on which adsorption takes place is adsorbate (Grassi et al., 2012). Adsorption occurs because of raised free surface energy of the solids because of their enhanced surface area. Adsorption is of two types: physical adsorption and chemical adsorption. Physical adsorption (also known as physisorption) occurs due to Van der Waals forces of attraction between the adsorbate and the adsorbent. Physical adsorption combines many layers of adsorbate on the adsorbent and takes place at low temperature. Chemical adsorption (also known as chemisorption) occurs due to chemical forces of attraction or a chemical bond between the adsorbate and the adsorbent. Chemical adsorption combines a single

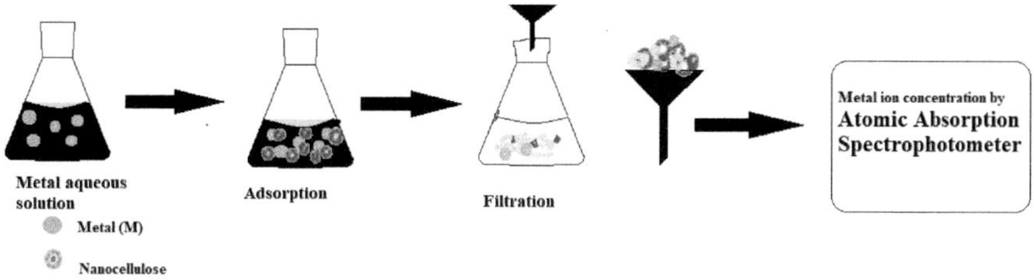

Fig. 16.4. Schematic representation of adsorption process.

layer of adsorbate on the adsorbent and takes place at all temperatures. Adsorption mechanism is divided into three steps:

i) Transportation and movement of adsorbate from the bulk solution across the outer surface sites on the adsorbent.
ii) Migration of adsorbate within the pores of the adsorbent by intraparticle diffusion.
iii) Adsorption of adsorbate at inner surface sites.

The adsorption phenomenon is a sustainable technology for heavy metals sequestration from aqueous solution and wastewater. The adsorption process is combatant in terms of efficiency, design simplicity, and operations. However, nanocellulose (with outstanding surface area, pore size, and adhered functional groups) prepared from natural materials, i.e., agricultural waste and transformed as nanoadsorbents have pulled in appreciable attentiveness to remove heavy metals. The adsorption capacity of such nanoadsorbents gets enhanced after chemical modification with functional groups and addition of more functional groups which is due to increase in active binding sites that promote the higher uptake of heavy metal ions depicted in Fig. 16.5.

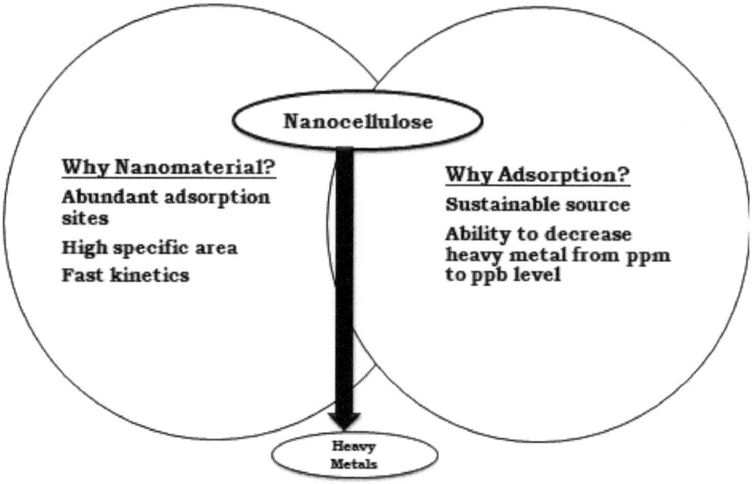

Fig. 16.5. Prominence of adsorption and nanocellulose.

4. Nanocellulose as Nanoadsorbent for Heavy Metal Removal

Cellulose from agricultural waste stands for a huge reservoir to derive nanocellulose due to its low cost and extensive availability. Urbanization and rapid industrialization have made environment intensely unpleasant by generating troubles like runoff from urban areas and discharge of industrial wastewater into aquatic bodies. The metal loaded wastewater of industries thus causes severe contamination (Cheraghi et al., 2016). Therefore, nanocellulose which is vital and flexible with large surface area holds more potential as adsorbent for treating the wastewater (Kaur et al., 2017).

It is manifested from the reported literature below that these adsorbents derived from nanocellulose and other nanomaterials on account of technical relevance and cost effectiveness have conspicuous potential for removal of heavy metals from wastewater. Indigenously synthesized carbon nanomaterials (CNMs) can be used for removing cadmium, lead, nickel, and zinc (Ruparelia et al., 2008). Chemical vapour deposition (CVD) set up with turpentine oil was used in synthesizing nanocarbon (NC) and nanoporous carbon (NPC) of different surface morphology along with commercially available activated carbon (AC). Nitric acid and potassium hydroxide were used post treatment for catalyst removal. The AC, NC, and NPC samples were characterized using XRD. NPC was found to be most suitable among the three mentioned adsorbents to remove lead, cadmium, nickel, and zinc.

Nanocellulose hybrids containing reactive polyhedral oligomeric silsesquioxane (R-POSS) were extracted and used as a novel biosorbent to remove copper and nickel from a synthetic solution (Xie et al., 2011). R-POSS bearing multi-N-methylol, a reactive monomer, had a high tendency for adsorption of heavy metals. Equilibrium isotherm followed the Langmuir isotherm and adsorption kinetic model followed the pseudo second-order equation in the study. Yu et al. (2012) reported carboxylated cellulose nanocrystals as adsorbent for the adsorption of heavy metals from synthetic solution. Chemically modified cotton nanocrystals with succinic anhydride and with saturated $NaHCO_3$ forming NaSCNCs were used for batch experiments to remove lead and cadmium. With increase in pH, the adsorption capacities of lead and cadmium on SCNCs and NaSCNCs increased. The Langmuir adsorption model fitted well into the data. Maximum adsorption capacities for lead were 367.6 mg/g, 259.7 mg/g, and for cadmium were 465.1 mg/g, 344.8 mg/g. The preparation and characterization of nanocellulose fibers using vinyl sulphonic acid for cationic toxic metals remediation was investigated by Kardam et al. (2012). The modified NCFs exhibited increase in sorption efficiency and stability in regeneration cycles for Pb (II), Ni (II), Cd (II), and Cr (III) ions. Lead and zinc ion removal from synthetic solution and effluent by using iron oxide nanoparticles was observed by Sidhaarth et al. (2012). Nanoparticles were prepared by employing the co-precipitation method. SEM, XRD, and AFM techniques were used for characterization of the material. The Scherrer formula was used to determine the particle size which was 18.14 nm. The Langmuir, Freundlich, and Temkin models fitted well into the data. 89% of lead and 95% zinc were removed. Multi-walled carbon nanotubes were successfully used by Salam (2013) for removing copper (II), zinc (II), lead (II), and cadmium (II) from synthetic solution. The Fraction power function, Elovich model, Lagergren pseudo first-order, and pseudo second-order models were employed to study the kinetics of the metals. The result also showed that increase in temperature was the key factor in increasing the percent adsorption and the process was found to be endothermic. The inference from the thermodynamic studies indicated the spontaneity of the process with negative free energies for all metal ions. A positive entropy value showed increase in randomness due to physical adsorption.

Hoakken et al. (2013) used succinic anhydride modified mercerized nanocellulose to study the sequestration of divalent form of heavy metal ions such as nickel, cadmium, cobalt, copper, and zinc from a synthetic solution. Nanocellulose was regenerated with nitric acid and ultrasonication. FTIR and SEM techniques were used to characterize the modified adsorbent. The Langmuir and Sip model represented the adsorption phenomena. The maximum metal uptake capacity was 0.72 mmol/g–0.95 mmol/g. Synthesis of nanocellulose fibers using rice straw and then exploration of fibers for remediation of cadmium, nickel, and lead from wastewater was recommended by Kardam et al. (2014). Batch experiments were conducted on prepared nanocellulose fibers of average size 6 nm. Removal efficiencies calculated were 9.7 mg/g for cadmium, 8.55 mg/g for nickel, and 9.42 mg/g for lead. The adsorption process fitted well into the Langmuir isotherm and the Freundlich isotherm. Suopajarvi et al. (2014) derived an adsorbent from wheat straw pulp by nanofibrillation and sulfonation pre-treatments and removed divalent lead ions from aqueous solution. Effects of pH, contact time, and initial lead concentration were the parameters analysed. The isotherm data was modeled with the Freundlich and Langmuir models. The Langmuir isotherm model was well fitted to the data. The adsorption capacity of synthesized adsorbent at pH 5 was 1.2 mmol/g. Pulp and paper residue and crab shells were used to prepare cellulose and chitin nanofibers with carboxylate functional groups (Sehaqui et al., 2014). The raw material was chemically modified with TEMPO-mediated oxidation (2,2,6,6-tetramethyl-1-piperidinyloxy) followed by mechanical disintegration. Addressed heavy metals were copper, zinc, chromium, and nickel. Characterization tools employed were X-ray photoelectron spectroscopy, UV spectrophotometry, and wavelength dispersive X-ray analysis. Sheikhi et al. (2015) reported the removal of copper metal by using electrosterically stabilized nanocrystalline cellulose (ENCC) which was prepared from wood fibers by periodate/chlorite oxidation with adsorption capacity 185 mg/g. Straws of different agricultural crops, viz., maize, rice, and wheat were used for the sequestration of different heavy metals. Regenerable agricultural waste material as adsorbent was developed by using maize straw modified with succinic anhydride in xylene (Guo et al., 2015). Cd (II) was removed from NaS-MS by batch mode experiments. Different parameters, viz., pH, contact time, initial concentration, adsorbent dose, and temperature were investigated. Total of 97% adsorbed Cd (II) ions were retrieved in the oxide form. Mautner et al. (2016) used fiber sludge, a paper industry waste to synthesize phosphorylated cellulose nanofibrils. The nanofibrils were refined into nanopaper ion exchangers. The diameter of the obtained CNFs was between 10 nm and 100 nm. The CNFs were further used for the adsorption of copper.

Ram et al. (2018) reported the synthesis of spherical nanocellulose by acid hydrolysis and lipase catalyzed esterification with 3-mercaptopropionic acid. The adsorbent prepared from nanocelluose was used for the adsorption of Hg (II) ions from their aqueous solution with approximately 98.6% removal. Regeneration of the adsorbent was investigated for nine successive cycles with NaCl, HNO_3, 0.1 N HCl, and CH_3COONa; 0.1 N HCl proved to be the best. Maximum adsorption capacity was found to be 404.95 mg/g. Sugarcane bagasse fibers were used to prepare cellulose nanocrystals for polymer reinforcement and further used for adsorption of Pb (II) ions from synthetic solution (Khoo et al., 2018). Remediation of heavy metals from wastewater using modified nanocellulose/microcellulose has been reviewed

(Varghese et al., 2019). The synthesis of carboxycellulose nanofibres from untreated Australian spinifex grass and effective removal of Cd^{2+} ions by nitro-oxidation method were reported by Sharma et al. (2018). The nanofibers obtained had high surface area and low crystallinity.

Heavy metals were also removed from a synthetic solution and industrial wastewater using nanocellulose as nanoadsorbent prepared from agricultural waste rice husk. The synthesis of nanocellulose was accomplished by chemo-mechanical treatment (acid hydrolysis and ultrasonication) as presented in Fig. 16.6 (Kaur et al., 2018). Nanocellulose was characterized by different techniques, viz., Atomic Force Microscopy, Field Emission Microscopy, X-Ray Diffraction, Fourier Transform Infrared Spectroscopy, Thermogravimetric Analysis, etc. The nanocellulose obtained was used as a nanoadsorbent for the removal of Cu (II), Pb (II), and Zn (II) by the low cost and highly efficient adsorption process. Nanoadsorbent to metal interaction was accomplished in batch mode to optimize different parameters, viz., pH, adsorbent dose, contact time, metal ion concentration, and temperature by adsorption process. The Cu (II) adsorption on the nanoadsorbent surface was modeled with Freundlich, Langmuir, and Temkin isotherm models with R^2 values 0.993, 0.968, and 0.960, respectively (Kaur et al., 2019a). In another study, central composite design (CCD) a subset of response surface methodology (RSM) was applied for mathematical model equation evaluation and optimization of parameters for the removal of Cu (II) and Pb (II) from a synthetic solution using rice husk nanoadsorbent. The interactive effect of the three independent variables such as pH, adsorbent dose, and initial metal ion concentration on Cu (II) and Pb (II) adsorption were investigated. Analysis of variance (ANOVA) was used to depict the relative significance of variables in the removal process. pH 6.7, adsorbent dose 0.7 g/L, and initial concentration 21 mg/L for Cu (II) and pH 4, adsorbent dose 0.4 g/L and initial concentration 50 mg/L for Pb (II) were found to be influential for 82% and 76% metal removal, respectively. The removal efficiency was found out by building a second-order regression equation model using Design Expert Software. Adsorption mechanism was explained by hydrolysis/precipitation and electrostatic interactions as presented in Fig. 16.7 (Kaur et al., 2019b).

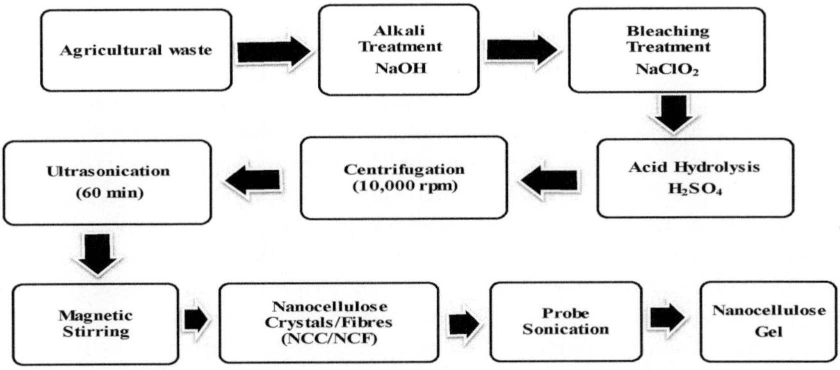

Fig. 16.6. Schematic representation of nanocellulose extraction.

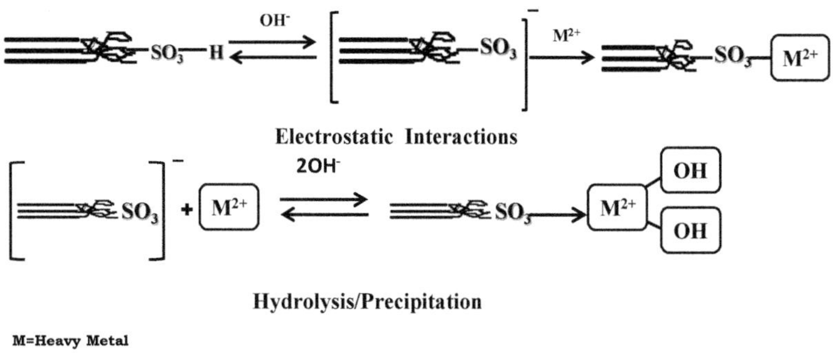

Fig. 16.7. Adsorption mechanism.

5. Conclusions

In this chapter, nanocellulose-based adsorbents developed from agricultural waste and their usage in the removal of heavy metals from synthetic water and wastewater by adsorption process have been extensively reviewed till date. It can be concluded that among different technologies available to remediate heavy metals which are costly, having high energy requirements, and generate toxic pollutants, adsorption is more efficient and environmentally friendly in terms of disposal of heavy metal laden wastewater. Moreover, heavy metal as a pollutant is a distress to the environment which requires a strenuous task to investigate the suitable adsorbent for its removal from wastewater. Owing to abundant availability of agricultural waste and its ease of modification, it is of special significance to synthesize nanocellulose "wealth from waste". Nanocellulose has diverse applications in various sectors due to its remarkable properties. Nanocellulose can be potentially used as a nanoadsorbent due to its prominent surface area and plentiful active sites for the sequestration of heavy metals from wastewater and its proven record to be an eco-friendly and a cost-effective product.

References

Abraham, E., Deepa, B., Pothan, L.A., Jacob, M., Thomas, S., Cvelbar, U. and Anandjiwala, R. (2011). Extraction of nanocellulose fibrils from lignocellulosic fibres: A novel approach. *Carbohydr. Polym.*, 86(4): 1468–1475.

Azizi Samir, M.A.S., Alloin, F. and Dufresne, A. (2005). Review of recent research into cellulosic whiskers, their properties, and their application in nanocomposite field. *Biomacromolecules*, 6(2): 612–626.

Azizi Haghighat, Z. and Ameri, E. (2016). Synthesis and characterization of nano magnetic wheat straw for lead adsorption. *Desalination and Water Treatment*, 57(21): 9813–9823.

Babel, S. and Kurniawan, T.A. (2003). Low-cost adsorbents for heavy metals uptake from contaminated water: A review. *J. Hazard. Mater.*, 97(1-3): 219–243.

Barakat, M.A. (2011). New trends in removing heavy metals from industrial wastewater. *Arabian Journal of Chemistry*, 4(4): 361–377.

Berglund, L.A. and Peijs, T. (2010). Cellulose biocomposites: from bulk moldings to nanostructured systems. *MRS Bulletin*, 35(3): 201–207.

Bishop, P.L. (2002). *Pollution Prevention: Fundamentals and Practice*. Beijing: Tsinghua University Press, 768pp.

Chen, N., Zhu, J.Y. and Tong, Z. (2016). Fabrication of microfibrillated cellulose gel from waste pulp sludge via mild maceration combined with mechanical shearing. *Cellulose*, 23(4): 2573–2583.

Cheraghi, E., Ameri, E. and Moheb, A. (2016). Continuous biosorption of Cd (II) ions from aqueous solutions by sesame waste: Thermodynamics and fixed-bed column studies. *Desalination and Water Treatment*, 57(15): 6936–6949.

Chirayil, C.J., Mathew, L. and Thomas, S. (2014). Review of recent research in nano cellulose preparation from different lignocellulosic fibers. *Reviews on Advanced Materials Science*, 37: 20–28.

Chowdhury, I.H., Chowdhury, A.H., Bose, P., Mandal, S. and Naskar, M.K. (2016). Effect of anion type on the synthesis of mesoporous nanostructured MgO, and its excellent adsorption capacity for the removal of toxic heavy metal ions from water. *RSC Advances*, 6: 6038–6047.

de Amorim, J.D.P., de Souza, K.C., Duarte, C.R., da Silva Duarte, I., Ribeiro, F.D.A.S., Silva, G.S. and Sarubbo, L.A. (2020). Plant and bacterial nanocellulose: Production, properties and applications in medicine, food, cosmetics, electronics, and engineering. A review. *Environ. Chem. Let.*, 1–19.

Dufresne, A. and Castaño, J. (2016). Polysaccharide nanomaterial reinforced starch nanocomposites: A review. *Starch-Stärke*, 69(1–2): 1–19.

Duran, N., Lemes, A.P., Duran, M., Freer, J. and Baeza, J. (2011). A minireview of cellulose nanocrystals and its potential integration as co-product in bioethanol production. *J. Chil. Chem. Soc.*, 56(2): 672–677.

Ferrer, A., Pal, L. and Hubbe, M. (2017). Nanocellulose in packaging: Advances in barrier layer technologies. *Indu. Crop Prod.*, 95: 574–582.

Fu, F. and Wang, Q. (2011). Removal of heavy metal ions from wastewaters: A review. *J. Environ. Manag*, 92(3): 407–418.

Gardner, D.J., Oporto, G.S., Mills, R. and Samir, M.A.S.A. (2008). Adhesion and surface issues in cellulose and nanocellulose. *J. Adhe. Sci. Tech.*, 22(5–6): 545–567.

Grassi, M., Kaykioglu, G., Belgiorno, V. and Lofrano, G. (2012). Removal of emerging contaminants from water and wastewater by adsorption process. pp. 15–37. *In*: *Emerging Compounds Removal from Wastewater*, Springer.

Guo, H., Zhang, S., Kou, Z., Zhai, S., Ma, W. and Yang, Y. (2015). Removal of cadmium (II) from aqueous solutions by chemically modified maize straw. *Carbohyd. Polym.*, 115: 177–185.

Habibi, Y. (2014). Key advances in the chemical modification of nanocelluloses. *Chem. Soc. Rev.*, 43(5): 1519–1542.

Harvey, P.J., Handley, H.K. and Taylor, M.P. (2015). Identification of the sources of metal (lead) contamination in drinking waters in north-eastern Tasmania using lead isotopic compositions. *Environ. Sci. Pollu. Res.*, 22(16): 12276–12288.

Hokkanen, S., Repo, E. and Sillanpää, M. (2013). Removal of heavy metals from aqueous solutions by succinic anhydride modified mercerized nanocellulose. *Chem. Engg. J.*, 223: 40–47.

Hossain, M.A., Ngo, H.H., Guo, W.S. and Setiadi, T. (2012). Adsorption and desorption of copper (II) ions onto garden grass. *Biores. Technol.*, 121: 386–395.

Johansson, C., Bras, J., Mondragon, I., Nechita, P., Plackett, D., Simon, P. and Aucejo, S. (2012). Renewable fibers and bio-based materials for packaging applications: A review of recent developments. *BioResources*, 7(2): 2506–2552.

Jonoobi, M., Oladi, R., Davoudpour, Y., Oksman, K., Dufresne, A., Hamzeh, Y. and Davoodi, R. (2015). Different preparation methods and properties of nanostructured cellulose from various natural resources and residues: A review. *Cellulose*, 22(2): 935–969.

Kamel, S. (2007). Nanotechnology and its applications in lignocellulosic composites, a mini review. *Express Polym. Let.*, 1(9): 546–575.

Kara, İ., Yilmazer, D. and Akar, S.T. (2017). Metakaolin based geopolymer as an effective adsorbent for adsorption of zinc (II) and nickel (II) ions from aqueous solutions. *App. Clay Sci.*, 139: 54–63.

Kardam, A., Raj, K.R., Arora, J.K. and Srivastava, S. (2012). Artificial neural network modeling for biosorption of Pb (II) ions on nanocellulose fibers. *Bionanoscience*, 2(3): 153–160.

Kardam, A., Raj, K.R., Srivastava, S. and Srivastava, M.M. (2014). Nanocellulose fibers for biosorption of cadmium, nickel, and lead ions from aqueous solution. *Clean Technol. Environ. Pol.*, 16(2): 385–393.

Kargarzadeh, H., Ioelovich, M., Ahmad, I., Thomas, S. and Dufresne, A. (2017). Methods for extraction of nanocellulose from various sources. pp. 1–49. *In*: *Handbook of Nanocellulose and Cellulose Nanocomposites*.

Kaur, M., Sharma, P. and Kumari, S. (2017). Extraction, preparation and characterization: Nanocellullose. *Int. J. Environ. Sci. Res.*, 6: 1881–86.

Kaur, M., Sharma, P. and Kumari, S. (2018). Chemically modified nanocellulose from rice husk: Synthesis and characterisation. *Advan. in Res.*, 13: 1–11.

Kaur, M., Sharma, P. and Kumari, S. (2019a). Equilibrium studies for copper removal from aqueous solution using nanoadsorbent synthesized from rice husk. *SN App. Sci.*, 1: 1–9.

Kaur, M., Sharma, P. and Kumari, S. (2019b). Response surface methodology approach for optimization of Cu^{2+} and Pb^{2+} removal using nanoadsorbent developed from rice husk. *Mater. Today Commun.*, 21: 1–10.

Kaur, P., Sharma, N., Munagala, M., Rajkhowa, R., Aallardyce, B., Shastri, Y. and Agrawal, R. (2021). Nanocellulose: Resources, physio-chemical properties, current uses and future applications. *Front. Nanotechnol.*, 1–17.

Khalil, H.A., Davoudpour, Y., Islam, M.N., Mustapha, A., Sudesh, K., Dungani, R. and Jawaid, M. (2014). Production and modification of nanofibrillated cellulose using various mechanical processes: A review. *Carbohydr. Polym.*, 99: 649–665.

Khoo, R.Z., Chow, W.S. and Ismail, H. (2018). Sugarcane bagasse fiber and its cellulose nanocrystals for polymer reinforcement and heavy metal adsorbent: A review. *Cellulose*, 1–28.

Kumar, A., Negi, Y.S., Choudhary, V. and Bhardwaj, N.K. (2014). Characterization of cellulose nanocrystals produced by acid-hydrolysis from sugarcane bagasse as agro-waste. *J. Mater. Phy. Chem.*, 2(1): 1–8.

Li, Z., Yang, R., Yang, F., Zhang, M. and Wang, B. (2015). Structure and properties of chitin whisker reinforced papers for food packaging application. *BioResources*, 10(2): 2995–3004.

Lasrado, D., Ahankari, S. and Kar, K. (2020). Nanocellulose-based polymer composites for energy applications: A review. *J. App. Polym. Sci.*, 137(27): 48959.

Malik, D.S., Jain, C.K. and Yadav, A.K. (2017). Removal of heavy metals from emerging cellulosic low-cost adsorbents: A review. *App. Wat. Sci.*, 7(5): 2113–2136.

Mautner, A., Maples, H.A., Kobkeatthawin, T., Kokol, V., Karim, Z., Li, K. and Bismarck, A. (2016). Phosphorylated nanocellulose papers for copper adsorption from aqueous solutions. *Int. J. Environ. Sci. Technol.*, 13(8): 1861–1872.

Mautner, A. (2020). Nanocellulose water treatment membranes and filters: A review. *Polym. Int.*, 1–11.

Peng, Y., Gardner, D.J. and Han, Y. (2012). Drying cellulose nanofibrils: In search of a suitable method. *Cellulose*, 19(1): 91–102.

Rajinipriya, M., Nagalakshmaiah, M., Robert, M. and Elkoun, S. (2018). Importance of agricultural and industrial waste in the field of nanocellulose and recent industrial developments of wood based nanocellulose: A review. *ACS Sus. Chem. Engg.*, 6(3): 2807–2828.

Ram, B. and Chauhan, G.S. (2018). New spherical nanocellulose and thiol-based adsorbent for rapid and selective removal of mercuric ions. *Chem. Engg. J.*, 331: 587–596.

Ruparelia, J.P., Duttagupta, S.P., Chatterjee, A.K. and Mukherji, S. (2008). Potential of carbon nanomaterials for removal of heavy metals from water. *Desalination*, 232(1-3): 145–156.

Salam, M.A. (2013). Removal of heavy metal ions from aqueous solutions with multi-walled carbon nanotubes: Kinetic and thermodynamic studies. *Int. J. Environ. Sci. Technol.*, 10(4): 677–688.

Sayago, U.F.C., Castro, Y.P., Rivera, L.R.C. and Mariaca, A.G. (2020). Estimation of equilibrium times and maximum capacity of adsorption of heavy metals by *E. crassipes*. *Environ. Monit. Assess.*, 192(2): 1–16.

Sehaqui, H., de Larraya, U.P., Liu, P., Pfenninger, N., Mathew, A.P., Zimmermann, T. and Tingaut, P. (2014). Enhancing adsorption of heavy metal ions onto biobased nanofibers from waste pulp residues for application in wastewater treatment. *Cellulose*, 21(4): 2831–2844.

Shah, G.M., Imran, M., Bakhat, H.F., Hammad, H M., Ahmad, I., Rabbani, F. and Khan, Z.U.H. (2018). Kinetics and equilibrium study of lead bio-sorption from contaminated water by compost and biogas residues. *Int. J. Environ. Sci. Technol.*, 1–12.

Sharma, P.R., Chattopadhyay, A., Sharma, S.K., Geng, L., Amiralian, N., Martin, D. and Hsiao, B.S. (2018). Nanocellulose from spinifex as an effective adsorbent to remove cadmium (II) from water. *ACS Sustainable Chemistry and Engineering*, 6(3): 3279–3290.

Sheikhi, A., Safari, S., Yang, H. and van de Ven, T.G. (2015). Copper removal using electrosterically stabilized nanocrystalline cellulose. *ACS App. Mater. Inter.*, 7(21): 11301–11308.

Sidhaarth, K.A., Jeyanthi, J. and Suryanarayana, N. (2012). Comparative studies of removal of lead and zinc from industrial wastewater and aqueous solution by iron oxide nanoparticle: Performance and mechanisms. *Euro. J. Sci. Res.*, 70(2): 169–184.

Siró, I. and Plackett, D. (2010). Microfibrillated cellulose and new nanocomposite materials: A review. *Cellulose*, 17(3): 459–494.

Suopajärvi, T., Liimatainen, H., Karjalainen, M., Upola, H. and Niinimäki, J. (2015). Lead adsorption with sulfonated wheat pulp nanocelluloses. *J. Wat. Pro. Engg.*, 5: 136–142.

Thakur, S.K., Tomar, N.K. and Pandeya, S.B. (2006). Influence of phosphate on cadmium sorption by calcium carbonate. *Geoderma*, 130: 240–249.

Thakur, V., Guleria, A., Kumar, S., Sharma, S. and Singh, K. (2021). Recent advances in nanocellulose processing, functionalization, and applications: A review. *Materials Advances*, 2(6): 1872–1895.

Trache, D., Hussin, M.H., Haafiz, M.M. and Thakur, V.K. (2017). Recent progress in cellulose nanocrystals: Sources and production. *Nanoscale*, 9(5): 1763–1786.

Varghese, A.G., Paul, S.A. and Latha, M.S. (2019). Remediation of heavy metals and dyes from wastewater using cellulose-based adsorbents. *Environ. Chem. Let.*, 17(2): 867–877.

Volesky, B. (1990). Removal and recovery of heavy metals by biosorption. *Biosorption of Heavy Metals*, 7: 3–403.

Volesky, B. and Holan, Z.R. (1995). Biosorption of heavy metals. *Biotechnology Progress*, 11(3): 235–250.

Wegner, T.H. and Jones, P.E. (2006). Advancing cellulose-based nanotechnology. *Cellulose*, 13(2): 115–118.

Xie, K., Zhao, W. and He, X. (2011). Adsorption properties of nano-cellulose hybrid containing polyhedral oligomeric silsesquioxane and removal of reactive dyes from aqueous solution. *Carbohydr. Polym.*, 83(4): 1516–1520.

Yang, J., Hou, B., Wang, J., Tian, B., Bi, J., Wang, N. and Huang, X. (2019). Nanomaterials for the removal of heavy metals from wastewater. *Nanomaterials*, 9(3): 1–39.

Yu, M., Yang, R., Huang, L., Cao, X., Yang, F. and Liu, D. (2012). Preparation and characterization of bamboo nanocrystalline cellulose. *BioResources*, 7(2): 1802–1812.

Zhan, W., Xu, C., Qian, G., Huang, G., Tang, X. and Lin, B. (2018). Adsorption of Cu (ii), Zn (ii), and Pb (ii) from aqueous single and binary metal solutions by regenerated cellulose and sodium alginate chemically modified with polyethyleneimine. *RSC Advances*, 8(33): 18723–18733.

Zvinowanda, C.M., Okonkwo, J.O., Shabalala, P.N. and Agyei, N.M. (2009). A novel adsorbent for heavy metal remediation in aqueous environments. *Int. J Environ. Sci. Technol.*, 6(3): 425–434.

Ecosafety and Phytotoxicity

Chapter 17

Ecosafety and Phytotoxicity Associated with Titania Nanoparticles

Nupur Bahadur[1,*] and *Paromita Das*[1,2]

1. Introduction

Titanium dioxide (TiO_2) nanoparticles, a major constituent of many commercial goods, are produced in large amounts throughout the world. Usage as well as disposal of commercial goods containing TiO_2 nanoparticles causes the release of these nanoparticles into the ecosystem. In a terrestrial ecosystem, plants are one of the most important living components. When exposed to excess nanoparticles, parameters such as seed germination and growth in a plant may be affected. The mobility of TiO_2 nanoparticles in a plant depends upon their size, phase, concentration, and surface modification (Nowack and Bucheli, 2007). The level of accumulation of the TiO_2 nanoparticles in the plant is dependent on factors such as plant species and physico-chemical characteristics of the nanoparticles. For example, accumulation of TiO_2 nano-conjugate in plant roots can be attributed to their small size which allows them to pass through the cell wall pores (Larue et al., 2011). The phytotoxicity effects of nanoparticles are dependent on the type of nanoparticles and plant species. To assess the difference from species to species, usually tests are conducted at two stages of plant growth and development (Nowack and Bucheli, 2007; Gottschalk et al., 2009; Larue et al., 2011; Das et al., 2020): during seed germination, that is, when germination rate and root elongation are usually measured, and during plant growth, in which root and shoot lengths are mostly assessed to see the nanoparticle exposure effects.

[1] TADOX Technology Centre for Water Reuse, Water Resources Division, The Energy and Resources Institute, TERI Gram, Gual Pahari, Gurgaon Faridabad Road, Gurgaon, Haryana 122 001, India.

[2] Department of Biotechnology, TERI School of Advanced Studies, Plot No. 10 Institutional Area, Vasant Kunj, New Delhi - 110070, India.

* Corresponding author: nupur.bahadur@teri.res.in

For treating industrial wastewater, including heavy metal and dye degradation and synthetic and real textile effluents, TiO_2 is used as a semiconductor mediated adsorbent and photocatalyst in an indigenously developed technology (Bahadur and Bhargava, 2019; Das et al., 2020). The TiO_2 nanoparticles used in this technology are recyclable and can be completely recovered for further use; however, the eco-safety and phytotoxicity associated with these titania nanoparticles in terms of their potential risk towards environment and ecosystem are being further explored. The effect of TiO_2 nanoparticles was observed on *Vigna radiata* (Garden pea) at a concentration range from 0 to 1000 mg L^{-1}. The synthesized TiO_2 nanoparticles were safe till 500 mg L^{-1} concentration for plant growth parameters (Das et al., 2020).

Also discussed are other factors related to plant development including chlorophyll content, antioxidant enzymes such as sodium oxide dismutase (SOD) and reactive oxygen species (ROS) which finally leads to high or low photosynthetic rate that decides the fate of plant growth. Chlorophyll is the key factor in plant growth, as it absorbs solar light required to convert carbon dioxide and water into glucose, which speeds up the photosynthesis process. The leaves having higher chlorophyll content can efficiently absorb the required amount of solar light for the photosynthesis process. Plants exposed to TiO_2 nanoparticles exhibited higher content of chlorophyll a, b, and (a+b) total, leading to higher plant growth (Gogoi and Basumatary, 2018).

This chapter reviews various factors affecting both aspects of eco-safety and phytotoxicity of seed germination and plant growth and development with TiO_2 nanoparticle–exposure such as size, shape, phase, concentration, and surface modification of TiO_2 nanoparticles. Various plant growth parameters are discussed including effects on seed germination, root growth, shoot growth, root laterals, vigor index (I), number of leaves, and colour of leaves. Also, other factors related to plant development are discussed such as chlorophyll content, antioxidant enzymes (SOD), and ROS which finally leads to high or low photosynthetic rate deciding the fate of plant growth.

2. Factors Affecting Eco-safety and Phytotoxicity with TiO_2 NMs Exposure

2.1 Size

Plant cells are surrounded by a cell wall which has an additional barrier having size range from 5 to 30 nm pore diameter (Auffan et al., 2009). TiO_2 nanoparticles with size less than 30 nm can effortlessly enter plants through these additional cell wall barriers. Therefore, a plant's overall photosynthetic and biochemical processes are affected by the size of the TiO_2 nanoparticles. From an eco-safety and phytotoxicity perspective, surface area and particle size are the two most important nanoparticle characteristics; as the nanoparticle size decreases, its surface area increases and a larger percentage of its atoms or molecules exist on the surface of the TiO_2 nanoparticles. As the surface area increases, the number of reactive group on the particle's surface also increases (Begum et al., 2011). As the size of these nanoparticles decreases, the structural and physico-chemical properties of TiO_2 nanoparticles transform, which can result in toxicological effects (Lee et al., 2012). Various studies have shown that the effects of TiO_2 exposure can be both positive (Raliya et al., 2015; Jiang et al., 2017; Faraji and

Sepehri, 2018; Das et al., 2020) and negative (Castiglione et al., 2014; Fellmann and Eichert, 2017; Liu et al., 2017) towards plant growth and development.

Though TiO_2 nanoparticles are extensively used in products of common utilities, there is limited research on their uptake and movements in plants (Nevius et al., 2012; Nadiminti et al., 2013). Easy translocation and tissue-specific distribution could take place due to small size of TiO_2 nanoparticles which form a covalent bond with most of the non-conjugate natural organic matter (Feizi et al., 2013). Furthermore because of the small size of these nanoparticles, the penetration into the seeds becomes easier. In one of the studies, Larue et al. (2016) observed that up to 100 nm nanoparticles can be accumulated in plant roots by exposure of wheat plants with TiO_2 nanoparticles. They also observed translocation of these nanoparticles from roots to leaves below 30 nm (Larue et al., 2016). It could be justified that the nanoparticles sometimes can easily enlarge the root pores or create new ones and easily penetrate through the roots. TiO_2 nanoparticles also showed enhanced nitrogen and protein contents (Asli and Neumann 2009); considerable increase in leaf pigments (Morteza et al., 2013) and enhanced photosynthesis process of plants (Zheng et al., 2005). Further, in a study done by Das et al. (2020) it is seen that very small sized TiO_2 nanoparticles, 14 nm (Bare TiO_2) and 8 nm (SDS modified TiO_2), showed no negative effect on mung bean plants till concentration as high as 500 mg L^{-1} (Das et al., 2020). In a study by Zheng et al. (2005) it was seen that smaller nanoparticles can lead to better germination of spinach seeds at 2.5% nano-TiO_2 (commercial) concentration. Li et al. (2014) showed that smaller sized TiO_2 nanoparticles can stimulate more ROS production when photo-excited, which will further produce higher oxidative stress to the biological systems (Li et al., 2014). In contrast, other studies have shown how nanoparticles travel through a plant and affect the transpiration rate and overall growth. Kurepa et al. (2010) studied the conjugation of TiO_2-Alizarin red having particle size of 3 nm which successfully crossed the cell walls and entered the plant cells very easily and accumulated in particular subcellular sites of *Arabidopsis* roots and leaves (Kurepa et al., 2010). On the other hand, Asli et al. (2009) shows that TiO_2 nanoparticles of particle size 25 nm accumulated on root surface of maize (*Z. mays* L.) plant such that the root hydraulic conductivity and water availability were reduced, which further reduced the transpiration rate, affecting the plant growth (Asli and Neumann, 2009).

2.2 Phase

TiO_2 occurs in three main crystal phases: anatase, brookite, and rutile. Among these, rutile is the most stable phase whereas anatase is the most favoured for the solution phase method of preparation for TiO_2 (Bahadur et al., 2010). Predominantly commercially available TiO_2 are usually of anatase and rutile phases. The TiO_2 nanoparticles properties vary with the crystal phases. The different crystal phases of TiO_2 nanoparticles would lead to different kinds of interactions with plants. For example, anatase TiO_2 shows various effects (morphological) on different plants species. For example, there was increased plant growth and biomass in spinach (Yang et al., 2007), parsley (Dehkourdi and Mosavi, 2013), and mung bean (Das et al., 2020). On the contrary, in some plants, such as in rice, anatase TiO_2 delayed and reduced the germination rates (Jalill and Yousef, 2015). Other factors leading to plant growth and development are also affected by anatase TiO_2. These include decreased photosynthetic rate in the case of long raceme elms (Gao et al., 2013) and induced chromosomal aberrations in onions (Pakrashi et al., 2014). Likewise, rutile phase of TiO_2 has been reported to increase photosynthetic rate and plant growth in spinach, indicated by increased plant dry weight and oxygenase activity (Zheng et al., 2005).

Other studies showed that the mixture of anatase and rutile (82: 18) in TiO_2 nanoparticles (P-25) when used in a concentration of 1 g L^{-1} leads to reduced root pore size and hydraulic potential in corns (Asli and Neumann, 2009). From these studies it could be concluded that different plants show diverse response towards TiO_2 nanoparticles with different crystal phases which might be due to variations in particle properties such as zeta potential, band gaps, and stability. The band gaps of rutile phase and anatase phases are 3.0 eV and 3.2 eV, respectively. To the best of our knowledge till date, there are no literature studies explaining the effect of brookite on plant growth and development so far.

2.3 Concentration

Concentration of TiO_2 applied to plants also plays an important role in expressing their effects. In a study done by Song et al. (2012), it was observed that TiO_2 nanoparticles improved nitrogen metabolism and photosynthetic rate, leading to higher growth of spinach. When the TiO_2 nanoparticles concentration was lower than 200 mg L^{-1} in the culture media, the plant cells showed increased antioxidant enzyme (SOD, APX, GPX, and CAT) defence activity to remove the accumulated ROS (Song et al., 2012). It could be further said that there can be a strong antioxidant defence capable of overcoming the oxidative stress and further recover the redox balance in the cells at low nanoparticle concentrations. Other studies have shown that at TiO_2 nanoparticle concentrations till 500 mg L^{-1} the *Vigna radiata* seeds showed very good root and shoot length (Das et al., 2020). On the contrary, in a study on *Vicia narbonensis* L. and *Zea mays* L. it was seen that at higher concentrations (0.4% concentration) of TiO_2 nanoparticles, the plants showed delayed germination, reduced mitotic index, and reduced root elongation (Castiglione et al., 2011). Thus, it is evident from the studies that when the concentration of nanoparticle is low, the production of ROS leads to activation of antioxidant enzymes, which helps in protecting the plants from further biotic and abiotic stresses. Basically, it boosts the plant defence system by degrading the ROS. However, when the concentration of nanoparticles is high, the plant's defence system and growth get affected. A plant's defence system is also a very important factor in this respect (Fu et al., 2014). It can also be seen that antimicrobial properties of TiO_2 nanoparticles improves the strength and resistance of plants towards stress. In fact, from most of the studies it could be seen that the presence of TiO_2 positively impacts a plant's growth.

2.4 Modifications

TiO_2 nanoparticles can be modified with inorganic or organic compounds to evade aggregation and improve dispersion which enhances their optical performance (Chen and Mao, 2007). The surface properties of TiO_2 nanoparticles such as zeta potential, surface area, and surface affinity would be considerably modified to a better performing nanoparticles (Du et al., 2017). For example, doping TiO_2 with metal/metalloid would improve the optical properties of these nanoparticles (Chen and Mao, 2007)

Table 17.1. Effects of particle size, phases, and concentration of TiO_2 nanoparticles on different plant species.

S. No.	Plant Species	Particle Size of Nanoparticle (nm)	TiO_2 Phase	Concentration of TiO_2 Nanoparticles ($mg\ L^{-1}$)	Effects on Plants	References
1.	Tomato	25 ± 3.5	Anatase	10, 100, 250, 500, 750, 1000	• Nanoparticles transported through the phloem • High chlorophyll content • Root and shoot length increased • At high concentration of 1000 $mg\ L^{-1}$ seed germination reduced	Raliya et al. (2015)
2.	Lettuce	4	Anatase	10, 100, 1000	• Nanoparticles were trapped on the root surface • Due to nanoparticle exposure, there was iron, calcium, and phosphorus aggregation in roots	Larue et al. (2016)
3.	Thale cress	5–15	Rutile	50, 100, 200, 300, 500	• Tetracycline toxicity increased • Biomass increased at 50 and 100 $mg\ L^{-1}$ concentrations	Liu et al (2017)
4.	Spinach	N/A	Rutile	2.5, 5, 10, 15, 20, 25, 40, 60	• Increased chlorophyll content and photosynthetic rate	Zheng et al. (2005)
5.	Corn	27 ± 4	Mixture of Anatase: Rutile (82: 18)	0.3, 1	• Smaller root cell wall pore size • Less leaf growth • Low transpiration rate	Asli et al. (2009)
6.	Tomato	27 ± 4	Mixture of Anatase: Rutile (82:18)	50, 100, 1000, 2500, 5000	• Longer roots • High sodium oxide dismutase activity	Song et al. (2013)

and addition of surfactants such as cetyl trimethyl ammonium bromide (CTAB) and sodium dodecyl sulphate (SDS), will stabilize nano-TiO_2 (Das et al., 2020). Thus, it is believed that the interaction between modified TiO_2 nanoparticles and plants would be different from that of non-modified TiO_2 nanoparticles. Some of the literature studies have clearly shown varied effects on plants with variations in TiO_2 nanoparticle modification. For example, anatase TiO_2 nanoparticle modified with an activated carbon resulted in improved seed germination of tomato and mung bean (Singh et al., 2016). Further, Raliya et al. (2015) studied TiO_2 nanoparticle biosynthesis using fungi *Aspergillus flavus* TFR 7. Foliar spray of TiO_2 nanoparticles at a concentration of 10 mg L^{-1} on the leaves of 14-day old mung bean plants showed noteworthy enhancement in root length (49.6%), root area (43%), root nodule (67.5%), shoot length (17.02%), total soluble leaf protein (94%), and chlorophyll content (46.4%) (Raliya et al., 2015). Increase in the overall plant growth in mung bean was seen when exposed with TiO_2 nanoparticle modified with SDS at 500 mgL^{-1} (Das et al., 2020). From the above studies, it can be said that the biomass or surfactant modified TiO_2 nanocomposites not only improve seed germination in plants but also the shoot length and root length depending on the concentration of nanoparticles. Even at higher concentration such as 500 mg L^{-1}, germination is higher, but the growth of root and shoot either reduced or remained constant as compared to plants, unexposed to TiO_2 based nanoparticles (Das et al., 2020). Table 17.1 summarizes key studies describing the effect of various properties and different aspects of TiO_2 NPs on a variety of plant species.

From the literature it could be seen that there are various unresolved effects of TiO_2 nanoparticles when exposed to plants. Hence, there is a crucial need to gather more information regarding the eco-safety and phytotoxicity effects of TiO_2 nanoparticles to support government's regulatory efforts. However, the future viewpoints on the TiO_2 nanoparticles and plant interaction will totally depend on an understanding of the molecular mechanisms responsible for any effect escalated by these nanoparticles towards the plants. The reported literature studies are encouraging enough to take this work forward in a detailed manner in future.

3. Conclusion

Titanium dioxide is one of the most produced and consumed nanoparticles worldwide. When plants are exposed to these nanoparticles, studies have shown both positive and negative response in plant eco-safety and phytotoxicity. The TiO_2 nanoparticles' size, phase, concentration, and modification affect various factors. It can be said that particles as small as 30 nm could easily penetrate a plant cell as the pore size diameter of the plant cell wall ranges from 5 to 30 nm. But particles larger than 30 nm accumulate in the cells and create blockage leading to cell death and finally plant death. Further, anatase and rutile phases of TiO_2 show both positive and negative effects on plant development such that each crystal phase has different particle properties, including zeta potential, stability, and band gaps (such that rutile phase and anatase phase have 3.0 eV and 3.2 eV, respectively). The TiO_2 nanoparticle concentration also plays a very important role. At low concentration (40 mg L^{-1}) of TiO_2 nanoparticles, the plants can overcome oxidative stress whereas at high concentration as high as 1000 mg L^{-1}, plant growth is reduced. Further, TiO_2 nanoparticles are modified with inorganic or organic compounds to avoid aggregation and accumulation and achieve better dispersion; it also enhances the optical performance of these nanoparticles. TiO_2 nanoparticles when modified by carbon-based materials can improve the growth of plants whereas when TiO_2 is doped with other inorganic materials such as aluminium and galium, it can lead to reduced plant growth as these new modified nanoparticles negatively affect the plants. Some other important factors affecting plant growth are ROS generated by TiO_2 and plants as well. Depending on the TiO_2 particle size and concentration, the plants release the antioxidant enzyme named SOD and the ROS levels are regulated by the SOD in plants. Further, the levels of ROS are in controlled levels such that they do not lead to cell death and instead even in stress the plant growth remains high. The lower the particle size and the concentration, the higher the SOD and it may regulate the ROS efficiently. When the concentration of nanoparticles is high, they overcome a plant's defence system and its growth. Thus, both eco-safety and phytotoxicity effects are associated with TiO_2 nanoparticles and various factors govern this behaviour. Hence, a further deep understanding is required to reach to any conclusion.

Acknowledgement

The part of experimental work reported in this MS was carried out at TERI-Deakin Nano Biotechnology Centre (TDNBC), TERI and was presented by Ms. Paromita Das in the 3rd NanoforAgri 2019 Conference. Authors are grateful to The Director, TDNBC for providing required infrastructural support to carry out the work. The joint financial support from the Department of Science & Technology, GoI, under the DST-Water Mission, Water Technology Initiative Program (WTI), and ONGC Energy Centre, New Delhi as Industry partner for the project [Ref No: DST/WTI/2K16/78(G)] is gratefully acknowledged.

References

Asli, S. and Neumann, P.M. (2009). Colloidal suspensions of clay or titanium dioxide nanoparticles can inhibit leaf growth and transpiration via physical effects on root water transport. *Plant, Cell Environ.*, 32: 577–584. https://doi.org/10.1111/j.1365-3040.2009.01952.x.

Auffan, M., Rose, J., Bottero, J.Y. et al. (2009). Towards a definition of inorganic nanoparticles from an environmental, health, and safety perspective. *Nat. Nanotechnol.*, 4: 634–641. https://doi.org/10.1038/nnano.2009.242.

Bahadur, N. and Bhargava, N. (2019). Novel pilot scale photocatalytic treatment of textile & dyeing industry wastewater to achieve process water quality and enabling zero liquid discharge. *J. Water Process Eng.*, 32: 100934. https://doi.org/10.1016/j.jwpe.2019.100934.

Bahadur, N., Jain, K., Srivastava, A.K. et al. (2010). Effect of nominal doping of Ag and Ni on the crystalline structure and photo-catalytic properties of mesoporous titania. *Mater. Chem. Phys.*, 124: 600–608. https://doi.org/10.1016/j.matchemphys.2010.07.020.

Begum, P., Ikhtiari, R. and Fugetsu, B. (2011). Graphene phytotoxicity in the seedling stage of cabbage, tomato, red spinach, and lettuce. *Carbon N. Y.*, 49: 3907–3919. https://doi.org/10.1016/j.carbon.2011.05.029.

Castiglione, M.R., Lucia, C., Geri, C. and Cremonini, R. (2011). The effects of nano-TiO_2 on seed germination, development and mitosis of root tip cells of *Vicia narbonensis* L. and *Zea mays* L. *J. Nanopart. Res.*, 13: 2443–2449. doi: 10.1007/s11051-010-0135-8.

Castiglione, M.R., Giorgetti, L. and Cremonini, R. (2014). Impact of TiO_2 nanoparticles on *Vicia narbonensis* L.: potential toxicity effects. *Protoplasma*, 251(6): 1471–9. doi: 10.1007/s00709-014-0649-5.

Chen, X. and Mao, S.S. (2007). Titanium dioxide nanomaterials: Synthesis, properties, modifications and applications. *Chem. Rev.*, 107: 2891–2959. https://doi.org/10.1021/cr0500535.

Das, P., Bahadur, N. and Dhawan, V. (2020). Surfactant-modified titania for cadmium removal and textile effluent treatment together being environmentally safe for seed germination and growth of *Vigna radiata. Environ. Sci. Pollut. Res.*, 27: 7795–7811. https://doi.org/10.1007/s11356-019-07480-1.

Dehkourdi, E.H. and Mosavi, M. (2013). Effect of anatase nanoparticles (TiO_2) on parsley seed germination (*Petroselinum crispum*) *in vitro. Biol. Trace Elem. Res.*, Nov; 155(2): 283–6. doi: 10.1007/s12011-013-9788-3.

Du, W., Tan, W., Peralta-Videa, J.R. et al. (2017). Interaction of metal oxide nanoparticles with higher terrestrial plants: Physiological and biochemical aspects. *Plant Physiol. Biochem.*, 110: 210–225. https://doi.org/10.1016/j.plaphy.2016.04.024.

Faraji, J. and Sepehri, A. (2018). Titanium dioxide nanoparticles and sodium nitroprusside alleviate the adverse effects of cadmium stress on germination and seedling growth of wheat (*Triticum aestivum* L.). *Univ. Sci.*, 23: 61–87. https://doi.org/10.11144/Javeriana. SC23-1.tdna.

Feizi, H., Kamali, M., Jafari, L. and Rezvani Moghaddam, P. (2013). Phytotoxicity and stimulatory impacts of nanosized and bulk titanium dioxide on fennel (*Foeniculum vulgare* Mill). *Chemosphere*, 91: 506–511. https://doi.org/10.1016/j.chemosphere.2012.12.012.

Fellmann, S. and Eichert, T. (2017). Acute effects of engineered nanoparticles on the growth and gas exchange of *Zea mays* L.: What are the underlying causes? https://doi.org/10.1007/s11270-017-3364-y.

Fu, P.P., Xia, Q., Hwang, H.M. et al. (2014). Mechanisms of nanotoxicity: Generation of reactive oxygen species. *J. Food Drug Anal.*, 22: 64–75. https://doi.org/10.1016/j.jfda.2014.01.005.

Gao, J., Xu, G., Qian, H. et al. (2013). Effects of nano-TiO_2 on photosynthetic characteristics of *Ulmus elongata* seedlings. *Environ, Pollut.*, 176: 63–70. https://doi.org/10.1016/j.envpol.2013.01.027.

Gogoi, M. and Basumatary, M. (2018). Estimation of the chlorophyll concentration in seven Citrus species of Kokrajhar district, BTAD, Assam, India. *Tropical Plant Research*, 5: 83–87. https://doi.org/10.22271/tpr.2018.v5.i1.012.

Gottschalk, F., Sonderer, T., Scholz, R.W. and Nowack, B. (2009). Modeled environmental concentrations of engineered nanomaterials (TiO_2, ZnO, Ag, CNT, fullerenes) for different regions. *Environ. Sci. Technol.*, 43: 9216–9222. https://doi.org/10.1021/es9015553.

Jalill, R.D.A. and Yousef, A.M. (2015). Comparison the phytotoxicity of TiO_2 nanoparticles with bulk particles on Amber 33 variety of rice (*Oryza sativa*) *in vitro. Sch. Acad. J. Biosci.*, 3(3): 254–262.

Jiang, F., Shen, Y., Ma, C. et al. (2017). Effects of TiO_2 nanoparticles on wheat (*Triticum aestivum* L.) seedlings cultivated under super-elevated and normal CO_2 conditions. *PLoS One*, 12: e0178088. https://doi.org/10.1371/journal.pone.0178088.

Kurepa, J., Paunesku, T., Vogt, S. et al. (2010). Uptake and distribution of ultrasmall anatase TiO_2 alizarin red s nanoconjugates in *Arabidopsis thaliana. Nano Lett.*, 10: 2296–2302. https://doi.org/10.1021/nl903518f.

Larue, C., Khodja, H., Herlin-Boime, N. et al. (2011). Investigation of titanium dioxide nanoparticles toxicity and uptake by plants. *J. Phys. Conf. Ser.*, 304: https://doi.org/10.1088/1742-6596/304/1/012057.

Larue, C., Castillo-Michel, H., Stein, R.J. et al. (2016). Innovative combination of spectroscopic techniques to reveal nanoparticle fate in a crop plant. *Spectrochim Acta - Part B At Spectrosc.*, 119: 17–24. https://doi.org/10.1016/j.sab.2016.03.005.

Lee, W.M., Kwak, J.I.l. and An, Y.J. (2012). Effect of silver nanoparticles in crop plants *Phaseolus radiatus* and Sorghum bicolor: Media effect on phytotoxicity. *Chemosphere*, 86: 491–499. https://doi.org/10.1016/j.chemosphere.2011.10.013.

Li, M., Yin, J.J., Wamer, W.G. and Lo, Y.M. (2014). Mechanistic characterization of titanium dioxide nanoparticle-induced toxicity using electron spin resonance. *J. Food Drug Anal.*, 22: 76–85. https://doi.org/10.1016/j.jfda.2014.01.006.

Liu, H., Ma, C., Chen, G. et al. (2017). Titanium dioxide nanoparticles aAlleviate tetracycline toxicity to *Arabidopsis thaliana* (L.). *ACS Sustain. Chem. Eng.*, 5: 3204–3213. https://doi.org/10.1021/acssuschemeng.6b02976.

Morteza, E., Moaveni, P., Farahani, H.A. and Kiyani, M. (2013). Study of photosynthetic pigments changes of maize (*Zea mays* L.) under nano TiO_2 spraying at various growth stages. *Springerplus*, 2: 1–5. https://doi.org/10.1186/2193-1801-2-247.

Nadiminti, P.P., Dong, Y.D., Sayer, C. et al. (2013). Nanostructured liquid crystalline particles as an alternative delivery vehicle for plant agrochemicals. *ACS Appl. Mater. Interfaces*, 5: 1818–1826. https://doi.org/10.1021/am303208t.

Nevius, B.A., Chen, Y.P., Ferry, J.L. and Decho, A.W. (2012). Surface-functionalization effects on uptake of fluorescent polystyrene nanoparticles by model biofilms. *Ecotoxicology*, 21: 2205–2213. https://doi.org/10.1007/s10646-012-0975-3.

Nowack, B. and Bucheli, T.D. (2007). Occurrence, behavior and effects of nanoparticles in the environment. *Environ. Pollut.*, 150: 5–22. https://doi.org/10.1016/j.envpol.2007.06.006.

Pakrashi, S., Jain, N., Dalai, S. et al. (2014). *In vivo* genotoxicity assessment of titanium dioxide nanoparticles by *Allium cepa* root tip assay at high exposure concentrations. *PLoS One*, 9: https://doi.org/10.1371/journal.pone.0087789.

Raliya, R., Biswas, P. and Tarafdar, J.C. (2015). TiO_2 nanoparticle biosynthesis and its physiological effect on mung bean (*Vigna radiata* L.). *Biotechnol Reports*, 5: 22–26. https://doi.org/10.1016/j.btre.2014.10.009.

Singh, P., Singh, R., Borthakur, A. et al. (2016). Effect of nanoscale TiO_2 -activated carbon composite on *Solanum lycopersicum* (L.) and *Vigna radiata* (L.) seeds germination. *Energy, Ecol. Environ.*, 1: 131–140. https://doi.org/10.1007/s40974-016-0009-8.

Song, G., Gao, Y., Wu, H. et al. (2012). Physiological effect of anatase TiO_2 nanoparticles on *Lemna minor*. *Environ. Toxicol. Chem.*, 31: 2147–2152. https://doi.org/10.1002/etc.1933.

Yang, F., Liu, C., Gao, F. et al. (2007). The improvement of spinach growth by nano-anatase TiO_2 treatment is related to nitrogen photoreduction. *Biol. Trace Elem. Res.*, 119: 77–88. https://doi.org/10.1007/s12011-007-0046-4.

Zheng, L., Hong, F., Lu, S. and Liu, C. (2005). Effect of Nano-TiO_2 on strength of naturally aged seeds and growth of spinach. *Biol. Trace Elem. Res.*, 104: 83–91.

Index

O

P

Q

R

About the Editors

Sunil Kumar Deshmukh

Dr. Sunil Kumar Deshmukh is Scientific Advisor to Greenvention Biotech, Uruli-Kanchan, Pune, India and Agpharm Bioinnovations LLP, Patiala, Punjab, India. Veteran industrial mycologist, spent a substantial part of his career in drug discovery at Hoechst Marion Roussel Limited [now Sanofi India Ltd.], Mumbai, and Piramal Enterprises Limited, Mumbai. He has also served TERI-Deaken Nano Biotechnology Centre, TERI, New Delhi, and as an Adjunct Associate Professor at Deakin University, Australia. He has to his credit 8 patents, 140 publications, and 16 books on various aspects of fungi and natural products of microbial origin. He is a president of the Association of Fungal Biologists (AFB) and a past president of the Mycological Society of India (MSI). Dr. Deshmukh serves as a referee for more than 20 national and international journals. He has approximately four decades of research experience in getting bioactives from fungi and keratinophilic fungi.

Mandira Kochar

Dr. Mandira Kochar is the Area Convenor for the Nanobiotechnology Centre which is a part of the Sustainable Agriculture Division at TERI. She has a PhD Degree in Genetics from Delhi University South Campus alongside extensive research experience over the past 17 years focused on Molecular Microbiology, Microbial Physiology and Genomics to support thematic areas of Agriculture and Environment Sustainability. Her research work over the last decade has concentrated on the genomics, molecular, and functional interactions of plant growth regulators crosstalk in plant-associated bacteria and symbiotic interactions with plants. Since 2011 she has been independently leading research projects in TERI's Sustainable Agriculture Division and leads the team on Nanobiotechnology interventions for sustainable agriculture and the environment. She works on nano agri inputs, carbon nanomaterials for agriculture and enzyme nanocomposites for environment remediation. Besides these areas her work is also focused on delivering unique advanced biological solutions towards achieving sustainability for agriculture. She has published 27 peer reviews paper/review articles/ book chapters including one in Science. She is a regular reviewer for many Springer and Elsevier Journals.

Pawan Kaur

Dr. Pawan Kaur is an Agri-Nanotechnologist with 10 years of research experience and 4 years of teaching experience at renowned institutes and universities, with expertise in the development of nanoproducts in particular nanofungicides using microbial interventions for their application in the agriculture field. Her current research involvements are biosynthesis of nanopesticides using indigenous resources; development of nanomaterials from bio-waste and natural mineral sources; eco-friendly nanocarriers for smart and efficient delivery of bioactives; nanoparticles for post-harvest management. She also worked as a scientist at Teri-Deakin Nanobiotechnology Centre, TERI Gram, Gurugram and was actively involved in the development of polymeric nano-based smart delivery systems for agricultural applications. Her additional activities included the development of novel biological approaches for the synthesis of nano pesticides and nano fertilizers and guiding PhD students at Deakin University, Australia for their PhD work. Her work has been published in over 30 research papers in reputed international journals, 3 book chapters, and 2 edited books (Springer Nature Publisher). She is also a member of editorial boards and acts as an invited reviewer for several reputed journals.

Pushplata Prasad Singh

Dr. Pushplata Prasad Singh is the Director (Acting) of the TERI-Deakin Nanobiotechnology Centre, Sustainable Agriculture Division, The Energy and Resources Institute (TERI). Her expertise is in the development of NanoAgri-inputs, studying nano-bio interactions, nano-toxicity, life-cycle assessment. She is an active member of the International initiative for Safe and Sustainable Nanotechnology (INISS) and Bio-Nano-Network (BNN). She along with other members from TERI, multiple government organizations and industries has contributed to the development of Indian "Guidelines for assessment of NanoAgri-inputs and NanoAgro-products". She is also a member of BIS-Nanotechnology (Bureau of Indian Standards). With a PhD in Human Genetics and Genomics and Post-PhD experiences in Genomics of human diseases, plants and beneficial microbes; and industrial experience in development of advanced nanomaterials, she collaborates with disparate partners to drive alignment on business objectives and deliverables to achieve complete development of safe and biogenic Agro-products in a set time frame. She has been significantly contributing towards major research and network projects "Centre of Excellence for Advanced Research in Agricultural Nanotechnology" (CEARAN) and "DBT-TDNBC-DEAKIN Research Network Across continents for learning and innovation" (DTD-RNA) funded by Department of Biotechnology, Govt. of India.